Birkhäuser

Frontiers in Mathematics

More information about this series at http://www.springer.com/series/5388

Adnan Tercan • Canan C. Yücel

Module Theory, Extending Modules and Generalizations

With the cooperation of Patrick F. Smith, University of Glasgow

Birkhäuser

Adnan Tercan
Department of Mathematics
Hacettepe University
Ankara, Turkey

Canan C. Yücel
Department of Mathematics
Pamukkale University
Denizli, Turkey

ISSN 1660-8046 ISSN 1660-8054 (electronic)
Frontiers in Mathematics
ISBN 978-3-0348-0950-4 ISBN 978-3-0348-0952-8 (eBook)
DOI 10.1007/978-3-0348-0952-8

Library of Congress Control Number: 2016940427

Mathematics Subject Classification (2010): 16D10, 16D40, 16D50, 16D70, 16L30, 16L60, 16P20, 16P60, 16S50, 16S90

Printed on acid-free paper

This book is published under the trade name Birkhäuser.
The registered company is Springer International Publishing AG (www.birkhauser-science.com)

Dedicated to our families, teachers, and students

Preface

A module is *extending* (or CS, C_1) if every complement submodule is a direct summand or, equivalently, every submodule is essential in a direct summand. The terms extending and *lifting* (dual to extending) were firstly used by Harada and Oshiro ([Har82], [Oshi84]).

The importance of extending modules and rings in Ring and Module Theory, and more generally in Algebra, became obvious in the 1990's, but not exclusively, through the impact of the publication of monographs of S.H. Mohammed and B.J. Müller [MM90] and of N.V. Dung, D.V. Huynh, P.F. Smith, and R. Wisbauer [DHSW94]. Since that time there has been a continuing interest in such rings and modules and their various generalizations which arose not only directly from the study of CS concept, but also from work concerning the dual notion to extending, namely, lifting. Many results obtained for extending modules and rings were transfered to lifting modules and rings [CLVW06]. At a first glance, the extending and generalized extending concepts appear to be too similar to expect many differences in their application to the structure theory of rings and modules. However, we have shown many "surprising" differences, as we indicate throughout this monograph. To this end, we classified generalized extending modules into two groups. The first group consists of generalized extending modules such that either for every submodule or a kind of special submodules, there exists a direct summand with the property that the direct sum of the aforementioned submodule with the direct summand is essentially contained in the module (inner type generalization). The second group consists of generalizations of extending modules which are based on a technical condition like a homomorphism into a direct summand or an equivalence relation in the lattice of submodules, etc. (outer type generalization). We also apply our former equivalence relation idea to the dual extending case. We believe that this last application will foster research on dual extending modules and related classes of modules, which in turn, will greatly broaden the scope of the theory.

The purpose of this monograph is to give an up-to-date presentation of known as well as new results on generalized extending modules and some complementary results on extending matrix rings, and also to provide standard background material, but with somewhat selective topics, on Ring and Module Theory. A number of open research problems are listed at the end of the book to generate interest in research on generalized extending and related duals. Each section includes exercises of varying degrees of difficulty for graduate students. To keep the book to a reasonable length, some results have been included as exercises in various sections and appropriate references have been included in remarks at the end of each chapters. We are very thankful to Patrick F. Smith and Gary F. Birkenmeier for encouragement and constructive comments and suggestions for improvements.

We are thankful to our colleagues who helped by proof-reading various parts of the book and provided technical advise, among them Yeliz Kara, Uğur Yücel, Ramazan Yaşar, Talha Arıkan, Hacer İlhan, Hatice Zeybek, Esma Dirican, Nazife Erkurşun Özcan, Aslı Pekcan, Sema Yayla, Mesut Şahin, and Seçil Tokgöz.

The errors that still may remain in the book are our own fault.

We are also thankful to Thomas Hempfling, Anna Mätzener, Sylvia Lotrovsky, Sarah Annette Goob, and the other staff at Springer Birkhäuser who have been so helpful with our book project.

Ankara, TURKEY
Denizli, TURKEY
February 28, 2015

Adnan Tercan
Canan Celep Yücel

Introduction

The first chapter covers introductory material to parts of ring and module theory. We deal with standard, yet somewhat selective background topics. We hope that the chapter will prove useful for a general audience. We add numerous exercises that provide supplementary results. The chapter does not contain all of the aspects (e.g., category-theoretical concepts) on rings and modules, being too narrowly focused for this, but it could serve as a basis for independent study in rings and modules following a basic introduction to modern algebra.

Chapter 2 deals with certain type of modules, including those possessing the C_2 and C_3 properties, the summand intersection property (SIP), and the summand sum property (SSP) conditions. The first two sections present basic properties of complement submodules and relative injective modules which are used in the rest of the book. Section 3 introduces lifting submodules. Here, given two modules A and X over a ring R, we say that a submodule N of A is a *lifting submodule for X in A* if for any $\varphi \in \mathrm{Hom}_R(N, X)$ there exists $\theta \in \mathrm{Hom}_R(A, X)$ such that the restriction of θ on N is equal to φ. So, we build up the class

$$\mathrm{Lift}_X(A) = \{N : N \leq A \text{ and } N \text{ is a lifting submodule for } X \text{ in } A\}.$$

For instance, any direct summand of A belongs to $\mathrm{Lift}_X(A)$. This new class of submodules is examined in detail. Section 4 introduces the notion of *ejectivity* and the class $\mathrm{Elift}_X(A)$, which are interesting in their own right. We say that X is *A-ejective* if, for each $K \leq A$ and each homomorphism $\varphi : K \to X$ there exist a homomorphism $\theta : A \to X$ and an essential submodule E of K such that $\theta(x) = \varphi(x)$ for all $x \in E$. In this case, K is said to be an *elifting submodule* for X in A. So we set $\mathrm{Elift}_X(A) = \{K : K \leq A \text{ and } K \text{ is an elifting submodule for } X \text{ in } A\}$. This class is also examined in detail. The last section looks at module properties which eventually give direct summands. Especially, C_2 and C_3 properties are characterized in terms of lifting submodules.

Chapter 3 collects several results on extending modules and rings and continuous, quasi-continuous modules which cannot be found in other monographs. The direct sum of two extending modules need not be extending, so Section 1 examines when a direct sum of an arbitrary number of extending modules is again an extending module. The section ends with a look at the rational hull and members of some distinguished class of submodules of an extending module which enjoy the extending property.

Section 2 mainly concerns extending modules over commutative domains, but also somewhat generalized versions of the main result to the noncommutative setting. To this end we give a fundamental decomposition theorem on extending modules which states that a torsion-free extending module over a commutative domain is a finite direct sum of injective modules and uniform modules. We apply this fact to obtain more information about extending modules over Dedekind domains or principal ideal domains. Moreover, we show that any nonsingular reduced extending module over a commutative ring with only a finite number of minimal prime ideals has a finite uniform dimension; and any nonsingular extending module over a semiprime Goldie ring is a finite direct sum of injective modules and uniform modules.

Section 3 considers generalized triangular matrix (also called formal triangular matrix) CS-rings and the split null extension (also called trivial extension) CS-rings. Chapter 3 ends with a section on the notions of continuous and quasi-continuous modules; in particular, it is shown that continuous and quasi-continuous modules can be characterized in terms of lifting homomorphisms from certain submodules to the module itself.

The opening section of Chapter 4 introduces class of a generalized extending modules which are called weak CS-modules. A module A is *weak* CS if every semisimple submodule of A is essential in a direct summand. Any direct sum of a semisimple submodule and an injective module is a weak CS-module, but such a module is not, in general, a CS-module, even over a Dedekind domain. Moreover, over a Dedekind domain, every module with finite uniform dimension is a weak CS-module. In fact, weak CS-modules share some of the properties of CS-modules. For example, if a weak CS-module satisfies the ascending chain condition on essential submodules, then it is a direct sum of a semisimple module and a Noetherian module.

Section 2 introduces C_{11}-modules. A module A *satisfies* C_{11} (or is a C_{11}-module) if every (complement) submodule has a complement which is a direct summand of A. Any direct sum of modules with C_{11} satisfies C_{11}. Furthermore, a module A satisfies C_{11} if and only if $A = Z_2(A) \oplus K$ for some (nonsingular) submodule K of A and $Z_2(A)$ and K both satisfy C_{11}. This result shows that the study of modules with property C_{11} reduces to the case of Goldie torsion modules and nonsingular modules. In contrast to extending modules, direct summands of a C_{11}-module need not be C_{11}, in general. The following example is quite interesting in its own right. Let $S = \mathbb{R}[x, y, z]$ and $R = S/Ss$, where $s = x^2 + y^2 + z^2 - 1$. Then the R-module $A = R \oplus R \oplus R$ satisfies C_{11} but contains an indecomposable direct summand K with uniform dimension 2 and K does not satisfy C_{11}. For instance, K corresponds to regular sections of the tangent bundle of the real 2-sphere S^2.

However, it is shown in Section 3 that the C_{11} condition with a conditional direct summand property ensure that C_{11} is inherited by direct summands. This section also provides a decomposition into uniform submodules for C_{11}-modules that satisfy chain conditions on left annihilators or essential submodules.

In Section 4, we continue the study of structural properties of C_{11}-modules and connections between the C_{11} condition and various other generalizations of CS condition. A module A is FI-*extending* if every fully invariant submodule of A is essential in a direct summand of A [BCFG01], [BMR02], [BPR02]. Since the C_{11} property lies strictly between the CS and FI-extending properties, it is natural to seek conditions which ensure that a C_{11}-module is CS, or that a FI-extending module is C_{11}. Such conditions are given in this section.

The focus in Section 5 is on extensions of C_{11}-rings and modules. We study the transference of the C_{11} condition from a given ring or module to various ring or module extensions. In particular, we show that if R is a right C_{11}-*ring* (i.e., R_R is a C_{11}-module), then the ring of column finite matrices of size Γ over R, the ring of m-by-m upper triangular matrices over R, and any right essential overring T of R are all right C_{11}-rings. We also provide necessary and sufficient conditions under which the generalized triangular matrix ring is a right C_{11}-ring. For a module A, we prove that all essential extensions of A satisfying C_{11} are essential extensions of C_{11}-modules constructed from A and certain subsets of idempotent elements of the ring of endomorphisms of the injective hull of A. Finally, we prove that if A is a C_{11}-module, then so is its rational hull.

Section 6 introduces a framework which encompasses most of the generalizations of the CS property and allows us to target specific sets of submodules of a module for application of the CS property. Let $\varnothing \neq \underline{C} \subseteq L(A)$, where $L(A)$ denotes the set of submodules of A. We say A is $\underline{C}$-*extending* if for each $X \in \underline{C}$ there exists a direct summand D of A such that X is essential in D. This concept was introduced by Oshiro in [Oshi83] with slightly different terminology and notation. Oshiro assumes that $\underline{C}$ is closed under isomorphisms and essential extensions. In [DS98], the authors introduce type 1 χ-extending and type 2 χ-extending modules, where χ is a class of modules containing the zero submodule and closed under isomorphisms. In contrast to [Oshi83] and [DS98], we do not implicitly assume that $\underline{C}$ is closed under isomorphisms or essential extensions. Our investigation focuses on the behavior of $\underline{C}$-extending modules with respect to direct sums and direct summands. We obtain various well-known results about extending modules and generalizations as corollaries of our results.

In Section 7 we continue our investigation on $\underline{C}$-extending modules with respect to essential extensions. This section explains how to construct essential extensions of a module A which are $\underline{C}$-extending by using a set of representatives of an equivalence relation $\gamma_{\underline{C}}$ on $\{e = e^2 \in \underline{E}_{E(A)}\}$, where $\underline{E}_{E(A)}$ denotes the endomorphisms of the injective hull of A; and characterize when the rational hull of A is $\underline{C}$-extending in terms of such a set of representatives. The section ends with several well-known types of $\underline{C}$-extending conditions transfer from the module to its rational hull.

Section 8 concerns $\underline{P}\underline{I}$-extending rings. A ring R is *right* $\underline{P}\underline{I}$-*extending* if every projection invariant right ideal of R is essential in a direct summand of R. This section provides the transfer of the $\underline{P}\underline{I}$-extending condition from ring R to its various ring extensions. More specifically, one derives when a generalized upper triangular matrix is right $\underline{P}\underline{I}$-extending. The last section of Chapter 4 looks at weak version of the C_{11} property. A module A is *weak* C_{11} if every semisimple submodule has a complement which is a direct summand of A. Weak C_{11}-modules are a proper generalization of both C_{11}-modules and weak CS-modules.

The first section of Chapter 5 concerns the class of C_{12}-modules. This class properly contains the class of C_{11}-modules. A module A *satisfies* C_{12} if, for every submodule N of A, there exist a direct summand K of A and a monomorphism $\alpha : N \to K$ such that $\alpha(N)$ is an essential submodule of K. This section presents structural properties of C_{12}-modules and relationships with the other generalized extending conditions. The section ends with results on a weak version of C_{12}-modules.

Section 2 looks at modules A such that every homomorphism from a complement submodule of A to A can be lifted to A. Although such modules share some of the properties of CS-modules, it is shown that they form a substantially bigger class.

Section 3 focuses on the class of CLS-modules, which is another proper generalization of the class of extending modules. A module A is CLS if every z-closed submodule of A is a direct summand of A. Here, if N is a submodule of A, we say that N is *z-closed* if $Z(A/N) = 0$, i.e., A/N is nonsingular.

In Section 4, we defined a module A to be G-*extending* (or *Goldie extending*) if for each submodule X of A there exists a direct summand D of A such that $X \cap D$ is essential in both X and D. We develop several characterizations of the G-extending condition and we locate the G-extending condition with respect to several generalizations of the extending condition. Moreover, we obtain various conditions for a direct sum of G-extending modules to be G-extending or for a direct summand of a G-extending module to be G-extending. Our results enable us to obtain a characterization of the G-extending Abelian groups.

Uniform modules are often considered to be fundamental objects in the structure theory of various classes of modules, e.g., the class of modules with finite uniform dimension, or the class of extending modules. Hence, for a given class, solving the following problems is important.

(1) Determine conditions which force a module from a given class to be a direct sum of uniform submodules.
(2) Determine when a direct sum of uniform modules is in a given class.

Section 5 begins with results providing answers to the first problem and continues with solutions to the second problem, where the given class of modules is the G-extending class. Section 5 finishes with results on G-extending essential extensions of a module.

Section 6 looks at the G-extending condition under various ring extensions. Amongst other results, it shows that if R_R is G-extending and S is a right essential overring, then S_R and S_S are G-extending.

Section 7 is mainly devoted to the characterization of G-extending modules over Dedekind domains and principal ideal domains. Furthermore, the finite direct sums and direct summands of G-extending modules over such domains are examined.

In Section 8, we obtain a characterization of the right G-extending generalized triangular matrix rings. This result and its corollaries improve and generalize the existing results on right extending generalized triangular matrix rings. Connections to operator theory and a characterization of the class of right extending right SI-rings are also established.

Chapter 6 is devoted to dual Goldie and ec-complement versions of the extending property. It is well known that lifting modules are dual to extending modules and there are many works on this subject in the literature.

Section 1 introduces Goldie*-supplemented modules, which are dual to Goldie extending modules. We investigate this new class of modules and the class of H-supplemented modules. These classes are located among various well-known classes of modules related to the class of lifting modules.

Section 2 deals with the classes of Goldie-rad-supplemented and rad-H-supplemented modules which are based on the β^{**} equivalence relation. A module A is *Goldie*-rad-*supplemented* if for every submodule N of A, there exists a rad-supplement submodule S in A such that $N\beta^{**}S$. A module A is rad-H-*supplemented* if for every submodule N of A, there exists a direct summand D of A such that $N\beta^{**}D$. Section 3 concerns ECS and EC_{11}-modules as weaker forms of extending and C_{11}-modules, respectively. A module A is an ECS-*module* if every ec-complement submodule of A is a direct summand. A module A is an EC_{11}-*module* if every ec-submodule of A has a complement which is a direct summand. Here, given a submodule N of A, we say that N is an ec-*complement* if N is a complement in A that contains essentially a cyclic submodule.

Chapter 7 formulates a number of open problems and questions. We think that our problems are legitimate and will provide a basis for further research which will in turn greatly broaden the scope of the theory. Here we should mention recent developments on the latticial counterparts of conditions C_i $(i = 1, 2, 3)$ and C_{1j} $(j = 1, 2)$ for modules and applications to Grothendieck categories, module categories equipped with a torsion theory [AIT16].

The book ends with an Appendix. Since we do not treat in detail the rings of quotients in the previous chapters, it is better to construct these kind of rings for graduate students in mathematics. We believe that this appendix will allow to build a bridge between our theory and notions treated in the monograph [BPR13].

A Partial List of Symbols

$\mathbb{N}$	The set of positive integers
$\mathbb{Z}$	The ring of integers
$\mathbb{Z}_n$ or $\mathbb{Z}/\mathbb{Z}n$ $(n > 1)$	The ring of integers modulo n
$\mathbb{Z}(p^\infty)$	The Prüfer p-group
$\prod \mathbb{Z}$	The Specker group
$\mathbb{Q}$	The field of rational numbers
$\mathbb{R}$	The field of real numbers
$\mathbb{C}$	The field of complex numbers
$\mathrm{M}_n(R)$	The $n \times n$ matrix ring over R
$T_n(R)$	The $n \times n$ upper triangular matrix ring over R
$\mathbf{I}\ (R)$	The set of all idempotents of R
$\mathrm{Cen}(R)$	The center of R
$\langle X \rangle_R$	The subring of R generated by $X \subseteq R$
$Z(_RA)$ or $Z(A)$	The singular submodule of $_RA$
$Z_2(_RA)$ or $Z_2(A)$	The second singular submodule of $_RA$
$\mathrm{soc}(A)$ or $\mathrm{soc}(_RA)$	The socle of $_RA$
$\mathrm{rad}(A)$ or $\mathrm{rad}(_RA)$	The Jacobson radical of $_RA$
A^n or $A^{(n)}$	The direct sum of n copies of A
$J(R)$	The Jacobson radical of R
u-dim	Uniform dimension
$\mathrm{S}_l(R)$	The set of all left semicentral idempotents of R
$\mathrm{S}_r(R)$	The set of all right semicentral idempotents of R
$E(_RA)$ or $E(A)$	The injective hull of $_RA$
$\widetilde{E}(_RA)$ or $\widetilde{E}(A)$	The rational hull of $_RA$
$\mathcal{Q}(R)$	The maximal right ring of quotients of R
R-Mod	The category of left R-modules
acc (dcc)	The ascending (descending) chain condition
$N \trianglelefteq A$	N is a fully invariant submodule of A
$N \leq_e A$	N is an essential (large) submodule of A
$N \leq_c A$	N is a complement submodule of A

$N \leq_d A$	N is a direct summand of A
$N \leq_s A$	N is a small submodule of A
$N \overset{cs}{\hookrightarrow} L$	$N \subseteq L$ is cosmall in A
Δ	$\{f \in \mathrm{End}A \mid \ker f \leq_e A\}$
$CFM_\Gamma(R)$	The $\Gamma \times \Gamma$ column finite matrix ring over R
$CRFM_\Gamma(R)$	The $\Gamma \times \Gamma$ column and row finite matrix ring over R
$\|I\|$	The cardinal of I

Chapter 1

Introducing Modules

In this chapter we introduce modules, explore some of their general properties and then go on to look at particular types of modules. Why are modules so important? There are two basic reasons. In the first place, the concept of a module is a very general one and examples abound, including some very familiar ones. Secondly, the theory of rings is intimately connected with that of modules and it is inconceivable to try to study the former without some reference to the latter. Without further ado let us begin with the definition of a module.

1.1 Modules

Let A be an Abelian group with binary operation $+$. Let $\text{End}A$ denote the collection of endomorphisms θ of A, i.e., $\theta : A \to A$ satisfies

$$\theta(a+b) = \theta(a) + \theta(b) \quad (a, b \in A).$$

Define addition and multiplication in $\text{End}A$ by

$$(\theta + \phi)(a) = \theta(a) + \phi(a)$$
$$(\theta\phi)(a) = \theta(\phi(a))$$

for all $\theta, \phi \in \text{End}A$, $a \in A$. With these definitions it can be checked that $\text{End}A$ is a ring, called the *ring of endomorphisms of* A, with zero element the *zero mapping* $z_A : A \to A$ given by $z_A(a) = 0$ $(a \in A)$ and identity element the *identity mapping* $i_A : A \to A$ given by $i_A(a) = a$ $(a \in A)$. For example, suppose that A is an infinite cyclic group generated by an element a. For each $n \in \mathbb{Z}$ define $\theta_n \in \text{End}A$ by $\theta_n(ka) = nka$ $(k \in \mathbb{Z})$. It is easy to verify that

(i) $\text{End}A = \{\theta_n : n \in \mathbb{Z}\}$, and
(ii) $\theta_n + \theta_m = \theta_{n+m}$ and $\theta_n\theta_m = \theta_{nm}$ $(n, m \in \mathbb{Z})$.

Thus the mapping $f : \mathbb{Z} \to \text{End}A$ defined by $f(n) = \theta_n$ $(n \in \mathbb{Z})$ is a ring isomorphism, i.e., $\text{End}A \cong \mathbb{Z}$.

Now let R be a ring. *Throughout, without further notice, R will always denote a ring with identity 1.* An Abelian group A is called a *left R-module* if there exists a homomorphism $f : R \to \text{End}A$. In particular, f satisfies $f(1) = i_A$. A homomorphism $f : R \to \text{End}A$ is sometimes called a *representation of R as a ring of endomorphisms of A*. In particular, A is a left $(\text{End}A)$-module.

Let A be a left R-module and $f : R \to \text{End}A$ the associated homomorphism. Given any elements $r \in R$ and $a \in A$ we define the "product" ra as

$$ra = f(r)(a).$$

Note the following elementary properties:

$$\begin{array}{ll} \text{(i)} & ra \in A, \\ \text{(ii)} & r(a_1 + a_2) = ra_1 + ra_2, \\ \text{(iii)} & (r_1 + r_2)a = r_1a + r_2a, \\ \text{(iv)} & (r_1r_2)a = r_1(r_2a), \text{ and} \\ \text{(v)} & 1a = a \end{array} \tag{1.1}$$

for all $a, a_1, a_2 \in A$ and $r, r_1, r_2 \in R$. Conversely, suppose that A is an Abelian group and R a ring for which the product ra can be defined for all $r \in R$, $a \in A$, in such a way that (1.1) holds. For each r in R define $\theta_r \in \text{End}A$ by $\theta_r(a) = ra$ $(a \in A)$. Define $g : R \to \text{End}A$ by $g(r) = \theta_r$ $(r \in R)$. Then g is clearly a homomorphism and thus A is a left R-module. To summarise, an Abelian group A is a left R-module if and only if the product ra can be defined for all $r \in R$ and $a \in A$ that (1.1) holds.

The product in the last paragraph is very reminiscent of something so familiar, namely, the product of a vector by a scalar in a vector space, and for this reason we shall refer to it as *multiplication by a scalar*. In fact it is now clear that if R is a field, then left R-modules are precisely vector spaces over R. On the other hand, every Abelian group A is a $\mathbb{Z}$-module with multiplication by scalars given by

$$na = \begin{cases} a + \cdots + a \quad \text{(n times)}, & \text{if } n \in \mathbb{N}, \\ 0, & \text{if } n = 0, \\ (-a) + \cdots + (-a) \quad \text{(n times)}, & \text{if } -n \in \mathbb{N}, \end{cases}$$

for all $a \in A$. Vector spaces and Abelian groups are good examples of modules to keep in mind, and it will be helpful to consider, each time a new definition is introduced in the sequel, what it means for these two classes of examples. Indeed a great deal of what follows is motivated by them.

Before proceeding any further, let us make several remarks. Given a left R-module A, we know that there is a homomorphism $f : R \to \text{End}A$ but of course in general there may be many such homomorphisms. For example, if h is an endomorphism of R then $fh : R \to \text{End}A$ is a homomorphism. For each homomorphism from R to $\text{End}A$ there is a corresponding multiplication by scalars defined as above, so that, in general A will have many such products. However, the important thing to remember is that A has one such product and we tend to forget that it has others. Secondly, if S is a subring of R, then by considering the appropriate restriction of f we see at once that A is a left S-module. More generally, if T is a ring and $g : T \to R$ a homomorphism, then A is a left T-module. Thus the same Abelian group A will be a left module for many different rings.

If A is a left R-module, then there is a homomorphism $f : R \to \mathrm{End}A$ and as always in this kind of situations we are interested in how much information about R is lost under this homomorphism. The normal way to measure this loss is by considering the kernel $\ker f$ of f, which is a very important ideal as far as the module A is concerned. The ideal $\ker f$ is called the *annihilator of A in R*, and is denoted by $\mathrm{ann}_R A$. Why is it given this name? Recall that

$$\mathrm{ann}_R A = \{r \in R : f(r) = z_A\} = \{r \in R : ra = 0 \ (a \in A)\},$$

so that each element r in $\mathrm{ann}_R A$ "annihilates" each element a in A in the sense that $ra = 0$. Now let I be an ideal of R such that $I \subseteq \mathrm{ann}_R A$. Then $f : R \to \mathrm{End}A$ induces a homomorphism $f_1 : R/I \to \mathrm{End}A$ so that A is a non-trivial left (R/I)-module. Note further that in terms of the multiplication by scalars defined above we have

$$(r + I)a = ra \quad (r \in R,\ a \in A).$$

Borrowing from the language of representation theory, a left R-module A is called *faithful* if $\mathrm{ann}_R A = 0$, i.e., with the above notation $f : R \to \mathrm{End}A$ is a monomorphism, and so a faithful representation of R.

Right modules can be defined in analogy with left modules. An Abelian group A is called a *right R-module* if there exists an anti-homomorphism $g : R \to \mathrm{End}A$. This just means that g preserves addition and

$$g(r_1 r_2) = g(r_2)g(r_1)$$

for all $r_1, r_2 \in R$. In practice this allows one to define the product ar for all $a \in A$, $r \in R$ so that the analogous properties to (1.1) are satisfied. If S and T are rings, then by a *left S-, right T-bimodule* is meant an Abelian group A such that A is a left S-module, a right T-module, and

$$s(at) = (sa)t$$

for all $a \in A$, $s \in S$ and $t \in T$. We often denote the fact that an Abelian group is a left R-module, a right R-module, or a left S-, right T-bimodule simply by

$${}_R A, \ A_R, \ {}_S A_T,$$

respectively. Let S and T be subrings of a ring R and I an ideal of R. Then I is a left S-, right T-bimodule, i.e., ${}_S I_T$, and in particular ${}_R R_R$. Moreover, left ideals of

R are left R-modules and right ideals of R are right R-modules, where in all of these cases the product by a scalar on either side is given by the multiplication of R.

Let $n \in \mathbb{N}$ and $M_n(R)$ denote the ring consisting of all $n \times n$ matrices with entries in R, addition and multiplication of matrices being given by the usual rules. If R is a field, then the ring $M_n(R)$ is familiar. In the sequel matrix rings over rings other than fields will be considered in various places. At this point merely note that ${}_RM_n(R)_R$. For, if $m \in M_n(R)$ is a matrix with (i,j)th entry m_{ij} for $1 \leq i,j \leq n$ and $r \in R$, define rm and mr to be the matrices with (i,j)th entry rm_{ij} and $m_{ij}r$ for $1 \leq i,j \leq n$, respectively, and check that ${}_RM_n(R)_R$.

Now we introduce a method of producing modules from given ones. Let Λ be any non-empty index set (i.e., Λ is just a set of labels). Let R be a ring and A_λ $(\lambda \in \Lambda)$ a collection of left R-modules. *The direct product* $A = \prod_{\lambda\in\Lambda} A_\lambda$ is defined as follows. The set A consists of all lists $\{a_\lambda\}$, where $a_\lambda \in A_\lambda$ $(\lambda \in \Lambda)$, and addition and multiplication by a scalar are defined by

$$\{a_\lambda\} + \{b_\lambda\} = \{a_\lambda + b_\lambda\},$$

and

$$r\{a_\lambda\} = \{ra_\lambda\},$$

for all $a_\lambda, b_\lambda \in A_\lambda$ $(\lambda \in \Lambda)$, $r \in R$. It is not hard to check that A is also a left R-module. For each $\lambda \in \Lambda$ the module A_λ is called a *direct factor of* A. In the special case when there exists ${}_RC$ such that $A_\lambda = C$ $(\lambda \in \Lambda)$, then A is written as C^Λ. Perhaps it is worth spelling out what this definition means when Λ is a finite set. Suppose that k is a positive integer and A_i is a left R-module for each $1 \leq i \leq k$ (so we are taking $\Lambda = \{1,\ldots,k\}$). Then

$$A = \{(a_1,\ldots,a_k) \ : \ a_i \in A_i \ (1 \leq i \leq k)\},$$

with equality, addition, and multiplication being given by the rules:

$$\begin{aligned}(a_1,\ldots,a_k) &= (b_1,\ldots,b_k) \text{ if and only if } a_i = b_i \ (1 \leq i \leq k),\\ (a_1,\ldots,a_k) + (b_1,\ldots,b_k) &= (a_1+b_1,\ldots,a_k+b_k), \text{ and}\\ r(a_1,\ldots,a_k) &= (ra_1,\ldots,ra_k),\end{aligned}$$

for all $a_i, b_i \in A_i$ $(1 \leq i \leq k)$ and $r \in R$. In case $A_i = C$ $(1 \leq i \leq k)$, A is written as C^k.

Let R_λ $(\lambda \in \Lambda)$ be a non-empty collection of rings and $R = \prod_{\lambda\in\Lambda} R_\lambda$ their direct product as $\mathbb{Z}$-modules. The $\mathbb{Z}$-module R can be given the structure of a ring by defining multiplication as

$$\{r_\lambda\} \cdot \{s_\lambda\} = \{r_\lambda s_\lambda\}$$

for all $r_\lambda, s_\lambda \in R_\lambda$ $(\lambda \in \Lambda)$. The ring R is called the *direct product of the rings* R_λ $(\lambda \in \Lambda)$. Note that R has zero element $\{0_\lambda\}$ and identity element $\{1_\lambda\}$, where 0_λ is the zero and 1_λ the identity of R_λ $(\lambda \in \Lambda)$.

Notation. The symbol 0 will have a number of different roles in the sequel. For example, 0 will denote the zero element of a ring or of a module. In addition, the subset $\{0\}$ of a ring or module will also be denoted by 0. No confusion should arise since the context should make it clear which meaning to attach to 0.

Exercises

1.1. Given ${}_RA$, prove that

(i) $0a = 0$ and $r0 = 0$ for all $a \in A$ and $r \in R$,
(ii) $(-r)a = -(ra) = r(-a)$ for all $r \in R$, $a \in A$, and
(iii) $r(na) = n(ra)$ for all $r \in R$, $n \in \mathbb{Z}$, and $a \in A$.

1.2. Let R be a commutative ring (i.e., $rs = sr$ for all $r, s \in R$) and A an Abelian group. Prove that the following statements are equivalent:

(i) A is a left R-module,
(ii) A is a right R-module,
(iii) A is a left R-, right R-bimodule.

(In this case A is called simply an R-*module.*)

1.3. Prove that if ${}_RA$, then ${}_RA_{\mathbb{Z}}$.

1.4. Let I be an ideal of a ring R and A a left (R/I)-module. Prove that

(i) A is a left R-module such that $I \subseteq \text{ann}_R A$, and
(ii) $\text{ann}_{R/I} A = (\text{ann}_R A)/I$.
Given ${}_RB$, deduce that B is a faithful left (R/I)-module if and only if $I = \text{ann}_R B$.

1.5. Give an example of a ring R and a subring S of R such that S is not a left R-module. Give an example of a ring T for which there exists a left T-module A which is not a right T-module.

1.6. Let R and S be rings and ${}_RA_S$. Let T denote the set of "matrices" of the form

$$\begin{bmatrix} r & a \\ 0 & s \end{bmatrix},$$

where $r \in R$, $a \in A$, and $s \in S$. Define

$$\begin{bmatrix} r & a \\ 0 & s \end{bmatrix} = \begin{bmatrix} r' & a' \\ 0 & s' \end{bmatrix} \text{ if } r = r',\ a = a',\text{ and } s = s',$$

$$\begin{bmatrix} r & a \\ 0 & s \end{bmatrix} + \begin{bmatrix} r' & a' \\ 0 & s' \end{bmatrix} = \begin{bmatrix} r + r' & a + a' \\ 0 & s + s' \end{bmatrix} \text{ and}$$

$$\begin{bmatrix} r & a \\ 0 & s \end{bmatrix} \begin{bmatrix} r' & a' \\ 0 & s' \end{bmatrix} = \begin{bmatrix} rr' & ra' + as' \\ 0 & ss' \end{bmatrix},$$

for all $r, r' \in R$, $s, s' \in S$, and $a, a' \in A$. Prove that with respect to these definitions T is a ring. The ring T is usually denoted by

$$\begin{bmatrix} R & A \\ 0 & S \end{bmatrix}.$$

1.7. Let R be a ring and ${}_RA_R$. Prove that the set of "matrices"

$$\begin{bmatrix} r & a \\ 0 & r \end{bmatrix},$$

where $r \in R$ and $a \in A$, forms a ring which is usually denoted by

$$\begin{bmatrix} R & & A \\ & \diagdown & \\ 0 & & R \end{bmatrix}.$$

Further, prove that if R is a commutative ring and A is an R-module, then the ring

$$\begin{bmatrix} R & & A \\ & \diagdown & \\ 0 & & R \end{bmatrix}$$

is also commutative.

1.2 Homomorphisms

Recall that in the theory of vector spaces the key idea is that of a linear mapping. The corresponding concept in module theory is called a homomorphism, and this will be defined next. Let ${}_RA$ and ${}_RB$. A mapping $\theta : A \to B$ is called an R-*homomorphism* (or simply a *homomorphism* when there is no ambiguity about the ring being considered) provided

$$\begin{aligned} \theta(a_1 + a_2) &= \theta(a_1) + \theta(a_2), \text{ and} \\ \theta(ra) &= r\theta(a) \end{aligned}$$

for all $a, a_1, a_2 \in A$, $r \in R$. The first of these properties shows that θ is an Abelian group homomorphism in particular, and thus we know at once from the properties of such homomorphisms that

$$\begin{aligned} \text{(i)} \quad & \theta(0_A) = 0_B, \\ \text{(ii)} \quad & \theta(-a) = -\theta(a), \text{ and} \\ \text{(iii)} \quad & \theta(na) = n\theta(a), \end{aligned} \tag{1.2}$$

for all $a \in A$, $n \in \mathbb{Z}$, where $0_A, 0_B$ denote the zero elements of A and B, respectively. In particular, (1.2) shows that $\mathbb{Z}$-homomorphisms are precisely Abelian group homomorphisms. Moreover, (1.2) also gives that if a_1 and a_2 are elements of A such that $\theta(a_1) = \theta(a_2)$, then $\theta(a_1 - a_2) = 0_B$. The *kernel* $\ker\theta$ and the *image* $\operatorname{im}\theta$ of θ are defined as follows:

$$\ker\theta = \{a \in A : \theta(a) = 0_B\},$$

and

$$\operatorname{im}\theta = \{\theta(a) : a \in A\}.$$

The homomorphism θ is called a *monomorphism* provided it is one-to-one, i.e., $\theta(a_1) \neq \theta(a_2)$ for all $a_1 \neq a_2$ in A, equivalently $\ker\theta = \{0_A\}$. On the other hand, θ is called an *epimorphism* provided it is onto, i.e., for each element b in B there exists an element a in A such that $b = \theta(a)$, equivalently $B = \operatorname{im}\theta$. If there is an epimorphism $\theta : A \to B$, then B is sometimes called a *homomorphic image of* A. The mapping θ is called an *isomorphism* if it is both a monomorphism and an epimorphism, and in this case A and B are called *isomorphic*, B is called an *isomorphic copy of* A, and we write $A \cong B$. If $\theta : A \to B$ is an isomorphism, then θ is a bijection in particular, and hence has an inverse mapping $\theta^{-1} : B \to A$, which is also an R-homomorphism, as a moment's thought will show. If $\theta : A \to B$ is a monomorphism, then θ induces an isomorphism $A \cong \operatorname{im}\theta$, and for this reason θ will sometimes be called an *embedding* and we say that A *is embedded in* B.

Given Abelian groups A, B, the collection of all $\mathbb{Z}$-homomorphisms $\theta : A \to B$ will be denoted by $\operatorname{Hom}(A, B)$. If $\theta, \phi \in \operatorname{Hom}(A, B)$ then their sum $\theta + \phi$ is the mapping from A to B defined by

$$(\theta + \phi)(a) = \theta(a) + \phi(a) \quad (a \in A).$$

It is easy to check that with this definition of addition $\operatorname{Hom}(A, B)$ is an Abelian group. The zero of the group $\operatorname{Hom}(A, B)$ is the mapping which maps each element of A to the zero of B, and will also be denoted by 0. Now suppose that ${}_RA$ and ${}_RB$. Then it can be checked that the collection $\operatorname{Hom}_R(A, B)$ of all R-homomorphisms $\theta : A \to B$ is a subgroup of $\operatorname{Hom}(A, B)$. Note that in this notation $\operatorname{Hom}_{\mathbb{Z}}(A, B) = \operatorname{Hom}(A, B)$. Sometimes $\operatorname{Hom}_R(A, B)$, or even $\operatorname{Hom}(A, B)$, is a left or right R-module, as the following result shows (see also Exercises 1.9 and 1.24).

Proposition 1.1. *Let A be a right R-module and B an Abelian group. Then $\operatorname{Hom}(A, B)$ is a left R-module.*

Proof. Given $r \in R$ and $\theta \in \operatorname{Hom}(A, B)$, define the mapping $r.\theta : A \to B$ by

$$(r.\theta)(a) = \theta(ar) \ (a \in A).$$

Then $r.\theta \in \operatorname{Hom}(A, B)$. If $r_1, r_2 \in R$ and $\theta \in \operatorname{Hom}(A, B)$ then, for any $a \in A$,

$$\begin{aligned}[(r_1 + r_2).\theta](a) &= \theta(a(r_1 + r_2)) = \theta(ar_1 + ar_2) = \theta(ar_1) + \theta(ar_2) \\ &= (r_1.\theta)(a) + (r_2.\theta)(a) = (r_1.\theta + r_2.\theta)(a).\end{aligned}$$

It follows that $(r_1 + r_2).\theta = r_1.\theta + r_2.\theta$ and hence (1.1)(iii) is satisfied. The remaining properties in (1.1) can be verified in a similar way for this product, and it follows that $\mathrm{Hom}(A, B)$ is a left R-module. □

For any ${}_RA$, a homomorphism $\theta : A \to A$ is called an *R-endomorphism* (or simply *endomorphism*) *of* A. Denote the collection of all R-endomorphisms of A by $\mathrm{End}_R(A)$. Note that

$$\mathrm{End}_{\mathbb{Z}}(A) = \mathrm{End}A \quad \text{and} \quad \mathrm{End}_R(A) = \mathrm{Hom}_R(A, A).$$

Moreover, $\mathrm{End}_R(A)$ is a subring of $\mathrm{End}A$. An endomorphism θ of A is called an *automorphism* provided it is a bijection, i.e., automorphisms are endomorphisms which are also isomorphisms.

Given left R-modules A, B, C, D, let $\theta \in \mathrm{Hom}_R(A, B)$, $\phi \in \mathrm{Hom}_R(B, C)$ and $\psi \in \mathrm{Hom}_R(C, D)$. Define $\phi\theta : A \to C$ by

$$\phi\theta(a) = \phi(\theta(a)) \quad (a \in A).$$

It can easily be checked that $\phi\theta \in \mathrm{Hom}_R(A, C)$ and that $\psi(\phi\theta) = (\psi\phi)\theta$, so that the mapping $\psi(\phi\theta)$ is simply written $\psi\phi\theta$. There is a $\mathbb{Z}$-homomorphism $\theta^\star$: $\mathrm{Hom}_R(C, A) \to \mathrm{Hom}_R(C, B)$, defined by

$$\theta^\star(\alpha) = \theta\alpha \quad (\alpha \in \mathrm{Hom}_R(C, A));$$

likewise, there is a $\mathbb{Z}$-homomorphism $\theta_\star : \mathrm{Hom}_R(B, C) \to \mathrm{Hom}_R(A, C)$ defined by

$$\theta_\star(\beta) = \beta\theta \quad (\beta \in \mathrm{Hom}_R(B, C)).$$

Let A_λ $(\lambda \in \Lambda)$ be a non-empty collection of left R-modules and A their direct product $\prod_{\lambda\in\Lambda} A_\lambda$. For each $\mu \in \Lambda$ there exists an R-epimorphism $\pi_\mu : A \to A_\mu$ (called *the canonical* or *natural projection*) defined by

$$\pi_\mu(\{a_\lambda\}) = a_\mu$$

for all $a_\lambda \in A_\lambda$ $(\lambda \in \Lambda)$. The mapping π_μ is sometimes called the *μ-projection.* On the other hand, for each $\nu \in \Lambda$ and $b \in A_\nu$ let $\iota_\lambda(b)$ denote the element $\{b_\lambda\}$ of A defined by $b_\lambda = b$ if $\lambda = \nu$ and $b_\lambda = 0$ otherwise. Then ι_ν is an R-monomorphism (called *the canonical* or *natural injection*) for each $\nu \in \Lambda$. Note that

(i) $\pi_\mu\iota_\nu = 0$ for all $\mu \neq \nu$ in Λ, and
(ii) $\pi_\mu\iota_\mu = i_{A_\mu}$ $(\mu \in \Lambda)$.

In this context, in the sequel any mappings π_μ and ι_μ introduced without comment should be understood to be these R-homomorphisms.

Given left R-modules A, B, C, to say that

$$A \xrightarrow{\theta} B \xrightarrow{\phi} C$$

is an *exact sequence* means that θ and ϕ are R-homomorphisms such that

$$\operatorname{im}\theta = \ker\phi.$$

Note, in particular, that if $A = 0$, then ϕ is a monomorphism and if $C = 0$ then θ is an epimorphism. More generally, an exact sequence consists of a collection of left R-modules A_n $(n \in \mathbb{Z})$ and R-homomorphisms $\theta_n : A_n \to A_{n+1}$ such that $\operatorname{im}\theta_n = \ker\theta_{n+1}$ $(n \in \mathbb{Z})$ and this is denoted by

$$\cdots \longrightarrow A_{-2} \xrightarrow{\theta_{-2}} A_{-1} \xrightarrow{\theta_{-1}} A_0 \xrightarrow{\theta_0} A_1 \xrightarrow{\theta_1} A_2 \xrightarrow{\theta_2} \cdots \text{ exact.} \tag{1.3}$$

If $A_m = 0$ for all $m \geq n+1$, then (1.3) is written

$$\cdots \longrightarrow A_{-2} \xrightarrow{\theta_{-2}} A_{-1} \xrightarrow{\theta_{-1}} A_0 \xrightarrow{\theta_0} A_1 \xrightarrow{\theta_1} \cdots \longrightarrow A_n \xrightarrow{\theta_n} 0 \qquad \text{exact.}$$

An exact sequence

$$0 \longrightarrow A \xrightarrow{\theta} B \xrightarrow{\phi} C \longrightarrow 0 \tag{1.4}$$

is called a *short exact sequence.* A short exact sequence (1.4) it is said to *split* or to be *split exact* provided there exists $\psi \in \operatorname{Hom}_R(C, B)$ such that $\phi\psi = i_C$.

A set-up of the type

$$\begin{array}{ccc} A & \xrightarrow{\theta} & B \\ {\scriptstyle\alpha}\downarrow & & \downarrow{\scriptstyle\beta} \\ C & \xrightarrow{\phi} & D \end{array}$$

where A, B, C, D are all left R-modules and $\theta, \phi, \alpha, \beta$ R-homomorphisms is called a *diagram of R-modules.* The diagram is said to *commute* or to be *commutative* provided $\beta\theta = \phi\alpha$. Using such diagrams as building blocks, bigger diagrams of R-modules can be produced, e.g.,

$$\begin{array}{ccccccccl} & & B' & \xrightarrow{\phi'} & C' & \longrightarrow & 0 & & \text{exact} \\ & & {\scriptstyle\beta'}\downarrow & & \downarrow{\scriptstyle\gamma'} & & & & \\ A & \xrightarrow{\theta} & B & \xrightarrow{\phi} & C & \longrightarrow & 0 & & \text{exact} \\ {\scriptstyle\alpha}\downarrow & & {\scriptstyle\beta}\downarrow & & \downarrow{\scriptstyle\gamma} & & & & \\ A'' & \xrightarrow{\theta''} & B'' & \xrightarrow{\phi''} & C'' & \longrightarrow & 0 & & \\ \downarrow & & \downarrow & & \downarrow & & & & \\ 0 & & 0 & & 0 & & & & \\ \text{exact} & & \text{exact} & & \text{exact} & & & & \end{array} \tag{1.5}$$

where A, B, C, B', C', A'', B'', C'' are left R-modules and α, β, γ, β', γ', θ, ϕ, ϕ', θ'', ϕ'' R-homomorphisms. To say that such a bigger diagram commutes is to mean that each of the squares commutes, i.e., $\gamma'\phi' = \phi\beta'$, $\gamma\phi = \phi''\beta$ and $\beta\theta = \theta''\alpha$. The diagram (1.5) has *rows* and *columns* in the obvious sense. For example the *first column* of (1.5) is

$$\begin{array}{c} A \\ \alpha \downarrow \\ A'' \\ \downarrow \\ 0 \end{array}$$

exact

and the *top row* of (1.5) is

$$B' \xrightarrow{\phi'} C' \longrightarrow 0 \qquad \text{exact.}$$

Proving results involving diagrams is called *diagram chasing* and is without question one of the most entertaining aspects of the subject. To illustrate the technique we prove:

Proposition 1.2. *Suppose that diagram* (1.5) *commutes. Then the bottom row of* (1.5) *is exact.*

Proof. ϕ'' is onto. Let $c'' \in C''$. Since γ and ϕ are epimorphisms, there exists $b \in B$ such that $c'' = \gamma\phi(b)$. But $\gamma\phi = \phi''\beta$ (because (1.5) commutes) so that $c'' = \phi''\beta(b) = \phi''(\beta(b))$.

$\operatorname{im}\theta'' = \ker\phi''$. Let $a'' \in A''$. Since α is an epimorphism, $a'' = \alpha(a)$ for some $a \in A$. Then the fact that (1.5) is commutative gives

$$\phi''\theta''(a'') = \phi''\theta''\alpha(a) = \phi''\beta\theta(a) = \gamma\phi\theta(a) = 0$$

because the second row is exact. It follows that $\operatorname{im}\theta'' \subseteq \ker\phi''$. On the other hand, let $b'' \in \ker\phi''$. Since β is an epimorphism, $b'' = \beta(b_1)$ for some $b_1 \in B$. Thus $0 = \phi''(b'') = \phi''\beta(b_1) = \gamma\phi(b_1)$ and hence $\phi(b_1) \in \ker\gamma = \operatorname{im}\gamma'$ by exactness of the third column. Let $c' \in C'$ satisfy $\gamma'(c') = \phi(b_1)$. There exists $b' \in B'$ such that $c' = \phi'(b')$ and hence

$$\phi(b_1) = \gamma'(c') = \gamma'\phi'(b') = \phi\beta'(b').$$

Thus $b_1 - \beta'(b') \in \ker\phi = \operatorname{im}\theta$ by the exactness of the second row, and hence there exists $a_1 \in A$ such that $b_1 - \beta'(b') = \theta(a_1)$. But

$$b'' = \beta(b_1) = \beta[\beta'(b') + \theta(a_1)] = \beta\beta'(b') + \beta\theta(a_1) = \beta\theta(a_1) = \theta''\alpha(a_1),$$

because $\beta\beta' = 0$ by the exactness of the second column. Thus $b'' = \theta''(\alpha(a_1)) \in \operatorname{im}\theta''$. It follows that $\ker\phi'' \subseteq \operatorname{im}\theta''$. Thus $\ker\phi'' = \operatorname{im}\theta''$. □

If A and A' are right R-modules, then a $\mathbb{Z}$-homomorphism $\theta : A \to A'$ is called an *R-homomorphism* if

$$\theta(ar) = \theta(a)r \ (a \in A,\ r \in R)$$

and the above discussion applies equally well to homomorphisms of right R-modules.

Exercises

1.8. Given ${}_RA$ and $\theta \in \operatorname{End}_R(A)$, the mapping $\theta^2 \in \operatorname{End}_R(A)$ is defined by $\theta^2 = \theta\theta$, i.e., $\theta^2(a) = \theta(\theta(a))$ for all $a \in A$. More generally, define $\theta^{n+1} = \theta^n\theta$ for any $n \in \mathbb{N}$. Prove that

(i) $\ker\theta \subseteq \ker\theta^2 \subseteq \ker\theta^3 \subseteq \cdots$, and
(ii) $\operatorname{im}\theta \supseteq \operatorname{im}\theta^2 \supseteq \operatorname{im}\theta^3 \supseteq \cdots$.

Further, prove that the following statements are equivalent for a positive integer n:

a) $\theta^n = 0$.
b) $\ker\theta^n = A$.
c) $\operatorname{im}\theta^n = 0$.

(An element r of a ring R is called *nilpotent* if $r^n = 0$ for some $n \in \mathbb{N}$, where $r^n = r \cdots r$ (n times).)

1.9. Let A be an Abelian group and B_R. Prove that $\operatorname{Hom}(A, B)$ is a right R-module.

1.10. Given the commutative diagram of left R-modules

$$\begin{array}{ccccccccc} 0 & \longrightarrow & A & \longrightarrow & B & \longrightarrow & C & \longrightarrow & 0 \\ & & \downarrow{\scriptstyle\alpha} & & \downarrow{\scriptstyle\beta} & & \downarrow{\scriptstyle\gamma} & & \\ 0 & \longrightarrow & A' & \longrightarrow & B' & \longrightarrow & C' & \longrightarrow & 0 \end{array} \tag{1.6}$$

assume that the rows are exact and β is an isomorphism. Prove that

(i) α is a monomorphism and γ an epimorphism, and
(ii) α is an epimorphism if and only if γ is a monomorphism.

1.11. Given the commutative diagram (1.6), assume that α, β and γ are isomorphisms. Prove that the top row is (split) exact if and only if the bottom row is (split) exact.

1.12. Given the commutative diagram of left R-modules

$$\begin{array}{cccccl} A & \xrightarrow{\theta} & B & \xrightarrow{\phi} & C & \text{exact} \\ \downarrow{\scriptstyle\alpha} & & \downarrow{\scriptstyle\beta} & & \downarrow{\scriptstyle\gamma} & \\ D & \xrightarrow{\xi} & E & \xrightarrow{\eta} & F & \text{exact} \end{array}$$

prove

(i) if α, γ and ξ are monomorphisms, then so is β,
(ii) if α, γ and ϕ are epimorphisms, then so is β,
(iii) if α and ϕ are epimorphisms and β is a monomorphism, then γ is a monomorphism, and
(iv) if ξ and γ are monomorphisms and β an epimorphism, then α is an epimorphism.

1.13. Given the commutative diagram

$$\begin{array}{ccccccccc} A & \longrightarrow & B & \longrightarrow & C & \longrightarrow & D & \longrightarrow & E \quad \text{exact} \\ {\scriptstyle\alpha}\downarrow & & {\scriptstyle\beta}\downarrow & & {\scriptstyle\gamma}\downarrow & & {\scriptstyle\delta}\downarrow & & {\scriptstyle\varepsilon}\downarrow \\ A' & \longrightarrow & B' & \longrightarrow & C' & \longrightarrow & D' & \longrightarrow & E' \quad \text{exact} \end{array}$$

prove

(i) if α is an epimorphism and β, δ monomorphisms, then γ is a monomorphism,
(ii) if ε is a monomorphism and β, δ epimorphisms, then γ is an epimorphism, and
(iii) if α, β, δ and ε are all isomorphisms, then so is γ.

(This result is called the *Five Lemma.*)

1.14. Given ${}_{\mathbb{Z}}D$ and $0 \longrightarrow {}_{\mathbb{Z}}A \xrightarrow{\theta} {}_{\mathbb{Z}}B \xrightarrow{\phi} {}_{\mathbb{Z}}C \longrightarrow 0$ exact, prove

(i) $0 \longrightarrow \mathrm{Hom}(D, A) \xrightarrow{\theta^*} \mathrm{Hom}(D, B) \xrightarrow{\phi^*} \mathrm{Hom}(D, C)$ is exact, and
(ii) $0 \longrightarrow \mathrm{Hom}(C, D) \xrightarrow{\phi_*} \mathrm{Hom}(B, D) \xrightarrow{\theta_*} \mathrm{Hom}(A, D)$ is exact.

1.15. Let $k \in \mathbb{N}$ and ${}_RA$. Let $a_i \in A$ $(1 \leq i \leq k)$ and define $\phi : {}_RR \to A^k$ by

$$\phi(r) = (ra_1, \ldots, ra_k) \quad (r \in R).$$

Prove that ϕ is an R-homomorphism with kernel $\{r \in R : ra_i = 0\ (1 \leq i \leq k)\}$. Extend this result to arbitrary subsets of A. Prove that, in particular, there is an R-homomorphism $\psi : {}_RR \to {}_RA$ with kernel $\mathrm{ann}_R A$. Deduce that A is faithful if and only if ${}_RR$ can be embedded in A^Λ for some non-empty set Λ.

1.16. Let A_λ $(\lambda \in \Lambda)$ be a non-empty collection of left R-modules. Prove that a left R-module B is R-isomorphic to $\prod_{\lambda \in \Lambda} A_\lambda$ if and only if there exists a family of R-homomorphisms $\phi_\lambda : B \to A_\lambda$ $(\lambda \in \Lambda)$ such that for all ${}_RC$ and homomorphisms $\theta_\lambda : C \to A_\lambda$ $(\lambda \in \Lambda)$ there exists a unique R-homomorphism $\theta : C \to B$ such that the diagram

$$\begin{array}{ccc} C & \xrightarrow{\theta} & B \\ {\scriptstyle\theta_\lambda}\searrow & & \swarrow{\scriptstyle\phi_\lambda} \\ & A_\lambda & \end{array}$$

commutes for all $\lambda \in \Lambda$.

1.3 Submodules

Let A be an Abelian group and X a non-empty subset of A. Let $\theta \in \mathrm{End}A$. Let $\theta(X) = \{\theta(x) : x \in X\}$. If S is a subring of $\mathrm{End}A$, then X is said to be *S-invariant* if $\theta(X) \subseteq X$ for all $\theta \in S$. Now suppose that A is a left R-module, so there exists a homomorphism $f : R \to \mathrm{End}A$. A subgroup B of A is called an *R-submodule* (as usual the "R" tends to get dropped unless there is danger of confusion) provided B is $(\mathrm{im}f)$-invariant. Why are such subgroups of A called submodules? Recall that f provides a multiplication by scalars given by

$$ra = f(r)(a) \quad (r \in R,\ a \in A).$$

Thus a subgroup B of A is an R-submodule if and only if $rb \in B$ for all $r \in R$ and $b \in B$. In this case B inherits the properties listed in (1.1) and hence B is itself a left R-module with respect to the multiplication by scalars and addition of A restricted to B. If R is a field, then A is a vector space and the R-submodules of A are precisely the subspaces of A. Note also that any subgroup B of A is a $\mathbb{Z}$-submodule of A. Moreover, the R-submodules of ${}_RR$ are the left ideals of R. To summarise, the following result gives a convenient test for submodules:

Proposition 1.3. *A subset B of a left R-module A is an R-submodule of A if and only if*

$$0 \in B,\ b_1 - b_2 \in B \text{ and } rb \in B$$

for all $b, b_1, b_2 \in B, r \in R$, where 0 is the zero of A.

Note that the zero of ${}_RA$ belongs to every submodule of A because each submodule is a subgroup of A. Proposition 1.3 has the following consequence, whose proof is left to the reader. It will be convenient to understand by $\theta \in \mathrm{Hom}_R(A, B)$ that ${}_RA, {}_RB$ and of course θ is an R-homomorphism from A to B.

Corollary 1.4. *Let $\theta \in \mathrm{Hom}_R(A, B)$. Then $\ker\theta$ is an R-submodule of A and $\mathrm{im}\,\theta$ is an R-submodule of B.*

Let A be a left R-module. Then A is an R-submodule of A. A submodule B of A is called *proper* if $B \neq A$. On the other hand, if 0 is the zero of A, then $\{0\}$ is an R-submodule of A, called the *zero submodule* and denoted by 0 as well. A submodule B of A is called *non-zero* if $B \neq 0$. There are other ways of producing submodules. For example let ${}_RA$ and $a \in A$. Define

$$Ra = \{ra : r \in R\}.$$

Proposition 1.3 shows that Ra is an R-submodule of A, and in fact Ra is the smallest submodule of A containing the element a. Submodules of the form Ra are called *cyclic* and correspond to one-dimensional vector spaces and cyclic Abelian groups. In particular, $R = R1 = 1R$ shows that ${}_RR$ and R_R are both cyclic. Let A_λ $(\lambda \in \Lambda)$ be any non-empty collection of R-submodules of A. Then the set-theoretic intersection $\bigcap_{\lambda\in\Lambda} A_\lambda$ is also an R-submodule of A by Proposition 1.3.

Now define

$$\sum_{\lambda \in \Lambda} A_\lambda$$

to be the set of all elements of A which can be written in the form $a_1 + \cdots + a_n$, where for each $1 \leq i \leq n$ there exists $\lambda \in \Lambda$ such that $a_i \in A_\lambda$. By Proposition 1.3, $\sum_{\lambda \in \Lambda} A_\lambda$ is an R-submodule of A and it is called the *sum* of the submodules A_λ ($\lambda \in \Lambda$). In case Λ is a finite set, say $\Lambda = \{1, \ldots, k\}$ for some $k \in \mathbb{N}$, then $\sum_{\lambda \in \Lambda} A_\lambda$ is usually denoted by $\sum_{i=1}^{k} A_i$ or $A_1 + \cdots + A_k$. In particular, for any $k \in \mathbb{N}$ and elements $a_i \in A_i$ ($1 \leq i \leq k$) there is a corresponding submodule, namely

$$Ra_1 + \cdots + Ra_k,$$

which consists of all elements of A which can be written in the form

$$r_1 a_1 + \cdots + r_k a_k$$

for some $r_i \in R$ ($1 \leq i \leq k$), and this submodule corresponds to the linear span of a finite collection of elements of a vector space. Any submodule of the form $Ra_1 + \cdots + Ra_k$ is called *finitely generated.*

Given any $\theta \in \mathrm{Hom}_R(A, A')$, we have seen that $\ker \theta$ is a submodule of A. Conversely, any submodule B of ${}_RA$ is the kernel of an R-homomorphism $\phi : A \to A_1$. To see this, it is of course necessary to produce a module A_1 and homomorphism ϕ. Form the factor Abelian group A/B consisting of cosets

$$a + B = \{a + b : b \in B\} \quad (a \in A)$$

in the usual way, and define multiplication by scalars by the rule

$$r(a + B) = ra + B \quad (r \in R,\ a \in A).$$

Note that this product is well defined, because if a_1 and a_2 are elements of A representing the same coset, i.e., $a_1 + B = a_2 + B$, then $a_1 - a_2 \in B$ and hence so too does $ra_1 - ra_2 = r(a_1 - a_2)$, which implies $r(a_1 + B) = r(a_2 + B)$, i.e., the products coincide. It is a simple matter to check that (1.1) is satisfied and hence A/B is a left R-module (called the *factor* (or *quotient*) *module of* A *by* B). Define $\phi : A \to A/B$ by

$$\phi(a) = a + B \quad (a \in A).$$

A moment's thought shows that ϕ is an R-epimorphism with kernel B.

Let B be an R-submodule of ${}_RA$. Define $\iota : B \to A$ by $\iota(b) = b$ ($b \in B$). Then ι is an R-monomorphism called the *inclusion mapping* or *canonical injection*; in the sequel a mapping ι introduced without comment will mean this mapping. The mapping $\pi : A \to A/B$ defined by $\pi(a) = a + B$ ($a \in A$) is called the *canonical projection* of A onto A/B; π used without comment will always mean this mapping. Note also that if ${}_RC$ and $\theta \in \mathrm{Hom}_R(A, C)$, then $\theta|_B$ will denote the *restriction of* θ *to* B, namely $\theta|_B : B \to C$ is defined by $\theta|_B(b) = \theta(b)$ ($b \in B$). Clearly, $\theta|_B \in \mathrm{Hom}_R(B, C)$.

The next two results are sometimes called the *First and Second Isomorphism Theorems.*

Proposition 1.5. *Let $\theta : A \to A'$ be an R-homomorphism. Then* $\operatorname{im}\theta \cong A/\ker\theta$.

Proof. Let $K = \ker\theta$. Define a mapping $\phi : A/K \to \operatorname{im}\theta$ by

$$\phi(a + K) = \theta(a) \quad (a \in A).$$

If $a_1, a_2 \in A$ and satisfy $a_1 + K = a_2 + K$, then $a_1 - a_2 \in K$ and hence $\theta(a_1 - a_2) = 0$, i.e., $\theta(a_1) = \theta(a_2)$, so that the mapping ϕ is well defined. Since θ is an R-homomorphism, so is ϕ. Moreover, $\operatorname{im}\phi = \operatorname{im}\theta$ and

$$\ker\phi = \{a + K : a \in A \text{ and } \theta(a) = 0\} = \{0 + K\},$$

the zero submodule of A/K. Thus $\phi : A/K \to \operatorname{im}\theta$ is an isomorphism. □

Proposition 1.6. *Let B and C be R-submodules of a left R-module A. Then $B \cap C$ is an R-submodule of B, C is an R-submodule of $B + C$, and* $B/(B \cap C) \cong (B + C)/C$.

Proof. Define a mapping $\theta : B \to (B + C)/C$ by

$$\theta(b) = b + C \quad (b \in B).$$

It is not difficult to check that θ is an R-epimorphism with kernel $B \cap C$. By Proposition 1.5, it follows that $B/(B \cap C) \cong \operatorname{im}\theta = (B + C)/C$. □

Let A_λ $(\lambda \in \Lambda)$ denote a non-empty collection of left R-modules and let A be their direct product $\prod_{\lambda\in\Lambda} A_\lambda$. If $a = \{a_\lambda\} \in A$, where $a_\lambda \in A_\lambda$ $(\lambda \in \Lambda)$, then the *support* of a is defined to be the set of all λ in Λ such that $a_\lambda \neq 0$. The zero element of A has empty support, but the support of every other element is non-empty. Let B be the subset of A consisting of all elements of A with finite support. It can be checked that B is an R-submodule of A. The left R-module B is called the *direct sum* or *direct coproduct of the modules* A_λ $(\lambda \in \Lambda)$ and will be denoted $\bigoplus_{\lambda\in\Lambda} A_\lambda$. In the particular case that there exists ${}_RC$ such that $A_\lambda = C$ $(\lambda \in \Lambda)$ the module $\bigoplus_{\lambda\in\Lambda} A_\lambda$ is written $C^{(\Lambda)}$. Note that if Λ is a finite set, then the direct product A and direct sum B coincide. If $k \in \mathbb{N}$ and $\Lambda = \{1, \dots, k\}$, then B (and A) is denoted by $A_1 \oplus \cdots \oplus A_k$. Incidentally, if R_i $(1 \leq i \leq k)$ are rings, then their direct product will be denoted by $R_1 \oplus \cdots \oplus R_k$ and this ring is often referred to as the *direct sum of the rings* R_i $(1 \leq i \leq k)$. Since we are assuming that every ring considered has an identity element, the direct sum of an infinite family of rings (the definition is the obvious one!) will not be a ring since it will not have an identity element.

Notation. It will be convenient to denote the fact that B is an R-submodule of a left (respectively right) R-module A by

$$B \leq {}_RA \quad (\text{respectively } B \leq A_R).$$

Exercises

1.17. Prove that if $B \leq {}_RA$ and $C \leq {}_RB$, then $C \leq {}_RA$.

1.18. Let $B, C, D \leq {}_RA$ satisfy $B \subseteq C$, $B \cap D = C \cap D$, and $B + D = C + D$. Prove that $B = C$. Give an example to show that this result is false if the condition $B \subseteq C$ is removed.

1.19. Let $B, C, D \leq {}_RA$ and $C \subseteq B$. Prove that $B \cap (C + D) = C + (B \cap D)$. (This is called the *Modular Law*). Give an example to show that the condition $C \subseteq B$ is necessary.

1.20. Given submodules $C \subseteq B$ of ${}_RA$, prove that $B/C \leq {}_R(A/C)$ and $A/B \cong (A/C)/(B/C)$. (This is called the *Third Isomorphism Theorem.*)

1.21. Let $\theta \in \operatorname{Hom}_R(A, A')$. Prove that

(i) $B \leq {}_RA$ implies $\theta(B) \leq {}_RA'$ and $\theta^{-1}(\theta(B)) = B + \ker\theta$,
(ii) $B' \leq {}_RA'$ implies $\theta^{-1}(B') \leq {}_RA$ and $\theta(\theta^{-1}(B')) = B' \cap \operatorname{im}\theta$.
Deduce that if $C \leq {}_RA$, then

$$D \leq {}_R(A/C) \text{ if and only if } D = B/C \text{ for some } B \leq {}_RA \text{ with } C \subseteq B.$$

1.22. Let $\theta \in \operatorname{Hom}_R(A, B)$. Prove that the sequence

$$0 \longrightarrow \ker\theta \xrightarrow{\ i\ } A \xrightarrow{\ \theta\ } B \xrightarrow{\ \pi\ } B/(\operatorname{im}\theta) \longrightarrow 0$$

is exact. (The left R-module $B/(\operatorname{im}\theta)$ is called the *cokernel* of θ and is denoted by $\operatorname{coker}\theta$. Note that θ is an epimorphism if and only if $\operatorname{coker}\theta = 0$.)

1.23. Let $\theta \in \operatorname{Hom}_R(A, B)$, $C \leq {}_RA$, and $D \leq {}_RB$ such that $\theta(C) \subseteq D$. Prove that there exists a unique homomorphism $\phi : A/C \to B/D$ such that the diagram

$$\begin{array}{ccccccccc}
0 & \longrightarrow & C & \xrightarrow{\ i\ } & A & \xrightarrow{\ \pi\ } & A/C & \longrightarrow & 0 \quad \text{exact} \\
 & & \Big\downarrow{\scriptstyle \theta|_C} & & \Big\downarrow{\scriptstyle \theta} & & \Big\downarrow{\scriptstyle \phi} & & \\
0 & \longrightarrow & D & \xrightarrow{\ i\ } & B & \xrightarrow{\ \pi\ } & B/D & \longrightarrow & 0 \quad \text{exact}
\end{array}$$

commutes.

1.24. Let $\theta \in \operatorname{Hom}_R(A, B)$ and ${}_RC_R$. Prove that

(i) $\operatorname{Hom}_R(C, A) \leq {}_R\operatorname{Hom}(C, A)$,
(ii) $\operatorname{Hom}_R(R, A) \cong A$ as left R-modules,
(iii) $\operatorname{Hom}_R(A, C) \leq \operatorname{Hom}(A, C)_R$,
(iv) $\theta^\star : \operatorname{Hom}_R(C, A) \to \operatorname{Hom}_R(C, B)$ is an R-homomorphism, and
(v) $\theta_\star : \operatorname{Hom}_R(B, C) \to \operatorname{Hom}_R(A, C)$ is an R-homomorphism.

1.25. Let $A_i \leq {}_RA$ $(1 \leq i \leq k)$ for some positive integer k. Then every element of $B = A_1 + \cdots + A_k$ has the form $a_1 + \cdots + a_k$ for some $a_i \in A_i$ $(1 \leq i \leq k)$. Prove that the following statements are equivalent:

(i) every element $a \in B$ has a unique expression $a = a_1 + \cdots + a_k$ where $a_i \in A_i$ $(1 \leq i \leq k)$,
(ii) if $a_i \in A_i$ $(1 \leq i \leq k)$ and $a_1 + \cdots + a_k = 0$, then $a_i = 0$ $(1 \leq i \leq k)$, and
(iii) $A_i \cap (A_1 + \cdots + A_{i-1} + A_{i+1} + \cdots + A_k) = 0 \quad (1 \leq i \leq k)$.

(In this case we say that B is the *direct sum of the submodules* A_i $(1 \leq i \leq k)$ or that B is an *internal direct sum* to distinguish it from the earlier kind, which is sometimes called *external*. Note that for given submodules A_i of A their external direct sum $A_1 \oplus \cdots \oplus A_k$ always exists but they need not form an internal direct sum. In practice it is usually clear which direct sum is being considered.) Now extend this result to arbitrary collections of submodules of A.

1.26. Let A_i $(1 \leq i \leq k)$ be a finite collection of left R-modules and let A be their external direct sum $A_1 \oplus \cdots \oplus A_k$. For each $1 \leq i \leq k$, define

$$B_i = \{(a_1, \ldots, a_k) \in A : a_j = 0 \text{ for all } 1 \leq j \leq k \text{ with } j \neq i\}.$$

Prove that $B_i \leq {}_RA$ $(1 \leq i \leq k)$ and A is the direct sum $B_1 \oplus \cdots \oplus B_k$ of the submodules B_i $(1 \leq i \leq k)$. Extend this fact to arbitrary collections of left R-modules. (This example shows that every external direct sum is also an internal direct sum.)

1.27. Let A_λ $(\lambda \in \Lambda)$ be a non-empty collection of left R-modules. Prove that a left R-module B is isomorphic to $\bigoplus_{\lambda \in \Lambda} A_\lambda$ if and only if there exist homomorphisms $\phi_\lambda : A_\lambda \to B$ $(\lambda \in \Lambda)$ such that for all ${}_RC$ and homomorphisms $\theta_\lambda : A_\lambda \to C$ $(\lambda \in \Lambda)$ there exists a unique homomorphism $\theta : B \to C$ such that the diagram

$$\begin{array}{ccc} B & \xrightarrow{\theta} & C \\ \phi_\lambda \nwarrow & & \nearrow \theta_\lambda \\ & A_\lambda & \end{array}$$

commutes for all $\lambda \in \Lambda$. (Compare Exercise 1.16.)

1.28. Let Λ' and Λ'' be disjoint non-empty subsets of an index set Λ such that $\Lambda = \Lambda' \cup \Lambda''$. Let A_λ $(\lambda \in \Lambda)$ be a collection of left R-modules. Prove

(i) $\prod_{\lambda \in \Lambda} A_\lambda \cong \left(\prod_{\lambda \in \Lambda'} A_\lambda\right) \oplus \left(\prod_{\lambda \in \Lambda''} A_\lambda\right)$, and
(ii) $\bigoplus_{\lambda \in \Lambda} A_\lambda \cong \left(\bigoplus_{\lambda \in \Lambda'} A_\lambda\right) \oplus \left(\bigoplus_{\lambda \in \Lambda''} A_\lambda\right)$.

Further prove that if B_λ $(\lambda \in \Lambda)$ is a collection of left R-modules such that $A_\lambda \cong B_\lambda$ $(\lambda \in \Lambda)$ then

$$\prod_{\lambda \in \Lambda} A_\lambda \cong \prod_{\lambda \in \Lambda} B_\lambda \quad \text{and} \quad \bigoplus_{\lambda \in \Lambda} A_\lambda \cong \bigoplus_{\lambda \in \Lambda} B_\lambda.$$

1.29. Let A_λ $(\lambda \in \Lambda)$ be a non-empty collection of left R-modules and let ${}_RB_S$. Prove

(i) $\mathrm{Hom}_R\left(B, \prod_{\lambda\in\Lambda} A_\lambda\right) \cong \prod_{\lambda\in\Lambda} \mathrm{Hom}_R(B, A_\lambda)$ as left S-modules, and
(ii) $\mathrm{Hom}_R(\bigoplus_{\lambda\in\Lambda} A_\lambda, B) \cong \prod_{\lambda\in\Lambda} \mathrm{Hom}_R(A_\lambda, B)$ as right S-modules.

1.30. Given an exact sequence $0 \longrightarrow A \xrightarrow{\theta} B \xrightarrow{\phi} C \longrightarrow 0$ of left R-modules, prove that the following statements are equivalent.

(i) The sequence splits.
(ii) There exists $\chi \in \mathrm{Hom}_R(B, A)$ such that $\chi\theta = i_A$.
(iii) There exists an R-isomorphism $\beta : B \to A \oplus C$ such that the diagram

$$\begin{array}{ccccccccc} 0 & \longrightarrow & A & \xrightarrow{\theta} & B & \xrightarrow{\phi} & C & \longrightarrow & 0 \quad \text{exact} \\ & & \downarrow{\scriptstyle i_A} & & \downarrow{\scriptstyle \beta} & & \downarrow{\scriptstyle i_C} & & \\ 0 & \longrightarrow & A & \xrightarrow{i} & A\oplus C & \xrightarrow{\pi} & C & \longrightarrow & 0 \quad \text{exact} \end{array}$$

commutes.

1.31. A submodule B of ${}_RA$ is called a *direct summand of* A provided there exists $C \leq {}_RA$ such that $A = B + C$ and $B \cap C = 0$, in which case we write $B \leq_{\mathrm{d}} {}_RA$ and call C a *direct complement* of B. Prove that the following statements are equivalent.

(i) $B \leq_{\mathrm{d}} {}_RA$.
(ii) There exists $C \leq {}_RA$ such that A is the direct sum $B \oplus C$ of its submodules B, C.
(iii) There exists $\theta \in \mathrm{Hom}_R(A, B)$ such that $\theta|_B = i_B$.
(iv) Any short exact sequence $0 \longrightarrow B \xrightarrow{\imath} A \longrightarrow D \longrightarrow 0$ splits.
(v) The short exact sequence $0 \longrightarrow B \xrightarrow{\imath} A \xrightarrow{\pi} A/B \longrightarrow 0$ splits.

1.32. Suppose that ${}_RA = B_1 \oplus C_1 = B_2 \oplus C_2$ is the direct sum of its submodules B_1, C_1 and of its submodules B_2, C_2. Prove that if $B_1 = B_2$, then $C_1 \cong C_2$. Give an example to show that $B_1 = B_2$ does not imply $C_1 = C_2$. Deduce that if $B \leq_{\mathrm{d}} {}_RA$, then any two direct complements of B are isomorphic.

1.33. Let $B, C \leq {}_RA$ and $B \oplus C$ be their external direct sum. Prove that there exists an exact sequence

$$0 \longrightarrow B \cap C \longrightarrow B \oplus C \longrightarrow B + C \longrightarrow 0.$$

1.34. Let $B \leq {}_RA = A_1 \oplus A_2$. Prove that in the usual notation

$$\pi_1(B)/(B \cap A_1) \cong \pi_2(B)/(B \cap A_2).$$

Conversely, suppose that there exist $D_i \leq {}_RC_i \leq {}_RA$ $(i = 1, 2)$ such that $C_1/D_1 \cong C_2/D_2$. Prove that there exists $C \leq {}_RA$ such that $C_i = \pi_i(C)$ and $D_i = C \cap A_i$ $(i = 1, 2)$.

1.35. Let $\theta \in \mathrm{Hom}_R(A, B)$ and $\phi \in \mathrm{Hom}_R(B, A)$. Prove that $\phi\theta$ is an automorphism of A if and only if

(i) θ is a monomorphism,
(ii) ϕ is an epimorphism, and
(iii) $B = \mathrm{im}\,\theta \oplus \ker \phi$.

Further, prove that if $\phi\theta$ is an automorphism of A and $\theta\phi$ an automorphism of B, then $A \cong B$. Give an example to show that $\phi\theta$ being an automorphism of A does not imply that $\theta\phi$ is an automorphism of B.

1.36. Let ${}_RA$ and $\theta \in \mathrm{End}_R(A)$. Prove that $A = \mathrm{im}\,\theta \oplus \ker\theta$ if and only if $\ker\theta = \ker\theta^2$ and $\mathrm{im}\,\theta = \mathrm{im}\,\theta^2$. (Here to make sense the direct sum must be of the internal variety.)

1.37. An element e of a ring R is called *idempotent* provided $e^2 = e$. A left R-module A is called *decomposable* provided A is the direct sum $B \oplus C$ of non-zero submodules B and C, otherwise A is *indecomposable*. Prove that the following statements are equivalent for ${}_RA$.

(i) A is indecomposable.
(ii) $0, A$ are the only direct summands of A.
(iii) z_A and i_A are the only idempotents in $\mathrm{End}_R(A)$.

1.38. Let A_λ $(\lambda \in \Lambda)$ be a collection of submodules of ${}_RA$ and $L \leq {}_RR$ (this just means that L is a left ideal of R). Define LA to be the set of all elements of A of the form

$$r_1a_1 + \cdots + r_ka_k,$$

where k is a positive integer and $r_i \in L$, $a_i \in A_i$ $(1 \leq i \leq k)$. Prove that

(i) $LA \leq {}_RA$,
(ii) $L(\sum_{\lambda\in\Lambda} A_\lambda) = \sum_{\lambda\in\Lambda} (LA_\lambda)$, and
(iii) $L(\bigcap_{\lambda\in\Lambda} A_\lambda) \subseteq \bigcap_{\lambda\in\Lambda} (LA_\lambda)$.

1.39. Let L_λ $(\lambda \in \Lambda)$ be a collection of left ideals of R and ${}_RA$. Prove that

(i) $(\sum_{\lambda\in\Lambda} L_\lambda)A = \sum_{\lambda\in\Lambda} (L_\lambda A)$, and
(ii) $(\bigcap_{\lambda\in\Lambda} L_\lambda)A \subseteq \bigcap_{\lambda\in\Lambda} (L_\lambda A)$.

1.40. Prove that $\mathbb{Q}$ is not a finitely generated $\mathbb{Z}$-module.

1.41. Let $B \leq {}_RA$. Prove that the factor module A/B is finitely generated if and only if there exists a finitely generated submodule C of A such that $A = B + C$. Deduce that

(i) if A is finitely generated, then so is A/B, and
(ii) if B and A/B are both finitely generated, then so is A.

Prove the same results with "cyclic" in place of "finitely generated". Give an example of a ring R for which there exists a cyclic left R-module containing a non-finitely generated submodule.

1.42. Let p be any prime integer. Let $A = \{m/p^n : m, n \in \mathbb{Z}, n \geq 0\}$. Prove that A is a subgroup of the additive Abelian group $\mathbb{Q}$ and that $\mathbb{Z}$ is a subgroup of A. Let $\mathbb{Z}(p^\infty)$ denote the factor group $A/\mathbb{Z}$. For each $n \geq 0$ let a_n denote the element $(1/p^n) + \mathbb{Z}$ of $\mathbb{Z}(p^\infty)$ and $A_n = \mathbb{Z}a_n$. Prove that

(i) $pa_n = a_{n-1}$ $(n \geq 1)$,
(ii) $A_n \subset A_{n+1}$ $(n \geq 0)$,
(iii) B is a proper $\mathbb{Z}$-submodule of $\mathbb{Z}(p^\infty)$ if and only if $B = A_n$ for some $n \geq 0$, and
(iv) the $\mathbb{Z}$-submodules of $\mathbb{Z}(p^\infty)$ can be listed thus:

$$0 = A_0 \subset A_1 \subset \cdots \subset A_n \subset A_{n+1} \subset \cdots \subset \bigcup_{n \in \mathbb{N}} A_n = \mathbb{Z}(p^\infty).$$

Deduce that $\mathbb{Z}(p^\infty)$ is not finitely generated although every proper submodule is cyclic. ($\mathbb{Z}(p^\infty)$ is called the *Prüfer p-group.*)

1.43. Let R be a commutative ring and $A = Ra_1 + \cdots + Ra_k$ be a finitely generated R-module. Prove that $\operatorname{ann}_R A = \{r \in R : ra_i = 0 \ (1 \leq i \leq k)\}$.

1.44. A ring R is called a *domain* provided $rs \neq 0$ for all $0 \neq r, s \in R$. Let R be a commutative domain and A any R-module. Let $T = \{a \in A : ra = 0$ for some $0 \neq r \in R\}$. Prove that $T \leq {}_R A$. (The submodule T is called the *torsion submodule of* A. A is called *torsion* if $T = A$ and *torsion-free* if $T = 0$.) Prove

(i) $\operatorname{ann}_R T \neq 0$ if T is finitely generated, and
(ii) A/T is torsion free.

Give an example to show that (i) is false if T is not finitely generated. Now suppose $0 \to {}_R B \to {}_R C \to {}_R D \to 0$ is exact. Prove that

(iii) C is a torsion module if and only if both B and D are torsion modules,
(iv) if B and D are torsion-free, then so is C,
(v) if C is torsion-free, then so is B.

Give an example to show that C can be torsion-free although D is torsion.

1.4 Projective modules

Every vector space has a basis, but this is not true for modules in general. For example if $A = \mathbb{Z}_2$, then A is a $\mathbb{Z}$-module such that $2A = 0$ and hence A cannot have a basis. Modules which have a basis are called *free.* A left R-module F is called *free* provided there exists a non-empty collection of elements b_ω $(\omega \in \Omega)$ of F such that

(i) $F = \sum_{\omega \in \Omega} Rb_\omega$, and
(ii) for any finite subset Ω' of Ω and elements r_ω $(\omega \in \Omega')$ of R with $\sum_{\omega \in \Omega'} r_\omega b_\omega = 0$ it follows that $r_\omega = 0$ $(\omega \in \Omega')$.

In the language of vector space theory, property (i) says that the elements b_ω ($\omega \in \Omega$) span F and property (ii) says that they are linearly independent. Expressed in another way, the elements b_ω ($\omega \in \Omega$) have the property

> *for each $x \in F$ there exist a unique finite subset Ω' of Ω and unique elements r_ω ($\omega \in \Omega'$) of R such that $x = \sum_{\omega \in \Omega'} r_\omega b_\omega$.*

The set $\{b_\omega : \omega \in \Omega\}$ is called a *basis* of F. The usual convention is that the zero module, i.e., the module which has only one element namely 0, is a free left R-module with empty basis. Free right modules are defined in analogous fashion. We have already remarked that vector spaces over a field K are free K-modules but not every $\mathbb{Z}$-module is free. For any ring R, ${}_RR$ is free with basis $\{1\}$ and R_R is free with the same basis. Moreover, if Λ and Ω_λ ($\lambda \in \Lambda$) are non-empty index sets and F_λ ($\lambda \in \Lambda$) a collection of free left R-modules such that for each $\lambda \in \Lambda, F_\lambda$ has basis $\{b_\omega : \omega \in \Omega_\lambda\}$, then $\bigoplus_{\lambda \in \Lambda} F_\lambda$ is a free left R-module with basis $\{\iota_\lambda(b_\omega) : \omega \in \Omega_\lambda, \lambda \in \Lambda\}$. In particular, combining these last two facts we see that if $F_\lambda = {}_RR$ ($\lambda \in \Lambda$) then $({}_RR)^{(\Lambda)}$ is a free left R-module for any non-empty set Λ. Let F be a free left R-module with basis $\{b_\lambda : \lambda \in \Lambda\}$. If $\theta : F \to G$ is an R-isomorphism then G is a free left R-module with basis $\{\theta(b_\lambda) : \lambda \in \Lambda\}$. On the other hand define $\phi : ({}_RR)^{(\Lambda)} \to F$ by

$$\phi(\{r_\lambda\}) = \Sigma_{\lambda \in \Lambda} r_\lambda b_\lambda$$

for all $\{r_\lambda\}$ in $({}_RR)^{(\Lambda)}$. Note that ϕ is well defined because for any element $\{r_\lambda\}$ of $({}_RR)^{(\Lambda)}$, $r_\lambda = 0$ for all but a finite number of elements λ of Λ. It is elementary to check that ϕ is an R-isomorphism. If we make the convention that $({}_RR)^{(\Lambda)} = 0$ when Λ is the empty set, then we have proved:

Proposition 1.7. *A left R-module F is free if and only if $F \cong ({}_RR)^{(\Lambda)}$ for some set Λ.*

Proposition 1.8. *Let A be any left R-module. Then there exist a free left R-module F and an R-epimorphism $\theta : F \to A$.*

Proof. Let $F = ({}_RR)^{(A)}$ and note that any element of F is a sequence $\{r_a\}$ of elements r_a of R, all but a finite number zero, indexed by the elements a of A. Define $\theta : F \to A$ by

$$\theta(\{r_a\}) = \sum_{a \in A} r_a a$$

for all $\{r_a\} \in ({}_RR)^{(A)}$. It can be checked that θ is an R-epimorphism. □

We shall say that ${}_RA$ has property $(\mathcal{P})$ provided that given any diagram

$$\begin{array}{ccccc} & & A & & \\ & & \theta\downarrow & & \\ {}_RB & \xrightarrow{\phi} & {}_RC & \longrightarrow & 0 \quad \text{exact} \end{array} \tag{1.7}$$

there exists an R-homomorphism $\psi : A \to B$ such that $\phi\psi = \theta$, i.e., the diagram

$$\begin{array}{ccccc} & & A & & \\ & \psi \swarrow & \theta\downarrow & & \\ {}_RB & \xrightarrow{\phi} & {}_RC & \longrightarrow & 0 \quad \text{exact} \end{array}$$

commutes.

Proposition 1.9. *Any free left R-module satisfies ($\mathcal{P}$).*

Proof. If $F = 0$, then there is nothing to prove. Let ${}_RF$ be free with non-empty basis $\{b_\omega : \omega \in \Omega\}$ and consider the diagram

$$\begin{array}{ccccc} & & F & & \\ & & \theta\downarrow & & \\ {}_RB & \xrightarrow{\phi} & {}_RC & \longrightarrow & 0 \quad \text{exact} \end{array}$$

For each $\omega \in \Omega$, $\theta(b_\omega) \in C$ and hence there exists $x_\omega \in B$ such that $\theta(b_\omega) = \phi(x_\omega)$. Define $\psi : F \to B$ by

$$\psi\left(\sum_{\omega\in\Omega} r_\omega b_\omega\right) = \sum_{\omega\in\Omega} r_\omega x_\omega,$$

for all $r_\omega \in R$ ($\omega \in \Omega$) with $r_\omega = 0$ for all but a finite number of elements $\omega \in \Omega$. (Note that ψ is well defined because $\{b_\omega : \omega \in \Omega\}$ is a basis of F and hence every element of F has a unique expression $\sum_{\omega\in\Omega} r_\omega b_\omega$.) It can be checked that ψ is an R-homomorphism and $\phi\psi = \theta$. □

Are the free modules the only modules that satisfy ($\mathcal{P}$)? Let A be a direct summand of a free left R-module F, so that $F = A \oplus A'$ for some submodule A' of F. Let $\iota : A \to F$ and $\pi : F \to A$ denote the injection and projection mappings. Consider the diagram (1.7). Now consider the bigger diagram

$$\begin{array}{ccccc} & & F & & \\ & & \pi\downarrow\ \uparrow\iota & & \\ & & A & & \\ & & \theta\downarrow & & \\ {}_RB & \xrightarrow{\phi} & {}_RC & \longrightarrow & 0 \quad \text{exact.} \end{array}$$

By Proposition 1.9, there exists an R-homomorphism $\psi : F \to B$ such that $\phi\psi = \theta\pi$. Then $\psi\iota : A \to B$ is an R-homomorphism and

$$\phi(\psi\iota) = \phi\psi\iota = \theta\pi\iota = \theta i_A = \theta.$$

It follows that A satisfies ($\mathcal{P}$). Thus any direct summand of a free left R-module satisfies ($\mathcal{P}$). The converse is also true as the following result shows.

Proposition 1.10. *The following statements are equivalent for a left R-module P.*

(i) *P satisfies ($\mathcal{P}$).*
(ii) *For any R-epimorphism $\phi : {}_RA \to P$ there exists an R-homomorphism $\psi : P \to A$ such that $\phi\psi = i_P$.*
(iii) *Any short exact sequence $0 \to {}_RA \to {}_RB \to {}_RP \to 0$ splits.*
(iv) *P is isomorphic to a direct summand of a free left R-module.*
(v) *P is a direct summand of a free left R-module.*

Proof. (i) $\Rightarrow$ (ii) Let $\phi : A \to P$ be an R-epimorphism. Consider the diagram

$$\begin{array}{ccccc} & & P & & \\ & & i_P\downarrow & & \\ A & \xrightarrow{\phi} & P & \longrightarrow & 0 \quad \text{exact.} \end{array}$$

By hypothesis, there exists an R-homomorphism $\psi : P \to A$ such that $\phi\psi = i_P$, as required.

(ii) $\Rightarrow$ (iii) Obvious.

(iii) $\Rightarrow$ (iv) Proposition 1.8 shows that there exist a free ${}_RF$ and an R-epimorphism $\phi : F \to P$. Consider the exact sequence

$$0 \longrightarrow \ker\phi \xrightarrow{i} F \xrightarrow{\phi} P \longrightarrow 0.$$

By hypothesis, there exists an R-homomorphism $\psi : P \to F$ such that $\phi\psi = i_P$. Applying Exercise 1.35, it follows that $P \cong \operatorname{im}\psi$ and $F = (\operatorname{im}\psi) \oplus (\ker\phi)$.

(iv) $\Rightarrow$ (v) Let G be a free left R-module such that G is the direct sum $G_1 \oplus G_2$ of its submodules G_1, G_2, and $P \cong G_1$. Then $P \oplus G_2 \cong G$ by Exercise 1.28, and hence $P \oplus G_2$ is free.

(v) $\Rightarrow$(i) This is proved above. □

A left R-module is called *projective* if it satisfies the equivalent properties of Proposition 1.10. Clearly free left R-modules are projective, but in general not every projective left R-module is free. For example, let R be the commutative ring $\mathbb{Z} \oplus \mathbb{Z} = \{(m, n) : m, n \in \mathbb{Z}\}$ and e the idempotent $(1, 0)$ of R. Then $e = e^2$ and $R = Re \oplus R(1 - e)$. But $0 \neq R(1 - e) = \operatorname{ann}_R Re$ shows that Re is not free. The next two results are very useful in the study of projective modules. If ${}_RA$, then define $A^* = \operatorname{Hom}_R(A, R)$ which is a right R-module by Exercises 1.9 and 1.24.

Theorem 1.11 (The Dual Basis Lemma). *A left R-module P is projective if and only if there exist a non-empty index set Λ, elements $p_\lambda \in P$, and $\psi_\lambda \in P^*$ such that*

(i) *$p = \sum_{\lambda\in\Lambda} \psi_\lambda(p)p_\lambda$, and*
(ii) *$\psi_\lambda(p) = 0$ for all but a finite number of elements $\lambda \in \Lambda$, for each $p \in P$.*

Proof. Suppose first that sets $\{p_\lambda : \lambda \in \Lambda\}$ and $\{\psi_\lambda : \lambda \in \Lambda\}$ exist satisfying (i) and (ii). Let $F = ({}_RR)^{(\Lambda)}$. Define $\phi : F \to P$ and $\psi : P \to F$ as follows:

$$\phi(\{r_\lambda\}) = \sum_{\lambda\in\Lambda} r_\lambda p_\lambda, \text{ and}$$
$$\psi(p) = \{\psi_\lambda(p)\},$$

for all $\{r_\lambda\} \in F$ and $p \in P$. Then ϕ and ψ are R-homomorphisms and $\phi\psi = i_P$. It follows that P is isomorphic to a direct summand of the free left R-module F and hence P is projective.

Conversely, suppose that P is projective. By Proposition 1.10, there exists a free left R-module G with basis $\{b_\lambda : \lambda \in \Lambda\}$ such that G is the direct sum $P \oplus A$ of its submodules P and A. For each $\lambda \in \Lambda$ define $\pi_\lambda \in \operatorname{Hom}_R(G, R)$ as follows: if $g \in G$, then $\pi_\lambda(g) = r_\lambda$, where $r_\lambda \in R$ $(\lambda \in \Lambda)$ is obtained from the unique expression $g = \sum_{\lambda\in\Lambda} r_\lambda b_\lambda$. Note that

$$g = \sum_{\lambda\in\Lambda} \pi_\lambda(g) b_\lambda \quad (g \in G).$$

For each $\lambda \in \Lambda$ there exist elements $p_\lambda \in P$ and $a_\lambda \in A$ such that $b_\lambda = p_\lambda + a_\lambda$. Define $\phi_\lambda = \pi_\lambda|_P \in \operatorname{Hom}_R(P, R)$ for each $\lambda \in \Lambda$. If $p \in P$, then

$$p = \sum_{\lambda\in\Lambda} \phi_\lambda(p) b_\lambda = \sum_{\lambda\in\Lambda} \phi_\lambda(p)(p_\lambda + a_\lambda) = \sum_{\lambda\in\Lambda} \phi_\lambda(p) p_\lambda + \sum_{\lambda\in\Lambda} \phi_\lambda(p) a_\lambda,$$

so that

$$p - \sum_{\lambda\in\Lambda} \phi_\lambda(p) p_\lambda = \sum_{\lambda\in\Lambda} \phi_\lambda(p) a_\lambda \in P \cap A = 0.$$

It follows that

$$p = \sum_{\lambda\in\Lambda} \phi_\lambda(p) p_\lambda \quad (p \in P).$$

The result follows. □

The next result concerns the following diagram of left R-modules and R-homomorphisms

$$\begin{array}{ccccccccccl} 0 & \longrightarrow & A & \xrightarrow{\theta} & B & \xrightarrow{\phi} & C & \longrightarrow & 0 & & \text{exact} \\ & & & & & & \gamma\downarrow & & & & \\ 0 & \longrightarrow & A' & \xrightarrow{\theta'} & B' & \xrightarrow{\phi'} & C' & \longrightarrow & 0 & & \text{exact} \end{array} \qquad (1.8)$$

Theorem 1.12 (Schanuel's Lemma). *Suppose that in* (1.8) γ *is an isomorphism and* B *and* B' *are projective. Then* $A \oplus B' \cong A' \oplus B$.

Proof. Since B is projective, there exists an R-homomorphism $\beta : B \to B'$ such that $\phi'\beta = \gamma\phi$. Define

$$D = \{(b, b') : \gamma\phi(b) = \phi'(b')\},$$

and note that D is a left R-module. Define $\psi : A' \oplus B \to D$ by

$$\psi(a', b) = (b, \theta'(a') + \beta(b)) \quad (a' \in A', b \in B).$$

(Note that

$$\phi'[\theta'(a') + \beta(b)] = \phi'\theta'(a) + \phi'\beta(b) = 0 + \gamma\phi(b) = \gamma\phi(b),$$

and hence $(b, \theta'(a') + \beta(b)) \in D$ for all $a' \in A'$ and $b \in B$.) A simple check shows that ψ is an R-homomorphism. Moreover, ψ is a monomorphism because θ' is a monomorphism. Let $(b, b') \in D$. Then

$$\phi'(b') = \gamma\phi(b) = \phi'\beta(b)$$

implies that $b' - \beta(b) \in \ker\phi' = \operatorname{im}\theta'$, and hence there exists $a' \in A'$ such that $b' = \beta(b) + \theta'(a')$. This shows that $(b, b') = \psi(a', b)$. It follows that ψ is an isomorphism and thus $A' \oplus B \cong D$. In a similar way, $A \oplus B' \cong D$. Thus $A \oplus B' \cong A' \oplus B$, as required. □

Exercises

1.45. Prove that a non-zero left R-module F is free if and only if there exists a non-empty collection of elements b_λ $(\lambda \in \Lambda)$ of F such that for every $_RA$ and elements $a_\lambda \in A$ $(\lambda \in \Lambda)$ there exists a unique R-homomorphism $\phi : F \to A$ with $\phi(b_\lambda) = a_\lambda$ $(\lambda \in \Lambda)$. In this case prove that $\{b_\lambda : \lambda \in \Lambda\}$ is a basis of F.

1.46. Let $n \in \mathbb{N}$. For each $1 \le i, j \le n$ let e_{ij} denote the $n \times n$ matrix with (i, j)th entry 1, the identity of R, and all other entries zero. Prove

(i) $M_n(R)$ is a free left (right) R-module with basis $\{e_{ij} : 1 \le i, j \le n\}$, and
(ii)
$$e_{ij}e_{st} = \begin{cases} 0, & \text{if } j \ne s, \\ e_{it}, & \text{if } j = s, \end{cases}$$
for all $1 \le i, j, s, t \le n$.

1.47. Let K be a field, $n \in \mathbb{N}$ with $n \ge 2$, and $R = M_n(K)$. Let $_RA$. Prove

(i) A is a vector space over the field K,
(ii) $_RA$ is finitely generated if and only if $_KA$ is finite dimensional,
(iii) if A is finitely generated and free then $A = 0$ or A has dimension $4m$ over K for some $m \in \mathbb{N}$, and
(iv) if $L = Re_{11}$ then L is a projective left R-module but L is not free.

1.48. Prove that $\mathrm{Hom}(\mathbb{Q},\mathbb{Z}) = 0$. Deduce that $\mathbb{Q}$ is not a projective $\mathbb{Z}$-module.

1.49. Let ${}_RA$ be cyclic, say $A = Ra$. Prove that $A \cong R/L$, where L is the left ideal $\{r \in R : ra = 0\}$ of R. Deduce that ${}_RA$ is projective if and only if $L = Re$ for some idempotent $e \in R$.

1.50. Let I be an ideal of R. Prove

(i) if F is a free left R-module with basis $\{b_\omega : \omega \in \Omega\}$, then F/IF is a free left (R/I)-module with basis $\{b_\omega + IF : \omega \in \Omega\}$,
(ii) if ${}_RA \cong {}_RB \oplus {}_RC$ then ${}_{R/I}(A/IA) \cong {}_{R/I}(B/IB) \oplus {}_{R/I}(C/IC)$, and
(iii) if P is a projective left R-module, then P/IP is a projective left (R/I)-module.

1.51. Let A_λ $(\lambda \in \Lambda)$ be a non-empty collection of left R-modules. Prove that $\bigoplus_{\lambda\in\Lambda} A_\lambda$ is projective if and only if A_λ is projective for all $\lambda \in \Lambda$.

1.52. Let $0 \to {}_RA \to {}_RB \to {}_RC \to 0$ be exact. Prove that if C is projective, then A is projective if and only if B is projective. Give an example to show that C need not be projective even when both A and B are projective.

1.53. Prove that

(i) a free left R-module is finitely generated if and only if it has a finite basis,
(ii) given $n \in \mathbb{N}$, any ${}_RA$ is generated by n elements if and only if there exists an epimorphism $\phi : ({}_RR)^n \to A$.

Further prove that the following statements are equivalent for ${}_RP$:

(i) P is finitely generated projective.
(ii) There exists $n \in \mathbb{N}$ such that $({}_RR)^n \cong P \oplus Q$ for some ${}_RQ$.
(iii) There exist $m \in \mathbb{N}$, $p_i \in P$, $\phi_i \in P^*$ $(1 \le i \le m)$ such that $p = \sum_{i=1}^m \phi_i(p)p_i$ for all $p \in P$.

In this case prove that there exists ${}_RQ$ such that $P \oplus Q$ is free and Q is finitely generated.

1.54. Let P be a projective left R-module. Prove that $P^* = \mathrm{Hom}_R(P,R)$ is a projective right R-module. Now suppose that in addition P is finitely generated. Prove that P^* is finitely generated and $P^{**} = \mathrm{Hom}_R(P^*,R) \cong P$.

1.55. Let ${}_RA$, $S = \mathrm{End}_R(A)$ and $B = A^n$ for some $n \in \mathbb{N}$. Let $f : \mathrm{End}_R(B) \to M_n(S)$ be the mapping that sends each $\theta \in \mathrm{End}_R(B)$ into the $n \times n$ matrix $f(\theta)$ with (i,j)th entry $\pi_i\theta\iota_j$ for all $1 \le i,j \le n$. Prove that f is a ring isomorphism. Deduce that $\mathrm{End}_R(({}_RR)^k) \cong M_k(R)$ for all $k \in \mathbb{N}$. Further, prove that if ${}_RP$ is finitely generated and projective, then there exist $t \in \mathbb{N}$ and an idempotent $e \in M_t(R)$ such that $\mathrm{End}_R(P) \cong eM_t(R)e$.

1.56. Let ${}_RP$ be projective. Define $\tau(P) = \sum \mathrm{im}\,\theta$, where the sum is taken over all $\theta \in P^*$. Prove that

(i) $\tau(P)$ is an ideal of R,

(ii) $\tau(P) = \tau(P)^2$,
(iii) $P = \tau(P)P$,
(iv) $\tau(P) = R$ if and only if $P^n \cong R \oplus A$ for some positive integer n and ${}_RA$.

Now suppose that in addition P is finitely generated. Prove that there exist a positive integer k and elements $t_i \in \tau(P)$ $(1 \leq i \leq k)$ such that every element of $\tau(P)$ can be written in the form

$$r_1t_1s_1 + \cdots + r_kt_ks_k$$

for some $r_i, s_i \in R$ $(1 \leq i \leq k)$.

(Ideals with the property *(ii)* are called *idempotent.* The ideal $\tau(P)$ is called the *trace ideal* of P.)

1.57. A left ideal L of a ring R is called *principal* provided there exists $a \in L$ such that $L = Ra = \{ra : r \in R\}$. Let R be a domain (c.f. Exercise 1.44). Prove that any principal left ideal of R is a free left R-module. Suppose further that R is commutative (and a domain). Prove that every projective R-module is torsion-free.

1.58. Let R be a commutative ring. Prove that every ideal of R is a free R-module if and only if R is a domain and every ideal of R is principal. (Such domains are called *principal ideal domains.*)

1.59. Let R be a ring such that every left ideal is projective (e.g., principal ideal domains have this property). Let F be a free left R-module with basis $\{b_n : n \in \mathbb{N}\}$ and let $P \leq {}_RF$. For each $m \in \mathbb{N}$, let $F_m = Rb_1 + \cdots + Rb_m$ and $\phi_m : F_m \to R$ be the mapping defined by

$$\phi_m(r_1b_1 + \cdots + r_mb_m) = r_m,$$

for all $r_i \in R$ $(1 \leq i \leq m)$. Prove that, for all $m \in \mathbb{N}$,

(i) F_m is a free left R-module with basis $\{b_1, \ldots, b_m\}$,
(ii) $\phi_m \in \mathrm{Hom}_R(F_m, R)$,
(iii) $L_m = \phi_m(P \cap F_m)$ is a left ideal of R, and
(iv) there exists $P_m \leq {}_RF_m$ such that $P \cap F_m = (P \cap F_{m-1}) \oplus P_m$ and $P_m \cong L_m$.

Deduce that $P = \bigoplus_{m \in \mathbb{N}} P_m$ and hence P is projective. (This argument can be adapted using ordinal numbers to prove *Kaplansky's Theorem*: if R is a ring in which every left ideal is projective, then every submodule of a free left R-module is isomorphic to a direct sum of left ideals of R and hence is projective. One important consequence of this theorem is that if R is a principal ideal domain then every ideal is free and hence every submodule of a free R-module is free. In particular, every submodule of a free $\mathbb{Z}$-module is free.)

1.60. Consider the following diagram of left R-modules

$$\begin{array}{cccccccccccccl} 0 & \longrightarrow & C & \longrightarrow & P_n & \longrightarrow & \cdots & \longrightarrow & P_0 & \longrightarrow & A & \longrightarrow & 0 & \text{exact} \\ & & & & & & & & & & \alpha\downarrow & & & \\ 0 & \longrightarrow & D & \longrightarrow & Q_n & \longrightarrow & \cdots & \longrightarrow & Q_0 & \longrightarrow & B & \longrightarrow & 0 & \text{exact} \end{array}$$

where $n \in \mathbb{N}$, α is an isomorphism, and P_i and Q_i are projective for all $0 \leq i \leq n$. Prove that there exist projective left R-modules X and Y such that $C \oplus X \cong D \oplus Y$. Suppose further that n is even. Let

$$X_1 = Q_0 \oplus P_1 \oplus Q_2 \oplus P_3 \oplus \cdots \oplus Q_n,$$

and

$$Y_1 = P_0 \oplus Q_1 \oplus P_2 \oplus Q_3 \oplus \cdots \oplus P_n.$$

Prove that $C \oplus X_1 \cong D \oplus Y_1$. State and prove the corresponding result when n is odd. (This result extends Schanuel's Lemma.)

1.5 Zorn's lemma

Let X be a non-empty set. A relation $\leq$ on X is called a *partial order* provided

(i) $x \leq x$,
(ii) if $x \leq y$ and $y \leq x$, then $x = y$, and
(iii) if $x \leq y$ and $y \leq z$, then $x \leq z$, for all $x, y, z \in X$.

Suppose X has a partial order $\leq$ (X is called a *partially ordered set*). A non-empty subset Y of X is called *a chain in* X or *totally ordered* if $x \leq y$ or $y \leq x$ for all $x, y \in Y$. An element x of a non-empty subset Z of X is called a *maximal element of* Z if whenever $z \in Z$ and $x \leq z$, then $x = z$. On the other hand, an element x of X is called an *upper bound of* Z if $z \leq x$ for all $z \in Z$. We shall *assume*:

Theorem 1.13 (Zorn's Lemma). *Let X be a partially ordered set such that every chain in X has an upper bound. Then X contains a maximal element.*

Let A be a left R-module and $\underline{S}$ a non-empty collection of submodules of A. A partial order $\leq$ can be defined on $\underline{S}$ as follows:

$$B \leq C \text{ if } B \subseteq C \text{ for all } B, C \in \underline{S}.$$

A non-empty subset $\underline{C}$ of $\underline{S}$ is a *chain* provided $B \subseteq C$ or $C \subseteq B$ for all $B, C \in \underline{C}$. Note that

$$\bigcup_{B \in \underline{C}} B \leq {}_RA \text{ for any chain } \underline{C}. \tag{1.9}$$

Thus if $\underline{C}$ is a chain in $\underline{S}$ and $U = \bigcup_{B \in \underline{C}} B \in \underline{S}$, then U is an upper bound for $\underline{C}$. By a *maximal member of* $\underline{S}$ we mean of course a submodule $M \in \underline{S}$ such that whenever $M \subseteq B$ and $B \in \underline{S}$, then $M = B$. In case $\underline{S}$ consists of all proper submodules of A, then the maximal members of $\underline{S}$ are called *maximal submodules of* A. The notation "$M \leq_{\mathrm{m}} {}_RA$" will be used to describe the fact that M is a maximal submodule of ${}_RA$. The maximal submodules of ${}_RR$ and of R_R are called *maximal left ideals* and *maximal right ideals*, respectively. Note that the $\mathbb{Z}$-module $\mathbb{Z}(p^\infty)$ has no maximal submodules (see Exercise 1.42). However, Theorem 1.13 gives

Theorem 1.14. *Let $\underline{S}$ be a non-empty collection of submodules of a left R-module A such that the union of any chain in $\underline{S}$ belongs to $\underline{S}$. Then $\underline{S}$ contains a maximal member.*

Theorems 1.13 and 1.14 have many applications. We first illustrate the usefulness of Theorem 1.14 and at the end of the section give an application of Theorem 1.13.

Proposition 1.15. *A left R-module A is finitely generated if and only if A is not the union of a chain of proper submodules.*

Proof. Suppose first that ${}_RA$ is finitely generated. There exist a positive integer k and elements $a_i \in A$ $(1 \leq i \leq k)$ such that $A = Ra_1 + \cdots + Ra_k$. Suppose that $A = \bigcup_{B \in \underline{C}} B$ for some chain $\underline{C}$ of submodules of A. For each $1 \leq i \leq k$ there exists $B_i \in \underline{C}$ such that $a_i \in B_i$. Consider the finite collection of submodules B_i $(1 \leq i \leq k)$ of A. For each $1 \leq i, j \leq k$ either $B_i \subseteq B_j$, or $B_j \subseteq B_i$. Hence, there exists $1 \leq m \leq k$ such that $B_i \subseteq B_m$ $(1 \leq i \leq k)$. But this implies that $a_i \in B_m$ $(1 \leq i \leq k)$ and hence

$$A = Ra_1 + \cdots + Ra_k \subseteq B_m \subseteq A,$$

so that $A = B_m$, contradicting the fact that B_m is proper. This proves the necessity.

Conversely, suppose that A is not the union of a chain of proper submodules. Suppose that A is not finitely generated. Let

$$\underline{S} = \{B \leq {}_RA : {}_R(A/B) \text{ is not finitely generated}\}.$$

Then $0 \in \underline{S}$ because $A/0 \cong A$. Let $\underline{T}$ be a chain of submodules in $\underline{S}$ and $U = \bigcup_{C \in \underline{T}} C$. Suppose that ${}_R(A/U)$ is finitely generated. By Exercise 1.41, there exists $D \leq {}_RA$ such that D is finitely generated and $A = D + U$. Then

$$A = D + U = D + \bigcup_{C \in \underline{T}} C = \bigcup_{C \in \underline{T}} (U + C).$$

If C_1 and $C_2 \in \underline{T}$, then $C_1 \subseteq C_2$ or $C_2 \subseteq C_1$, so that $U + C_1 \subseteq U + C_2$ or $U + C_2 \subseteq U + C_1$. Therefore, by hypothesis, $A = U + C$ for some $C \in \underline{T}$. But in this case A/C is finitely generated, by Exercise 1.41, which contradicts the fact that $C \in \underline{T} \subseteq \underline{S}$. Thus A/U is not finitely generated and hence $U \in \underline{S}$. By Theorem 1.14, $\underline{S}$ contains a maximal member M. Clearly $M \in \underline{S}$ implies that $M \neq A$. Let $a \in A, a \notin M$. Then $M + Ra \notin \underline{S}$ by the choice of M and hence $A/(M + Ra)$ is finitely generated. By Exercise 1.41, there exists a finitely generated submodule E of A such that $A = (M + Ra) + E = M + (Ra + E)$. But this implies, by Exercise 1.41, that A/M is finitely generated, contradicting the fact that $M \in \underline{S}$. It follows that A is finitely generated. □

Corollary 1.16. *Let A be a finitely generated left R-module. Then any proper submodule of A is contained in a maximal submodule of A.*

Proof. Let $B \leq {}_RA$ with $B \neq A$. Let

$$\underline{S} = \{C \leq {}_RA : B \subseteq C \neq A\}.$$

The collection $\underline{S}$ is non-empty because $B \in \underline{S}$. Moreover, the union of any chain of submodules in $\underline{S}$ belongs to $\underline{S}$ by Proposition 1.15. Thus $\underline{S}$ contains a maximal member M by Theorem 1.14. But clearly $M \leq_{\mathrm{m}} {}_RA$ and the result follows. □

In particular this corollary applies to the modules ${}_RR$ and R_R, both of which are cyclic. Any proper left ideal of R is contained in a maximal left ideal and any proper right ideal of R is contained in a maximal right ideal. The second application of Theorem 1.14 concerns essential submodules. A submodule E of a left R-module A is called *essential* (or *large*) provided $E \cap B \neq 0$ for each non-zero $B \leq {}_RA$, and in this case we write $E \leq_{\mathrm{e}} {}_RA$. In particular $A \leq_{\mathrm{e}} {}_RA$. On the other hand, $0 \leq_{\mathrm{e}} {}_RA$ if and only if $A = 0$. The next result shows that each submodule of A is a direct summand of an essential submodule of A.

Proposition 1.17. *Let B be a submodule of a left R-module A. Then there exists a submodule C of A such that $B \cap C = 0$ and $B \oplus C$ is an essential submodule of A.*

Proof. Let

$$\underline{S} = \{D \leq {}_RA : B \cap D = 0\}.$$

Note that $\underline{S}$ is non-empty because $0 \in \underline{S}$. Let $\underline{C}$ be a chain of submodules in $\underline{S}$ and $U = \bigcup_{X \in \underline{C}} X$. Note that

$$B \cap U = B \cap [\bigcup_{X \in \underline{C}} X] = \bigcup_{X \in \underline{C}} (B \cap X) = 0,$$

which implies that $U \in \underline{S}$. By Theorem 1.14, $\underline{S}$ contains a maximal member C. Clearly, $B \cap C = 0$ because $C \in \underline{S}$. Suppose that $0 \neq Y \leq {}_RA$. If $Y \subseteq C$, then

$$0 \neq Y = C \cap Y \subseteq (B + C) \cap Y,$$

so that $(B + C) \cap Y \neq 0$. If $Y \nsubseteq C$, then $C \subset C + Y$ implies that $C + Y \notin \underline{S}$ by the choice of C. Hence, $B \cap (C + Y) \neq 0$, i.e., there exist elements $b \in B$, $c \in C$, and $y \in Y$ such that $0 \neq b = c + y$. If $y = 0$, then $b = c = 0$, a contradiction. Thus $0 \neq y = b - c \in (B + C) \cap Y$. In any case $(B + C) \cap Y \neq 0$. It follows that $B \oplus C \leq_{\mathrm{e}} A$. □

A left R-module A is called *subdirectly irreducible* (Philip Hall proposed the term *"monolithic"*) provided the intersection of all non-zero submodules of A is itself non-zero. Every module has many subdirectly irreducible homomorphic images, as the next result shows.

Proposition 1.18. *Let A be a non-zero left R-module. Then there exists a collection of submodules B_λ $(\lambda \in \Lambda)$ of A such that $\bigcap_{\lambda \in \Lambda} B_\lambda = 0$ and A/B_λ is subdirectly irreducible for all $\lambda \in \Lambda$.*

Proof. Let $0 \neq a \in A$. Define

$$\underline{S} = \{C \leq {}_R A : a \notin C\}.$$

Note that $\underline{S}$ is non-empty because $0 \in \underline{S}$. Clearly, the union of any chain in $\underline{S}$ also belongs to $\underline{S}$ by (1.9) and hence $\underline{S}$ contains a maximal member B_a by Theorem 1.14. Note that $a \notin B_a$. Consider the module A/B_a. If $0 \neq X \leq {}_R(A/B_a)$, then $X = D/B_a$ for some $B_a \subset D \leq {}_R A$ by Exercise 1.21. The choice of B_a dictates that $a \in D$ and hence $Ra + B_a \subseteq D$, so that $(Ra + B_a)/B_a \subseteq X$. It follows that the non-zero module $(Ra + B_a)/B_a$ is the intersection of all non-zero submodules of A/B_a and hence A/B_a is subdirectly irreducible. Clearly, $\bigcap_{a \in A} B_a = 0$ and the result is proved. □

Finally, we give an application of Theorem 1.13.

Proposition 1.19. *Let $\underline{S}$ be a non-empty collection of submodules of a left R-module A. Then there exists a submodule B of A which is maximal with respect to the property of being a direct sum of members of $\underline{S}$.*

Proof. Let $\underline{T}$ denote the collection of non-empty subsets $\{C_\lambda : \lambda \in \Lambda\}$ of $\underline{S}$ such that $\sum_{\lambda \in \Lambda} C_\lambda$ is a direct sum. Clearly, $\{C\} \in \underline{T}$ for each $C \in \underline{S}$ and hence $\underline{T}$ is non-empty. Let $\underline{C} = \{C_\omega : \omega \in \Omega\}$ be a chain in $\underline{T}$ and let $U = \bigcup_{\omega \in \Omega} C_\omega$. There exist an index set Γ and submodules $D_\gamma \in \underline{S}$ such that $U = \{D_\gamma : \gamma \in \Gamma\}$. Suppose there exist a positive integer k, elements $\gamma(i) \in \Gamma$, and elements $d_i \in D_{\gamma(i)}$ $(1 \leq i \leq k)$ such that $d_1 + \cdots + d_k = 0$. For each $1 \leq i \leq k$ there exists $\omega(i) \in \Omega$ such that $D_{\gamma(i)} \in C_{\omega(i)}$. However, because $\underline{C}$ is a chain there exists $1 \leq j \leq k$ such that $C_{\omega(i)} \subseteq C_{\omega(j)}$ for all $1 \leq i \leq k$. In particular, this means that $D_{\gamma(i)} \in C_{\omega(j)}$ $(1 \leq i \leq k)$ and hence

$$D_{\gamma(1)} + \cdots + D_{\gamma(k)}$$

is a direct sum. Therefore $d_i = 0$ $(1 \leq i \leq k)$. It follows that $\sum_{\gamma \in \Gamma} D_\gamma$ is a direct sum and this means that $U \in \underline{T}$. By Theorem 1.13, $\underline{T}$ has a maximal member M and we take B to be the sum of the submodules in M. □

Let V be a non-zero vector space over a field K and $\underline{S}$ denote the collection of one-dimensional subspaces of V. By Proposition 1.19, there exists a subspace U of V which is a direct sum $\bigoplus_{v \in B} Kv$ of one-dimensional subspaces Kv for some subset B of non-zero elements of V and which is maximal with respect to this property. It is easy to show that the maximality implies that $U = V$, and hence B is a basis of V. Thus every vector space has a basis.

Exercises

1.61. Prove that if A is a finitely generated faithful left R-module, then there exists $B \leq {}_RA$ such that A/B is faithful, but A/C is unfaithful for all $B \subset C \leq {}_RA$.

1.62. Let A be a finitely generated left R module which is not cyclic. Prove that there exists $B \leq {}_RA$ maximal with respect to the property that A/B is not cyclic.

1.63. Let B be a non-essential submodule of ${}_RA$. Prove that there exists $C \leq {}_RA$ such that $B \leq_{\text{e}} {}_RC$ but $B \not\leq_{\text{e}} {}_RD$ for any $C \subset D \leq {}_RA$.

1.64. Prove that

(i) if $C \leq_{\text{e}} {}_RA$ and $C \subseteq B \leq {}_RA$ then $B \leq_{\text{e}} {}_RA$,
(ii) if $C \leq_{\text{e}} {}_RB$ and $B \leq_{\text{e}} {}_RA$ then $C \leq_{\text{e}} {}_RA$,
(iii) if $B \leq_{\text{e}} {}_RA$ then $B \cap C \leq_{\text{e}} {}_RC$ for any $C \leq {}_RA$,
(iv) if $B_i \leq_{\text{e}} {}_RA$ $(1 \leq i \leq k)$ then $B_1 \cap \cdots \cap B_k \leq_{\text{e}} {}_RA$, for any positive integer k,
(v) if $C \leq_{\text{e}} {}_RB$ and $\theta : {}_RA \to {}_RB$ is a homomorphism then $\theta^{-1}(C) \leq_{\text{e}} {}_RA$,
(vi) if $C \subseteq B$ are submodules of ${}_RA$ such that $B/C \leq_{\text{e}} {}_R(A/C)$ then $B \leq_{\text{e}} {}_RA$, and
(vii) if $B \leq_{\text{e}} {}_RA$ and $a \in A$ then $L = \{r \in R : ra \in B\} \leq_{\text{e}} {}_RR$.

1.65. Prove that ${}_RA$ is finitely generated if and only if whenever there exists a family of submodules A_λ $(\lambda \in \Lambda)$ of A with $A = \sum_{\lambda\in\Lambda} A_\lambda$ then there exists a finite subset Λ' of Λ such that $A = \sum_{\lambda\in\Lambda'} A_\lambda$.

1.66. Let $\underline{S}$ be a non-empty collection of submodules of ${}_RA$. A submodule $M \in \underline{S}$ is called a *minimal member of* $\underline{S}$ if $B \in \underline{S}$ and $B \subseteq M$ together imply $B = M$. Prove that if $\underline{S}$ has the property that $\bigcap_{C\in\underline{C}} C \in \underline{S}$ for each chain $\underline{C}$ in $\underline{S}$, then $\underline{S}$ contains a minimal member.

1.67. A submodule M of ${}_RA$ is called *minimal* if it is a minimal member of the collection of all non-zero submodules of A. Prove that ${}_{\mathbb{Z}}\mathbb{Z}$ does not contain a minimal submodule, but ${}_{\mathbb{Z}}\mathbb{Z}(p^\infty)$ has a unique minimal submodule.

1.68. A left R-module is called *finitely cogenerated* or *finitely embedded* provided for every non-empty collection of submodules A_λ $(\lambda \in \Lambda)$ of A such that $\bigcap_{\lambda\in\Lambda} A_\lambda = 0$ there exists a finite subset Λ' of Λ such that $\bigcap_{\lambda\in\Lambda'} A_\lambda = 0$. Suppose that ${}_RA$ is finitely cogenerated. Prove that

(i) any submodule of A is finitely cogenerated,
(ii) any non-zero submodule of A contains a minimal submodule,
(iii) A contains only a finite number of minimal submodules.

Prove further that if $0 \to {}_RB \to {}_RC \to {}_RD \to 0$ is an exact sequence with B and D finitely cogenerated, then C is finitely cogenerated. Prove that if C is finitely cogenerated, then B is finitely cogenerated, but D need not be finitely cogenerated.

1.69. Prove that ${}_RA$ is finitely cogenerated if and only if the intersection of every chain of non-zero submodules of A is non-zero.

1.70. Prove that every left R-module A contains a submodule F which is maximal with respect to the property of being free. Now suppose that in addition R is a commutative domain. Prove that A/F is a torsion module.

1.71. Let R be a commutative domain and A a finitely generated torsion-free R-module. Prove that there exist a free submodule F of A and a non-zero element r of R such that $rA \subseteq F$. Deduce that A is isomorphic to a submodule of F.

1.72. Let R be a commutative principal ideal domain (See Exercise 1.58) and A a finitely generated R-module with torsion submodule T. Prove that there exists a free submodule F of A such that A is the direct sum $T \oplus F$ of its submodules T and F.

1.6 Injective modules

In this section we introduce and study the basic properties of injective modules. The proof of the first result requires Zorn's Lemma. A left R-module X is called *injective* provided for any diagram

$$\begin{array}{ccccc} 0 & \longrightarrow & {}_RA & \xrightarrow{\theta} & {}_RB \quad \text{exact} \\ & & \phi\downarrow & & \\ & & {}_RX & & \end{array} \tag{1.10}$$

there exists an R-homomorphism $\psi : B \to X$ such that $\psi\theta = \phi$, i.e., the diagram

$$\begin{array}{ccccc} 0 & \longrightarrow & {}_RA & \xrightarrow{\theta} & {}_RB \quad \text{exact} \\ & & \phi\downarrow & \swarrow \psi & \\ & & {}_RX & & \end{array}$$

commutes. In this situation we say that ϕ *can be lifted to* ψ. Injective modules are the *duals* of projective modules in the sense that arrows have been reversed and epimorphisms replaced by monomorphisms. The first result gives a very useful test for injectivity and for this reason is sometimes called the *Injective Test Lemma.*

Theorem 1.20 (Baer's Lemma). *A left R-module X is injective if and only if for each left ideal L of R and each R-homomorphism $\phi : L \to X$ there exists $x \in X$ such that $\phi(r) = rx$ $(r \in L)$.*

Proof. Suppose first that X is injective. Let L be a left ideal of R and $\phi : L \to X$ be an R-homomorphism. Consider the diagram

$$\begin{array}{ccccc} 0 & \longrightarrow & L & \xrightarrow{\iota} & R \quad \text{exact} \\ & & \phi\downarrow & & \\ & & X & & \end{array}$$

where ι denotes the inclusion map. By hypothesis, there exists an R-homomorphism $\psi : R \to X$ such that $\psi\iota = \phi$. Let $x = \psi(1)$. For any $r \in L$,

$$\phi(r) = \psi\iota(r) = \psi(r) = r\psi(1) = rx.$$

This proves the necessity.

Conversely, suppose that X has the stated property. Consider the diagram (1.10). First suppose that $A \leq {}_RB$ and $\theta = \iota$, the inclusion map. Let

$$\underline{S} = \{(C, \gamma) : A \subseteq C \leq {}_RB,\ \gamma : C \to B \text{ is a homomorphism and } \gamma|_A = \phi\}.$$

Note that $\underline{S}$ is non-empty because $(A, \phi) \in \underline{S}$. Define a partial order $\leq$ on the set $\underline{S}$ as follows: given (C, γ), $(D, \delta) \in \underline{S}$ then

$$(C, \gamma) \leq (D, \delta) \text{ if and only if } C \subseteq D \text{ and } \delta|_C = \gamma.$$

(It can be checked that $\leq$ is a partial order!) Let $\{(C_\lambda, \gamma_\lambda) : \lambda \in \Lambda\}$ be a chain in the partially ordered set $\underline{S}$. Define $C = \bigcup_{\lambda \in \Lambda} C_\lambda$, which is a submodule of ${}_RB$ by (1.9). Further define $\gamma : C \to X$ as follows: for each $c \in C$ there exists $\lambda \in \Lambda$ such that $c \in C_\lambda$ and in this case $\gamma(c) = \gamma_\lambda(c)$. Note that the definition of γ is independent of the choice of λ in Λ because if $c \in C_\lambda \cap C_\mu$ for distinct elements $\lambda, \mu \in \Lambda$, then $(C_\lambda, \gamma_\lambda) \leq (C_\mu, \gamma_\mu)$ or $(C_\mu, \gamma_\mu) \leq (C_\lambda, \gamma_\lambda)$, which implies that $\gamma_\lambda(c) = \gamma_\mu(c)$. Clearly, $\gamma \in \mathrm{Hom}_R(C, E)$ and $\gamma|_A = \phi$. Thus $(C, \gamma) \in \underline{S}$. It follows that every chain in $\underline{S}$ has an upper bound. By Theorem 1.13, $\underline{S}$ contains a maximal member (E, ψ).

Suppose that $E \neq B$. Let $b \in B$, $b \notin E$. Define $L = \{r \in R : rb \in E\}$. It can easily be checked that L is a left ideal of R. Define $\alpha : L \to X$ by

$$\alpha(r) = \psi(rb) \quad (r \in L).$$

Again some routine checking shows that $\alpha \in \mathrm{Hom}_R(L, X)$. By hypothesis, there exists $x \in X$ such that $\alpha(r) = rx$ for all $r \in L$. Now define $\beta : E + Rb \to X$ by

$$\beta(e + rb) = \psi(e) + rx \quad (e \in E, r \in R).$$

We have to check that β is well defined because we have no guarantee that a given element of $E + Rb$ can be uniquely expressed in the form $e + rb$ for some $e \in E$ and $r \in R$. Therefore, suppose that $e + rb = e' + r'b$ for some $e, e' \in E$ and $r, r' \in R$. Then $(r - r')b = e' - e \in E$ implies $r - r' \in L$ and hence

$$\begin{aligned} rx - r'x = (r - r')x = \alpha(r - r') &= \psi((r - r')b) \\ = \psi(rb - r'b) = \psi(e' - e) &= \psi(e') - \psi(e), \end{aligned}$$

i.e., $\psi(e) + rx = \psi(e') + r'x$. To summarise, if $e, e' \in E$ and $r, r' \in R$, then $e + rb = e' + r'b$ implies $\beta(e + rb) = \beta(e' + r'b)$. Thus β is well defined. It is left to the reader to check that β is an R-homomorphism such that $\beta|_A = \psi|_A = \phi$. But this means that $(E + Rb, \beta) \in \underline{S}$, which contradicts the choice of (E, ψ) because $(E, \psi) \leq (E + Rb, \beta)$ and $(E, \psi) \neq (E + Rb, \beta)$. This contradiction shows that $E = B$ and hence $\psi : B \to X$ is an R-homomorphism such that $\psi|_A = \phi$.

In general note that given (1.10), $\theta : A \to \operatorname{im}\theta$ is an R-isomorphism. Consider the diagram

$$\begin{array}{ccccc} 0 & \longrightarrow & \operatorname{im}\theta & \xrightarrow{\iota} & B \qquad \text{exact} \\ & & {\scriptstyle \phi\theta^{-1}}\downarrow & & \\ & & X & & \end{array}$$

where ι is the inclusion mapping. The above argument shows that there exists $\psi \in \operatorname{Hom}_R(B, X)$ such that $\psi\iota = \phi\theta^{-1}$. Therefore, $\psi\theta = \psi\iota\theta = \phi$. It follows that X is injective. □

The theorem has the following immediate consequence.

Corollary 1.21. *Let K be a field. Then every K-module is injective.*

Theorem 1.20 can be used to determine which Abelian groups are injective $\mathbb{Z}$-modules. It turns out that these are precisely the divisible groups. An Abelian group D is called *divisible* provided for each $d \in D$ and non-zero integer k there exists $d' \in D$ such that $kd' = d$ (so that in this sense each element of D is divisible by each non-zero integer). For any $k \in \mathbb{Z}$ let $kD = \{kd : d \in D\}$. Then D is divisible if and only if $D = kD$ for any $0 \neq k \in \mathbb{Z}$. Note that the zero group is divisible. It is also clear that $\mathbb{Q}$ is a divisible group.

Lemma 1.22.

(i) *Every factor group of a divisible group is divisible.*
(ii) *Every direct sum and direct product of divisible groups is divisible.*
(iii) *Every Abelian group can be embedded in a divisible group.*

Proof. (i) Suppose D is divisible and $C \leq_{\mathbb{Z}} D$. Then

$$k(D/C) = (kD + C)/C = (D + C)/C = D/C,$$

for any non-zero integer k.

(ii) Suppose that D_λ $(\lambda \in \Lambda)$ is a non-empty collection of divisible groups and $D = \prod_{\lambda\in\Lambda} D_\lambda$. Then

$$kD = k\prod_{\lambda\in\Lambda} D_\lambda = \prod_{\lambda\in\Lambda}(kD_\lambda) = \prod_{\lambda\in\Lambda} D_\lambda = D,$$

for any non-zero integer k. The argument for direct sums is similar.

(iii) Let A be any Abelian group. Proposition 1.8 shows that there exist a set Λ and a $\mathbb{Z}$-epimorphism $\theta : F \to A$, where $F = \mathbb{Z}^{(\Lambda)}$. If $K = \ker\theta$, then K is a subgroup of F and $A \cong F/K$. Let $D = \mathbb{Q}^{(\Lambda)}$. By (ii) D is a divisible group and clearly F/K can be embedded in the group D/K which is divisible by (i). □

In order to apply Lemma 1.22 we first prove a result about the ring $\mathbb{Z}$.

Lemma 1.23. *$\mathbb{Z}$ is a principal ideal domain.*

Proof. It is clear that $\mathbb{Z}$ is a domain. Let I be an ideal of $\mathbb{Z}$. If $I = 0$, then I is the principal ideal $\mathbb{Z}0$. Now suppose that $I \neq 0$. If $n \in I$, then also $-n \in I$ and hence I contains a positive integer. Let k be the least positive integer belonging to I. If $m \in I$ then the Division Algorithm gives integers a, b such that $m = ak + b$ and $0 \leq b \leq k - 1$. But the choice of k implies that $b = 0$ because $b = m - ak \in I$. Hence $m = ak$. It follows that $I = \mathbb{Z}k$, and so $\mathbb{Z}$ is a principal ideal domain. □

Proposition 1.24. *An Abelian group A is an injective $\mathbb{Z}$-module if and only if A is a divisible group.*

Proof. Suppose first that A is an injective $\mathbb{Z}$-module. Let $a \in A$, $0 \neq n \in \mathbb{Z}$. Define $\phi : \mathbb{Z}n \to A$ by $\phi(kn) = ka$ $(k \in \mathbb{Z})$. Note that ϕ is well defined because if $kn = k'n$ for some $k, k' \in \mathbb{Z}$, then $k = k'$ and hence $ka = k'a$. Clearly, ϕ is a homomorphism and hence by Theorem 1.20 there exists $b \in A$ such that $\phi(r) = rb$ $(r \in \mathbb{Z}n)$. In particular, $a = \phi(n) = nb$. It follows that A is divisible.

Conversely, suppose that A is divisible. Let I be a non-zero ideal of $\mathbb{Z}$ and $\theta : I \to A$ an homomorphism. Lemma 1.23 shows that there exists $0 \neq m \in I$ such that $I = \mathbb{Z}m$. By hypothesis there exists $c \in A$ such that $\theta(m) = mc$. If $s \in I$ then there exists $t \in \mathbb{Z}$ such that $s = tm$ and

$$\theta(s) = \theta(tm) = t\theta(m) = tmc = sc.$$

Applying Theorem 1.20 again, it follows that A is an injective $\mathbb{Z}$-module. □

Combining Lemma 1.22 (iii) and Proposition 1.24 it is now clear that any $\mathbb{Z}$-module can be embedded in an injective $\mathbb{Z}$-module. In fact this is true not only for the ring $\mathbb{Z}$, but for any ring. This is a consequence of the next result.

Lemma 1.25. *Let D be a divisible group. Then* $\mathrm{Hom}(R, D)$ *is an injective left R-module.*

Proof. Let $X = \mathrm{Hom}(R, D)$ and consider the diagram (1.10). Associated with (1.10) is the following diagram of $\mathbb{Z}$-modules:

$$\begin{array}{ccccc} 0 & \longrightarrow & A & \xrightarrow{\theta} & B \qquad \text{exact} \\ & & \alpha\downarrow & & \\ & & D & & \end{array}$$

where $\alpha : A \to D$ is defined by $\alpha(a) = \phi(a)(1)$ $(a \in A)$. (Check that α is a $\mathbb{Z}$-homomorphism.) By Proposition 1.24, there exists a $\mathbb{Z}$-homomorphism $\beta : B \to D$ such that $\alpha = \beta\theta$. Define $\psi : B \to X$ as follows: for each $b \in B$, $\psi(b)$ is the mapping from R to D defined by $\psi(b)(r) = \beta(rb)$ $(r \in R)$. Note that for each $b \in B$, $\psi(b)$ is a $\mathbb{Z}$-homomorphism (because β is a $\mathbb{Z}$-homomorphism), so that it is indeed the case that $\psi(b) \in X$. Now check that ψ is an R-homomorphism. Finally, note that $\psi\theta = \phi$. To see this, remember that we have to prove that $\psi\theta(a) = \phi(a)$ $(a \in A)$, and in order to prove that for any $a \in A$ the mappings $\psi\theta(a)$ and $\phi(a)$ are equal it is necessary to show:

$$\psi\theta(a)(r) = \phi(a)(r) \ (r \in R). \tag{1.11}$$

Let $r \in R$. Then

$$\begin{aligned}\psi\theta(a)(r) &= \psi[\theta(a)(r)] = \beta(r\theta(a)) = \beta\theta(ra) = \alpha(ra) \\ &= \phi(ra)(1) = (r\phi(a))(1) = \phi(a)(1r) = \phi(a)(r).\end{aligned}$$

This proves (1.11) and hence that $\psi\theta = \phi$. It follows that X is injective. □

Corollary 1.26. *For any ring R, any left R-module can be embedded in an injective left R-module.*

Proof. Let A be a left R-module. By Lemma 1.22 (iii), there exist a divisible Abelian group D and a $\mathbb{Z}$-monomorphism $\theta : A \to D$. By Exercises 1.14 and 1.24, $\theta^* : \mathrm{Hom}(R, A) \to \mathrm{Hom}(R, D)$ is an R-monomorphism. But

$${}_RA \cong {}_R\,\mathrm{Hom}_R(R, A) \leq {}_R\,\mathrm{Hom}(R, A)$$

by Exercise 1.24. Thus gathering together these facts we see that there is an R-monomorphism $\phi : A \to \mathrm{Hom}(R, D)$, and $\mathrm{Hom}(R, D)$ is injective by Lemma 1.25. □

A left R-module B is called an *extension* of a left R-module A if A is a submodule of B. An extension B of A is called *proper* if $A \neq B$, and is called *essential* if $A \leq_e {}_RB$. There is an intimate connection between injective modules and extensions which we aim to demonstrate. First we prove:

Lemma 1.27. *Let $\theta : {}_RA \to {}_RB$ be a monomorphism. Then there exist an extension C of A and an R-isomorphism $\phi : C \to B$ such that $\phi|_A = \theta$.*

Proof. Let $D = \{(0, b) : b \in B,\ b \notin \operatorname{im}\theta\} \leq {}_R(R \oplus B)$. Let $C = A \cup D$. Define a mapping $\phi : C \to B$ by

$$\phi(c) = \begin{cases} \theta(c), & \text{if } c \in A, \\ \pi(c), & \text{if } c \in D, \end{cases}$$

where $\pi : R \oplus B \to B$ denotes the canonical projection. The mapping ϕ is a bijection and hence has an inverse $\phi^{-1} : B \to C$. The set C can be given the structure of a left R-module by defining:

$$\begin{aligned} c_1 + c_2 &= \phi^{-1}[\phi(c_1) + \phi(c_2)], \text{ and} \\ rc &= \phi^{-1}[\phi(rc)], \end{aligned}$$

for all $r \in R$, c, c_1, $c_2 \in C$. It can be checked that with these definitions C is a left R-module, $A \leq {}_RC$, $\phi : C \to B$ is an R-isomorphism and $\phi|_A = \theta$. □

Corollary 1.28. *Suppose that $\theta : {}_RA \to {}_RB$ is an isomorphism and ${}_RA$ has no proper essential extension. Then B has no proper essential extension.*

Proof. Suppose that $B \leq_e {}_RC$. By Lemma 1.27, there exist an extension D of A and an R-isomorphism $\phi : D \to C$ such that $\phi|_A = \theta$. Let $0 \neq E \leq {}_RD$. Then $0 \neq \phi(E) \leq {}_RC$, so that $B \cap \phi(E) \neq 0$. But $B \cap \phi(E) \subseteq \phi(A \cap E)$. Thus $A \cap E \neq 0$. It follows that $A \leq_e {}_RD$. By hypothesis, $A = D$ and hence

$$C = \phi(D) = \phi(A) = \theta(A) = B.$$

It follows that B has no proper essential extension. □

Combining Corollary 1.26 and Lemma 1.27 gives at once:

Proposition 1.29. *Any left R-module has an injective extension.*

The next result gives two characterizations of injective modules.

Proposition 1.30. *The following statements are equivalent for a left R-module E:*

(i) *E is injective.*
(ii) *E is a direct summand of each of its extensions.*
(iii) *E has no proper essential extension.*

Proof. (i) $\Rightarrow$ (ii) Suppose that $E \leq {}_RA$. Consider the diagram

$$\begin{array}{ccccc} 0 & \longrightarrow & E & \stackrel{\iota}{\longrightarrow} & A \\ & & {\scriptstyle i_E}\downarrow & & \\ & & E & & \end{array} \qquad \text{exact}$$

By hypothesis, there exists $\theta : A \to E$ such that $\theta\iota = i_E$. By Exercises 1.30 and 1.31, E is a direct summand of A.

(ii) $\Rightarrow$ (iii) Suppose that $E \leq_e {}_RB$. By hypothesis, there exists $C \leq {}_RB$ such that $B = E \oplus C$ and, in particular, $E \cap C = 0$. Thus $C = 0$ and hence $E = B$.

(iii) $\Rightarrow$ (i) Proposition 1.29 shows that there exists an injective left R-module X such that $E \leq {}_RX$. By Proposition 1.17, there exists $D \leq {}_RX$ maximal with respect to the property $E \cap D = 0$ and in this case $E \oplus D \leq_e {}_RX$. The aim now is to prove

$$X = E \oplus D. \tag{1.12}$$

Suppose on the contrary that $X \neq E \oplus D$. Note that

$$E \cong E/(E \cap D) \cong (E \oplus D)/D \leq {}_R(X/D),$$

and $(E \oplus D)/D \neq X/D$ (see Proposition 1.6). By Corollary 1.28, it follows that $(E \oplus D)/D \not\leq_e {}_R(X/D)$. Therefore, there exists $D \subset H \leq {}_RX$ such that

$$(E \oplus D)/D \cap (H/D) = 0.$$

But this implies that $(E \oplus D) \cap H = D$ and hence $E \cap H \subseteq E \cap D = 0$. Thus $E \cap H = 0$, contradicting the choice of D. This proves (1.12). Therefore, E is a direct summand of the injective module X and it is an easy exercise to show that E is injective. □

Theorem 1.31. *Any left R-module A has an essential extension E which is an injective left R-module. Moreover, if E' is any essential extension of A such that E' is injective, then there exists an R-isomorphism $\theta : E \to E'$ such that $\theta|_A = i_A$.*

Proof. Proposition 1.29 shows that A has an injective extension X. Apply Exercise 1.63 to obtain a submodule E of X maximal with respect to the property that $A \leq_{\mathrm{e}} {}_RE$. Suppose that F is an essential extension of E. Consider the diagram

$$\begin{array}{ccccc} 0 & \longrightarrow & E & \overset{\iota}{\longrightarrow} & F \\ & & \iota\downarrow & & \\ & & X & & \end{array} \qquad \text{exact}$$

where in each case ι denotes the inclusion mapping. Since X is injective, there exists an R-homomorphism $\theta : F \to X$ such that $\theta\iota = \iota$. If $\ker\theta \neq 0$ then $A \cap \ker\theta \neq 0$ (see Exercise 1.64). Thus $\ker\theta = 0$, i.e., θ is a monomorphism. It follows that $E \leq_{\mathrm{e}} {}_R(\operatorname{im}\theta)$ because $E \leq_{\mathrm{e}} {}_RF$. But $E \subseteq \operatorname{im}\theta$ and hence the choice of E gives $E = \operatorname{im}\theta$. If $f \in F$, then $\theta(f) \in E$ so that

$$\theta(f) = \iota(\theta(f)) = \theta(\theta(f)),$$

which in turn implies $f = \theta(f) \in E$. It follows that $F = E$. Hence E has no proper essential extension. By Proposition 1.30, E is an injective module. Moreover, from its definition we have at once that $A \leq_{\mathrm{e}} {}_RE$.

Now suppose that E' is an essential extension of A such that E' is injective. Consider the diagram

$$\begin{array}{ccccc} 0 & \longrightarrow & A & \overset{\iota}{\longrightarrow} & E \\ & & \iota\downarrow & & \\ & & E' & & \end{array} \qquad \text{exact}$$

where ι denotes the corresponding inclusion mapping. Since E' is injective, there exists a homorphism $\theta : E \to E'$ such that $\theta\iota = \iota$. Note that $A \cap \ker\theta = 0$ and that $A \leq_{\mathrm{e}} {}_RE$. Hence $\ker\theta = 0$. But this implies that $\operatorname{im}\theta \cong E$ is an injective submodule of E' and hence a direct summand of E' (Proposition 1.30). But $A \leq_{\mathrm{e}} E'$ and $A \subseteq \operatorname{im}\theta$ combine to give that $\operatorname{im}\theta \leq_{\mathrm{e}} {}_RE'$. It follows that $\operatorname{im}\theta = E'$, i.e., $\theta : E \to E'$ is an isomorphism. Clearly $\theta|_A = i_A$. □

Let A be any left R-module. Theorem 1.31 shows that there exists an injective module E such that $A \leq_{\mathrm{e}} {}_RE$; E is called an *injective envelope* or *injective hull of A* and will be denoted by $E({}_RA)$ or simply $E(A)$. Note that Theorem 1.31 also shows that $E(A)$ is *unique up to isomorphism* in the sense that if ${}_RE'$ is injective and $A \leq_{\mathrm{e}} {}_RE'$, then $E' \cong E(A)$.

Exercises

1.73. Prove that any direct summand of an injective module is injective.

1.74. Let A_λ $(\lambda \in \Lambda)$ be a non-empty collection of left R-modules. Prove that $\prod_{\lambda\in\Lambda} A_\lambda$ is injective if and only if A_λ is injective for all $\lambda \in \Lambda$.

1.75. Prove that the following statements are equivalent for ${}_RE$.

(i) E is injective.

(ii) For any R-monomorphism $\phi : E \to A$ there exists $\psi \in \mathrm{Hom}_R(A, E)$ such that $\psi\phi = i_E$.

(iii) Any short exact sequence $0 \to E \to {}_RB \to {}_RC \to 0$ splits.

1.76. Let $\phi : {}_RA \to {}_RB$ be a monomorphism. Prove that there exists an R-monomorphism $\psi : E(A) \to E(B)$ such that the diagram

$$\begin{array}{ccccccc} 0 & \longrightarrow & A & \xrightarrow{\phi} & B & & \text{exact} \\ & & \iota\downarrow & & \iota\downarrow & & \\ 0 & \longrightarrow & E(A) & \xrightarrow{\psi} & E(B) & & \text{exact} \end{array}$$

commutes.

1.77. Prove that a left R-module X is injective if and only if for any *essential* left ideal L of R and R-homomorphism $\phi : L \to X$, there exists $x \in X$ such that $\phi(r) = rx$ $(r \in R)$.

1.78. Let A_i $(1 \le i \le k)$ be any finite collection of left R-modules. Prove that $E(A_1 \oplus \cdots \oplus A_n) \cong E(A_1) \oplus \cdots \oplus E(A_n)$.

1.79. Let $A \le {}_RB$ and let $E(B)$ be an injective envelope of B. Prove that there exists an injective envelope $E(A)$ of A such that $E(B) = E(A) \oplus C$ for some $C \le {}_RE(B)$. Prove further that if $A \le_e {}_RB$ then $E(A) = E(B)$.

1.80. Let A be a left R-module. Prove that the following statements are equivalent for a left R-module E:

(i) $E = E(A)$.

(ii) $A \le_e {}_RE$, but $A \not\le_e {}_RF$ for any $E \le {}_RF$ with $E \ne F$ (E is called a *maximal essential extension of* A).

(iii) $A \le {}_RE$, E is injective, but G is not injective for any $A \le {}_RG \le {}_RE$ with $G \ne E$ (E is called a *minimal injective extension of* A).

1.81. Let I be an ideal of R and let ${}_RX$ be injective. Prove that $Y = \{x \in X : Ix = 0\}$ is an injective left (R/I)-module. Prove further that if $X = E({}_RA)$, then $Y = E({}_{R/I}(A \cap Y))$.

1.82. For any Abelian group G, define $G^+ = \mathrm{Hom}(G, \mathbb{Q}/\mathbb{Z})$. ($G^+$ is called the *character group of* G.) Let ${}_RA$ and C_R. Prove that

(i) A^+ is a right R-module and C^+ a left R-module,
(ii) for each $0 \neq a \in A$ there exists $\alpha \in A^+$ such that $\alpha(a) \neq 0$,
(iii) there exists an R-monomorphism $\theta : A \to A^{++}$,
(iv) if $\phi : {}_RA \to {}_RB$ is an epimorphism, then $\phi^+ : B^+ \to A^+$ is a monomorphism, where $\phi^+(\psi) = \psi\phi$ $(\psi \in B^+)$,
(v) A can be embedded in F^+ for some free right R-module F, and
(vi) if F_R is free, then ${}_RF^+$ is injective.

Deduce that every left R-module can be embedded in an injective left R-module of the form F^+ for some free right R-module F.

1.83. Let X be an injective extension of ${}_RA$. Let $B \leq {}_RX$ be chosen maximal with respect to the property that $A \cap B = 0$. Let $C \leq {}_RX$ be chosen maximal with respect to $A \subseteq C$ and $C \cap B = 0$. Prove that $C = E(A)$.

1.84. A submodule A of ${}_RB$ is called *superfluous* or *small* provided $B \neq A + C$ for any proper submodule C of B, and in this case we write $A \leq_s {}_RB$. Prove

(i) if $A \leq_s {}_RB$ and $C \leq {}_RA$, then $C \leq_s {}_RB$,
(ii) if $A_i \leq_s {}_RB$ $(1 \leq i \leq k)$, then $A_1 + \cdots + A_k \leq_s {}_RB$, for any positive integer k,
(iii) if $A \leq_s {}_RB$ and $\phi : {}_RB \to {}_RC$ is a homomorphism, then $\phi(A) \leq_s {}_RC$, and
(iv) if $A \leq_s {}_RB$ and $B \leq {}_RC$, then $A \leq_s {}_RC$.

1.85. Consider the diagram of left R-modules

$$\begin{array}{ccccccccccc} 0 & \longrightarrow & K & \xrightarrow{\iota} & P & \xrightarrow{\phi} & A & \longrightarrow & 0 & & \text{exact} \\ & & & & & & \alpha\downarrow & & & & \\ 0 & \longrightarrow & K' & \xrightarrow{\iota} & P' & \xrightarrow{\phi'} & A' & \longrightarrow & 0 & & \text{exact} \end{array}$$

where α is an isomorphism, P, P' are projective, and $K \leq_s {}_RP$. Prove that there exist $P_1, P_2 \leq {}_RP'$ such that

(i) $P' = P_1 \oplus P_2$,
(ii) $P_1 \cong P$,
(iii) $P_2 \subseteq K'$, and
(iv) $P_1 \cap K' \leq_s {}_RP_1$.

Deduce that if $K' \leq_s {}_RP'$, then $P \cong P'$.

(The module P is called a *projective cover* of A and is the analogue of an injective envelope of A. This exercise should be compared with Theorem 1.31. Note however that for many rings R not every left R-module has a projective cover.)

1.86. Prove that $E({}_{\mathbb{Z}}\mathbb{Z}) = \mathbb{Q}$ and $E({}_{\mathbb{Z}}\mathbb{Z}_p) = \mathbb{Z}(p^\infty)$ for any prime p.

1.87. Suppose that ${}_RX$ and ${}_RY$ are injective and $\theta : X \to Y$ and $\phi : Y \to X$ are R-monomorphisms. Prove that $X \cong Y$.

1.88. Prove that the following statements are equivalent for an injective left R-module E.

(i) E is indecomposable.
(ii) $E = E(A)$ for every $0 \neq A \leq {}_R E$.
(iii) $B \cap C \neq 0$ for all $0 \neq B,\ C \leq {}_R E$.

1.89. An element u of a ring R is called a *unit* provided there exists $v \in R$ such that $uv = vu = 1$, otherwise u is called a *non-unit.* Let E be an indecomposable injective left R-module. Prove that

(i) if $\theta \in \mathrm{End}_R(E)$ and $\ker\theta = 0$ then θ is an automorphism of E,
(ii) if $\theta,\ \phi \in \mathrm{End}_R(E)$ and $\theta\phi$ is an automorphism of E, then both θ and ϕ are automorphisms of E,
(iii) if $\theta,\ \phi \in \mathrm{End}_R(E)$ and $\theta + \phi$ is an automorphism of E, then either θ or ϕ is an automorphism of E, and
(iv) the set of non-units of the ring $\mathrm{End}_R(E)$ forms an ideal.

Rings R for which the set of non-units forms an ideal are called *local rings.* Prove that if A is a left R-module and $\mathrm{End}_R(A)$ is a local ring, then A is indecomposable.

1.90. Consider the diagram of left R-modules

$$\begin{array}{ccccccccccc} 0 & \longrightarrow & A & \longrightarrow & E & \longrightarrow & B & \longrightarrow & 0 & & \text{exact} \\ & & \alpha\downarrow & & & & & & & & \\ 0 & \longrightarrow & A' & \longrightarrow & E' & \longrightarrow & B' & \longrightarrow & 0 & & \text{exact} \end{array}$$

where α is an isomorphism and both E and E' are injective. Prove that $E \oplus B' \cong E' \oplus B$. (This is *Schanuel's Lemma for Injectives.*)

1.91. Prove that if every left ideal of a ring R is projective, then every homomorphic image of an injective left R-module is injective.

1.92. Suppose given free left R-modules F, G and a homomorphism $\theta : F \to G$ with kernel K. Suppose further that $E(F)/K$ is an injective left R-module and consider the diagram

$$\begin{array}{ccccccc} & & 0 & \longrightarrow & F/K & \xrightarrow{\theta'} & G & & \text{exact} \\ & & & & \iota\downarrow & & & & \\ E(F) & \xrightarrow{\pi} & & & E(F)/K & \longrightarrow & 0 & & \text{exact} \end{array}$$

where θ' is the monomorphism induced by θ. Prove that

(i) there exists $\alpha \in \mathrm{Hom}_R(G, E(F))$ such that $\pi\alpha\theta' = \iota$,
(ii) $\mathrm{im}\,\alpha\theta' \subseteq F$, and
(iii) $F = K \oplus \mathrm{im}\ (\alpha\theta')$ and F/K is projective.

1.93. Prove that the following statements are equivalent for a ring R.

(i) Every left ideal of R is projective.

(ii) Every submodule of a projective left R-module is projective.

(iii) Every homomorphic image of an injective left R-module is injective.

(Rings which have these equivalent properties are called *left hereditary. Right hereditary* rings are defined in analogous fashion and rings which are both left and right hereditary are called *hereditary.* In particular, principal ideal domains are hereditary.)

1.94. Prove that a ring R is left hereditary if and only if the sum of any two injective submodules of a left R-module is injective.

1.95. A left R-module A is called *quasi-injective* if for any $B \leq {}_R A$ and homomorphism $\phi : B \to A$ there exists $\psi \in \mathrm{End}_R(A)$ such that $\phi = \psi|_B$, i.e., the diagram

$$\begin{array}{ccccc} 0 & \longrightarrow & B & \xrightarrow{\iota} & A \qquad \text{exact} \\ & & \phi\downarrow & \swarrow \psi & \\ & & A & & \end{array}$$

commutes. Prove that a left R-module X is quasi-injective if and only if $\theta(X) \subseteq X$ for all $\theta \in \mathrm{End}_R(E(X))$. (This is the *Johnson-Wong Theorem.*)

1.7 Flat modules

Having considered projective and injective modules, there is a third class of modules which arise naturally and have important properties. These are the so-called flat modules. Tensor products are considered first and to define them it is convenient to introduce some notation concerning free Abelian groups (in other words, free $\mathbb{Z}$-modules).

Let S be a non-empty set. The *free Abelian group on the set S* is defined to be the free $\mathbb{Z}$-module $\mathbb{Z}^{(S)}$. The *standard basis* of $\mathbb{Z}^{(S)}$ is the set $\{\iota_s(1)\}$ indexed by the elements $s \in S$. At the risk of causing some confusion, we denote $\iota_s(1)$ simply by s for each $s \in S$. In this notation, $\mathbb{Z}^{(S)}$ consists of all formal sums $\sum_{s\in S} n_s s$ where $n_s \in \mathbb{Z}$ $(s \in S)$ and $n_s \neq 0$ for at most a finite number of elements $s \in S$. Moreover, given k, n_s, $m_s \in \mathbb{Z}$ $(s \in S)$, the following hold:

$$\sum_{s\in S} n_s s = \sum_{s\in S} m_s s$$

if and only if $n_s = m_s$ $(s \in S)$

$$\sum_{s\in S} n_s s + \sum_{s\in S} m_s s = \sum_{s\in S} (n_s + m_s) s,$$

and

$$k\left(\sum_{s\in S} n_s s\right) = \sum_{s\in S} (kn_s)s.$$

Now suppose that A_R and ${}_RB$. Let $A \times B$ denote the Cartesian product of the sets A and B, i.e., $A \times B$ consists of all ordered pairs (a, b) with $a \in A$ and $b \in B$. Let $F(A, B)$ denote the free Abelian group on the set $A \times B$. Thus $F(A, B)$ consists of all finite sums

$$\sum_{i=1}^{k} n_i(a_i, b_i)$$

where k is a positive integer, $n_i \in \mathbb{Z}$, $a_i \in A$ and $b_i \in B$ for all $1 \le i \le k$. Let T be the subset of $F(A, B)$ consisting of all elements of the form

$$\begin{gathered}(a, b_1 + b_2) - (a, b_1) - (a, b_2),\\ (a_1 + a_2, b) - (a_1, b) - (a_2, b), \text{ or}\\ (ar, b) - (a, rb)\end{gathered}$$

for all a, a_1, $a_2 \in A$, b, b_1, $b_2 \in B$ and $r \in R$. Define

$$G(A, B) = \sum_{t\in T} \mathbb{Z}t.$$

Then $G(A, B)$ is a subgroup of the Abelian group $F(A, B)$ and the factor group $F(A, B)/G(A, B)$ is called the *tensor product of A and B* and is denoted by $A \otimes_R B$, i.e.,

$$A \otimes_R B = F(A, B)/G(A, B).$$

Moreover, for each $a \in A$ and $b \in B$ we define

$$a \otimes b = (a, b) + G(A, B).$$

These definitions have certain immediate consequences, namely

$$\begin{aligned}
&\text{(i)} && A \otimes_R B \text{ is an Abelian group for all } A_R, {}_RB,\\
&\text{(ii)} && a \otimes (b_1 + b_2) = a \otimes b_1 + a \otimes b_2,\\
&\text{(iii)} && (a_1 + a_2) \otimes b = a_1 \otimes b + a_2 \otimes b,\\
&\text{(iv)} && (ar) \otimes b = a \otimes (rb),\\
&\text{(v)} && a \otimes 0 = 0 \otimes b = 0, \text{ and}\\
&\text{(vi)} && (na) \otimes b = n(a \otimes b) = a \otimes (nb),
\end{aligned} \tag{1.13}$$

for all a, a_1, $a_2 \in A$, b, b_1, $b_2 \in B$, $r \in R$ and $n \in \mathbb{Z}$.

In particular, if $x \in A \otimes_R B$, then there exist a positive integer k and elements $a_i \in A$, $b_i \in B$ such that

$$\begin{aligned} x &= \sum_{i=1}^{k} n_i(a_i, b_i) + G(A, B) = \sum_{i=1}^{k} n_i[(a_i, b_i) + G(A, B)] \\ &= \sum_{i=1}^{k} n_i(a_i \otimes b_i) = \sum_{i=1}^{k} (n_i a_i) \otimes b_i, \end{aligned}$$

so that each element of $A \otimes_R B$ can be written as a finite sum of elements $a \otimes b$ where the $a's \in A$ and the $b's \in B$. Note however that although every element of $F(A, B)$ can be expressed uniquely in terms of the elements (a, b), *this ceases to be true* when we pass to the factor group $A \otimes_R B$, i.e., in general elements of $A \otimes_R B$ have more than one expression in terms of the elements $a \otimes b$. Thus great care has to be exercised in defining mappings on the group $A \otimes_R B$.

If A_R and ${}_RB$, then $A \otimes_R B$ is an Abelian group and not a left R-module in general. However, in certain circumstances $A \otimes_R B$ can be given the structure of a left R-module. To be slightly more general, let R and S be rings, and let ${}_SA_R$ and ${}_RB$. Define

$$s\left(\sum_{i=1}^{k} n_i(a_i, b_i)\right) = \sum_{i=1}^{k} n_i(sa_i, b_i)$$

for all positive integers k, $n_i \in \mathbb{Z}$, $a_i \in A$, $b_i \in B$ $(1 \le i \le k)$ and $s \in S$. (Note that the product is well defined because each element of $F(A, B)$ has a unique expression in terms of elements of $A \times B$.) With this definition of multiplication by a scalar, $F(A, B)$ becomes a left S-module. Moreover, it can be checked that $G(A, B) \le {}_SF(A, B)$ and hence $A \otimes_R B = F(A, B)/G(A, B)$ is a left S-module, where

$$s\left(\sum_{i=1}^{k} a_i \otimes b_i\right) = \sum_{i=1}^{k} (sa_i) \otimes b_i$$

for all $k \ge 1$, $a_i \in A$, $b_i \in B$ $(1 \le i \le k)$ and $s \in S$. Similarly, if C_R and ${}_RD_S$, then $C \otimes_R D$ is a right S-module where

$$\left(\sum_{i=1}^{k} c_i \otimes d_i\right) s = \sum_{i=1}^{k} c_i \otimes (d_i s)$$

for all $k \ge 1$, $c_i \in C$, $d_i \in D$ $(1 \le i \le k)$ and $s \in S$.

Let $\theta : A_R \to C_R$ and $\phi : {}_RB \to {}_RD$ be R-homomorphisms. Define $(\theta, \phi) : F(A, B) \to F(C, D)$ as follows:

$$(\theta, \phi)\left[\sum_{i=1}^{k} n_i(a_i, b_i)\right] = \sum_{i=1}^{k} n_i(\theta(a_i), \phi(b_i))$$

for all positive integers k, $n_i \in \mathbb{Z}$, $a_i \in A$ and $b_i \in B$ $(1 \le i \le k)$. Because of the uniqueness of expression for elements in $F(A,B)$, (θ,ϕ) is well defined and it is easily seen that (θ,ϕ) is a $\mathbb{Z}$-homomorphism. Moreover,

$$\begin{aligned}
&(\theta,\phi)[(a,b_1+b_2)-(a,b_1)-(a,b_2)]\\
&\quad = (\theta(a),\phi(b_1+b_2))-(\theta(a),\phi(b_1))-(\theta(a),\phi(b_2))\\
&\quad = (\theta(a),\phi(b_1)+\phi(b_2))-(\theta(a),\phi(b_1))-\theta(a),\phi(b_2))\\
&\quad \in G(C,D),
\end{aligned}$$

for all $a \in A$, b_1, $b_2 \in B$. Similarly

$$(\theta,\phi)[(a_1+a_2,b)-(a_1,b)-(a_2,b)] \in G(C,D),$$

and

$$(\theta,\phi)[(ar,b)-(a,rb)] \in G(C,D)$$

for all a, a_1, $a_2 \in A$, $b \in B$ and $r \in R$. In terms of the above notation

$$(\theta,\phi)(T) \subseteq G(C,D),$$

and hence $(\theta,\phi)(G(A,B)) \subseteq G(C,D)$. Exercise 1.23 is applicable here and it gives the existence of a unique $\mathbb{Z}$-homomorphism $\psi : A \otimes_R B \to C \otimes_R D$ such that the diagram

$$\begin{array}{ccc}
F(A,B) & \xrightarrow{(\theta,\phi)} & F(C,D)\\
\pi\downarrow & & \pi\downarrow\\
A\otimes_R B & \xrightarrow{\psi} & C\otimes_R D
\end{array}$$

commutes, where π is the canonical epimorphism in each case. The above mapping ψ will be denoted by $\theta \otimes \phi$.

Suppose that $0 \to A_R \xrightarrow{\theta} B_R \xrightarrow{\phi} C_R \to 0$ is an exact sequence and D is a left R-module. Let $\varepsilon = i_D$ and note that in the commutative diagram

$$\begin{array}{ccccccccc}
 & & 0 & & 0 & & & & \\
 & & \downarrow & & \downarrow & & & & \\
 & & G(B,D) & \longrightarrow & G(C,D) & \longrightarrow & 0 & & \text{exact}\\
 & & \iota\downarrow & & \iota\downarrow & & & & \\
F(A,D) & \xrightarrow{(\theta,\varepsilon)} & F(B,D) & \xrightarrow{(\phi,\varepsilon)} & F(C,D) & \longrightarrow & 0 & & \text{exact}\\
\pi\downarrow & & \pi\downarrow & & \pi\downarrow & & & & \\
A\otimes_R D & \xrightarrow{\theta\otimes\varepsilon} & B\otimes_R D & \xrightarrow{\phi\otimes\varepsilon} & C\otimes_R D & \longrightarrow & 0 & & \\
\downarrow & & \downarrow & & \downarrow & & & & \\
0 & & 0 & & 0 & & & &
\end{array}$$

the columns are exact. Proposition 1.2 gives the following result.

Proposition 1.32. *Let* $0 \to A_R \xrightarrow{\theta} B_R \xrightarrow{\phi} C_R \to 0$ *be an exact sequence and let* D *be a left* R*-module. Then*

$$A \otimes_R D \xrightarrow{\alpha} B \otimes_R D \xrightarrow{\beta} C \otimes_R D \longrightarrow 0$$

is exact, where $\alpha = \theta \otimes i_D$ *and* $\beta = \phi \otimes i_D$.

A left R-module D is called *flat* if for every monomorphism $\theta : A_R \to B_R$ the homomorphism $\theta \otimes i_D : A \otimes_R D \to B \otimes_R D$ is a monomorphism. In other words, ${}_RD$ is flat if and only if for every exact sequence

$$0 \longrightarrow A_R \xrightarrow{\theta} B_R \xrightarrow{\phi} C_R \longrightarrow 0$$

the corresponding sequence

$$0 \longrightarrow A \otimes_R D \xrightarrow{\alpha} B \otimes_R D \xrightarrow{\beta} C \otimes_R D \longrightarrow 0$$

is also exact where $\alpha = \theta \otimes i_D$ and $\beta = \phi \otimes i_D$.

Our next objective is to prove that every projective left R-module is flat, and as a first instalment we shall prove

$${}_RR \text{ is flat.} \tag{1.14}$$

This is a consequence of the next result.

Lemma 1.33. *For any right* R*-module* A, $A \otimes_R R$ *and* A *are isomorphic as right* R*-modules.*

Proof. Define $\alpha : A \to A \otimes_R R$ by

$$\alpha(a) = a \otimes 1 \quad (a \in A).$$

By (1.13), $\alpha \in \operatorname{Hom}_R(A, A \otimes_R R)$. In fact, α is a bijection. Of course, one way of proving this is to check directly that α is both one-to-one and onto. However, it is almost always easier to produce the inverse mapping α^{-1} and this is what we do here. We want a mapping from $A \otimes_R R$ to A and because of the lack of uniqueness for elements of $A \otimes_R R$ we proceed as follows. Define $\theta : F(A, R) \to A$ by

$$\theta\left(\sum_{i=1}^{k} n_i(a_i, r_i)\right) = \sum_{i=1}^{k} n_i(a_i r_i)$$

for all $k \geq 1$, $a_i \in A$, $r_i \in R$ $(1 \leq i \leq k)$. Note that $\theta \in \operatorname{Hom}_R(F(A,R), A)$ and $G(A,R) \subseteq \ker\theta$. By Exercise 1.23, θ induces an R-homomorphism $\gamma : A \otimes_R R \to A$ such that

$$\gamma\left(\sum_{i=1}^{k} a_i \otimes r_i\right) = \sum_{i=1}^{k} a_i r_i$$

for all $k \geq 1$, $a_i \in A$, $r_i \in R$ $(1 \leq i \leq k)$. (What is important is that γ exists, not that it is an R-homomorphism.) But $\alpha\gamma$ and $\gamma\alpha$ are the identity mappings on $A \otimes_R R$ and A, respectively. Thus $\gamma = \alpha^{-1}$ and α is a bijection, as claimed. □

Let $\phi : A_R \to B_R$ be a monomorphism. Consider the diagram

$$\begin{array}{ccccc} 0 & \longrightarrow & A & \xrightarrow{\phi} & B & \text{exact} \\ & & \alpha\downarrow & & \beta\downarrow & \\ & & A \otimes_R R & \xrightarrow{\phi\otimes\varepsilon} & B \otimes_R R & \end{array} \tag{1.15}$$

where α and β are the isomorphisms given by Lemma 1.33 and $\varepsilon = i_R$. Since (1.15) commutes, $\phi \otimes \varepsilon = \beta\phi\alpha^{-1}$, which is a monomorphism. This proves (1.14).

Having established (1.14), in order to prove that every projective left R-module is flat, a relationship between flat left R-modules and certain injective right R-modules will be established (note how the different ideas link together here), namely

$${}_RD \text{ is flat if and only if } D^+ \text{ is an injective right } R\text{–module}, \tag{1.16}$$

where D^+ is the so-called *character module* $\mathrm{Hom}(D, \mathbb{Q}/\mathbb{Z})$ of D.

The proof of (1.16) is a little involved and it is helpful to give it in outline before becoming involved in the details. Consider any exact sequence

$$0 \longrightarrow A_R \xrightarrow{\theta} B_R.$$

The right R-module D^+ is injective if and only if

$$\mathrm{Hom}_R(B, D^+) \xrightarrow{\theta_*} \mathrm{Hom}_R(A, D^+) \longrightarrow 0 \text{ is exact}, \tag{1.17}$$

where $\theta_*(\beta) = \beta\theta$ ($\beta \in \mathrm{Hom}_R(B, D^+)$). On the other hand, the left R-module D is flat if and only if

$$0 \longrightarrow A \otimes_R D \xrightarrow{\phi} B \otimes_R D \text{ is exact}, \tag{1.18}$$

where $\phi = \theta \otimes i_D$. So far we have just used definitions. We shall prove that (1.18) holds if and only if

$$(B \otimes_R D)^+ \xrightarrow{\phi_*} (A \otimes_R D)^+ \longrightarrow 0 \text{ is exact}. \tag{1.19}$$

The next step is to show that there exist $\mathbb{Z}$-isomorphisms α and β such that the diagram

$$\begin{array}{ccccc} \mathrm{Hom}_R(B, D^+) & \xrightarrow{\theta_*} & \mathrm{Hom}_R(A, D^+) & \longrightarrow & 0 \\ \beta\downarrow & & \alpha\downarrow & & \\ (B \otimes_R D)^+ & \xrightarrow{\phi_*} & (A \otimes_R D)^+ & \longrightarrow & 0 \end{array} \quad \text{commutes.} \tag{1.20}$$

From (1.20) it follows at once that (1.17) holds if and only if (1.19), and hence (1.18), holds. Now we proceed to implement this programme.

Lemma 1.34. *Let $\psi : G \to H$ be a homomorphism of Abelian groups G, H. Then ψ is a monomorphism if and only if $\psi_* : H^+ \to G^+$ is an epimorphism.*

Proof. Suppose first that ψ is a monomorphism. Let $\gamma \in G^+ = \text{Hom}\ (G, \mathbb{Q}/\mathbb{Z})$. Consider the diagram

$$\begin{array}{ccccc} 0 & \longrightarrow & G & \overset{\psi}{\longrightarrow} & H \qquad \text{exact} \\ & & \gamma\downarrow & & \\ & & \mathbb{Q}/\mathbb{Z} & & \end{array}$$

and recall that $\mathbb{Q}/\mathbb{Z}$ is an injective $\mathbb{Z}$-module (Proposition 1.24). There exists $\delta \in \text{Hom}(H, \mathbb{Q}/\mathbb{Z})$ such that $\delta\psi = \gamma$, i.e., $\psi_*(\delta) = \gamma$. It follows that ψ_* is an epimorphism.

Conversely, suppose that ψ_* is an epimorphism. Let $0 \neq g \in G$. If $kg \neq 0$ for all $0 \neq k \in \mathbb{Z}$, define $\xi' : \mathbb{Z}g \longrightarrow \mathbb{Q}/\mathbb{Z}$ by

$$\xi'(kg) = k/2 + \mathbb{Z} \quad (k \in \mathbb{Z}).$$

Note that ξ' is well defined because for $k_1, k_2 \in \mathbb{Z}$, $k_1 g = k_2 g$ implies that $k_1 = k_2$ and hence $\xi'(k_1 g) = \xi'(k_2 g)$. Since ξ' is a $\mathbb{Z}$-homomorphism, it can be lifted to a $\mathbb{Z}$-homomorphism $\xi : G \longrightarrow \mathbb{Q}/\mathbb{Z}$ such that

$$\xi(g) = \xi'(g) = 1/2 + \mathbb{Z} \neq 0,$$

again using the fact that $\mathbb{Q}/\mathbb{Z}$ is injective. On the other hand, suppose that

$$I = \{k \in \mathbb{Z} : kg = 0\} \neq 0.$$

By Lemma 1.23, there exists an integer $n \geq 2$ such that the ideal $I = \mathbb{Z}n$. In this case define $\xi'' : \mathbb{Z}g \longrightarrow \mathbb{Q}/\mathbb{Z}$ by

$$\xi''(kg) = k/n + \mathbb{Z}\ (k \in \mathbb{Z}).$$

Again ξ'' is well defined because of the choice of n and ξ'' can be lifted to a $\mathbb{Z}$-homomorphism $\xi : G \longrightarrow \mathbb{Q}/\mathbb{Z}$ such that $\xi(g) \neq 0$. Thus in any case there exists a $\mathbb{Z}$-homomorphism $\xi \in G^+$ such that $\xi(g) \neq 0$. By hypothesis, ψ_* is onto and hence there exists $\eta \in H^+$ such that $\xi = \psi_*(\eta) = \eta\psi$. But $\eta\psi(g) = \xi(g) \neq 0$ implies that $\psi(g) \neq 0$. It follows that ψ is a monomorphism. □

Lemma 1.35. *For any right R-module A and left R-module D there exists a $\mathbb{Z}$-isomorphism $\alpha : \text{Hom}_R(A, D^+) \to (A \otimes_R D)^+$.*

Proof. First we define α. Let $\nu \in \text{Hom}_R(A, D^+)$, i.e., for each $a \in A$, $\nu(a) \in \text{Hom}(D, \mathbb{Q}/\mathbb{Z})$. Define $\mu : F(A, D) \longrightarrow \mathbb{Q}/\mathbb{Z}$ by

$$\mu\left(\sum_{i=1}^{k} n_i(a_i, d_i)\right) = \sum_{i=1}^{k} n_i \nu(a_i)(d_i),$$

for all $k \in \mathbb{N}$, $a_i \in A$, $d_i \in D$, $n_i \in \mathbb{Z}$ $(1 \le i \le k)$. It can be checked that μ is a $\mathbb{Z}$-homomorphism and $G(A, D) \subseteq \ker \mu$. Thus μ induces a $\mathbb{Z}$-homomorphism $\alpha(\nu) : A \otimes_R D \longrightarrow \mathbb{Q}/\mathbb{Z}$ defined by

$$\alpha(\nu)\left(\sum_{i=1}^{k} a_i \otimes d_i\right) = \sum_{i=1}^{k} \nu(a_i)(d_i),$$

for all $k \in \mathbb{N}$, $a_i \in A$, $d_i \in D$ $(1 \le i \le k)$. Moreover,

$$\alpha(\nu_1 + \nu_2) = \alpha(\nu_1) + \alpha(\nu_2) \quad (\nu_1, \nu_2 \in \mathrm{Hom}_R(A, D^+)),$$

and thus α is a $\mathbb{Z}$-homomorphism.

The proof is completed by showing that α is a bijection and this is done by producing its inverse mapping. Let $\delta \in (A \otimes_R D)^+$. Define $\delta' : A \longrightarrow D^+$ as follows: for each $a \in A$, $\delta'(a)$ is the mapping from D to $\mathbb{Q}/\mathbb{Z}$ defined by

$$\delta'(a)(d) = \delta(a \otimes d) \quad (d \in D).$$

It can be checked that $\delta'(a) \in D^+$ for each $a \in A$ and hence δ' is a mapping from A to D^+. Moreover, by combining (1.13) with the fact that δ is a $\mathbb{Z}$-homomorphism it can be shown easily that $\delta' \in \mathrm{Hom}_R(A, D^+)$. Define $\alpha' : (A \otimes_R D)^+ \to \mathrm{Hom}_R(A, D^+)$ by

$$\alpha'(\delta) = \delta' \quad (\delta \in (A \otimes_R D)^+).$$

A final check shows that the mapping α' is the inverse of α. This proves that α is an isomorphism. □

We are now in a position to prove (1.20). Suppose that $0 \longrightarrow A_R \xrightarrow{\theta} B_R$ is exact and ${}_RD$. Define $\mathbb{Z}$-isomorphisms

$$\alpha : \mathrm{Hom}_R(A, D^+) \to (A \otimes_R D)^+ \quad \text{and} \quad \beta : \mathrm{Hom}_R(B, D^+) \to (B \otimes_R D)^+$$

as in Lemma 1.35. To prove (1.20) we have to show that $\alpha\theta_* = \phi_*\beta$. Let $\xi \in \mathrm{Hom}_R(B, D^+)$. Then $\alpha\theta_*(\xi) = \alpha(\xi\theta)$ which is the mapping γ from $A \otimes_R D$ to $\mathbb{Q}/\mathbb{Z}$ such that

$$\gamma\left(\sum_{i=1}^{k} a_i \otimes d_i\right) = \sum_{i=1}^{k} \xi\theta(a_i)(d_i),$$

for all $k \in \mathbb{N}$, $a_i \in A$, $d_i \in D$ $(1 \le i \le k)$. On the other hand, $\phi_*\beta(\xi) = \beta(\xi)\phi$ which is the mapping δ from $A \otimes_R D$ to $\mathbb{Q}/\mathbb{Z}$ such that

$$\delta\left(\sum_{i=1}^{k} a_i \otimes d_i\right) = \beta(\xi)\left(\sum_{i=1}^{k} \theta(a_i) \otimes d_i\right) = \sum_{i=1}^{k} \xi\theta(a_i)(d_i)$$

for all $k \in \mathbb{N}$, $a_i \in A$, $d_i \in D$ $(1 \le i \le k)$. It follows that $\alpha\theta_* = \phi_*\beta$. This proves (1.20) and hence the following result.

Theorem 1.36 (Lambek's Theorem). *A left R-module D is flat if and only if its character module D^+ is an injective right R-module.*

Corollary 1.37. *Let A_λ ($\lambda \in \Lambda$) be a non-empty collection of left R-modules. Then $\bigoplus_{\lambda\in\Lambda} A_\lambda$ is flat if and only if A_λ is flat for each $\lambda \in \Lambda$.*

Proof. Let $A = \bigoplus_{\lambda\in\Lambda} A_\lambda$. Then

$$A^+ = \operatorname{Hom}(A, \mathbb{Q}/\mathbb{Z}) \cong \prod_{\lambda\in\Lambda} \operatorname{Hom}(A_\lambda, \mathbb{Q}/\mathbb{Z}) = \prod_{\lambda\in\Lambda} A_\lambda^+,$$

by Exercise 1.29. By Exercise 1.74 the module A^+ is injective if and only if ${A_\lambda}^+$ is injective for each $\lambda \in \Lambda$. The result now follows by Theorem 1.36. □

Recall that our objective is to prove that every projective left R-module is flat. This is an immediate consequence of (1.14) and Corollary 1.37, so we have:

Proposition 1.38. *Every projective module is flat.*

Let E be a right ideal of R and D be a left R-module, i.e., $E \leq R_R$ and ${}_RD$. Then there is a $\mathbb{Z}$-epimorphism $\varepsilon : E \otimes_R D \to ED$ which can be defined in a natural way. First define $\varepsilon' : F(E, D) \to ED$ by

$$\varepsilon'\left(\sum_{i=1}^{k} n_i(e_i, d_i)\right) = \sum_{i=1}^{k} n_i e_i d_i,$$

for all $k \geq 1$, $n_i \in \mathbb{Z}$, $e_i \in E$, $d_i \in D$ ($1 \leq i \leq k$). Note that ε' is a $\mathbb{Z}$-epimorphism and that $G(E, D) \subseteq \ker \varepsilon'$. Thus ε' induces a $\mathbb{Z}$-epimorphism $\varepsilon : E \otimes_R D \to ED$ such that

$$\varepsilon\left(\sum_{i=1}^{k} e_i \otimes d_i\right) = \sum_{i=1}^{k} e_i d_i,$$

for all $k \geq 1$, $e_i \in E$, $d_i \in D$ ($1 \leq i \leq k$). Another consequence of Theorem 1.36 is the following result which is the analogue of Baer's Lemma (Theorem 1.20) and is sometimes called the *Flat Test Lemma.*

Proposition 1.39. *The following statements are equivalent for a left R-module D.*

(i) *D is flat.*

(ii) *Given any right ideal E of R and any monomorphism $\theta : E \to R$, the mapping $\theta \otimes i_D : E \otimes_R D \to R \otimes_R D$ is a monomorphism.*

(iii) *Given any right ideal E of R, the $\mathbb{Z}$-epimorphism $\varepsilon : E \otimes_R D \to ED$ is an isomorphism.*

Proof. (i) ⇒ (ii) Clear.

(ii) ⇒ (iii) By (ii), the mapping $\iota \otimes i_D : E \otimes_R D \to R \otimes_R D$ is a $\mathbb{Z}$-monomorphism, where $\iota : E \to R$ denotes the inclusion map. By Lemma 1.33, there exists a $\mathbb{Z}$-isomorphism $\delta : R \otimes_R D \to D$. Let δ' denote the restriction of δ to the subgroup $E \otimes_R D$. Then δ' is a $\mathbb{Z}$-monomorphism and hence so is $\varepsilon = \delta'(\iota \otimes i_D)$.

(iii) $\Rightarrow$ (i) If $\varepsilon : E \otimes_R D \to ED$ is a $\mathbb{Z}$- monomorphism, then

$$0 \longrightarrow E \otimes_R D \xrightarrow{\theta} R \otimes_R D$$

is exact, where $\theta = \iota \otimes i_D$. Thus

$$(R \otimes_R D)^+ \xrightarrow{\theta_*} (E \otimes_R D)^+ \longrightarrow 0$$

is exact by Lemma 1.34. By (1.20),

$$\mathrm{Hom}_R(R, D^+) \longrightarrow \mathrm{Hom}_R(E, D^+) \longrightarrow 0$$

is exact. But this last fact being true for every right ideal E of R implies that D^+ is injective by Theorem 1.20. By Theorem 1.36, D is flat. □

Suppose that R is a field. Then any R-module is a vector space, has a basis, is free and hence is flat by Proposition 1.38. Which $\mathbb{Z}$-modules are flat? The answer is given in the next result. Note that in particular it shows that $\mathbb{Z} \oplus \mathbb{Q}$ is a flat $\mathbb{Z}$-module which is neither projective nor injective. Recall from Exercises 1.44 and 1.57 that if R is a commutative domain, then an R-module A is called *torsion-free* provided $ra \neq 0$ for all $0 \neq r \in R$ and $0 \neq a \in A$ and any projective R-module is torsion-free.

Proposition 1.40. *An Abelian group A is a flat $\mathbb{Z}$-module if and only if A is torsion-free.*

Proof. Let E be a non-zero ideal of $\mathbb{Z}$. By Lemma 1.23, there exists a positive integer n such that $E = \mathbb{Z}n$. If $x \in E \otimes_{\mathbb{Z}} A$, then

$$x = \sum_{i=1}^{k} (m_i n) \otimes a_i = \sum_{i=1}^{k} n \otimes (m_i a_i) = n \otimes \sum_{i=1}^{k} m_i a_i$$

for some $k \in \mathbb{N}, m_i \in \mathbb{Z}, a_i \in A$ $(1 \leq i \leq k)$. Thus

$$E \otimes_{\mathbb{Z}} A = \{n \otimes a \ : a \in A\}.$$

Moreover, the natural mapping $\varepsilon : E \otimes_R A \to EA$ is given by

$$\varepsilon(n \otimes a) = na \quad (a \in A).$$

Suppose that A is torsion-free. If $a \in A$ and $\varepsilon(n \otimes a) = 0$, then $na = 0$ and hence $a = 0$, so that $n \otimes a = n \otimes 0 = 0$. Thus ε is a monomorphism and Proposition 1.39 gives that A is flat. Conversely, suppose that A is flat. Let m be any positive integer. Consider the $\mathbb{Z}$-isomorphism $\theta : \mathbb{Z}m \to \mathbb{Z}$ defined by $\theta(km) = k$ $(k \in \mathbb{Z})$. Since A is flat, $\theta \otimes i_A : \mathbb{Z}m \otimes_{\mathbb{Z}} A \to \mathbb{Z} \otimes_{\mathbb{Z}} A$ is a monomorphism. Combining this fact with Lemma 1.33 we see that there exists a $\mathbb{Z}$-monomorphism $\phi : \mathbb{Z}m \otimes_{\mathbb{Z}} A \to A$ such that $\phi(m \otimes a) = a$. Thus $m \otimes a \neq 0$ for all $0 \neq a \in A$. Because A is flat, it follows that $ma = \varepsilon(m \otimes a) \neq 0$ for all $0 \neq a \in A$. Hence, A is torsion-free. □

Exercise 1.72 shows that any finitely generated flat $\mathbb{Z}$-module is free. The final result of this section is an extension of this fact to general rings. But before coming to that we prove two lemmas.

Lemma 1.41. *Let K be a submodule of a flat left R-module F. Then the factor module F/K is flat if and only if $EF \cap K = EK$ for each right ideal E of R.*

Proof. Let $A = F/K$. Let E be any right ideal of R. Consider the exact sequence $0 \longrightarrow K \xrightarrow{\iota} F \xrightarrow{\pi} A \longrightarrow 0$, where ι is the inclusion mapping and π the canonical epimorphism. Consider the commutative diagram

$$\begin{array}{ccccccc} & & E \otimes_R K & \xrightarrow{\alpha} & E \otimes_R F & \xrightarrow{\beta} & E \otimes_R A & \longrightarrow & 0 \\ & & \varepsilon_K \downarrow & & \varepsilon_F \downarrow & & \varepsilon_A \downarrow & & \\ 0 & \longrightarrow & EF \cap K & \xrightarrow{\iota} & EF & \xrightarrow{\delta} & EA & & \end{array}$$

where ε_K, ε_F, and ε_A are the natural $\mathbb{Z}$-epimorphisms defined before Proposition 1.39, $\alpha = i_E \otimes \iota$, $\beta = i_E \otimes \pi$, and $\delta = \pi|_{EF}$. Note that the bottom row is exact by inspection and the top row is exact by Proposition 1.32. By Exercise 1.12, ε_A is a $\mathbb{Z}$-monomorphism if and only if ε_K is a $\mathbb{Z}$-epimorphism. But $\operatorname{im} \varepsilon_K = EK$. Thus ε_A is a monomorphism if and only if $EF \cap K = EK$. The result now follows by Proposition 1.39. □

Lemma 1.42. *Let B be a submodule of a left R-module A with the property that for each element $b \in B$ there exists an R-homomorphism $\theta : A \to B$ such that $\theta(b) = b$. Then for each finite subset $\{b_1, \ldots, b_n\}$ of B there exists an R-homomorphism $\phi : A \to B$ such that $\phi(b_i) = b_i$ $(1 \le i \le k)$.*

Proof. The proof is by induction on n. Suppose that $n \ge 2$. By hypothesis, there exists $\alpha \in \operatorname{Hom}_R(A, B)$ such that $\alpha(b_n) = b_n$. Let $c_i = b_i - \alpha(b_i)$ $(1 \le i \le n-1)$. By induction there exists $\beta \in \operatorname{Hom}_R(A, B)$ such that $\beta(c_i) = c_i$ $(1 \le i \le n-1)$. Define $\phi = \alpha + \beta - \beta\alpha$. Then $\phi(b_i) = b_i$ $(1 \le i \le k)$. □

Given a left R-module A, there exist a free left R-module F and an R-epimorphism $\phi : F \to A$ by Proposition 1.8. Thus $A \cong F/K$, where $K = \ker \phi$. The next result shows that if A is flat, then A is projective whenever K is finitely generated.

Proposition 1.43. *Let A be a flat left R-module such that $A = F/K$ for some free left R-module F and finitely generated submodule K of F. Then A is projective.*

Proof. Let $\{b_1, \ldots, b_n\}$ be a basis of F (Exercise 1.53). Let $x \in K$. There exist elements $r_i \in R$ $(1 \le i \le n)$ such that $x = r_1 b_1 + \cdots + r_n b_n$. Let E be the right ideal $r_1 R + \cdots + r_n R$ of R. Then

$$x \in EF \cap K = EK = r_1 K + \cdots + r_n K,$$

by Proposition 1.38 and Lemma 1.41. There exist elements $k_i \in K$ $(1 \le i \le n)$ such that $x = r_1 k_1 + \cdots + r_n k_n$. Define $\theta \in \operatorname{Hom}_R(F, K)$ such that $\theta(b_i) = k_i$

$(1 \le i \le n)$ (Exercise 1.45). Then $\theta(x) = x$. There exist $m \in \mathbb{N}$ and $x_i \in K$ $(1 \le i \le m)$ such that $K = Rx_1 + \cdots + Rx_m$. By Lemma 1.42, there exists $\phi \in \mathrm{Hom}_R(F, K)$ such that $\phi(x_i) = x_i$ $(1 \le i \le m)$ and hence $\phi|_K = i_K$. Finally, Exercise 1.31 shows that $K \le_{\mathrm{d}} F$ and hence A is projective. □

Exercises

1.96. Prove that if $A_R \cong C_R$ and ${}_RB \cong {}_RD$, then $A \otimes_R B \cong C \otimes_R D$.

1.97. Prove that $T \otimes_{\mathbb{Z}} D = 0$ for any torsion Abelian group T and divisible Abelian group D.

1.98. Let ${}_RA$ and $E \le R_R$. Prove that there exists a $\mathbb{Z}$-isomorphism $\theta : (R/E) \otimes_R A \to A/EA$ such that $\theta((r+E) \otimes a) = ra + EA$ $(r \in R, a \in A)$. Now suppose that in addition R is a commutative ring. Prove that θ is an R-isomorphism. Deduce that if E and F are ideals of R then as, R-modules, $(R/E) \otimes_R (R/F) \cong R/(E+F)$.

1.99. Let R be a commutative ring and F and G free R-modules with bases $\{f_\lambda : \lambda \in \Lambda\}$ and $\{g_\gamma : \gamma \in \Gamma\}$, respectively. Prove that $F \otimes_R G$ is a free R-module with basis $\{f_\lambda \otimes g_\gamma : \lambda \in \Lambda, \gamma \in \Gamma\}$.

1.100. Let K be a subfield of a field F and let V be a K-vector space with basis $\{v_\lambda : \lambda \in \Lambda\}$. Prove that $F \otimes_K V$ is an F-vector space with basis $\{1 \otimes v_\lambda : \lambda \in \Lambda\}$.

1.101. Let $0 \longrightarrow A \longrightarrow B \longrightarrow C \longrightarrow 0$ be an exact sequence of left R-modules such that A and C are both flat. Prove that B is flat.

1.102. Let R be a commutative ring and ${}_RA$, ${}_RB$. Prove that $A \otimes_R B$ and $B \otimes_R A$ are R-isomorphic.

1.103. Let R and S be rings. Prove that $(A \otimes_R B) \otimes_S C \cong A \otimes_R (B \otimes_S C)$ for all A_R, ${}_RB_S$, and ${}_SC$. Deduce that if ${}_RB$ and ${}_SC$ are flat, then so is ${}_R(B \otimes_S C)$.

1.104. Prove that ${}_RA$ is flat if and only if the natural $\mathbb{Z}$-epimorphism $\varepsilon : E \otimes_R A \to EA$ is a monomorphism for each *finitely generated* right ideal E of R. Prove the same result with "finitely generated" replaced by "essential".

1.105. Prove that if K is a submodule of a flat left R-module F, then the factor module F/K is flat if and only if $EF \cap K = EK$ for each *finitely generated* right ideal E of R. Deduce that if R is a principal ideal domain, then the following statements are equivalent.

(i) F/K is flat. (ii) $rF \cap K = rK$ for each $r \in R$. (iii) F/K is torsion-free.

1.106. Let $0 \longrightarrow {}_RA \longrightarrow {}_RB \longrightarrow {}_RC \longrightarrow 0$ be exact with B flat. Prove that the left R-modules $I \otimes_R C$ and IB/IA are R-isomorphic for any ideal I of R.

1.107. Prove that

$$\mathrm{Hom}_R(A, \mathrm{Hom}_S(B, C)) \cong \mathrm{Hom}_S(B \otimes_R A, C)$$

for any ${}_RA$, ${}_SB_R$, ${}_SC$.

1.108. Prove that

$$\mathrm{Hom}_R(A, \mathrm{Hom}_S(C, B)) \cong \mathrm{Hom}_S(C, \mathrm{Hom}_R(A, B))$$

for all ${}_RA$, ${}_RB_S$, C_S.

1.109. Suppose that ${}_RA$ is finitely generated and projective. Prove that

$$\mathrm{Hom}_R(A, B) \otimes_S C \cong \mathrm{Hom}_R(A, B \otimes_S C)$$

for all ${}_RB_S$ and ${}_SC$.

1.110. Suppose that A_R is finitely generated projective. Prove that

$$A \otimes_R \mathrm{Hom}_S(B, C) \cong \mathrm{Hom}_S(\mathrm{Hom}_R(A, B), C)$$

for all ${}_SB_R$ and ${}_SC$.

1.8 Chain conditions

Let A be a left R-module. Recall from § 1.5 that a non-empty collection $\underline{S}$ of submodules of A is a *chain* provided $B \subseteq C$ or $C \subseteq B$ for all $B, C \in \underline{S}$. A chain of the form

$$A_1 \subseteq A_2 \subseteq A_3 \subseteq \cdots , \tag{1.21}$$

where each $A_i \leq {}_RA$ ($i \geq 1$), is called an *ascending chain.* On the other hand a chain of the from

$$B_1 \supseteq B_2 \supseteq B_3 \supseteq \cdots , \tag{1.22}$$

where each $B_i \leq {}_RA$ ($i \geq 1$), is called a *descending chain.* The module A is called *Noetherian* if for any ascending chain (1.21) there exists a positive integer n such that $A_n = A_{n+1} = A_{n+2} = \cdots$, and in this case A is said to *satisfy the ascending chain condition*. Modules which satisfy the analogous *descending chain condition* are called *Artinian*, i.e., A is Artinian if for any descending chain (1.22) there exists a positive integer m such that $B_m = B_{m+1} = B_{m+2} = \cdots$. The first result in this section gives a useful property that Noetherian and Artinian modules share.

Lemma 1.44. *Let A be a Noetherian or Artinian left R-module. Then A does not contain an infinite direct sum of non-zero submodules.*

Proof. Let $A_1 \oplus A_2 \oplus A_3 \oplus \cdots$ be a direct sum of non-zero submodules of A. Consider the chains:

$$A_1 \subset A_1 \oplus A_2 \subset A_1 \oplus A_2 \oplus A_3 \subset \cdots$$

and

$$(A_1 \oplus A_2 \oplus A_3 \oplus \cdots) \supset (A_2 \oplus A_3 \oplus A_4 \oplus \cdots) \supset \cdots .$$

Since A is Noetherian or Artinian it follows that the direct sum $A_1 \oplus A_2 \oplus A_3 \oplus \cdots$ is finite. □

Suppose that R is a field and A is an R-module, i.e., A is a vector space over R. If A is Noetherian or Artinian, then Lemma 1.44 shows that A is finite dimensional. On the other hand, if A is finite dimensional and $B \subset C$ are subspaces of A, then B has smaller dimension than C and hence any chain of subspaces of A contains at most $n+1$ distinct members, where n is the dimension of A. Thus the following statements are equivalent:

(i) A is Noetherian,

(ii) A is Artinian,

(iii) A is finite dimensional,

(iv) A is finitely generated.

The situation for $R = \mathbb{Z}$ is a good deal more interesting. Clearly finite Abelian groups are both Noetherian and Artinian $\mathbb{Z}$-modules. Note also that ${}_{\mathbb{Z}}\mathbb{Z}$ is Noetherian, but is not Artinian; that it is Noetherian is a consequence of Proposition 1.45 below and Lemma 1.23. In contrast, the $\mathbb{Z}$-module $\mathbb{Z}(p^\infty)$ is Artinian, but not Noetherian (see Exercise 1.42).

A ring R is called *left Noetherian* (respectively *right Noetherian*) if the module ${}_RR$ (respectively R_R) is Noetherian. In other words, a ring R is left Noetherian if and only if for any ascending chain of left ideals $L_1 \subseteq L_2 \subseteq L_3 \subseteq \cdots$ there exists a positive integer n such that $L_n = L_{n+1} = L_{n+2} = \cdots$. Rings that are both left and right Noetherian will be called simply *Noetherian*. There are analogous definitions for *left Artinian, right Artinian*, and *Artinian rings*. Examples are given below.

Proposition 1.45. *The following statements are equivalent for a left R-module A.*

(i) *A is Noetherian.*

(ii) *Every non-empty collection of submodules of A contains a maximal member.*

(iii) *Every submodule of A is finitely generated.*

Proof. (i) $\Rightarrow$ (ii) Let $\underline{S}$ be a non-empty collection of submodules of A. Let $A_1 \in \underline{S}$. Then either A_1 is a maximal member of $\underline{S}$, or there exists $A_2 \in \underline{S}$ such that $A_1 \subset A_2$. In the same way, either A_2 is a maximal member of $\underline{S}$, or there exists $A_3 \in \underline{S}$ such that $A_2 \subset A_3$. Repeat this process to obtain a chain

$$A_1 \subset A_2 \subset A_3 \cdots ,$$

which must be finite because A is Noetherian. Thus there exists a positive integer n such that A_n is a maximal member of $\underline{S}$.

(ii) $\Rightarrow$ (iii) Let $B \leq {}_RA$. Let $\underline{T}$ denote the collection of finitely generated submodules of B. By hypothesis, $\underline{T}$ contains a maximal member M. There exist $k \in \mathbb{N}$ and $m_i \in M$ $(1 \leq i \leq k)$ such that $M = Rm_1 + \cdots + Rm_k$. Let $b \in B$. Then $M \subseteq M + Rb \in \underline{T}$ implies that $M = M + Rb$ by the choice of M, and hence $b \in M$. It follows that $B = M$ and B is finitely generated.

(iii) ⇒ (i) Let $C_1 \subseteq C_2 \subseteq C_3 \subseteq \cdots$ be an ascending chain of submodules of A. Then $C = \bigcup_{i \in \mathbb{N}} C_i \leq {}_R A$ and by hypothesis C is finitely generated. By Proposition 1.15, there exists $m \in \mathbb{N}$ such that $C_m = C$, and in this case

$$C_m \subseteq C_{m+1} \subseteq C_{m+2} \subseteq \cdots \subseteq C = C_m$$

shows that $C_m = C_{m+1} = C_{m+2} = \cdots$. It follows that A is Noetherian. □

The companion result to Proposition 1.45 for Artinian modules is the following one. Its proof is left as an exercise.

Proposition 1.46. *The following statements are equivalent for a left R-module A.*

(i) *A is Artinian.*

(ii) *Every non-empty collection of submodules of A contains a minimal member.*

(iii) *Every factor module of A is finitely cogenerated.*

Proposition 1.47. *Let B be a submodule of a left R-module A. Then A is Noetherian (respectively Artinian) if and only if the submodule B and factor module A/B are both Noetherian (respectively Artinian).*

Proof. We prove the Noetherian version of the result and leave the Artinian case as an exercise. Suppose first that A is Noetherian. Any submodule of B is a submodule of A and thus B is Noetherian. On the other hand, let $X_1 \subseteq X_2 \subseteq X_3 \subseteq \cdots$ be an ascending chain of submodules of A/B. For each $i \in \mathbb{N}$ let $C_i = \pi^{-1}(X_i)$, where $\pi : A \to A/B$ is the canonical epimorphism. Then $C_i \leq {}_R A$ $(i \in \mathbb{N})$ and $C_1 \subseteq C_2 \subseteq C_3 \subseteq \cdots$. But A is Noetherian, so there exists $n \in \mathbb{N}$ such that $C_n = C_{n+1} = C_{n+2} = \cdots$. It follows that $X_n = X_{n+1} = X_{n+2} = \cdots$. Hence A/B is Noetherian.

Conversely, suppose that B and A/B are both Noetherian. Let $D_1 \subseteq D_2 \subseteq D_3 \subseteq \cdots$ be an ascending chain of submodules of A. Consider the chains:

$$(D_1 \cap B) \subseteq (D_2 \cap B) \subseteq (D_3 \cap B) \subseteq \cdots \leq {}_R B$$

and

$$(D_1 + B)/B \subseteq (D_2 + B)/B \subseteq (D_3 + B)/B \subseteq \cdots \leq {}_R(A/B).$$

There exists $k \in \mathbb{N}$ such that

$$(D_k \cap B) = (D_{k+1} \cap B) = (D_{k+2} \cap B) = \cdots$$

and

$$(D_k + B)/B = (D_{k+1} + B)/B = (D_{k+2} + B)/B = \cdots .$$

It follows that $D_k = D_{k+1} = D_{k+2} = \cdots$ by Exercise 1.18. Hence A is Noetherian. □

Any isomorphic copy of a Noetherian module is Noetherian, similarly for an Artinian module, so that an immediate consequence of Proposition 1.47 reads:

Corollary 1.48. *Let* $0 \longrightarrow A \longrightarrow B \longrightarrow C \longrightarrow 0$ *be an exact sequence of left* R*-modules. Then* B *is Noetherian (respectively Artinian) if and only if* A *and* C *are both Noetherian (respectively Artinian).*

Proposition 1.49. *Let* A_i $(1 \le i \le n)$ *be a finite collection of left* R*-modules. Then* $A_1 \oplus \cdots \oplus A_n$ *is Noetherian (respectively Artinian) if and only if* A_i *is Noetherian (respectively Artinian) for each* $1 \le i \le n$.

Proof. Again we prove the Noetherian version and leave the other proof as an easy exercise. Let $A = A_1 \oplus \cdots \oplus A_n$. If A is Noetherian, then so is each A_i because it is a homomorphic image of A for $1 \le i \le n$. Conversely, suppose that each A_i $(1 \le i \le n)$ is Noetherian. We show that A is Noetherian by induction on n. The case $n = 1$ is clear, so suppose that $n \ge 2$. Consider the projection $\pi_n : A \to A_n$ with kernel $B = A_1 \oplus \cdots \oplus A_{n-1}$. By hypothesis, A_n is Noetherian and by induction B is Noetherian. Thus A is Noetherian by Corollary 1.48. □

Corollary 1.50. *Let* R *be a left Noetherian (respectively Artinian) ring and* A *a finitely generated left* R*-module. Then* A *is Noetherian (respectively Artinian).*

Proof. Suppose that R is left Noetherian. By Exercise 1.53, there exist $n \in \mathbb{N}$ and an epimorphism $\phi : ({}_RR)^n \to A$. Applying Proposition 1.49, the left R-module $({}_RR)^n$ is Noetherian and hence so is A by Proposition 1.47. The proof of the Artinian case is similar. □

This last corollary allows us to prove a simple fact about matrix rings. Recall from Exercise 1.46 that $M_n(R)$ is a free left (respectively right) R-module with basis $\{e_{ij} : 1 \le i, j \le n\}$.

Proposition 1.51. *If* R *is a left Noetherian (respectively Artinian) ring, then so is the matrix ring* $M_n(R)$ *for any positive integer* n.

Proof. Suppose that R is left Noetherian. By Corollary 1.50, $M_n(R)$ is a Noetherian left R-module. But any left ideal of $M_n(R)$ is an R-submodule of $M_n(R)$. Hence $M_n(R)$ is left Noetherian. The proof of the Artinian case is the same. □

Another application of Corollary 1.50 is that an Abelian group A is a Noetherian $\mathbb{Z}$-module if and only if it is finitely generated. To obtain more information about Noetherian $\mathbb{Z}$-modules we first prove:

Lemma 1.52. *Let* A *be a* $\mathbb{Z}$*-module which is either Noetherian or Artinian. Then* A *is finite if and only if* A *is unfaithful.*

Proof. Suppose first that A contains only a finite number n of elements. By Lagrange's Theorem, $na = 0$ $(a \in A)$ and hence $0 \ne n \in \operatorname{ann}_{\mathbb{Z}} A$, i.e., A is unfaithful. Conversely, suppose that A is unfaithful. By Lemma 1.23, there exists a positive integer m such that $mA = 0$. We prove that A is finite by induction on m. If $m = 1$, then $A = 0$. Suppose that $m \ge 2$. There exist a prime p and $k \in \mathbb{N}$ such that $m = pk$. Note that $p(kA) = 0$ and hence the subgroup kA of A is a vector space over the field $\mathbb{Z}/\mathbb{Z}p$. By Lemma 1.44, kA is finite dimensional and hence finite. Moreover, $k(A/kA) = 0$ implies that A/kA is finite by induction because $k \leq m - 1$. Thus A is finite. □

Proposition 1.53.

(i) *An Abelian group A is a Noetherian $\mathbb{Z}$-module if and only if A is finite or $A \cong \mathbb{Z}^n \oplus F$ for some positive integer n and finite group F.*

(ii) *An Abelian group A is an Artinian $\mathbb{Z}$-module if and only if A is finite or $A \cong \mathbb{Z}(p_1^\infty) \oplus \cdots \oplus \mathbb{Z}(p_k^\infty) \oplus G$ for some positive integer k, primes p_i $(1 \leq i \leq k)$, and finite group G.*

Proof. (i) The sufficiency is clear by Proposition 1.49. Conversely, suppose that A is a Noetherian $\mathbb{Z}$-module. By Exercise 1.72 and Lemma 1.23, there exists a free submodule H such that $A = H \oplus T$, where T is the torsion submodule of A. Since T is finitely generated and torsion, it follows that T is finite. Since H is finitely generated, we conclude that $H \cong ({}_{\mathbb{Z}}\mathbb{Z})^n$ for some positive integer n (Exercise 1.53).

(ii) The sufficiency is clear by Proposition 1.49. Conversely, suppose that A is Artinian. Let D denote the sum of all divisible subgroups of A. It is clear that D itself is divisible. By Propositions 1.24 and 1.30, there exists a subgroup G of A such that A is the direct sum $D \oplus G$ of its subgroups D, G. Since A is Artinian, it follows that A does not contain an infinite cyclic subgroup (because $\mathbb{Z}$ is not Artinian) and hence A is a torsion group. Suppose that $D \neq 0$. Let $0 \neq a_1 \in D$ with $pa_1 = 0$ for some prime p. There exists $a_2 \in D$ such that $pa_2 = a_1$. Moreover, there exists $a_3 \in D$ such that $pa_3 = a_2$. In this way we obtain a sequence $a_1, a_2, a_3, \ldots$ of elements of D such that $pa_{i+1} = a_i$ $(i \in \mathbb{N})$. Let $C_1 = \mathbb{Z}a_1 + \mathbb{Z}a_2 + \mathbb{Z}a_3 + \cdots$. By Exercise 1.42, $C_1 \cong \mathbb{Z}(p^\infty)$, which is an injective $\mathbb{Z}$-module, and hence $D = C_1 \oplus C'$ for some $C' \leq {}_{\mathbb{Z}}D$. If $C' \neq 0$, then this argument can be repeated to show that C' is the direct sum $C_2 \oplus C''$ of its subgroups $C_2 \cong \mathbb{Z}(q^\infty)$ for some prime q and C''. Repeat this argument and use Lemma 1.44 to obtain $D = \mathbb{Z}(p_1^\infty) \oplus \cdots \oplus \mathbb{Z}(p_k^\infty)$ for some positive integer k. Now consider G. Because $D \cap G = 0$, if $G \neq 0$, then there exists a positive integer n_1 such that $G \supset n_1 G$. Because $D \cap n_1 G = 0$, if $n_1 G \neq 0$ then there exists a positive integer n_2 such that $n_1 G \supset n_2(n_1 G) = n_1 n_2 G$. Repeating this argument gives a sequence $n_1, n_2, n_3, \ldots$ of positive integers such that $G \supset n_1 G \supset n_1 n_2 G \supset \cdots$. But A is Artinian. Thus there exists a positive integer t such that $n_1 n_2 \cdots n_t G = 0$. Then by Lemma 1.52 G is finite. □

There are a number of ways of characterising left Noetherian rings in terms of injective modules. The next theorem gives one of these; others can be found in the exercises and in the next section.

Theorem 1.54 (Bass–Papp Theorem). *A ring R is left Noetherian if and only if any direct sum of injective left R-modules is injective.*

Proof. Suppose first that R is left Noetherian. Let E_λ $(\lambda \in \Lambda)$ be a non-empty collection of injective left R-modules and $E = \bigoplus_{\lambda\in\Lambda} E_\lambda$. Let L be a left ideal of R and $\phi : L \longrightarrow E$ an R-homomorphism. There exists a finite collection of elements $a_i \in L$ $(1 \leq i \leq n)$ such that $L = Ra_1 + \cdots + Ra_n$. For each $1 \leq i \leq n$ there exists a finite subset Λ_i of Λ such that

$$\pi_\lambda(\phi(a_i)) = 0 \quad (\lambda \in \Lambda, \lambda \notin \Lambda_i).$$

Let $\Lambda' = \Lambda_1 \cup \cdots \cup \Lambda_n$. Then $\operatorname{im}\phi$ is contained in $E' = \bigoplus_{\lambda\in\Lambda'} E_\lambda$, thought of as a submodule of E. Since Λ' is finite it follows that E' is injective (Exercise 1.74) and hence there exists $e \in E' \subseteq E$ such that $\phi(r) = re$ $(r \in L)$. Therefore E is injective by Theorem 1.20.

Conversely, suppose that the direct sum of any non-empty collection of injective left R-modules is injective. Let $L_1 \subseteq L_2 \subseteq L_3 \subseteq \cdots$ be an ascending chain of left ideals of R. Let $N = \bigcup_{i\in\mathbb{N}} L_i$. Then N is a left ideal of R. Let $I = E(R/L_1) \oplus E(R/L_2) \oplus E(R/L_3) \oplus \cdots$ and define $\theta : N \to I$ by

$$\theta(n) = (n + L_1, n + L_2, n + L_3, \ldots) \quad (n \in N).$$

Note that if $n \in N$ then there exists $k \in \mathbb{N}$ such that $n \in L_k$ and hence $n + L_t = 0$ for all $t \geq k$, so that $\theta(n) \in I$. By hypothesis, I is injective and by Theorem 1.20 there exists an $x \in I$ such that $\theta(n) = nx$ $(n \in N)$. There exist $m \in \mathbb{N}$ and $x_i \in E(R/L_i)$ $(i \in \mathbb{N})$ such that $x = (x_1, x_2, x_3, \ldots)$ and $x_s = 0$ for all $s \geq m$. If $n \in N$ then

$$(n + L_1, n + L_2, n + L_3, \ldots) = \theta(n) = n(x_1, x_2, x_3, \ldots)$$

implies that $n + L_m = nx_m = 0$ and hence $n \in L_m$. It follows that $L \subseteq L_m$ and hence $L_m = L_{m+1} = L_{m+2} = \cdots$. Thus R is left Noetherian. □

If the ring R is left Noetherian and A is a finitely generated left R-module, then A is Noetherian. The next lemma gives a converse in the case of commutative rings.

Lemma 1.55. *Let R be a commutative ring and A a Noetherian R-module. Then the ring $R/(\operatorname{ann}_R A)$ is Noetherian.*

Proof. There exists a finite collection of elements a_i $(1 \leq i \leq n)$ of A such that $A = Ra_1 + \cdots + Ra_n$. Define $\phi : R \to A^n$ by $\phi(r) = (ra_1, \ldots, ra_n)$ $(r \in R)$. Then ϕ is an R-homomorphism with kernel $\operatorname{ann}_R A$. By Propositions 1.47 and 1.49, the R-module $R/(\operatorname{ann}_R A)$ is Noetherian. But any ideal of $R/(\operatorname{ann}_R A)$ is an R-submodule. Thus $R/(\operatorname{ann}_R A)$ is a Noetherian ring. □

The next result is included for fun. The ingenious proof is due to E. Formanek.

Theorem 1.56 (Eaken's Theorem). *Let R be a subring of a commutative ring S such that S is a finitely generated R-module. Then R is a Noetherian ring if and only if S is a Noetherian ring.*

Proof. Suppose first that R is a Noetherian ring. Then S is a Noetherian R-module by Corollary 1.50. But any ideal of S is an R-submodule. It follows that S is a Noetherian ring.

Conversely, suppose that S is a Noetherian ring. Note two facts:

(i) S is a faithful R-module, and

(ii) if I is an ideal of R, then IS is an ideal of S.

Thus in order to complete the proof it is certainly sufficient to prove the following stronger result:

Lemma 1.57. *Let R be a commutative ring for which there exists a finitely generated faithful R-module A such that for any non-empty collection $\underline{I}$ of ideals of R the set $\{IA : I \in \underline{I}\}$ contains a maximal member. Then R is a Noetherian ring.*

Proof. In view of Lemma 1.55, it is sufficient to prove that A is Noetherian. Suppose not and let $\underline{S} = \{IA : I \leq {}_R R$ and A/IA is not Noetherian$\}$. Note that $\underline{S}$ is non-empty because $0 \in \underline{S}$. By hypothesis, $\underline{S}$ contains a maximal member JA for some ideal J of R. Thus A/JA is not Noetherian, but A/KA is Noetherian for any ideal K of R such that $JA \subset KA$. Let $B = A/JA$ and $N = \mathrm{ann}_R B$. Then B is a finitely generated faithful (R/N)-module, and by Exercise 1.61 there exists $C \leq {}_{(R/N)}B$ such that B/C is faithful, but B/D is unfaithful for any $C \subset D \leq {}_{(R/N)}B$. Let $0 \neq X \leq {}_{(R/N)}(B/C)$. Then there exists $C \subset D \leq {}_{(R/N)}B$ such that $X = D/C$. By the choice of C, there exists an ideal M of R such that $N \subset M$ and $MB \subseteq D$. But

$$B/MB = (A/JA)/[M(A/JA)] = (A/JA)/[(MA + JA)/JA] \cong A/MA$$

and $JA \subset MA$. Thus B/MB, and hence B/D, are Noetherian. It follows that $(B/C)/X$ is Noetherian because $(B/C)/X = (B/C)/(D/C) \cong B/D$. Since this is true for any non-zero submodule of B/C it follows that B/C is a Noetherian (R/N)-module. But B/C is faithful. By Lemma 1.55, the ring R/N is Noetherian and hence the finitely generated (R/N)-module B is Noetherian by Corollary 1.50. But $N = \mathrm{ann}_R B$ implies that the R-submodules and (R/N)-submodules of B coincide, so that B is a Noetherian R-module. This contradiction shows that A is Noetherian. This completes the proof of Lemma 1.57 and hence also of Theorem 1.56. □□

The technique used in the proof of Theorem 1.56 of first replacing what is required to prove by something more general and then proving this new result (which might on the surface be considered more difficult) is extremely useful.

Exercises

1.111. Prove that ${}_RA$ is Noetherian if and only if A/B is Noetherian for each $B \leq {}_RA$. What is the corresponding result for Artinian modules?

1.112. Let $n \in \mathbb{N}$ and let $A = A_0 \supseteq A_1 \supseteq \cdots \supseteq A_n = 0$ be a descending chain of submodules of a left R-module A. Prove that A is Noetherian (respectively Artinian) if and only if the factor module A_{i-1}/A_i is Noetherian (respectively Artinian) for each $1 \leq i \leq n$.

1.113. Prove that if I is an ideal of a left Noetherian (respectively Artinian) ring R, then the factor ring R/I is left Noetherian (respectively Artinian).

1.114. Let R_i $(1 \leq i \leq n)$ be a finite collection of rings. Prove that the ring $R = R_1 \oplus \cdots \oplus R_n$ is left Noetherian if and only if the ring R_i is left Noetherian for each $1 \leq i \leq n$.

1.115. Let R be a left Artinian domain. Prove that every non-zero element of R is a unit. (Rings with the property that their non-zero elements are units are called *division rings* or *skew fields*.)

1.116. Let R be a ring such that R is finitely generated as an additive group. Prove that R is Noetherian. Deduce that for any $n \in \mathbb{N}$, any subring S of $M_n(\mathbb{Z})$ is Noetherian.

1.117. Let K be a field. A ring R is called an *algebra over* K provided R is a K-vector space such that $k(rs) = (kr)s = r(ks)$ for all r, $s \in R$ and $k \in K$. An algebra over K is called *finite-dimensional* if it is a finite-dimensional vector space over K. Prove that for any $n \in \mathbb{N}$, the ring $M_n(K)$ is a finite-dimensional algebra over K. A subring S of R which is also a subspace of R is called a *subalgebra* of R. Prove that any subalgebra of a finite-dimensional algebra over K is itself finite dimensional. Prove that any finite-dimensional algebra over a field K is an Artinian ring. Deduce that any subalgebra of $M_n(K)$ is an Artinian ring.

1.118. If L is a left ideal of a ring R, then $L^2 = LL$, the left ideal of R consisting of all finite sums $a_1b_1 + \cdots + a_nb_n$ where $n \in \mathbb{N}$, a_i, $b_i \in L$ $(1 \leq i \leq n)$ (see Exercise 1.38). Let N be an ideal of a ring R such that $N^2 = 0$. Prove that R is left Noetherian if and only if the factor ring R/N is left Noetherian and ${}_RN$ is finitely generated. Hence, or otherwise, prove that if S and T are rings and ${}_SM_T$, then the ring

$$\begin{bmatrix} S & M \\ 0 & T \end{bmatrix}$$

is left Noetherian if and only if S and T are both left Noetherian and ${}_SM$ is finitely generated. State and prove similar results for left Artinian rings. Deduce that the ring

$$\begin{bmatrix} \mathbb{R} & \mathbb{R} \\ 0 & \mathbb{Q} \end{bmatrix}$$

is left Noetherian, but not right Noetherian, and is left Artinian, but not right Artinian.

1.119. Let e be an idempotent of a left Noetherian ring R. Prove that the subring eRe is also left Noetherian. Deduce that if $n \in \mathbb{N}$ and $M_n(R)$ is left Noetherian, then R is left Noetherian. Prove the corresponding results for left Artinian rings.

1.120. A left R-module U is called *uniform* provided $X \cap Y \neq 0$ for all non-zero submodules X and Y of U. Suppose that ${}_RA$ is non-zero and does not contain an infinite direct sum of non-zero submodules. Prove

(i) A contains a uniform submodule,
(ii) there exist $n \in \mathbb{N}$ and uniform submodules U_i $(1 \leq i \leq n)$ of A such that $U_1 + \cdots + U_n$ is a direct sum and $U_1 \oplus \cdots \oplus U_n \leq_e {}_RA$,
(iii) $E(U_i)$ is indecomposable for each $1 \leq i \leq n$, and
(iv) $E(A) = E(U_1) \oplus \cdots \oplus E(U_n)$.

Deduce that if R is a left Noetherian ring, then every injective left R-module is a direct sum of indecomposable injective left R-modules. (This last fact is *Matlis' Theorem.*)

1.121. Prove that if A is a Noetherian left R-module, then $E(A)$ is a finite direct sum of indecomposable injective left R-modules.

1.122. Prove that a non-zero Abelian group A is an indecomposable injective $\mathbb{Z}$-module if and only if $A \cong \mathbb{Q}$ or $A \cong \mathbb{Z}(p^\infty)$ for some prime p. Deduce that a non-zero Abelian group B is an injective $\mathbb{Z}$-module if and only if there exists a non-empty collection of $\mathbb{Z}$-modules B_λ $(\lambda \in \Lambda)$ such that $B \cong \bigoplus_{\lambda\in\Lambda} B_\lambda$ and for each $\lambda \in \Lambda$, $B_\lambda = \mathbb{Q}$, or there exists a prime $p \in \mathbb{N}$ such that $B_\lambda = \mathbb{Z}(p^\infty)$.

1.123. Prove that a ring R is left Noetherian if and only if for each ${}_RA$ the union of any chain of injective submodules of A is injective. Deduce that if R is left Noetherian, then each left R-module contains a maximal injective submodule.

1.124. Given a non-empty collection $\underline{L}$ of left ideals of R, a left R-module E is called $\underline{L}$-injective provided for every $L \in \underline{L}$ and homomorphism $\theta : L \to E$ there exists an element $e \in E$ such that $\theta(r) = re$ $(r \in R)$. Prove that

(i) every $\underline{E}$-injective left R-module is injective, where $\underline{E}$ denotes the collection of essential left ideals of R,
(ii) R is left Noetherian if and only if every $\underline{F}$-injective left R-module is injective, where $\underline{F}$ denotes the collection of all finitely generated left ideals of R,
(iii) R is left hereditary if and only if every $\underline{P}$-injective left R-module is injective, where $\underline{P}$ denotes the collection of all projective left ideals of R, and
(iv) R is left hereditary left Noetherian if and only if every $(\underline{F} \cap \underline{P})$-injective left R-module is injective.

1.125. Prove that the following statements are equivalent for a ring R.

(i) R is left hereditary left Noetherian.

(ii) For each ${}_RA$, the sum of any non-empty collection of injective submodules of A is injective.
(iii) For each ${}_RA$, there exists an injective submodule of A containing every injective submodule of A.

Deduce that if R is left hereditary left Noetherian, then each left R-module contains a *unique* maximal injective submodule.

1.126. Suppose that ${}_RA$ is not Artinian. Prove that there exists $B \leq {}_RA$ such that B is Artinian and A/B is not finitely cogenerated.

1.127. Let ${}_RA$ and $\theta \in \mathrm{End}_R A$.

(i) Suppose A is Noetherian. Prove that there exists $m \in \mathbb{N}$ such that $\ker \theta^m = \ker \theta^{m+1}$ and hence $\ker \theta \cap \operatorname{im} \theta^m = 0$. Deduce that if θ is an epimorphism, then θ is also a monomorphism.
(ii) Suppose A is Artinian. Prove that there exists $n \in \mathbb{N}$ such that $\operatorname{im} \theta^n = \operatorname{im} \theta^{n+1}$ and hence $A = \operatorname{im} \theta + \ker \theta^n$. Deduce that if θ is a monomorphism, then θ is also an epimorphism.
(iii) Suppose that A is both Noetherian and Artinian. Prove that there exists $k \in \mathbb{N}$ such that A is the direct sum $\operatorname{im} \theta^k \oplus \ker \theta^k$ of its submodules $\operatorname{im} \theta^k$ and $\ker \theta^k$.
(iv) Suppose that A is Noetherian, Artinian, and indecomposable. Prove that θ is an automorphism or $\theta^k = 0$ for some $k \in \mathbb{N}$. Deduce that $\mathrm{End}_R(A)$ is a local ring.

1.9 Modules of finite length

A left R-module U is called *simple* or *irreducible* provided $U \neq 0$ and the only submodules of U are 0 and U. If R is a field K, then simple R-modules are one-dimensional K-vector spaces. On the other hand, if $R = \mathbb{Z}$, then simple R-modules are simple Abelian groups and hence are cyclic groups of prime order, i.e., an Abelian group A is a simple $\mathbb{Z}$-module if and only if there exists a prime $p \in \mathbb{N}$ such that $A \cong \mathbb{Z}_p$.

Lemma 1.58. *Let B be a submodule of a left R-module A. Then*

(i) *B is a maximal submodule of A if and only if the factor module A/B is simple, and*
(ii) *B is a minimal submodule of A if and only if B is simple.*

Proof. Apply Exercise 1.21 for (i); (ii) is obvious. □

A left R-module A *has a composition series* provided there exists a finite chain

$$A = A_0 \supset A_1 \supset \cdots \supset A_n = 0 \tag{1.23}$$

of submodules A_i $(0 \leq i \leq n)$ of A such that A_{i-1}/A_i is a simple module for each $1 \leq i \leq n$. The non-negative integer n is called the *length* of the composition series (1.23). In particular, according to this definition the zero module has a trivial composition series $0 = 0$. For non-trivial examples note that a vector space over a field has a composition series if and only if it is finite dimensional, and an Abelian group has a composition series as a $\mathbb{Z}$-module if and only if it is finite. In general we have:

Proposition 1.59. *A left R-module A has a composition series if and only if it is both Noetherian and Artinian.*

Proof. Suppose that A has a composition series. Clearly, simple modules are both Noetherian and Artinian. Thus A is both Noetherian and Artinian by Exercise 1.112. Conversely, suppose that A is both Noetherian and Artinian. Suppose $A \neq 0$. There exists $B_1 \leq_{\mathrm{m}} {}_R A$ because A is Noetherian. If $B_1 \neq 0$, then there exists $B_2 \leq_{\mathrm{m}} {}_R B_1$, again because A is Noetherian. Repeating this argument produces a descending chain $A = B_0 \supset B_1 \supset B_2 \supset \cdots$ of submodules such that for each $i \geq 0$, $B_{i+1} \leq_{\mathrm{m}} {}_R B_i$ and hence B_i/B_{i+1} is simple by Lemma 1.58. Since A is Artinian, it follows that $B_k = 0$ for some $k \in \mathbb{N}$ and hence A has a composition series $A = B_0 \supset B_1 \supset \cdots \supset B_k = 0$. □

Clearly, if a left R-module has a composition series, then so does every isomorphic copy of it. Thus Proposition 1.59 has the following consequence.

Corollary 1.60. *Let $0 \longrightarrow A \longrightarrow B \longrightarrow C \longrightarrow 0$ be an exact sequence of left R-modules. Then B has a composition series if and only if A and C both have composition series.*

Proof. Apply Proposition 1.59 and Corollary 1.48. □

Let A be a left R-module with a composition series (1.23). Suppose that

$$A = B_0 \supset B_1 \supset \cdots \supset B_m = 0 \tag{1.24}$$

is another composition series of A. The composition series (1.23) and (1.24) are said to be *equivalent* provided $n = m$ and there exists a permutation ρ of $\{1, \ldots, n\}$ such that $A_{i-1}/A_i \cong B_{\rho(i)-1}/B_{\rho(i)}$ for each $1 \leq i \leq n$. In other words, two composition series are equivalent if their associated simple factor modules can be completely paired off in such a way that corresponding factor modules are isomorphic. For example, if p and q are distinct primes in $\mathbb{N}$, then

$$\mathbb{Z}_{pq} = \mathbb{Z}/\mathbb{Z}pq \supset \mathbb{Z}p/\mathbb{Z}pq \supset 0,$$

and

$$\mathbb{Z}_{pq} = \mathbb{Z}/\mathbb{Z}pq \supset \mathbb{Z}q/\mathbb{Z}pq \supset 0$$

are equivalent composition series for the $\mathbb{Z}$-module $\mathbb{Z}_{pq}$. In fact, it can be checked that these are the only two composition series of $\mathbb{Z}_{pq}$, and the fact that these composition series are equivalent is no surprise in view of the following result.

Theorem 1.61 (Jordan–Hölder Theorem). *Let A be a left R-module with a composition series. Then any two composition series for A are equivalent.*

Proof. Suppose $A \neq 0$. Define $d(A)$ to be the least positive integer t such that A has a composition series of length t. The proof is by induction on $d(A)$. Let $d(A) = n$ and let (1.23) be a composition series of A. If $n = 1$, then A is simple and the result follows at once. Suppose $n \geq 2$. Note that $d(A_1) = n - 1$. Let (1.24) be another composition series of A. The proof splits into two cases.

Case 1. $A_1 = B_1$. By induction on $d(A)$, the composition series

$$A_1 = A_1 \supset A_2 \supset \cdots \supset A_n = 0$$

and

$$A_1 = B_1 \supset B_2 \supset \cdots \supset B_m = 0$$

of A_1 are equivalent, and hence (1.23) and (1.24) are equivalent.

Case 2. $A_1 \neq B_1$. In this case $A_1 \subset A_1 + B_1$, and hence $A = A_1 + B_1$ by Lemma 1.58. Let $C = A_1 \cap B_1$. Corollary 1.60 shows that the module C has a composition series

$$C = C_0 \supset C_1 \supset \cdots \supset C_k = 0,$$

for some $k \geq 0$. Consider the chains

$$A = A_0 \supset A_1 \supset C_0 \supset C_1 \supset \cdots \supset C_k = 0 \tag{1.25}$$

and

$$A = B_0 \supset B_1 \supset C_0 \supset C_1 \supset \cdots \supset C_k = 0. \tag{1.26}$$

By Proposition 1.6, (1.25) and (1.26) are equivalent composition series of A. But by Case 1, (1.23) and (1.25) are equivalent and hence $k = n - 2$. By Case 1 again, (1.24) and (1.26) are equivalent. Thus (1.23) and (1.24) are equivalent. The result follows by induction. □

In particular, Theorem 1.61 shows that if a module A has a composition series, then any two composition series have the same length, and this non-negative integer is called the *length of* A, denoted by $d(A)$. We say that a module *has finite length* if it has a composition series. Note that if A is a vector space over a field K then A has finite length if and only if A is finite dimensional, and in this case $d(A)$ is the dimension of A. On the other hand, if A is a finite Abelian group of order n then $d(A)$ is the number of prime factors of n. Theorem 1.61 has the following consequence.

Corollary 1.62. *Let A be a left R-module of finite length and let B be a submodule of A. Then $d(A) = d(A/B) + d(B)$.*

Proof. Let

$$A/B = X_0 \supset X_1 \supset \cdots \supset X_m = 0$$

and

$$B = B_0 \supset B_1 \supset \cdots \supset B_n = 0$$

be composition series of A/B and B, respectively (Corollary 1.60). For each $1 \leq i \leq m$, let $C_i = \pi^{-1}(X_i)$. Then

$$A = C_0 \supset C_1 \supset \cdots \supset C_m \supset B_1 \supset B_2 \supset \cdots \supset B_n = 0$$

is a composition series of A because $C_m = B$ and

$$C_{i-1}/C_i \cong (C_{i-1}/B)/(C_i/B) = X_{i-1}/X_i \quad (1 \leq i \leq m),$$

by Exercises 1.20 and 1.21. It follows that $d(A) = m + n = d(A/B) + d(B)$. □

One of the most intriguing questions in Module Theory is the following:

Given submodules A_i, B_i $(i = 1, 2)$ of a left R-module A such that $A = A_1 \oplus B_1 = A_2 \oplus B_2$, when does it follow that $A_1 \cong A_2$ and $B_1 \cong B_2$?

In general it does not follow that $B_1 \cong B_2$ even if $A_1 \cong A_2$. For example, let K be any field and $V = K^{(\mathbb{N})}$. The K-vector space V has a basis $\{v_i : i \in \mathbb{N}\}$. Let V_1 and V_2 denote the subspaces of V spanned by the sets $\{v_{2i-1} : i \in \mathbb{N}\}$ and $\{v_{2i} : i \in \mathbb{N}\}$, respectively. Clearly $V = V_1 \oplus V_2 = V \oplus 0$ and $V_1 \cong V$, but $V_2 \ncong 0$. However, if V is a finite-dimensional vector space with subspaces V_i, W_i $(i = 1, 2)$ such that $V = V_1 \oplus V_2 = W_1 \oplus W_2$ then it need not follow that $V_1 \cong W_1$ and $V_2 \ncong W_2$ (although it clearly does if these subspaces of V are one dimensional), but if $V_1 \cong W_1$ then $V_2 \cong W_2$ by counting dimensions.

Now consider the following more general situation. Let A be a left R-module and A_λ $(\lambda \in \Lambda)$ and B_γ $(\gamma \in \Gamma)$ non-empty collections of submodules of A such that

$$A = \bigoplus_{\lambda \in \Lambda} A_\lambda = \bigoplus_{\gamma \in \Gamma} B_\gamma.$$

For each $\lambda \in \Lambda$ and $\gamma \in \Gamma$ define the R-homomorphism $\theta_{\lambda\gamma} : A_\lambda \to B_\gamma$ to be the restriction to the submodule A_λ of the canonical projection $\pi_\gamma : A \to B_\gamma$ and define the R-homomorphism $\phi_{\gamma\lambda} : B_\gamma \to A_\lambda$ to be the restriction to the submodule B_γ of the canonical projection $\pi_\lambda : A \to A_\lambda$. For each $\gamma \in \Gamma$ let

$$\Lambda_\gamma = \{\lambda \in \Lambda : \theta_{\lambda\gamma}\phi_{\gamma\lambda} \text{ is an automorphism of } B_\gamma\}.$$

Of course, Λ_γ could be empty. However, in certain circumstances Λ_γ is non-empty for some $\gamma \in \Gamma$, as the next result shows. Recall from Exercise 1.89 that a ring R is called *local* if the sum of any two non-units is also a non-unit.

Lemma 1.63. *Let A_λ $(\lambda \in \Lambda)$, B_1 and B_2 be submodules of a left R-module A such that $A = \bigoplus_{\lambda \in \Lambda} A_\lambda = B_1 \oplus B_2$. Suppose further that B_1 is non-zero and $\mathrm{End}_R(B_1)$ is a local ring. Then Λ_1 is a non-empty finite subset of Λ.*

Proof. Let $0 \neq b \in B_1$. There exist $n \in \mathbb{N}$, elements $\lambda(i) \in \Lambda$ and elements $a_i \in A_{\lambda(i)}$ $(0 \leq i \leq n-1)$ such that $b = a_0 + \cdots + a_{n-1}$. Note that $b \in \ker \phi_{1\lambda}$ for all $\lambda \in \Lambda$ with $\lambda \neq \lambda(i)$ $(0 \leq i \leq n-1)$, and hence $\Lambda_1 \subseteq \{\lambda(0), \ldots, \lambda(n-1)\}$ by Exercise 1.35. Thus Λ_1 is finite. Let $\Lambda' = \{\lambda \in \Lambda : \lambda \neq \lambda(i)\ (0 \leq i \leq n-1)\}$, $A_i = A_{\lambda(i)}$ $(0 \leq i \leq n-1)$, and $A_n = \bigoplus_{\lambda \in \Lambda'} A_\lambda$. Then

$$A = A_0 \oplus \cdots \oplus A_n = B_1 \oplus B_2.$$

Note that $\theta_{j1}\phi_{1j} \in \mathrm{End}_R(B_1)$ $(0 \leq j \leq n)$ and

$$\theta_{01}\phi_{10} + \cdots + \theta_{n1}\phi_{1n} = i_{B_1}.$$

Since $\mathrm{End}_R(B_1)$ is a local ring, it follows that there exists $0 \leq k \leq n$ such that $\theta_{k1}\phi_{1k}$ is an automorphism of B_1. However $b \in \ker \phi_{1n}$ implies that $k \neq n$ by Exercise 1.35. Thus $0 \leq k \leq n-1$ and Λ_1 is non-empty. □

Theorem 1.64 (Azumaya Decomposition Theorem). *Let A_λ $(\lambda \in \Lambda)$ and B_γ $(\gamma \in \Gamma)$ be non-zero submodules of a left R-module A such that $\mathrm{End}_R(A_\lambda)$ and $\mathrm{End}_R(B_\gamma)$ are local rings for all $\lambda \in \Lambda$ and $\gamma \in \Gamma$ and $A = \bigoplus_{\lambda \in \Lambda} A_\lambda = \bigoplus_{\gamma \in \Gamma} B_\gamma$. Then there exists a bijection $\beta : \Lambda \longrightarrow \Gamma$ such that $A_\lambda \cong B_{\beta(\lambda)}$ for all $\lambda \in \Lambda$.*

Proof. Define a relation $\approx$ on Λ by

$$\lambda \approx \mu \text{ if } A_\lambda \cong A_\mu,$$

where λ, $\mu \in \Lambda$. Then $\approx$ is an equivalence relation and the equivalence class containing the element λ of Λ is denoted by $[\lambda]$. There is a corresponding equivalence relation on the set Γ and the equivalence class containing the element γ of Γ is denoted by $[\gamma]$. It is sufficient to prove:

$$\begin{array}{l}\text{for each } \delta \in \Gamma \text{ there exists } \mu \in \Lambda \text{ such that } A_\mu \cong B_\delta \\ \text{and there exists a bijection } \alpha : [\mu] \to [\delta].\end{array} \tag{1.27}$$

Let $\delta \in \Gamma$. The proof of (1.27) splits into two cases.

Case 1. $[\delta]$ is finite.

Suppose that $[\delta] = \{\delta = \delta(1), \delta(2), \ldots, \delta(n)\}$ for some $n \in \mathbb{N}$. Let $\Gamma' = \{\gamma \in \Gamma : \gamma \neq \delta(1)\}$ and $B' = \bigoplus_{\gamma \in \Gamma'} B_\gamma$. Then Lemma 1.63 applied to

$$A = \bigoplus_{\lambda \in \Lambda} A_\lambda = B_\delta \oplus B'$$

shows that Λ_δ is a non-empty finite subset of Λ. Let $\mu \in \Lambda_\delta$. Since A_μ is indecomposable (Exercise 1.89) and $\theta_{\mu\delta}\phi_{\delta\mu}$ is an automorphism of B_δ, it follows that $\phi_{\delta\mu} : B_\delta \longrightarrow A_\mu$ and $\theta_{\mu\delta} : A_\mu \longrightarrow B_\delta$ are isomorphisms (see Exercise 1.35). Let $a \in A$. Then $\pi_\delta(a) \in B_\delta$ and hence $\pi_\delta(a) = \theta_{\mu\delta}(a') = \pi_\delta(a')$ for some $a' \in A_\mu$.

Thus $a-a' \in \ker \pi_\delta = B'$ and hence $a \in A_\mu + B'$. It follows that $A = A_\mu + B'$. Moreover, if $c \in A_\mu \cap B'$, then $c = \phi_{\delta\mu}(b)$ for some $b \in B_\delta$ and $0 = \theta_{\mu\delta}(c) = \theta_{\mu\delta}\phi_{\delta\mu}(b)$, so that $b = 0$ and hence $c = 0$. We have proved

$$A = A_\mu \oplus B'.$$

Let $\Gamma'' = \{\gamma \in \Gamma : \gamma \neq \delta(1), \delta(2)\}$ and $B'' = \bigoplus_{\gamma \in \Gamma''} B_\gamma$. Thus

$$A = \bigoplus_{\lambda \in \Lambda} A_\lambda = A_\mu \oplus B_{\delta(2)} \oplus B'',$$

and by the same argument there exists $\mu(2) \in \Lambda$ such that $\phi_{\delta(2)\mu(2)}$ is an isomorphism and $A = A_{\mu(2)} \oplus A_\mu \oplus B''$, so that in particular $\mu \neq \mu(2)$. Repeating this argument we produce distinct elements $\mu(i)$ $(1 \leq i \leq n)$ of Λ such that $A_{\mu(i)} \cong B_\delta$ $(1 \leq i \leq n)$. Thus $\Lambda(\mu)$ contains at least n elements. But reversing the argument shows that $\Lambda(\mu)$ contains at most n elements, and hence (1.27) is proved in this case.

Case 2. $[\delta]$ is infinite.

By Lemma 1.63, Λ_γ is a non-empty finite subset of Λ for each $\gamma \in \Gamma(\delta)$. Let $\mu \in \Lambda_\delta$. Again by Lemma 1.63,

$$[\mu] = \bigcup_{\gamma \in [\delta]} \Lambda_\delta.$$

It follows that there exists an injection from $[\mu]$ to $[\delta]$. But reversing the argument gives an injection from $[\delta]$ to $[\mu]$. Thus there exists a bijection $\alpha : [\mu] \to [\delta]$ and (1.27) follows. □

Corollary 1.65. *Let A_λ $(\lambda \in \Lambda)$ and B_γ $(\gamma \in \Gamma)$ be non-zero indecomposable injective submodules of a left R-module A such that $A = \bigoplus_{\lambda \in \Lambda} A_\lambda = \bigoplus_{\gamma \in \Gamma} B_\gamma$. Then there exists a bijection $\beta : \Lambda \to \Gamma$ such that $A_\lambda \cong B_{\beta(\lambda)}$ for all $\lambda \in \Lambda$.*

Proof. Apply the theorem and Exercise 1.89. □

Another consequence of Theorem 1.64 is the Krull–Schmidt Theorem, which has the names of Remak and Wedderburn also associated with it.

Theorem 1.66 (Krull–Schmidt Theorem). *Let A be a non-zero left R-module of finite length. Then there exists a positive integer n such that A is the direct sum $A_1 \oplus \cdots \oplus A_n$ of non-zero indecomposable submodules A_i $(1 \leq i \leq n)$. Moreover, if there exists a positive integer m such that A is the direct sum $B_1 \oplus \cdots \oplus B_m$ of non-zero indecomposable submodules B_j $(1 \leq j \leq m)$, then $m = n$ and there exists a permutation ρ of $\{1, \ldots, n\}$ such that $A_i \cong B_{\rho(i)}$ for each $1 \leq i \leq n$.*

Proof. Either A is indecomposable, or A is the direct sum $B_1 \oplus C_1$ of non-zero submodules B_1 and C_1. Either B_1 is indecomposable, or B_1 is the direct sum $B_2 \oplus C_2$ of non-zero submodules B_2 and C_2. Repeating this argument gives a descending chain $A = B_0 \supset B_1 \supset B_2 \supset \cdots$. Since A is Artinian (Proposition 1.59), it follows that B_m is non-zero and indecomposable for some $m \geq 0$. Moreover, if $m \geq 1$ then $A = C_1 \oplus \cdots \oplus C_m \oplus B_m$. Thus A contains a non-zero indecomposable direct summand A_1. Let D_1 be a direct complement of A_1. Thus $A = A_1 \oplus D_1$. By the same argument, if $D_1 \neq 0$, then there exist submodules A_2 and D_2 of A such that $D_1 = A_2 \oplus D_2$ and A_2 is non-zero and indecomposable. Repeating this argument gives a descending chain $A \supset D_1 \supset D_2 \supset \cdots$. Since A is Artinian, it follows that $D_n = 0$ for some $n \in \mathbb{N}$. Thus $A = A_1 \oplus \cdots \oplus A_n$ and each A_i $(1 \leq i \leq n)$ is non-zero and indecomposable. The result now follows by Theorem 1.64 and Exercise 1.127. □

Yet another application of Theorem 1.64 can be found in the proof of the following result.

Theorem 1.67 (Matlis–Papp Theorem). *A ring R is left Noetherian if and only if each injective left R-module is a direct sum of indecomposable injective left R-modules.*

Proof. Suppose that R is a left Noetherian ring. Let E be an injective left R-module. The collection $\underline{S}$ of indecomposable injective submodules of E is non-empty because $0 \in \underline{S}$ and by Proposition 1.19 there exists a submodule F of E such that F is maximal among direct sums of members of $\underline{S}$. By the Bass–Papp Theorem (Theorem 1.54), F is an injective submodule of E and hence $E = F \oplus G$ for some $G \leq {}_R E$ by Proposition 1.30. Suppose $G \neq 0$. Let $0 \neq g \in G$. Then Rg is a Noetherian submodule of G by Corollary 1.50, and hence Rg contains a uniform submodule U by Exercise 1.120. Moreover, Exercise 1.79 shows that U has an injective envelope $E(U)$ contained in G. But $E(U)$ is indecomposable by Exercise 1.88 and $F \subset F \oplus E(U)$ gives a contradiction to the choice of F. Thus $G = 0$ and $E = F$. It follows that every injective left R-module is a direct sum of indecomposable injective left R-modules.

Conversely, suppose that every injective left R-module is a direct sum of indecomposable injective left R-modules. The proof of Theorem 1.54 shows that to show R is left Noetherian it is sufficient to show that any module of the form $\bigoplus_{n \in \mathbb{N}} E(R/L_n)$ is injective, where $L_n \leq {}_R R$ $(n \in \mathbb{N})$. Let $n \in \mathbb{N}$. By hypothesis, there exists a collection H_γ $(\gamma \in \Gamma)$ of indecomposable injective left R-modules such that $E(R/L_n) = \bigoplus_{\gamma \in \Gamma} H_\gamma$. There exists a finite subset Γ' of Γ such that the element $1 + L_n \in \bigoplus_{\gamma \in \Gamma'} H_\gamma$ and it follows that $H_\gamma = 0$ for all $\gamma \in \Gamma'$. Thus each module $E(R/L_n)$ $(n \in \mathbb{N})$ is a finite direct sum of indecomposable injective left R-modules. Thus it is sufficient to prove that

$$\begin{gathered}\text{if } E_n \text{ is indecomposable injective for} \\ \text{each } n \in \mathbb{N}, \text{ then } A = \bigoplus_{n \in \mathbb{N}} E_n \text{ is injective.}\end{gathered} \tag{1.28}$$

Let $B = A^{(\mathbb{N})}$ and $I = E(B)$. By hypothesis, there exists a non-empty collection of indecomposable injective left R-modules I_λ $(\lambda \in \Lambda)$ such that $I = \bigoplus_{\lambda \in \Lambda} I_\lambda$.

For each $n \in \mathbb{N}$ let $\Lambda_n = \{\lambda \in \Lambda : I_\lambda \cong E_n\}$. Suppose that there exists $k \in \mathbb{N}$ such that Λ_k is finite, say $|\Lambda_k| = m \in \mathbb{N}$. Note that

$$B = A^{(\mathbb{N})} \cong \bigoplus_{n \in \mathbb{N}} E_n^{(\mathbb{N})}.$$

Thus $(E_k)^{m+1}$ is isomorphic to a submodule J of I and J is injective by Exercise 1.74. Thus $I = J \oplus K$ for some $K \leq {}_R I$. Consider

$$I = \bigoplus_{\lambda \in \Lambda} I_\lambda = J \oplus K$$

and apply Theorem 1.64 to reach a contradiction. A similar argument shows that Λ_n is non-empty for each $n \in \mathbb{N}$. Thus Λ_n is infinite for each $n \in \mathbb{N}$. Choose $\lambda(1) \in \Lambda$ such that $I_{\lambda(1)} \cong E_1$, $\lambda(2) \in \Lambda$ such that $\lambda(2) \neq \lambda(1)$ and $I_{\lambda(2)} \cong E_2$, and so on. It follows that $A \cong \bigoplus_{n \in \mathbb{N}} I_{\lambda(n)}$, which is a direct summand of I and hence is injective. This proves (1.28) and the result follows. □

Finally we examine finitely generated Abelian groups again. If A is a finitely generated Abelian group, then A is finite or $A = \mathbb{Z}^m \oplus F$ for some non-negative integer m and finite group F (see Proposition 1.53). Each non-zero finite Abelian group F has finite length as a $\mathbb{Z}$-module and, by the Krull–Schmidt Theorem (Theorem 1.64), is a finite direct sum of non-zero indecomposable $\mathbb{Z}$-submodules in an essentially unique way. The next result shows that indecomposable $\mathbb{Z}$-modules are cyclic and thus every finitely generated Abelian group is a finite direct sum of cyclic subgroups.

Proposition 1.68. *A non-zero finite Abelian group A is an indecomposable $\mathbb{Z}$-module if and only if A is a cyclic group of order p^n for some prime p and positive integer n.*

Proof. Suppose first that $A = \mathbb{Z}a$ and a has order p^n for some prime p and $n \in \mathbb{N}$. Suppose that $A = B \oplus C$ where B, $C \leq {}_\mathbb{Z} A$. There exist $b \in B$ and $c \in C$ such that $a = b + c$. Since $0 = p^n a = p^n b + p^n c$ it follows that $p^n b = 0$ and $p^n c = 0$. Suppose that b and c have orders p^s and p^t, respectively, where $s, t \geq 0$. Suppose that $s \geq t$. Then $p^s a = p^s b + p^s c = 0$, so that $s = n$ and $A = \mathbb{Z}b = B$. Similarly if $s \leq t$, then $A = C$. It follows that A is indecomposable.

Conversely, suppose that A is indecomposable. Let q be a prime such that $qd = 0$ for some $0 \neq d \in A$. Define $\phi : A \longrightarrow A$ by $\phi(x) = qx$ $(x \in A)$. An easy check shows that $\phi \in \mathrm{End}_R(A)$ and $d \in \ker \phi$, so that $\phi^k = 0$ for some $k \in \mathbb{N}$ by Fitting's Lemma (Exercise 1.127). Thus $q^k A = \phi^k(A) = 0$. Choose $m \in \mathbb{N}$ such that $m \leq k$ and q^m is maximal among the orders of the non-zero elements of A. Let $y \in A$ have order q^m. We shall complete the proof by proving

$$A = \mathbb{Z}y. \tag{1.29}$$

Let $G \leq {}_{\mathbb{Z}}A$ be chosen maximal such that $G \cap \mathbb{Z}y = 0$ (see Proposition 1.17). Suppose $A \neq G + \mathbb{Z}y$. Clearly, $A/(G+\mathbb{Z}y)$ is a q-group, so that there exists $z \in A$ such that $z \notin G + \mathbb{Z}y$ and $qz \in G + \mathbb{Z}y$. There exist $g \in G$, $u \in \mathbb{Z}$ such that $qz = g + uy$ and hence $0 = q^m z = q^{m-1}g + q^{m-1}uy$ and $q^{m-1}uy \in G \cap \mathbb{Z}y = 0$. Since $q^{m-1}uy = 0$ and y has order q^m, it follows that q divides u, say $u = qv$ for some $v \in \mathbb{Z}$. Thus $qz = g + uy = g + qvy$ and $q(z - vy) = g$. Let $z_1 = z - vy$. Since $z \in G + \mathbb{Z}y$, it follows that $z_1 \in G$ and hence $G \subset G + \mathbb{Z}z_1$. Therefore $(G + \mathbb{Z}z_1) \cap \mathbb{Z}y \neq 0$ by the choice of G, i.e., $0 \neq n'y = g' + m'z_1$ for some m', $n' \in \mathbb{Z}$ and $g' \in G$. Note that $m'z_1 = -g' + n'y \in G + \mathbb{Z}y$. We conclude that q divides m', say $m' = qm''$ for some $m'' \in \mathbb{Z}$. But

$$n'y = g' + m'z_1 = g' + m''qz_1 = g' + m''g \in G$$

gives $n'y \in G \cap \mathbb{Z}y = 0$ and $n'y = 0$, a contradiction. Thus $A = G + \mathbb{Z}y$ and hence $A = G \oplus \mathbb{Z}y$. But A is indecomposable, and so $A = \mathbb{Z}y$, as needed. □

Exercises

1.128. Let $A_\lambda \leq {}_RA$ ($\lambda \in \Lambda$) and $A = \bigoplus_{\lambda\in\Lambda} A_\lambda$. Suppose that there exists a finitely generated essential submodule B of A. Prove that there exists a finite subset Λ' of Λ such that $A = \bigoplus_{\lambda\in\Lambda'} A_\lambda$.

1.129. Let ${}_RA$ have finite length and $B, C \leq {}_RA$. Prove that $d(B+C)+d(B\cap C) = d(B) + d(C)$.

1.130. Let $B_2 \subseteq B_1$ and $C_2 \subseteq C_1$ be submodules of a left R-module A. Prove that

$$[(B_1 \cap C_1) + B_2]/[(B_1 \cap C_2) + B_2] \cong [(B_1 \cap C_1) + C_2]/[(B_2 \cap C_1) + C_2].$$

(This is called *Zassenhaus' Lemma.*)

1.131. Let ${}_RA \neq 0$. A *series* for A is a finite chain of submodules of the form

$$A = A_0 \supset A_1 \supset \cdots \supset A_n = 0, \tag{1.30}$$

and a *refinement* of this series is a series

$$A = B_0 \supset B_1 \supset \cdots \supset B_m = 0$$

such that $m \in \mathbb{N}$, $m \geq n$, and there exist integers $1 \leq i(1) < \cdots < i(n) \leq m$ such that $B_{i(j)} = A_j$ $(1 \leq j \leq n)$. Two series, (1.30) and

$$A = C_0 \supset C_1 \supset \cdots \supset C_k = 0,$$

are called *equivalent* if $n = k$ and there exists a permutation ρ of $\{1, \ldots, n\}$ such that $A_{i-1}/A_i \cong C_{\rho(i)-1}/C_{\rho(i)}$ for all $1 \leq i \leq n$. Prove that any two series for A have equivalent refinements. (This is called the *Schreier Refinement Theorem.*) Deduce the Jordan–Hölder Theorem.

1.132. Prove that if U is a simple left R-module, then $\mathrm{End}_R(U)$ is a division ring. (This is called *Schur's Lemma.*) Deduce that if A is a left R-module which is the direct sum $\bigoplus_{\lambda\in\Lambda} A_\lambda$ and the direct sum $\bigoplus_{\gamma\in\Gamma} B_\gamma$ of simple submodules A_λ $(\lambda \in \Lambda)$ and B_γ $(\gamma \in \Gamma)$, then there exists a bijection $\beta : \Lambda \to \Gamma$ such that $A_\lambda \cong B_{\beta(\lambda)}$ for all $\lambda \in \Lambda$.

1.133. The *socle* $\mathrm{soc}_R\, A$, or simply $\mathrm{soc}\, A$, of A is defined to be the sum of all minimal submodules of A if A contains a minimal submodule, and 0 otherwise. Prove that

(i) $V = \mathrm{soc}_K V$ for any vector space V over a field K,
(ii) $\mathrm{soc}_{\mathbb{Z}}\mathbb{Z} = 0$,
(iii) $\mathrm{soc}_{\mathbb{Z}}\mathbb{Z}(p^\infty) = \mathbb{Z}(1/p + \mathbb{Z})$ for any prime p,
(iv) $\mathrm{soc}_R B = B \cap \mathrm{soc}_R\, A$, for any $B \leq {}_RA$, and
(v) $(\mathrm{soc}_R A + B)/B \subseteq \mathrm{soc}_R(A/B)$ for any $B \leq {}_RA$.

1.134. A left R-module A is called *semisimple* or *completely reducible* provided $A = \mathrm{soc}_R\, A$. Prove that submodules and homomorphic images of semisimple modules are semisimple.

1.135. Prove that the following statements are equivalent for a left R-module A.

(i) A is semisimple.
(ii) A is the only essential submodule of A.
(iii) Every submodule of A is a direct summand of A.
(iv) $A = 0$ or A is the direct sum $\bigoplus_{\lambda\in\Lambda} A_\lambda$ of simple submodules A_λ $(\lambda \in \Lambda)$. In this case prove that if $B \leq {}_RA$, then there exists a subset Λ' of Λ such that A is the direct sum $B \oplus A'$, where $A' = \bigoplus_{\lambda\in\Lambda'} A_\lambda$.

1.136. Prove that the socle of any ${}_RA$ is the intersection of all essential submodules of A.

1.137. The *radical* $\mathrm{rad}_R A$, or simply $\mathrm{rad}\, A$, of a left R-module A is defined to be the intersection of all maximal submodules of A if A contains a maximal submodule, and A otherwise. Prove

(i) $\mathrm{rad}_K V = 0$ for any vector space V over a field K,
(ii) $\mathrm{rad}_{\mathbb{Z}}\mathbb{Z} = 0$,
(iii) $\mathrm{rad}_{\mathbb{Z}}\mathbb{Z}(p^\infty) = \mathbb{Z}(p^\infty)$,
(iv) $\mathrm{rad}_R B \subseteq \mathrm{rad}_R A$ for any $B \leq {}_RA$,
(v) $(\mathrm{rad}_R A + B)/B \subseteq \mathrm{rad}_R(A/B)$ for any $B \leq {}_RA$,
(vi) $\mathrm{rad}_R A = 0$ for any semisimple ${}_RA$, and
(vii) $\mathrm{rad}_R(A/\mathrm{rad}_R A) = 0$ for any ${}_RA$.

1.138. Given any ${}_RA$, prove that

(i) $\mathrm{rad}_R A$ is the intersection of all submodules B such that A/B is semisimple, and
(ii) $\mathrm{rad}_R A$ is the sum of all superfluous submodules of A.

1.139. Let $\theta \in \operatorname{Hom}_R(A, B)$. Prove that $\theta(\operatorname{soc} A) \subseteq \operatorname{soc} B$ and $\theta(\operatorname{rad} A) \subseteq \operatorname{rad} B$.

1.140. Let A_λ $(\lambda \in \Lambda)$ be a non-empty collection of left R-modules. Prove that

$$\operatorname{soc}\left(\prod_{\lambda \in \Lambda} A_\lambda\right) = \operatorname{soc}\left(\bigoplus_{\lambda \in \Lambda} A_\lambda\right) = \bigoplus_{\lambda \in \Lambda} (\operatorname{soc}\ A_\lambda),$$

and

$$\operatorname{rad}\left(\bigoplus_{\lambda \in \Lambda} A_\lambda\right) = \bigoplus_{\lambda \in \Lambda} (\operatorname{rad}\ A_\lambda).$$

1.141. Let $_RA$ have finite length and $A_i \leq {_RA}$, $B_i \leq {_RA}$ $(i = 1, 2)$ such that $A = A_1 \oplus A_2 = B_1 \oplus B_2$ and $A_1 \cong B_1$. Prove that $A_2 \cong B_2$.

1.142. Let R be a principal ideal domain. Prove that any finitely generated R-module is a finite direct sum of cyclic R-modules.

Remarks. This chapter provides background with standard, but somewhat selective topics. Since we have attempted to make this chapter as self-contained, we provided very basic notions, but no references. The authors are grateful to Prof. Dr. Patrick F. Smith for his valuable help on this chapter. For a general background on Ring and Module Theory we refer to readers to consult the elegant books [AF74], [Lam99], [Wis91].

Chapter 2

Types of Relative Injectivity

Recall that vector spaces have typical properties, such as every submodule is a direct summand, and the space itself and any of its direct summands are injective modules (see, e.g., Corollary 1.21, Exercise 1.73). Having these facts as starting points, in this chapter we introduce complement submodules and prove their basic properties, and we also discuss chain conditions on complement submodules. Then we focus on relative injectivity and ejectivity of modules. By considering lifting homomorphisms, we build classes of lifting and elifting submodules. We provide detailed structural properties of these two classes. In the last section we deal with the conditions C_2, C_3, SIP, etc. in a module, that is, we study conditions of the type that eventually yield a direct summand of the module.

2.1 Complement submodules

Let A be a left R-module. Recall from § 1.5 that a submodule N of A is called an essential submodule, or is essential in A, written $N \leq_e A$, if $N \cap K \neq 0$ for all $0 \neq K \leq A$. We first give some properties of essential submodules without proof (see, also Exercise 1.64).

Proposition 2.1. *Let A be a module. Then*

(i) $N \leq_e A \Leftrightarrow N \cap Ra \neq 0$ *for all* $0 \neq a \in A$.

(ii) *Given* $K \leq N \leq A$, $K \leq_e A \Leftrightarrow K \leq_e N$ *and* $N \leq_e A$.

(iii) *For any* $t \geq 1$, $N_i \leq_e K_i$ $(1 \leq i \leq t) \Rightarrow (N_1 \cap \cdots \cap N_t) \leq_e (K_1 \cap \cdots \cap K_t)$.

(iv) *For any non-empty index set* Λ, $N_\lambda \leq_e K_\lambda$ $(\lambda \in \Lambda) \Rightarrow \bigoplus_\Lambda N_\lambda \leq_e \bigoplus_\Lambda K_\lambda$.

Given $L \leq A$, by a *complement* (*submodule*) of L in A, we mean a submodule K of A, maximal with respect to the property $K \cap L = 0$. Thus, K is a complement of L in $A \Leftrightarrow$ (i) $K \cap L = 0$, and (ii) $N \cap L \neq 0$ for all $K \subset N \leq A$.

Proposition 2.2. *Let $L, N \leq A$ with $N \cap L = 0$. Then there exists a complement K of L in A such that $N \subseteq K$.*

Proof. Use Zorn's Lemma (see also the proof of Proposition 1.17). □

Notice that Proposition 2.2 guarantees the existence of complements of submodules in a module. In other words, every submodule of a module has a complement (see Exercise 2.2). The following proposition and corollary were stated before as Proposition 1.17 and Exercise 1.136, respectively. We prove both of them in this section's context for the sake of completeness.

Proposition 2.3. *Let $L \leq A$ and let K be any complement of L in A. Then $K \oplus L \leq_{\mathrm{e}} A$.*

Proof. Suppose $N \leq A$ and $(K \oplus L) \cap N = 0$. Suppose $K \subset K + N$. Then $(K + N) \cap L \neq 0$. There exist $k \in K$, $n \in N$, $0 \neq x \in L$ such that $x = k + n$. Then $n \in (K \oplus L) \cap N$, so that $n = 0$ and hence $x = 0$, a contradiction. Thus $K = K + N$, so that $N \subseteq K$, and hence $N = 0$. □

Corollary 2.4. *For any module A, $\operatorname{soc} A = \bigcap\{N : N \leq_{\mathrm{e}} A\}$.*

Proof. Let $N \leq_{\mathrm{e}} A$ and U be a simple submodule of A. Then $N \cap U \neq 0$ implies $N \cap U = U$, so that $U \subseteq N$. Thus $\operatorname{soc} A \subseteq \bigcap\{N : N \leq_{\mathrm{e}} A\}$.
Conversely, suppose that $a \in \bigcap\{N : N \leq_{\mathrm{e}} A\}$. Suppose $\operatorname{soc}(Ra) \neq Ra$. Then there exists a maximal submodule L of Ra such that $\operatorname{soc}(Ra) \subseteq L$. Suppose $L \leq_{\mathrm{e}} Ra$. Let K be a complement of L in A. Then $Ra \cap K = 0$ and $L \oplus K \leq_{\mathrm{e}} Ra \oplus K \leq_{\mathrm{e}} A$, by Proposition 2.3. Thus $L \oplus K \leq_{\mathrm{e}} A$ (Proposition 2.1) and $a \in L \oplus K$, so that $a \in L$, a contradiction. Thus $L \leq_{\mathrm{e}} Ra$ does not hold, and there exists $0 \neq V \leq Ra$ such that $L \cap V = 0$. It follows that V is simple and hence $V \subseteq \operatorname{soc}(Ra) \subseteq L$, a contradiction. Thus $Ra = \operatorname{soc}(Ra)$, and $a \in \operatorname{soc} A$. The result follows. □

A submodule K of a module A will be called a *complement* (*submodule*) (in A), written $K \leq_{\mathrm{c}} A$, provided there exists $L \leq A$ such that K is a complement of L in A. Clearly, $0 \leq_{\mathrm{c}} A$, $A \leq_{\mathrm{c}} A$, and $K \leq_{\mathrm{c}} A$ for any direct summand K of A.

Proposition 2.5. *Let $N \leq A$. Then there exists $K \leq A$, containing N, such that $N \leq_{\mathrm{e}} K \leq_{\mathrm{c}} A$.*

Proof. Let N' be a complement of N in A. By Proposition 2.2, there exists a complement K of N' in A with $N \subseteq K$. Let $0 \neq L \leq K$. Then $N' \subseteq L + N'$ and hence $(L + N') \cap N \neq 0$. There exist $x \in L$, $n' \in N'$, $0 \neq n \in N$, such that $n = x + n'$. Then $n' \in K \cap N'$, so that $n' = 0$, and $n \in L \cap N$. Thus $N \leq_{\mathrm{e}} K$. □

Proposition 2.6. *Let $K \leq_{\mathrm{c}} A$ and $K \leq N \leq A$. Then $N \leq_{\mathrm{e}} A \Leftrightarrow N/K \leq_{\mathrm{e}} A/K$.*

Proof. ($\Leftarrow$) By Exercise 1.64 (vi).

($\Rightarrow$) Suppose $N \leq_{\mathrm{e}} A$. Set $A' = A/K$, $N' = N/K$. Let $L' \leq A'$ with $N' \cap L' = 0$. Then there exists $L \leq A$ such that $K \subseteq L$, $L' = L/K$ and $N \cap L = K$. Let $K' \leq A$ such that K is a complement of K' in A. Then $N \cap L \cap K' = 0$, and hence $L \cap K' = 0$. Since $K \subseteq L$, we have that, $K = L$ and so $L' = 0$. It follows that $N' \leq_{\mathrm{e}} A'$. □

Proposition 2.7. *Let $K \leq A$. Then $K \leq_c A$ if and only if whenever $K \leq_e L \leq A$, then $K = L$.*

Proof. Use Proposition 2.5. □

Proposition 2.8. *Let $K, L \leq_c A$. Then K is a complement of L in A if and only if L is a complement of K in A.*

Proof. Suppose that K is a complement of L in A. Let $L \subseteq L' \leq A$ and $L' \cap K = 0$. By Proposition 2.3, $K \oplus L \leq_e A$. Let $0 \neq y \in L'$. Then there exists $r \in R$ such that $0 \neq ry = k + x$ for some $k \in K$, $x \in L$. Then $k = ry - x \in K \cap L' = 0$, and hence $ry \in L$. Thus $L \leq_e L'$. By Proposition 2.7, $L = L'$. Thus L is a complement of K in A. □

Proposition 2.9. *Let $N \leq K \leq A$. Then*

(i) *$K \leq_c A \Rightarrow K/N \leq_c A/N$, and*
(ii) *$K/N \leq_c A/N$, $N \leq_c A \Rightarrow K \leq_c A$.*

Proof. (i) Suppose $L \leq A$ with $K \subseteq L$ and $K/N \leq_e L/N$. By Exercise 1.64 (iv), $K \leq_e L$ and, by Proposition 2.7, $K = L$. Thus $K/N = L/N$. Again, by Proposition 2.7, $K/N \leq_c A/N$.

(ii) There exist $K', N' \leq A$ such that $N \subseteq K'$, K/N is a complement of K'/N in A/N, and N is a complement of N' in A. Thus $K \cap K' = N$ and $N \cap N' = 0$, giving $K \cap (K' \cap N') = 0$. Suppose that $K \leq L \leq A$ and $L \cap (K' \cap N') = 0$. Since $N \subseteq L \cap K'$ and $(L \cap K') \cap N' = 0$, it follows that $L \cap K' = N$. Thus $(L/N) \cap (K'/N) = 0$ and hence $L/N = K/N$. We conclude that $L = K$, and hence that K is a complement of $(K' \cap N')$ in A. □

Proposition 2.10. *Let $K \leq_c N$ and $N \leq_c A$. Then $K \leq_c A$.*

Proof. There exists $K' \leq N$ such that K is a complement of K' in N, and there exists $N' \leq A$ such that N is a complement of N' in A. Note that $K \cap (K' + N') = 0$. Suppose that $K \leq_e L \leq A$. Then $L \cap (K' + N') = 0$, and hence $[N \cap (L + N')] \cap K' = K' \cap (L + N') = 0$. But $K \subseteq N \cap (L + N')$, so that $K = N \cap (L + N')$. Hence $(N + L) \cap N' = 0$. It follows that $L \subseteq N$ and hence $K = L$, by Proposition 2.7. Now Proposition 2.7 gives $K \leq_c A$. □

A module A satisfies the *ascending* (respectively, *descending*) *chain condition on complements*, written A satisfies acc-c (respectively dcc-c), if for any chain

$$K_1 \subseteq K_2 \subseteq K_3 \subseteq \cdots \quad (\text{respectively}, K_1 \supseteq K_2 \supseteq K_3 \supseteq \cdots)$$

of complements there exists a positive integer n such that

$$K_n = K_{n+1} = K_{n+2} = \cdots .$$

Proposition 2.11. *The following statements are equivalent for a module A.*

(i) *A satisfies* acc-c.

(ii) *For any ascending chain of submodules $N_1 \subseteq N_2 \subseteq N_3 \subseteq \cdots$ of A there exists $k \geq 1$ such that $N_i \leq_e N_{i+1}$ for all $i \geq k$.*

(iii) *A satisfies* dcc-c.

(iv) *For any descending chain of submodules $N_1 \supseteq N_2 \supseteq N_3 \supseteq \cdots$ of A there exists $k \geq 1$ such that $N_{i+1} \leq_e N_i$ for all $i \geq k$.*

(v) *A does not contain an infinite direct sum of non-zero submodules.*

(vi) *There exists $N \leq_e A$ such that N does not contain an infinite direct sum of non-zero submodules.*

(vii) *For each $N \leq A$ there exists a finitely generated $K \leq_e N$.*

(viii) *For each $N \leq_e A$ there exists a finitely generated $K \leq_e N$.*

Proof. (i) $\Rightarrow$ (v) Let $N_1 \oplus N_2 \oplus N_3 \oplus \cdots$ be an infinite direct sum of non-zero submodules of A. By Proposition 2.5, there exists $K_1 \leq_c A$ with $N_1 \leq_e K_1$. Note that $K_1 \cap (N_2 \oplus N_3 \oplus \cdots) = 0$. Again, by Proposition 2.5, there exists $K_2 \leq_c A$ such that $(K_1 \oplus N_2) \leq_e K_2$. Note that

$$K_2 \cap (N_3 \oplus N_4 \oplus \cdots) = 0.$$

Continuing this process produces a chain of complements: $K_1 \subseteq K_2 \subseteq K_3 \subseteq \cdots$. Thus A does not satisfy acc-c.

(v) $\Rightarrow$ (ii) Let $N_1 \subseteq N_2 \subseteq N_3 \subseteq \cdots$ be an ascending chain of submodules of A such that N_i is not an essential submodule of N_{i+1} for all $i \geq 1$. For each $i \geq 1$, there exists $0 \neq K_i \leq N_{i+1}$ such that $N_i \cap K_i = 0$. Then it is easy to check that $K_1 + K_2 + K_3 + \cdots$ is an infinite direct sum of non-zero submodules.

(ii) $\Rightarrow$ (i) By Proposition 2.7.

(iii) $\Leftrightarrow$ (iv) $\Leftrightarrow$ (v) Similar to (i) $\Leftrightarrow$ (ii) $\Leftrightarrow$ (v)

(v) $\Leftrightarrow$ (vi) Clear.

(ii) $\Rightarrow$ (vii) Let $0 \neq N \leq A$. Let $0 \neq n_1 \in N$. Then either $Rn_1 \leq_e N$, or there exists $0 \neq n_2 \in N$ such that $Rn_1 \cap Rn_2 = 0$. Next, either $Rn_1 \oplus Rn_2 \leq_e N$, or there exists $0 \neq n_3 \in N$ such that $(Rn_1 \oplus Rn_2) \cap Rn_3 = 0$. Repeat this process and note that, by (ii), it must stop after a finite number of steps. Thus there exists $k \geq 1$ such that $Rn_1 \oplus \cdots \oplus Rn_k \leq_e N$.

(vii) $\Rightarrow$ (viii) Clear.

(viii) $\Rightarrow$ (v) Suppose that (viii) holds. Let $N = N_1 \oplus N_2 \oplus N_3 \oplus \cdots$ be a direct sum of submodules of A. Let N' be a complement of N in A. By Proposition 2.3, $N \oplus N' \leq_e A$ and hence there exists $K \leq_e N \oplus N'$ with K finitely generated. Since K is finitely generated, it follows that there exists $t \geq 1$ such that $K \subseteq N_1 \oplus \cdots \oplus N_t \oplus N'$. Then, for all $i \geq t+1$, $K \cap N_i = 0$ and hence $N_i = 0$. It follows that A satisfies (v). $\square$

Corollary 2.12. *Suppose that A satisfies* acc-c *and $N \leq A$. Then*

(i) *N satisfies* acc-c.
(ii) *A/N satisfies* acc-c *provided $N \leq_c A$.*

Proof. (i) is clear and (ii) follows by Proposition 2.9 (ii). □

We shall say that a module A satisfies *acc-e* (respectively, *dcc-e*) if every ascending (descending) chain of essential submodules terminates. The next result is a consequence of Proposition 2.11.

Proposition 2.13. *The following statements are equivalent for a module A.*

(i) *A satisfies* acc-e.
(ii) *For all $K \leq_e N \leq A$, the module N/K is finitely generated.*
(iii) *$A/\operatorname{soc} A)$ is Noetherian.*

Proof. (i) ⇒ (ii) Suppose that A satisfies acc-e. Let $K \leq_e N \leq A$. Let L be a complement of K in A. Then $N \cap L = 0$ and, by Proposition 2.3, $K \oplus L \leq_e A$. Hence $A/(K \oplus L)$ is Noetherian and $N/K \cong (N \oplus L)/(K \oplus L) \leq A/(K \oplus L)$ implies N/K is finitely generated.

(ii) ⇒ (iii) Suppose that (ii) holds. Let $S = \operatorname{soc} A$. Let $S \subseteq N \leq A$. Let K be a complement of S in N. We prove first that K satisfies acc-c. Suppose not and let $K' = K_1 \oplus K_2 \oplus K_3 \oplus \cdots$ be a direct sum of non-zero submodules of K. For each $i \geq 1$, $S \cap K_i = 0$ and hence, by Corollary 2.4, there exists $L_i \leq_e K_i$ with $L_i \neq K_i$. Let

$$L = L_1 + L_2 + L_3 + \cdots = L_1 \oplus L_2 \oplus L_3 \oplus \cdots .$$

By Proposition 2.1 (iv), $L \leq_e K'$ and hence K'/L is finitely generated. But

$$K'/L \cong (K_1/L_1) \oplus (K_2/L_2) \oplus (K_3/L_3) \oplus \cdots ,$$

which is an infinite direct sum of non-zero submodules, a contradiction. Thus K satisfies acc-c.

By Proposition 2.11, there exists $P \leq_e K$ with P finitely generated. Moreover, $P \oplus S \leq_e N$ by Propositions 2.1 and 2.3. Then (ii) implies $N/(P \oplus S)$ is finitely generated and hence N/S is finitely generated. It follows that A/S is Noetherian.

(iii) ⇒ (i) Clear by Corollary 2.4. □

An analogous argument gives the following result.

Proposition 2.14. *The following statements are equivalent for a module A.*

(i) *A satisfies* dcc-e.
(ii) *For all $K \leq_e N \leq A$, the module N/K is finitely cogenerated.*
(iii) *$A/(\text{soc A})$ is Artinian.*

Corollary 2.15.

(i) *A module A is Noetherian if and only if A satisfies* acc-c *and* acc-e.
(ii) *A module A is Artinian if and only if A satisfies* dcc-c *and* dcc-e.

Proof. By Propositions 2.11, 2.13, and 2.14. □

Another consequence of Proposition 2.11 is the following result.

Proposition 2.16. *Let $N \leq A$ such that both N and A/N satisfy* acc-c. *Then A satisfies* acc-c.

Proof. Let $K \leq A$. Let L be a complement of $N \cap K$ in K. By Proposition 2.3, $(N \cap K) \oplus L \leq_{\mathrm{e}} K$. By Proposition 2.11 there exists $K' \leq_{\mathrm{e}} (N \cap K)$ with K' finitely generated. Moreover, $N \cap L = 0$ implies that L is isomorphic to a submodule of A/N, and hence L satisfies acc-c. Again, by Proposition 2.11, there exists $L' \leq_{\mathrm{e}} L$ with L' finitely generated. Then $K' \oplus L'$ is clearly finitely generated and $K' \oplus L' \leq_{\mathrm{e}} K$ by Proposition 2.1. Finally, Proposition 2.11 shows that A satisfies acc-c. □

A submodule U of A is called *uniform*, written $U \leq_{\mathrm{u}} A$, if $U \neq 0$ and $X \cap Y \neq 0$ for all $0 \neq X, Y \leq U$. In other words, $U \leq_{\mathrm{u}} A \Leftrightarrow X \leq_{\mathrm{e}} U$ for all $0 \neq X \leq U$. The next result gives more information about modules with acc-c.

Proposition 2.17. *Suppose that A is a non-zero module satisfying* acc-c. *Then*

(i) *A contains a uniform submodule.*
(ii) *There exist a positive integer n and uniform submodules U_i $(1 \leq i \leq n)$ of A such that $U_1 \oplus \cdots \oplus U_n \leq_{\mathrm{e}} A$.*
(iii) *Given $N \leq A$, $N \leq_{\mathrm{e}} A \Leftrightarrow N \cap U_i \neq 0$ $(1 \leq i \leq n)$.*
(iv) *For any direct sum $N_1 \oplus \cdots \oplus N_k$ of non-zero submodules of A, $k \leq n$.*
(v) *If $V_1 \oplus \cdots \oplus V_k \leq_{\mathrm{e}} A$, with $V_i \leq_{\mathrm{u}} A$ $(1 \leq i \leq k)$, then $k = n$.*

Proof. (i) If A is not uniform, then there exist $0 \neq L_1, L_1' \leq A$ such that $L_1 \cap L_1' = 0$. If L_1' is not uniform, then there exist $0 \neq L_2, L_2' \leq L_1'$ such that $L_2 \cap L_2' = 0$. Either L_2' is uniform or this process continues. Note that it produces a direct sum of non-zero submodules, namely $L_1 \oplus L_2 \oplus L_3 \oplus \cdots$. By Proposition 2.11, either A is uniform or there exists $t \geq 1$ such that L_t' is uniform.

(ii) By (i), A contains a uniform submodule U_1. Either $U_1 \leq_{\mathrm{e}} A$ or there exists $0 \neq K_1 \leq A$ such that $U_1 \cap K_1 = 0$. By (i) again, K_1 contains a uniform submodule U_2. Note that $U_1 \cap U_2 = 0$. Either $U_1 \oplus U_2 \leq_{\mathrm{e}} A$, or there exists $0 \neq K_2 \leq A$ such that $(U_1 \oplus U_2) \cap K_2 = 0$. In this case, K_2 contains a uniform submodule U_3, by (i). In this way we can produce a direct sum of uniform submodules of A. By Proposition 2.11, there exists a positive integer n such that $(U_1 \oplus \cdots \oplus U_n) \leq_{\mathrm{e}} A$.

(iii) If $N \leq_{\mathrm{e}} A$ then clearly $N \cap U_i \neq 0$ for each $1 \leq i \leq n$. Conversely, suppose that $N \cap U_i \neq 0$ $(1 \leq i \leq n)$. Let $0 \neq K \leq A$. Let k be a non-zero element of K. Then $Rk \cap (U_1 \oplus \cdots \oplus U_n) \neq 0$. Thus, there exist $r \in R$, $u_i \in U_i$ $(1 \leq i \leq n)$ such that

$$0 \neq rk = u_1 + \cdots + u_n.$$

Clearly, $u_i \neq 0$ for some $1 \leq i \leq n$. Note that, because U_i is uniform, we have $Ru_i \cap (U \cap N) \neq 0$. Thus, there exists $s \in R$ such that $0 \neq su_i \in N$. Then

$$0 \neq srk = su_1 + \cdots + su_i + \cdots + su_n.$$

Let $x = srk - su_i$. Let $V = \bigoplus_{j \neq i} U_j$. By induction on n, $N \cap V \leq_{\mathrm{e}} V$. Either $x = 0$, or there exists $t \in R$ such that $0 \neq tx \in N \cap V$. Thus, either srk is a non-zero element of N, or $tsrk$ is a non-zero element of N. In any case, $N \cap K \neq 0$. Thus $N \leq_{\mathrm{e}} A$.

(iv) Clearly $N_2 \oplus \cdots \oplus N_k$ is not an essential submodule of A and hence, without loss of generality, $U_1 \cap (N_2 \oplus \cdots \oplus N_k) = 0$, by (iii). Form the direct sum $U_1 \oplus N_2 \oplus \cdots \oplus N_k$. By the same argument we can suppose without loss of generality that $U_1 + U_2 + N_3 + \cdots + N_k$ is a direct sum. If $k > n$, then repeating this process would produce the direct sum $U_1 \oplus \cdots \oplus U_n \oplus N_{n+1} \oplus \cdots \oplus N_k$. But this implies

$$(U_1 \oplus \cdots \oplus U_n) \cap N_k = 0,$$

a contradiction. Thus $k \leq n$.

(v). By (iv). □

Let A be a module with acc-c. There exists a positive integer n such that n is the number of non-zero direct summands in any essential direct sum of uniform submodules; n is called the *uniform dimension* or *Goldie dimension* of A, and is denoted $u\text{–}\dim A$. Of course,

$$u\text{–}\dim A = 0 \Leftrightarrow A = 0.$$

Proposition 2.18. *Let A be a module which satisfies* acc-c *and let $N \leq A$. Then*

(i) *N satisfies* acc-c *and $u\text{–}\dim N \leq u\text{–}\dim A$. Moreover, $u\text{–}\dim N = u\text{–}\dim A$ if and only if $N \leq_{\mathrm{e}} A$.*
(ii) *If $N \leq_{\mathrm{c}} A$ then A/N satisfies* acc-c *and in this case $u\text{–}\dim A = u\text{–}\dim N + u\text{–}\dim (A/N)$.*

Proof. (i) Clear.

(ii) Suppose $N \leq_{\mathrm{c}} A$. Let N' be a complement of N in A. By Proposition 2.3, $N \oplus N' \leq_{\mathrm{e}} A$. Moreover, by Proposition 2.6, $N' \cong (N \oplus N')/N \leq_{\mathrm{e}} A/N$. Now N' has acc-c, by (i), and hence so does A/N by Proposition 2.12. Moreover, it is clear that

$$\begin{aligned} u\text{–}\dim A &= u\text{–}\dim N + u\text{–}\dim N' \\ &= u\text{–}\dim N + u\text{–}\dim (A/N). \end{aligned}$$ □

Exercises

2.1. Prove Proposition 2.1.

2.2. Find an example of a module A such that a submodule of A has more than one complement in A (see Proposition 2.2).

2.3. Let A be a left R-module and $0 \neq a \in A$. Let $E \leq_{\rm e} A$. Show that there is an $L \leq_{\rm e} {}_RR$ such that $La \neq 0$ and $La \subseteq E$.

2.4. Prove that the torsion submodule of a module over a commutative ring is a complement.

2.5. Let ${}_RA$ be a module and $N, K \leq A$ such that $K \leq N$. show that if $N/K \leq_{\rm c} A/K$ and $K \leq_{\rm c} A$, then $N \leq_{\rm c} A$.

2.6. Find an example of a module A such that $X, Y \leq_{\rm c} A$, but $X \cap Y$ is not a complement.

2.7. Let K be any field and $R = \begin{bmatrix} K & K \\ 0 & K \end{bmatrix}$. Show that complements and direct summands of R_R are the same.

2.8. Let A be a left R-module and $a \in A$. Let $l(a) = \{r \in R : ra = 0\}$ (resp., $r(a) = \{r \in R : ar = 0\}$) be the left (resp., right) *annihilator of a*. The *singular submodule* of A is defined by $Z(A) = \{a \in A : l(a) \leq_{\rm e} {}_RR\}$. Then the module A is called *singular* if $A = Z(A)$ and *nonsingular* if $Z(A) = 0$. Prove

(i) $Z(A) = \{a \in A : Ea = 0$ for some essential left ideal E of $R\}$,
(ii) $Z(N) = N \cap Z(A)$ for any $N \leq A$,
(iii) $Z(A)$ is a singular module,
(iv) If N is nonsingular and $N \leq_{\rm e} A$, then A is nonsingular,
(v) If $N \leq_{\rm e} A$ then A/N is singular,
(vi) If A/N is nonsingular, then $N \leq_{\rm c} A$,
(vii) If A is nonsingular and A/N singular, then $N \leq_{\rm e} A$.

2.9. Let ${}_RA$ be a module and $N \leq A$. A submodule K of A with the property that $N \leq_{\rm e} K \leq_{\rm c} A$ is called a *closure* (or *essential closure*) of N in A (see Proposition 2.5). Then:

(i) Give an example which shows that a closure of a submodule need not be unique.
(ii) Let $\mathrm{c}(N) = \{a \in A : Ea \leq A$ for some essential left ideal of $R\}$. Show that $\mathrm{c}(N) \leq_{\rm c} A$.
(iii) Assume A is nonsingular. Prove that $\mathrm{c}(N)$ is a unique complement in A such that $N \leq_{\rm e} \mathrm{c}(N)$, i.e, $\mathrm{c}(N)$ is the unique closure of N in A.

2.10. Let ${}_RA$ be a module. The *Goldie torsion* (or *second singular*) *submodule* $Z_2(A)$ of A is that submodule of A, containing $Z(A)$ such that $Z_2(A)/Z(A)$ is the singular submodule of $A/Z(A)$. Prove that

(i) $Z_2(A) = \{a \in A : Ea \subseteq Z(A)$ for some essential left ideal of $R\}$,
(ii) $A/Z_2(A)$ is nonsingular,
(iii) $Z_2(A) \leq_c A$.

2.11. Find an example such that $N \leq_c A$ is not superfluous in Proposition 2.18 (ii).

2.12. Let N, K be submodules of a module. Find an example which shows the failure of the equality,

$$u\text{--}\dim(N+K) = u\text{--}\dim(N) + u\text{--}\dim(K) - u\text{--}\dim(N \cap K).$$

2.13. Let ${}_RA$ be a module and $K, N \leq A$.

(i) Show that if $K \leq_c A$, then $Z_2(K) \leq_c Z_2(A)$.
(ii) Assume $K \leq N \leq A$ and $K \leq_c A$. Show that $K \leq_c N$.

2.14. Let R_R be a nonsingular module and $x \in R$. Prove that $r(x) \leq_c R_R$.

2.15. Let A be a left R-module. Prove that A is finitely cogenerated if and only if $\operatorname{soc} A$ is finitely generated and essential in A.

2.16. Let ${}_RA$ be a module and let $N \leq A$. We call N a *z-closed submodule* of A provided A/N is nonsingular. Prove that

(i) every z-closed submodule is a complement,
(ii) N is a z-closed submodule of the left R-module A if and only if $a \in A$, E is essential left ideal of R, and $Ea \leq N$ implies that $a \in N$.
(iii) in a nonsingular module complement and z-closed submodules coincide.

2.17. Let V be a vector space over a field F with $\dim(V) \geq 2$ and let $R_R = \begin{bmatrix} F & V \\ 0 & F \end{bmatrix} = \{\begin{bmatrix} a & v \\ 0 & a \end{bmatrix} : a \in F, v \in V\}$. Then

(i) Show that R_R is not finitely cogenerated.
(ii) For $v \in V$, consider the ideal $I_v = \begin{bmatrix} 0 & Fv \\ 0 & 0 \end{bmatrix}$ of R.

Show that for each $v \in V$, I_v is a complement submodule which is not z-closed submodule of R_R.

2.2 Relative injective modules

Let A and X be left R-modules. X is called *A-injective* if for each submodule N of A, every R-homomorphism $\varphi : N \to X$ can be lifted to an R-homomorphism $\theta : A \to X$, i.e., $\theta(n) = \varphi(n)$ for all $n \in N$. A module X is called *quasi-injective* (or *QI* or *self-injective*) provided X is X-injective.

Some of the material in this section can be found in [AF74], [Wis91], [Wis88].

First note that any injective module is A-injective, for any module A. On the other hand, any ${}_RR$-injective module is injective. Recall that for any module A, $E(A)$ stands for the injective hull of A.

Proposition 2.19. *Let K be any submodule of an R-module A. Then an R-module X is A-injective if and only if the following conditions are satisfied:*

(i) *X is K-injective.*
(ii) *X is (A/K)-injective.*
(iii) *Any homomorphism $\varphi : K \to X$ can be lifted to a homomorphism $\theta : A \to X$.*

Proof. Suppose that X is A-injective. Then (i) and (iii) clearly hold. Now suppose that $K \subseteq N \leq A$ and $\varphi \in \mathrm{Hom}_R(N/K, X)$. Define $\varphi' : N \to X$ by

$$\varphi'(n) = \varphi(n + K) \quad (n \in N).$$

Note that $\varphi' \in \mathrm{Hom}_R(N, X)$ and hence φ' can be lifted to $\theta' \in \mathrm{Hom}_R(A, X)$. Note that $\theta'(n) = \varphi'(n)$ $(n \in N)$ and, in particular,

$$\theta'(k) = \varphi'(k) = \varphi(k + K) = 0 \quad (k \in K).$$

Define $\theta : A/K \to X$ by $\theta(a+K) = \theta'(a)$ $(a \in A)$. Since $\theta'(K) = 0$, it follows that θ is well defined. Clearly, $\theta \in \mathrm{Hom}_R(A/K, X)$. Moreover,

$$\theta(n + K) = \theta'(n) = \varphi'(n) = \varphi(n + K) \quad (n \in N).$$

It follows that X is (A/K)-injective.

Conversely, suppose that X satisfies (i), (ii), and (iii). Let $N \leq A$ and $\varphi \in \mathrm{Hom}_R(N, X)$. Let φ' denote the restriction of φ to $N \cap K$. Since X is K-injective, there exists $\alpha \in \mathrm{Hom}_R(K, X)$ such that α lifts φ'. By (iii), there exists $\beta \in Hom_R(A, X)$ such that β lifts α. Thus

$$\beta(k) = \varphi(k) \quad (k \in K \cap N).$$

Let $\gamma = \varphi - \beta$. Then $\gamma \in \mathrm{Hom}_R(N, X)$ and $\gamma(K \cap N) = 0$. Define $\varphi'' : (N+K)/K \to X$ by $\varphi''(n+K) = \gamma(n)$ $(n \in N)$. Note that φ'' is well defined because $\gamma(K \cap N) = 0$. Clearly $\varphi'' \in \mathrm{Hom}_R((N + K)/K, X)$. By (ii), there exists $\theta' \in \mathrm{Hom}_R(A/K, X)$ such that θ' lifts φ''. Define $\theta'' \in \mathrm{Hom}_R(A, X)$ by $\theta''(a) = \theta'(a + K)$ $(a \in A)$. Let $\theta = \beta + \theta'' \in \mathrm{Hom}_R(A, X)$. Let $n \in N$. Then $\theta(n) = \beta(n) + \theta''(n) = \varphi(n) - \gamma(n) + \theta''(n) = \varphi(n) - \varphi''(n + K) + \theta'(n + K) = \varphi(n)$. Thus θ lifts φ. It follows that X is A-injective. □

Proposition 2.20. *Let R be a ring and A an R-module such that A is the sum $\sum_{\lambda \in \Lambda} A_\lambda$ of its submodules A_λ $(\lambda \in \Lambda)$. Then an R-module X is A-injective if and only if X is A_λ-injective for all $\lambda \in \Lambda$.*

Proof. Suppose that X is A-injective. Then X is A_λ-injective for all $\lambda \in \Lambda$, by Proposition 2.19. Conversely, suppose that X is A_λ-injective for all $\lambda \in \Lambda$. Let N be a submodule of A and $\varphi \in \mathrm{Hom}_R(N, X)$. Let $\underline{S}$ denote the collection of pairs (L, α), where $N \subseteq L \leq A$, $\alpha \in \mathrm{Hom}_R(L, X)$, and $\alpha|_N = \varphi$. If $(L, \alpha), (L', \alpha') \in \underline{S}$, then we define $(L, \alpha) \leq (L', \alpha')$ if $L \subseteq L'$ and $\alpha'|_L = \alpha$. A non-empty collection of elements (L_w, α_w) of $\underline{S}$, where w belongs to some index set Ω, is called a chain if for all w, w' in Ω, either $(L_w, \alpha_w) \leq (L_{w'}, \alpha_{w'})$, or $(L_{w'}, \alpha_{w'}) \leq (L_w, \alpha_w)$. Let $\{(L_w, \alpha_w) : w \in \Omega\}$ be a chain in $\underline{S}$. Let $L = \bigcup_{w\in\Omega} L_w$. Then $L \leq A$ and clearly $N \subseteq L$. Define $\alpha : L \to X$ by $\alpha(a) = \alpha_w(a)$, where $a \in L_w$. It is easy to check that $(L, \alpha) \in \underline{S}$. By Zorn's Lemma, $\underline{S}$ contains a maximal member (K, θ).

We shall now prove that $K = A$. Let $\lambda \in \Lambda$. Let $P = A_\lambda \cap K$ and $\beta = \theta|_P$. Note that $\beta \in \operatorname{Hom}_R(P, X)$ and, because X is A_λ-injective, it follows that β can be lifted to a homomorphism $\gamma : A_\lambda \to X$. Define $\theta' : A_\lambda + K \to X$ by

$$\theta'(a + k) = \gamma(a) + \theta(k) \quad (a \in A_\lambda, k \in K).$$

Suppose that $a \in A_\lambda$, $k \in K$, and $a + k = 0$. Then $a = -k \in A_\lambda \cap K = P$, so that

$$\gamma(a) = \beta(a) = \theta(a) = \theta(-k) = -\theta(k),$$

and hence $\theta'(a + k) = 0$. It follows that θ' is well defined, and it is easy to check that $\theta' \in \operatorname{Hom}_R((A_\lambda + K), X)$ and $\theta'|_K = \theta$. By the choice of K, we have that $A_\lambda + K = K$ and hence $A_\lambda \subseteq K$. Thus $A \subseteq K$ and hence $A = K$. It follows that X is A-injective. □

Corollary 2.21. *Let R be a ring and A an R-module. Then an R-module X is A-injective if and only if X is (Ra)-injective for all $a \in A$.*

Proof. Immediate by Proposition 2.20. □

Proposition 2.22. *Let R be a ring and A an R-module. An R-module X is A-injective if and only if $\varphi(A) \subseteq X$ for every $\varphi \in \operatorname{Hom}_R(E(A), E(X))$.*

Proof. Suppose first that $\varphi(A) \subseteq X$ for every $\varphi \in \operatorname{Hom}_R(E(A), E(X))$. Let $N \leq A$ and $\alpha \in \operatorname{Hom}_R(N, X)$. Consider the diagram

$$\begin{array}{ccccccc} 0 & \longrightarrow & N & \longrightarrow & A & \xrightarrow{\iota} & E(A) \\ & & {\scriptstyle\alpha}\downarrow & & & & \\ & & X & & & & \\ & & {\scriptstyle\iota}\downarrow & & & & \\ & & E(X) & & & & \end{array}$$

where each ι is an inclusion mapping. Since $E(X)$ is injective, there exists $\beta \in \operatorname{Hom}_R(E(A), E(X))$ such that β lifts α. By hypothesis, $\beta(A) \subseteq X$ and hence the restriction γ of β to A is a homomorphism from A to X which lifts α. Thus X is A-injective.

Conversely, suppose that X is A-injective. Let $\varphi \in \operatorname{Hom}_R(E(A), E(X))$. Let $N = \{a \in A : \varphi(a) \in X\}$. Clearly, $N \leq A$. Let φ' denote the restriction of φ to N. By hypothesis, there exists $\theta \in \operatorname{Hom}_R(A, X)$ such that θ lifts φ'. Thus

$$\theta(n) = \varphi(n) \quad (n \in N).$$

Let φ'' denote the restriction of φ to A. Then $\lambda = \theta - \varphi'' \in \mathrm{Hom}_R(A, E(X))$ and $\lambda(N) = 0$. Suppose that $\lambda(A) \neq 0$. Then $X \cap \lambda(A) \neq 0$. Let $a \in A$ satisfy $0 \neq \lambda(a) \in X$. Then $\varphi(a) = \varphi''(a) = \theta(a) - \lambda(a) \in X$, and hence $a \in N$. But, in this case, $\lambda(a) = 0$, a contradiction. Hence $\lambda(A) = 0$, and so $\varphi(a) = \theta(a) \in X$ $(a \in A)$. Thus $\varphi(A) \subseteq X$. This completes the proof. □

Using Proposition 2.22 we can give a second proof of Proposition 2.20 as follows.

Second proof of Proposition 2.20. Let $A = \sum_{\lambda \in \Lambda} A_\lambda$ be a sum of submodules A_λ $(\lambda \in \Lambda)$, and let X be an R-module such that X is A_λ-injective for each $\lambda \in \Lambda$. Then we shall prove that X is A-injective. Let $E(A)$ denote the injective hull of A. Then for each $\lambda \in \Lambda$, there exists an injective hull $E(A_\lambda)$ of A_λ in $E(A)$. Let $\varphi \in \mathrm{Hom}_R(E(A), E(X))$. For each $\lambda \in \Lambda$, let $\varphi_\lambda \in \mathrm{Hom}_R(E(A_\lambda), E(X))$, denote the restriction of φ to $E(A_\lambda)$. By Proposition 2.22, $\varphi_\lambda(A_\lambda) \subseteq X$ $(\lambda \in \Lambda)$. But $\varphi(A) = \sum_{\lambda \in \Lambda} \varphi_\lambda(A_\lambda) \subseteq X$. By Proposition 2.22, X is A-injective. □

Proposition 2.23. *Let R be a ring and A an R-module. Let Y be a complement submodule of an A-injective R-module X. Then Y is A-injective.*

Proof. We can suppose, without loss of generality, that $E(X)$ contains as submodules X, Y, and $E(Y)$. Since $X \cap E(Y)$ is an essential extension of Y, it follows that $Y = X \cap E(Y)$. Let $\varphi \in \mathrm{Hom}_R(E(A), E(Y))$. Since $E(Y) \leq E(X)$, we see that $\varphi \in \mathrm{Hom}_R(E(A), E(X))$. By Proposition 2.22, $\varphi(A) \subseteq X$. Hence $\varphi(A) \subseteq X \cap E(Y) = Y$. Thus $\varphi(A) \subseteq Y$ for every $\varphi \in \mathrm{Hom}_R(E(A), E(Y))$. By Proposition 2.22, Y is A-injective. □

Lemma 2.24. *Let $K \subseteq N$ be submodules of an R-module A such that N/K is A-injective. Then N/K is a direct summand of A/K.*

Proof. By Proposition 2.19, N/K is (A/K)-injective. The identity mapping $\iota : N/K \to N/K$ can be lifted to a homomorphism $\theta : A/K \to N/K$. It is easy to check that $A/K = (N/K) \oplus (\ker \theta)$. □

Corollary 2.25. *Let R be a ring and A a quasi-injective R-module. Then any complement in A is a direct summand of A.*

Proof. By Proposition 2.23 and Lemma 2.24. □

Proposition 2.26. *The following statements are equivalent for an R-module A.*

(i) *A is semisimple.*
(ii) *Every R-module is A-injective.*
(iii) *Every submodule of A is A-injective.*
(iv) *Every submodule of an A-injective R-module is A-injective.*

Proof. (i) $\Rightarrow$ (ii) Let X be any R-module. Let $N \leq A$ and $\varphi \in \mathrm{Hom}_R(N, X)$. There exists $N' \leq A$ such that $A = N \oplus N'$. Define $\theta : A \to X$ by

$$\theta(n + n') = \varphi(n) \quad (n \in N, \quad n' \in N').$$

Then $\theta \in \mathrm{Hom}_R(A, X)$ and θ lifts φ. Thus X is A-injective.

(ii) $\Rightarrow$ (iii), (iv) Obvious.

(iv) $\Rightarrow$ (iii) Suppose that (iv) holds. Let $E(A)$ denote an injective envelope of A. By (iv), every submodule of $E(A)$, and hence of A, is A-injective.

(iii) $\Rightarrow$ (i) By Lemma 2.24. □

Lemma 2.27. *Let R be a ring and A an R-module.*

(i) *Any direct summand of an A-injective R-module is A-injective.*
(ii) *Let X_λ $(\lambda \in \Lambda)$ be any non-empty collection of A-injective R-modules. Then $X = \prod_\lambda X_\lambda$ is A-injective.*

Proof. Easy. □

Let P denote some module property, e.g., "Noetherian", "Artinian", etc. A left R-module A is called *locally P* if every finitely generated submodule of A has P.

Theorem 2.28. *An R-module A is locally Noetherian if and only if the direct sum of any family of A-injective modules is A-injective.*

Proof. Suppose A is locally Noetherian. Let X_λ $(\lambda \in \Lambda)$ be any non-empty collection of A-injective R-modules and let $X = \bigoplus_\lambda X_\lambda$. Let N be any finitely generated submodule of A. Let K be any submodule of N and let $\varphi \in \mathrm{Hom}_R(K, X)$. Since K is finitely generated, it follows that $\mathrm{im}\,\varphi$ is finitely generated and hence $\mathrm{im}\,\varphi$ is contained in a submodule X' of X, where X' is the direct sum of a finite number of the submodules X_λ. By Lemma 2.27 (ii), X' is A-injective and hence φ can be lifted to a homomorphism $\theta : A \to X'$. Note that $\theta \in \mathrm{Hom}_R(N, X)$. It follows that X is N-injective, for any finitely generated submodule N of A. For each element $a \in A$, X is Ra-injective. By Corollary 2.21, X is A-injective.

Conversely, suppose that the direct sum of any family of A-injective modules is A-injective. Let L be any finitely generated submodule of A. Let $N_1 \subseteq N_2 \subseteq N_3 \subseteq \cdots$ be an ascending chain of submodules of L. Let N denote the submodule $\bigcup_i N_i$ of A. Let

$$X = E(A/N_1) \oplus E(A/N_2) \oplus E(A/N_3) \oplus \cdots .$$

Define a mapping $\varphi : N \to X$ by

$$\varphi(n) = (n + N_1, n + N_2, n + N_3, \dots).$$

Note that φ is well defined, because for any $n \in N$ there exists $k \geq 1$ such that $n \in N_k$. By hypothesis, X is A-injective, and hence φ can be lifted to $\theta \in \mathrm{Hom}_R(A, X)$. Since L is finitely generated, it follows that $\theta(L)$ is finitely generated and hence

$\theta(L)$ is contained in $E(A/N_1) \oplus \cdots \oplus E(A/N_t)$, for some positive integer t. Let $n \in N$. Then

$$(n + N_1, n + N_2, n + N_3, \dots) = \varphi(n) = \theta(n) = (e_1, \dots, e_t, 0, 0, \dots),$$

for some $e_i \in E(A/N_i)$ $(1 \leq i \leq t)$. Thus $n \in N_{t+1}$. It follows that $N = N_{t+1}$, and hence $N_{t+1} = N_{t+2} = N_{t+3} = \cdots$. Thus L is Notherian, and consequently A is locally Notherian. □

Given a left R-module X, a left R-module A is called *X-projective* if for every $Y \leq X$ and $\varphi \in \mathrm{Hom}_R(A, X/Y)$ there exists $\theta \in \mathrm{Hom}_R(A, X)$ such that $\varphi = \pi\theta$, where $\pi : X \to X/Y$ is the canonical epimorphism.

Proposition 2.29. *The following statements are equivalent for left R-modules A and X.*

(i) *X is A-injective and every submodule of A is X-projective.*

(ii) *A is X-projective and every homomorphic image of X is A-injective.*

Proof. (i) $\Rightarrow$ (ii) Suppose that (i) holds. Then A is X-projective. Let $Y \leq X$. Let $N \leq A$, $\varphi \in \mathrm{Hom}_R(N, X/Y)$, and consider the diagram

$$\begin{array}{ccccc} 0 & \longrightarrow & N & \xrightarrow{\iota} & A \qquad \text{exact} \\ & & \downarrow{\scriptstyle\varphi} & & \\ X & \xrightarrow{\pi} & X/Y & \longrightarrow & 0 \qquad \text{exact} \end{array}$$

where ι denotes the inclusion mapping. Since N is X-projective, there exists $\theta \in \mathrm{Hom}_R(N, X)$ such that $\pi\theta = \varphi$. But X is A-injective, and hence there exists $\psi \in \mathrm{Hom}_R(A, X)$ such that $\psi\iota = \theta$. Now $\pi\psi \in \mathrm{Hom}_R(A, X/Y)$ and $(\pi\psi)\iota = \varphi$. It follows that X/Y is A-injective.

(ii) $\Rightarrow$ (i) Suppose that (ii) holds. Then X is A-injective. Let $K \leq A$. Let $Z \leq X$, $\varphi \in \mathrm{Hom}_R(K, X/Z)$, and consider the diagram

$$\begin{array}{ccccc} 0 & \longrightarrow & K & \xrightarrow{\iota} & A \qquad \text{exact} \\ & & \downarrow{\scriptstyle\varphi} & & \\ X & \xrightarrow{\pi} & X/Z & \longrightarrow & 0 \qquad \text{exact} \end{array}$$

There exists $\theta \in \mathrm{Hom}_R(A, X/Z)$ such that $\theta\iota = \varphi$, because X/Z is A-injective. Since A is X-projective, there exists $\psi \in \mathrm{Hom}_R(A, X)$ such that $\pi\psi = \theta$. Now $\psi\iota \in \mathrm{Hom}_R(K, X)$ and $\pi(\psi\iota) = \varphi$. It follows that K is X-projective. □

By a *class* $\underline{\mathrm{X}}$ of R-modules we mean a collection of R-modules containing a zero module and closed under isomorphisms. Any member of $\underline{\mathrm{X}}$ is called an *$\underline{X}$-module.* A class $\underline{\mathrm{X}}$ is called *S-closed* if every submodule of an $\underline{\mathrm{X}}$-module is an $\underline{\mathrm{X}}$-module. Given a module A, we now consider the general question: "What properties does A have if every $\underline{\mathrm{X}}$-module is A-injective?"

Lemma 2.30. *Let A be an R-module and $\underline{X}$ an S-closed class of R-modules. Then the following statements are equivalent.*

(i) *Every $\underline{X}$-module is A-injective.*
(ii) *For all $K \leq N \leq A$ with $N/K \in \underline{X}$, N/K is a direct summand of A/K.*

Proof. (i) $\Rightarrow$ (ii) Let $K \leq N \leq A$ with $N/K \in \underline{X}$. Then N/K is A-injective by (i) and hence, by Lemma 2.24, N/K is a direct summand of A/K.

(ii) $\Rightarrow$ (i) Suppose that (ii) holds. Let $X \in \underline{X}$, $N \leq A$, and $\varphi \in \operatorname{Hom}_R(N, X)$. Let $K = \ker \varphi$. Then $N/K \cong \operatorname{im} \varphi \leq X$, so that $N/K \in \underline{X}$. By hypothesis, there exists $L \leq A$ such that $K \subseteq L$ and $A/K = (N/K) \oplus (L/K)$. Define $\theta : A \to X$ by $\theta(a + b) = \varphi(a)$ $(a \in N,\ b \in L)$. Note that if $a \in N$, $b \in L$, and $a + b = 0$, then $a = -b \in N \cap L = K$, so that $\varphi(a) = 0$. It follows that θ is well defined. Clearly θ is an R-homomorphism and $\theta|_N = \varphi$. Thus X is A-injective. □

Recall that a module A is called finitely cogenerated if it has finitely generated essential socle (Exercise 2.15). The (Jacobson) radical, $\operatorname{rad} A$, of an R-module A is defined to be A if A contains no maximal submodule, and to be the intersection of all the maximal submodules of A otherwise (see Exercise 1.137).

Theorem 2.31. *The following statements are equivalent for an R-module A.*

(i) *Every simple R-module is A-injective.*
(ii) *Every finitely cogenerated R-module is A-injective.*
(iii) *Every Artinian R-module is A-injective.*
(vi) $\operatorname{rad}(A/N) = 0$ *for each submodule N of A.*

Proof. (ii) $\Rightarrow$ (iii) $\Rightarrow$ (i) Obvious.

(i) $\Rightarrow$ (iv) Let $N \leq A$ and let K be the submodule of A containing N such that $K/N = \operatorname{rad}(A/N)$. Suppose that $K \neq N$. Let $x \in K$, $x \notin N$. Then $(Rx + N)/N$ is a non-zero cyclic module, so contains a maximal submodule. Let $P \leq A$ such that $N \subseteq P \subseteq Rx + N$ and P/N is a maximal submodule of $(Rx + N)/N$. Then $(Rx+N)/P$ is simple and hence A-injective by hypothesis. By Lemma 2.24, $(Rx+N)/P$ is a direct summand of A/P. Thus there exists $P \subseteq L \leq A$ such that $A/P = [(Rx + N)/P] \oplus (L/P)$. Now $A/L \cong (Rx + N)/P$, so that A/L is simple and hence L is a maximal submodule of A. Since $N \subseteq L$, it follows that $K \subseteq L$, and hence $Rx + N \subseteq L$. Thus $(Rx + N)/P = 0$, a contradiction. Thus $K = N$ and $\operatorname{rad}(A/N) = 0$.

(iv) $\Rightarrow$ (i) Suppose (iv) holds. Let $K \leq N \leq A$ such that N/K is simple. Since $\operatorname{rad}(A/K) = 0$, there exists a maximal submodule L of A such that $K \subseteq L$ but $N \nsubseteq L$. Then $K \subseteq N \cap L \subseteq N$ implies $N \cap L = K$ and hence $A/K = (N/K) \oplus (L/K)$. By Lemma 2.30, every simple R-module is A-injective.

(i) $\Rightarrow$ (ii) Suppose that (i) holds. Then every finitely generated semisimple R-module is A-injective by Lemma 2.27. Let $K \leq N \leq A$ with N/K finitely cogenerated. Let $L \leq A$ containing K such that L/K is the socle of N/K. Then L/K is finitely generated semisimple and hence A-injective. By Lemma 2.24, it follows that L/K is a direct summand of A/K and hence also of N/K. Thus $N/K = L/K$ (because L/K is essential in N/K) and N/K is a direct summand of A/K. By Lemma 2.30, (ii) follows. □

Corollary 2.32. *The following statements are equivalent for an R-module A.*

(i) *Every semisimple R-module is A-injective.*

(ii) *A is locally Noetherian and every simple R-module is A-injective.*

Proof. (ii) ⇒ (i) By Theorem 2.28.

(i) ⇒ (ii) Suppose (i) holds. To show that A is locally Noetherian it is necessary to adapt the proof of Theorem 2.28. Let N be a finitely generated submodule of A. Suppose that N is not Noetherian. Then there exists an infinite ascending chain $0 = L_0 \subset L_1 \subset L_2 \subset \cdots$ of finitely generated submodules L_i of N. For each $i \geq 1$, there exists $K_i \leq A$ such that $L_{i-1} \subseteq K_i \subset L_i$ and L_i/K_i is simple. Let $i \geq 1$ and let $\pi_i : L_i \to L_i/K_i$ denote the canonical projection. Since L_i/K_i is A-injective, π_i can be lifted to a homomorphism $\varphi_i : A \to L_i/K_i$. Let $L = \bigcup_{i\geq 1} L_i$ and define a mapping $\varphi : L \to (L_1/K_1) \oplus (L_2/K_2) \oplus (L_3/K_3) \oplus \cdots$ by

$$\varphi(x) = (\varphi_1(x), \varphi_2(x), \varphi_3(x), \dots) \quad (x \in L).$$

Note that for each $x \in L$ there exists a positive integer n such that $x \in L_n$ and hence $x \in K_t$ and $\varphi_t(x) = 0$ $(t \geq n+1)$. By hypothesis, φ can be lifted to a homomorphism $\theta : N \to (L_1/K_1) \oplus (L_2/K_2) \oplus (L_3/K_3) \oplus \cdots$, and since N is finitely generated this leads to a contradiction by the argument given in the proof of Theorem 2.28. □

Theorem 2.33. *The following statements are equivalent for an R-module A.*

(i) *Every singular R-module is A-injective.*

(ii) *A is the direct sum $A_1 \oplus A_2$ of a semisimple R-module A_1 and a nonsingular R-module A_2 such that A_2/N is semisimple for each essential submodule N of A_2.*

In this case A/K is semisimple for each essential submodule K of A and $A/(\operatorname{soc} A)$ is locally Noetherian.

Proof. (i) ⇒ (ii) Suppose that (i) holds. Let A_1 denote the singular submodule of A, i.e., $A_1 = \{a \in A : Ea = 0$ for some essential left ideal E of $R\}$. Then A_1 is singular, so that A_1 is A-injective and hence a direct summand of A by Lemma 2.24. Moreover, any submodule of A_1 is singular, and by hypothesis a direct summand of A, and hence also of A_1. Thus A_1 is semisimple. There exists $A_2 \leq A$ such that $A = A_1 \oplus A_2$. Clearly, A_2 is nonsingular. Let N be an essential submodule of A_2. Let $N \subseteq K \leq A_2$. Then K/N is singular, so that K/N is A-injective. By Lemma 2.24, K/N is a direct summand of A_2/N. It follows that A_2/N is semisimple. Thus (ii) holds.

(ii) ⇒ (i) Suppose that A is the direct sum $A_1 \oplus A_2$ of a semisimple submodule A_1 and a nonsingular submodule A_2 with the property that A_2/N is semisimple for each essential submodule N of A_2. Let $K \leq N \leq A_2$ with N/K singular. Since A_2 is nonsingular it follows that K is essential in N. Let L be a complement of K in A_2. Then $N \cap L = 0$ and $K \oplus L$ is an essential submodule of A_2. By hypothesis, $A_2/(K \oplus L)$ is semisimple and hence there exists a submodule P of A_2, containing $K \oplus L$, such that

$$A_2/(K \oplus L) = [(N \oplus L)/(K \oplus L)] \oplus [P/(K \oplus L)].$$

It is easy to check that $A_2/K = (N/K) \oplus (P/K)$. By Lemma 2.30, every singular R-module is A_2-injective. But, by Proposition 2.26, every singular R-module is A_1-injective. Hence every singular R-module is A-injective by Proposition 2.20.

Again supposing that (ii) holds, let Q be an essential submodule of A. Then $A_1 \subseteq Q$ and hence $Q = A_1 \oplus (A_2 \cap Q)$. Thus $A/Q \cong A_2/(A_2 \cap Q)$. But $A_2 \cap Q$ is an essential submodule of A_2, so that A/Q is semisimple. Let U be a finitely generated submodule of A. Let V be an essential submodule of U. Let W be a complement of V in A. Then $U \cap W = 0$ and $V \oplus W$ is an essential submodule of A. Thus $U/V \cong (U \oplus W)/(V \oplus W) \leq A/(V \oplus W)$. It follows that U/V is semisimple and hence Noetherian, since U is finitely generated. Thus U/V is Noetherian for each essential submodule V of U. It follows that U/S is Noetherian, where S denotes the socle of U, by [SHD90, Thm 2.4]. Thus $A/(\operatorname{soc} A)$ is locally Noetherian. □

Exercises

2.18. Prove the following statements.

(i) Any injective module is A-injective, for any module A.
(ii) Any R_R-injective module is injective.

2.19. Write out the proof of Lemma 2.27.

2.20. A ring R is called *(von Neumann) regular* if for each $a \in R$ there exists $x \in R$ such that $a = axa$. Prove that

(i) a ring R is regular if and only if each principal (finitely generated) right ideal of R is a direct summand of R,
(ii) any regular ring is nonsingular, and
(iii) if I, $J \leq_{\mathrm{d}} R$, where R is a regular ring, then $I \cap J \leq_{\mathrm{d}} R$.

2.21. Let A be a left R-module and $S = \operatorname{End}({}_RA)$. Prove the following statements.

(i) If ${}_RA$ is semisimple then S is regular.
(ii) If ${}_RA$ is nonsingular, quasi-injective then S is regular, self-injective.

2.22. Show that nonsingularity is not superfluous in Exercise 2.21 (ii).

2.23. Let A be a left R-module and left $S = \operatorname{End}({}_RA)$, $\triangle = \{f \in S\colon \ker f \leq_{\mathrm{e}} A\}$, and $\bigtriangledown = \{f \in S\colon f(A) \leq_{\mathrm{s}} A\}$. Prove

(i) ∇ and $\triangle$ are ideals of S.

(ii) If ${}_RA$ is quasi-injective, then $\triangle$ is the same the Jacobson radical of S and $S/\triangle$ is a regular ring.

2.24. Let R be a ring. Let $\underline{A}$, $\underline{N}$, $\underline{F}$, $\underline{G}$, and $\underline{Z}$ be classes of Artinian, Noetherian, finitely generated, have finite Goldie dimension and nonsingular modules over R, respectively. Determine whether each of these classes is S-closed or not?

2.25. Let A be a left R-module and $N \leq A$. Prove that

(i) If A is locally Noetherian, then N and A/N are both locally Noetherian.

(ii) If N is Noetherian and A/N is locally Noetherian, then A is locally Noetherian.

(ii) Show that the converse (even if A/N is Noetherian) of (i) is not true in general.

2.3 Lifting submodules

Let A and X be left R-modules. We deal with the class of submodules of A such that X is relative injective with respect to each member of it. A submodule N of A is called a *lifting submodule* for X in A if for any $\varphi \in \mathrm{Hom}_R(N, X)$ there exists $\theta \in \mathrm{Hom}_R(A, X)$ such that $\varphi = \theta|_N$. So we set

$$\mathrm{Lift}_X(A) = \{N : N \leq A \text{ and } N \text{ is a lifting submodule for } X \text{ in } A\}.$$

We examine properties of this new class of submodules. First observe that $0 \in \mathrm{Lift}_X(A)$ and $A \in \mathrm{Lift}_X(A)$. More generally, we have

Lemma 2.34. *Let N be a direct summand of the module A. Then $N \in \mathrm{Lift}_X(A)$.*

Proof. Let $A = N \oplus N'$ for some submodule N' of A. Suppose $\varphi \in \mathrm{Hom}_R(N, X)$. Define $\theta : A \to X$ by $\theta(n + n') = \varphi(n)$ $(n \in N,\ n' \in N')$. It is easy to see that $\theta \in \mathrm{Hom}_R(A, X)$ and $\varphi = \theta|_N$. □

Lemma 2.35. *The following statements are equivalent.*

(i) *X is A-injective.*

(ii) *Every submodule of A is a lifting submodule for X in A.*

(iii) *Every essential submodule of A is a lifting submodule for X in A.*

Proof. The implications (i) $\Rightarrow$ (ii) and (ii) $\Rightarrow$ (iii) are clear.

(iii) $\Rightarrow$ (i) Let N be a submodule of A. Let N' be a complement of N in A (see Proposition 2.2). Then $N \oplus N' \leq_e A$. Let $\varphi \in \mathrm{Hom}_R(N, X)$. By Lemma 2.34, there exists $\theta \in \mathrm{Hom}_R(N \oplus N', X)$ such that $\theta|_N = \varphi$. By (iii), there exists $\chi \in \mathrm{Hom}_R(A, X)$ such that $\chi|_{N \oplus N'} = \theta$. Thus $\chi|_N = \varphi$. It follows that X is A-injective. □

Lemma 2.36. *Let K, N be submodules of A such that $K \leq N$. Then*

(i) $K \in \mathrm{Lift}_X(N)$, $N \in \mathrm{Lift}_X(A)$ *implies that* $K \in \mathrm{Lift}_X(A)$.
(ii) $K \in \mathrm{Lift}_X(A)$ *implies that* $K \in \mathrm{Lift}_X(N)$.
(iii) $N \in \mathrm{Lift}_X(A)$ *implies that* $N/K \in \mathrm{Lift}_X(A/K)$.
(iv) $K \in \mathrm{Lift}_X(A)$, $N/K \in \mathrm{Lift}_X(A/K)$ *implies that* $N \in \mathrm{Lift}_X(A)$.

Proof. (i) and (ii) are clear.

(iii) Let $\varphi \in \mathrm{Hom}_R(N/K, X)$. Let $\pi : N \to N/K$ denote the canonical projection. Then $\varphi\pi : N \to X$ is a homomorphism. Since $N \in \mathrm{Lift}_X(A)$, there exists $\theta \in \mathrm{Hom}_R(A, X)$ such that $\theta(n) = \varphi\pi(n) = \varphi(n + K)$ for all $n \in N$. Define $\overline{\theta} : A/K \to X$ by $\overline{\theta}(a + K) = \theta(a)$ $(a \in A)$.

Suppose $a + K = a' + K$, where $a, a' \in A$. Then $a - a' \in K$ and hence $\varphi\pi(a - a') = 0$. Thus $\theta(a - a') = 0$, so that $\theta(a) = \theta(a')$. Hence $\overline{\theta}$ is well defined. Clearly, $\overline{\theta} \in \mathrm{Hom}_R(A/K, X)$. For any $n \in N$, $\overline{\theta}(n + K) = \theta(n) = \varphi(n + K)$. It follows that $N/K \in \mathrm{Lift}_X(A/K)$.

(iv) Let $\varphi \in \mathrm{Hom}_R(N, X)$. Then $\varphi|_K \in \mathrm{Hom}_R(K, X)$. There exists $\theta \in \mathrm{Hom}_R(A, X)$ such that $\varphi|_K = \theta|_K$. Define $\chi : N/K \to X$ by

$$\chi(n + K) = \varphi(n) - \theta(n) \quad (n \in N).$$

Note that χ is well defined and a homomorphism. There exists $\psi \in \mathrm{Hom}_R(A/K, X)$ such that $\psi|_{N/K} = \chi$. Let $\pi : A \to A/K$ denote the canonical projection. Let $\alpha = \psi\pi + \theta \in \mathrm{Hom}_R(A, X)$. For any $n \in N$,

$$\alpha(n) = \psi\pi(n) + \theta(n) = \psi(n + K) + \theta(n) = \chi(n + K) + \theta(n) = \varphi(n).$$

Thus $\alpha|_N = \varphi$, and so $N \in \mathrm{Lift}_X(A)$. □

Corollary 2.37. *For any* $N \in \mathrm{Lift}_X(A)$, $\mathrm{Lift}_X(N) = \{K \leq N : K \in \mathrm{Lift}_X(A)\}$.

Proof. Suppose $K \in \mathrm{Lift}_X(N)$. Then $K \leq N$ and, by Lemma 2.36 (i), $K \in \mathrm{Lift}_X(A)$. Therefore $\mathrm{Lift}_X(N) \subseteq \{K \leq N : K \in \mathrm{Lift}_X(A)\}$. Conversely, suppose $K \leq N$ and $K \in \mathrm{Lift}_X(A)$. By Lemma 2.36 (ii), $K \in \mathrm{Lift}_X(N)$. □

Let K, N be submodules of the module A such that $K \leq N$. Then $K \in \mathrm{Lift}_X(A)$ does not imply $N \in \mathrm{Lift}_X(A)$, as the following example illustrates.

Example 2.38. Let X be a non-injective left R-module. There exists $E \leq_e {}_RR$ such that $E \notin \mathrm{Lift}_X({}_RR)$. Let $A = R \oplus R$, $K = R \oplus 0$, $N = R \oplus E$. Then $K \in \mathrm{Lift}_X(A)$ by Lemma 2.34, but $N \notin \mathrm{Lift}_X(A)$, by Lemma 2.36.

Proposition 2.39. *Let $N, K \leq A$ such that $N + K$ and $N \cap K$ belong to* $\mathrm{Lift}_X(A)$. *Then both N and K belong to* $\mathrm{Lift}_X(A)$.

Proof. Let $\varphi \in \mathrm{Hom}_R(N, X)$. Then $\varphi|_{N\cap K} \in \mathrm{Hom}_R(N \cap K, X)$. There exists $\theta_1 \in \mathrm{Hom}_R(A, X)$ such that $\theta_1|_{N\cap K} = \varphi|_{N\cap K}$. Define $\chi : N + K \to X$ by

$$\chi(n + k) = \varphi(n) + \theta_1(k) \quad (n \in N, \quad k \in K).$$

Suppose $n, n' \in N$, $k, k' \in K$, and $n + k = n' + k'$. Then $n - n' = k' - k$, so that $k' - k \in N \cap K$. Hence $\theta_1(k') - \theta_1(k) = \theta_1(k' - k) = \varphi(k' - k) = \varphi(n - n') = \varphi(n) - \varphi(n')$, which implies $\varphi(n) + \theta_1(k) = \varphi(n') + \theta_1(k')$. Thus, χ is well defined. Clearly, $\chi \in \mathrm{Hom}_R(N + K, X)$, so by hypothesis, there exists $\theta \in \mathrm{Hom}_R(A, X)$ such that $\theta|_{N+K} = \chi$. For any $n \in N$,

$$\theta(n) = \chi(n) = \varphi(n).$$

Thus $\theta|_N = \varphi$. It follows that $N \in \mathrm{Lift}_X(A)$. Similarly $K \in \mathrm{Lift}_X(A)$. □

Corollary 2.40. *Let K, N be submodules of A.*

(i) *If $N \cap K = 0$ and $N \oplus K \in \mathrm{Lift}_X(A)$, then $N, K \in \mathrm{Lift}_X(A)$.*
(ii) *If $N + K = A$ and $N \cap K \in \mathrm{Lift}_X(A)$, then $N, K \in \mathrm{Lift}_X(A)$.*

Proof. Immediate by Proposition 2.39 □

Lemma 2.41. *Let $K \in \mathrm{Lift}_X(A)$, $N \leq A$. Suppose $N \cap K \in \mathrm{Lift}_X(K)$ and $(N + K)/K \in \mathrm{Lift}_X(A/K)$. Then $N \in \mathrm{Lift}_X(A)$.*

Proof. By Lemma 2.36 (i) and (iv), we deduce $N \cap K$ and $N + K$ both belong to $\mathrm{Lift}_X(A)$. Apply Proposition 2.39. □

Corollary 2.42. *Let $K \leq A$. Then X is A-injective if and only if* (i) *X is K-injective,* (ii) *X is (A/K)-injective, and* (iii) *$K \in \mathrm{Lift}_X(A)$.*

Proof. By Lemma 2.36 and Lemma 2.41. □

Theorem 2.43. *Let $X = \prod_{\lambda\in\Lambda} X_\lambda$. Then $\mathrm{Lift}_X(A) = \bigcap_{\lambda\in\Lambda} \mathrm{Lift}_{X_\lambda}(A)$, for any module A.*

Proof. Let $\lambda \in \Lambda$ and $Y = X_\lambda$. Let $N \in \mathrm{Lift}_X(A)$. Let $\varphi \in \mathrm{Hom}_R(N, Y)$. Let $i : Y \to X$ denote the inclusion mapping and $\pi : X \to Y$ the canonical projection. Then $i\varphi \in \mathrm{Hom}_R(N, X)$. By hypothesis, there exists $\theta \in \mathrm{Hom}_R(A, X)$ such that $\theta|_N = i\varphi$.

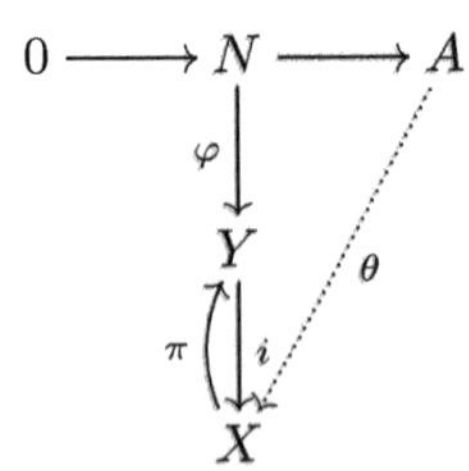

Now $\pi\theta \in \mathrm{Hom}_R(A, Y)$ and, for any $n \in N$, $\pi\theta(n) = \pi i\varphi(n) = \varphi(n)$. Thus $\varphi = \pi\theta|_N$. It follows that $N \in \mathrm{Lift}_Y(A)$. Hence $\mathrm{Lift}_X(A) \subseteq \mathrm{Lift}_Y(A)$. Therefore $\mathrm{Lift}_X(A) \subseteq \bigcap_{\lambda\in\Lambda} \mathrm{Lift}_{X_\lambda}(A)$.

Conversely, let $K \in \bigcap_{\lambda\in\Lambda} \mathrm{Lift}_{X_\lambda}(A)$. Let $\alpha \in \mathrm{Hom}_R(K, X)$. For each $\lambda \in \Lambda$, let $\pi_\lambda : X \to X_\lambda$ denote the canonical projection. Then $\pi_\lambda\alpha \in \mathrm{Hom}_R(K, X_\lambda)$, $\lambda \in \Lambda$. By hypothesis, for each $\lambda \in \Lambda$, there exists $\beta_\lambda \in \mathrm{Hom}_R(A, X_\lambda)$ such that $\beta_\lambda(k) = \pi_\lambda\alpha(k)$, $k \in K$. Define $\beta : A \to X$ by

$$\beta(a) = \{\beta_\lambda(a)\}_{\lambda\in\Lambda} \quad (a \in A).$$

For each $k \in K$, $\beta(k) = \alpha(k)$. Thus $K \in \mathrm{Lift}_X(A)$. □

Corollary 2.44. *Let $X = \prod_{\lambda\in\Lambda} X_\lambda$. Then X is A-injective if and only if X_λ is A-injective for all $\lambda \in \Lambda$.*

Proof. By Lemma 2.35 and Theorem 2.43. □

Proposition 2.45. *Consider the following conditions for a any submodule N of a module A:*

(i) *$\theta(A) \leq X$ for any $\theta \in \mathrm{Hom}_R(A, E(X))$ with $\theta(N) \leq X$.*

(ii) *$N \in \mathrm{Lift}_X(A)$.*

(iii) *$\theta(A) \leq X$ for any $\theta \in \mathrm{Hom}_R(A, E(X))$ with $\theta(N) \leq X$ and $\theta^{-1}(X) \in \mathrm{Lift}_X(A)$.*

Then (i) $\Rightarrow$ (ii) $\Rightarrow$ (iii).

Proof. (i) $\Rightarrow$ (ii) Let $\varphi \in \mathrm{Hom}_R(N, X)$. Then there exists $\theta \in \mathrm{Hom}_R(A, E(X))$ such that $\theta|_N = i\varphi$, where $i : X \to E(X)$ is the inclusion mapping. Thus $\theta(N) \leq X$. By hypothesis, $\theta(A) \leq X$, and hence $\theta \in \mathrm{Hom}_R(A, X)$. It follows that $N \in \mathrm{Lift}_X(A)$.

(ii) $\Rightarrow$ (iii) Suppose (ii) holds. Let $\theta \in \mathrm{Hom}_R(A, E(X))$ such that $N \leq \theta^{-1}(X) \in \mathrm{Lift}_X(A)$. There exists $\theta' \in \mathrm{Hom}_R(A, X)$ such that $\theta'(k) = \theta(k)$ ($k \in \theta^{-1}(X)$). Consider $\theta - \theta' : A \to E(X)$. If $(\theta - \theta')(A) \neq 0$, then $(\theta - \theta')(A) \cap X \neq 0$, and hence there exist $0 \neq x \in X$, $a \in A$ such that $x = (\theta - \theta')(a) = \theta(a) - \theta'(a)$. Thus $\theta(a) = x + \theta'(a) \in X$ and hence $a \in \theta^{-1}(X)$. In this case, $\theta'(a) = \theta(a)$, so that $x = 0$, a contradiction. We conclude that $(\theta - \theta')(A) = 0$, and so $\theta(A) = \theta'(A) \leq X$. □

We support previous result with the following two examples. The first example shows that (ii) does not imply (i) in the Proposition 2.45.

Example 2.46. Let R be the ring $\mathbb{Z}$ of integers. Take $X = A = \mathbb{Z}$ and $N = 0$ in Proposition 2.45. Then $N \in \mathrm{Lift}_X(A)$. Let $0 \neq a \in \mathbb{Z}$ and define $\theta : \mathbb{Z} \to \mathbb{Q}$ by $\theta(n) = \frac{n}{a}$ ($n \in \mathbb{Z}$). Then $\theta(N) \leq X$, but $\theta(A) \nleq X$. Hence (ii)$\not\Rightarrow$(i).

Example 2.47. Let $R = \mathbb{Z}$, $X = \mathbb{Z}/\mathbb{Z}p$, where p be any prime integer, $A = \mathbb{Q}$ and $N = \mathbb{Z}$ in Proposition 2.45.

Assume $N \in \mathrm{Lift}_X(A)$. Let $\pi : \mathbb{Z} \to \mathbb{Z}/\mathbb{Z}p$ denote the canonical epimorphism, defined by

$$\pi(n) = n + \mathbb{Z}p \quad (n \in \mathbb{Z}).$$

Then there exists a homomorphism $\alpha : \mathbb{Q} \to \mathbb{Z}/\mathbb{Z}p$ such that $\alpha|_{\mathbb{Z}} = \pi$:

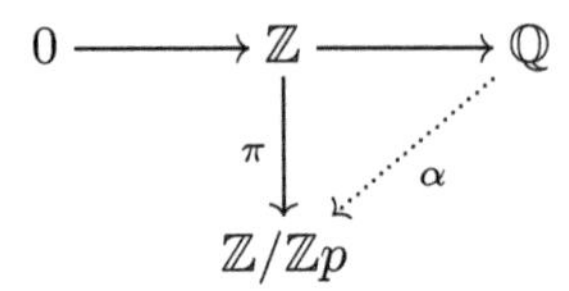

Now $\alpha(1/p) = x + \mathbb{Z}p$ for some $x \in \mathbb{Z}$. Thus

$$p\alpha(1/p) = \alpha(1) = \pi(1) = 1 + \mathbb{Z}p.$$

It follows that $px + \mathbb{Z}p = 1 + \mathbb{Z}p$, and hence $1 \equiv 0 \pmod{p}$, a contradiction. Thus $N \notin \mathrm{Lift}_X(A)$.

Observe that we obtained some results contained in § 2.2 as corollaries in this section (see Corollary 2.42 and Corollary 2.44). Now we consider a collection of submodules of a module to get some general results on certain type of modules in terms of the class of lifting submodules.

Theorem 2.48. *Let $\underline{A}$ be a non-empty collection of submodules of A. Then the following statements are equivalent.*

(i) *If $N \in \underline{A}$, then $N \leq_{\mathrm{d}} A$.*
(ii) *$\underline{A} \subseteq \mathrm{Lift}_X(A)$ for all left R-modules X.*
(iii) *$\underline{A} \subseteq \mathrm{Lift}_X(A)$ for all $X \in \underline{A}$.*

Proof. (i) $\Rightarrow$ (ii) By Lemma 2.34.

(ii) $\Rightarrow$ (iii) Obvious.

(iii) $\Rightarrow$ (i) Let $B \in \underline{A}$. Consider the identity mapping $i_B : B \to B$. Since $B \in \mathrm{Lift}_B(A)$ by (iii), there exists $\theta \in \mathrm{Hom}_R(A, B)$ such that $\theta(a) = a$, where $a \in A$. It can easily be checked that $A = B \oplus (\ker\theta)$. Thus $B \leq_{\mathrm{d}} A$. □

Corollary 2.49. *The following statements are equivalent for a module A.*

(i) *A is semisimple.*
(ii) *Every right R-module X is A-injective.*
(iii) *Every submodule of A is A-injective.*

Proof. Apply Theorem 2.48 to $\underline{A} = \{N : N \leq A\}$ and use Lemma 2.35. □

Corollary 2.50. *The following statements are equivalent for a module A.*

(i) *If $N \leq_{\mathrm{c}} A$, then $N \leq_{\mathrm{d}} A$.*
(ii) *If $N \leq_{\mathrm{c}} A$, then $N \in \mathrm{Lift}_X(A)$ for all left R-modules X.*
(iii) *If $N \leq_{\mathrm{c}} A$, then $N \in \mathrm{Lift}_X(A)$ for all $X \leq_{\mathrm{c}} A$.*

Proof. Apply Theorem 2.48 to $\underline{A} = \{N : N \leq_{\mathrm{c}} A\}$. □

Notice that the first part of Corollary 2.50 (the converse always holds) introduces an interesting class of modules and rings as well as the main concept in this book. Throughout this chapter we work with this kind of modules without having them. Incidentally, we will define and present details on these modules and rings in next chapter and in the sequel.

The equivalent conditions stated in Corollary 2.50 yield, in particular that for a module A and any complement submodule N in A, we have that $N \in \mathrm{Lift}_A(A)$. However, one cannot expect that $N \in \mathrm{Lift}_A(A)$ will imply N is a direct summand of A, as the following example illustrates.

Example 2.51. Let p be any prime integer and let R be the local ring $\mathbb{Z}_{(p)}$. Let A denote the $\mathbb{Z}$-module $(\mathbb{Z}/\mathbb{Z}p) \oplus \mathbb{Q}$. Then

(i) A is an R-module.
(ii) $K \leq_{\mathrm{c}} A$ if and only if $K \leq_{\mathrm{d}} A$ or $K = R(1+\mathbb{Z}p, q)$ for some non-zero element q in Q.
(iii) There exists a complement submodule in A which is not a direct summand of A.
(iv) If $K \leq_{\mathrm{c}} A$ then $K \in \mathrm{Lift}_A(A)$.

Proof. (i) Let $A_1 = (\mathbb{Z}/\mathbb{Z}p) \oplus 0$ and $A_2 = 0 \oplus \mathbb{Q}$, so that A is the direct sum $A_1 \oplus A_2$ of its submodules A_1 and A_2. The ring R is the subring of $\mathbb{Q}$ consisting of all rational numbers s/t such that $s, t \in \mathbb{Z}$, $t \neq 0$, and t is coprime to p. Note first that for any element $a \in A$ and any $s, t \in \mathbb{Z}$ such that p does not divide t, there exists a unique element $a' \in A$ such that $ta' = sa$; we denote a' by $(s/t)a$. In this way A is an R-module.

(ii) Let $q \in \mathbb{Q}$ and $K = R(1+\mathbb{Z}p, q)$. We show first that $K \leq_{\mathrm{c}} {}_{\mathbb{Z}}A$. Note that K is a uniform submodule of A. Suppose that N is a submodule of A such that $K \leq_{\mathrm{e}} N$. Let $x \in N$. Then $U = \mathbb{Z}x + \mathbb{Z}(1+\mathbb{Z}p, q)$ is a finitely generated uniform $\mathbb{Z}$-module, and hence U is cyclic (see Proposition 1.68). Suppose that $U = \mathbb{Z}(a+\mathbb{Z}p, b)$, where $a \in \mathbb{Z}$, $b \in \mathbb{Q}$. There exists $n \in \mathbb{Z}$ such that $(1 + \mathbb{Z}p, q) = n(a + \mathbb{Z}p, b)$. Note that $1 - na \in \mathbb{Z}p$, and hence n is coprime to p, and $(a + \mathbb{Z}p, b) \in R(1 + \mathbb{Z}p, q) = K$. Thus $x \in K$. It follows that $K = N$, and so $K \leq_{\mathrm{c}} A$.

Let $L \leq_{\mathrm{c}} A$ and suppose $L \neq A$. Note that A has uniform dimension 2, and hence L is uniform by Exercise 1.120 (see Proposition 2.18). We shall show first that L is an R-submodule of A. Let $L' = \{a \in A : ta \in L$ for some $t \in \mathbb{Z}$, t coprime to $p\}$. Then L' is a submodule of A containing L, in fact $L' = RL$. If

$0 \neq a \in L'$, then $ta \in L$ for some $t \in \mathbb{Z}$, coprime to p, and hence $ta \neq 0$. It follows that $L \leq_{\rm e} L'$. Thus $L = L'$, and L is an R-submodule of A.

Next we show that $L = 0, A, A_1, A_2$, or $R(1+\mathbb{Z}p, q)$ for some $q \in \mathbb{Q}$. Suppose that $L \neq 0, A, A_1$, and A_2. Note that A_1 and A_2 are both uniform, so that L is not contained in A_1 or A_2. Thus $(c + \mathbb{Z}p, d) \in L$ for some $c \in \mathbb{Z}$, coprime to p and $0 \neq d \in \mathbb{Q}$. Without loss of generality we can suppose that $c = 1$. Because L is an R-submodule of A, $R(1 + \mathbb{Z}p, d) \subseteq L$. But $R(1 + \mathbb{Z}p, d) \leq_{\rm c} A$, and so $L = R(1 + \mathbb{Z}p, d)$. This completes the proof of (ii).

(iii) Note that $K = R(1+\mathbb{Z}p, 1)$ is a complement in A. Suppose $K \leq_d A$. Then $A = K \oplus L$ for some $L \leq A$. Let $(a+\mathbb{Z}p, b) \in L$, where $a \in \mathbb{Z}$, $b = m/n \in \mathbb{Q}$. Then $p(a + \mathbb{Z}p, b) = (0 + \mathbb{Z}p, pm/n) \in L$. Therefore, $n(0 + \mathbb{Z}p, pm/n) = (0 + \mathbb{Z}p, pm) = (0+\mathbb{Z}p, 0)$, because $K \cap L = 0$. Hence $npb = pm = 0$ gives $b = 0$, so that if $x \in L$, then $x = (y + \mathbb{Z}p, 0)$ where $y \in \mathbb{Z}$. Thus $L \leq A_1$ which is simple, so $L = A_1$. Now $A = K \oplus A_1$. Hence

$$K \cong A/A_1 \cong A_2 \cong \mathbb{Q} \cong p\mathbb{Q}.$$

Then there exists an element $(c+\mathbb{Z}p, d) \in K$ such that $(1+\mathbb{Z}p, 1) = p(c+\mathbb{Z}p, d) = (0 + \mathbb{Z}p, pd)$ $(c \in \mathbb{Z}, d \in \mathbb{Q})$. Thus $1 \in \mathbb{Z}p$, a contradiction. We conclude that K is not a direct summand of A.

(iv) To show that if $K \leq_{\rm c} A$ then $K \in \mathrm{Lift}_A(A)$, it is sufficient to prove that for any non-zero $q \in \mathbb{Q}$ and any homomorphism $\varphi : R(1 + \mathbb{Z}p, q) \to A$, φ can be lifted to an endomorphism θ of A. Let $K = R(1 + \mathbb{Z}p, q)$. Suppose that $\varphi(1 + \mathbb{Z}p, q) = (a + \mathbb{Z}p, b)$ for some $a \in \mathbb{Z}, b \in \mathbb{Q}$. Define a mapping $\theta : A \to A$ by

$$\theta(c + \mathbb{Z}p, d) = (ca + \mathbb{Z}p, db/q) \quad (c \in \mathbb{Z}, d \in \mathbb{Q}).$$

It is clear that θ is well defined. It can be checked that $\theta : A \to A$ is a homomorphism and that φ is the restriction of θ to K. Thus $K \in \mathrm{Lift}_A(A)$. □

Example 2.52.

(i) Let A be any finitely generated torsion-free group (i.e., $\mathbb{Z}$-module). If $N \leq_{\rm c} A$, then $N \leq_{\rm d} A$.

(ii) Let p be any prime integer and ${}_{\mathbb{Z}}A = \mathbb{Z}/\mathbb{Z}p \oplus \mathbb{Z}$. Then there exists $L \leq_{\rm c} A$ such that $L \notin \mathrm{Lift}_A(A)$.

Proof. (i) Let $N \leq_{\rm c} A$. Let $N \leq K \leq A$ such that K/N is the torsion submodule of A/N. Then A/K is finitely generated and torsion-free, hence free, so $K \leq_{\rm d} A$. Also, $N \leq_{\rm e} K$ because K is torsion-free and K/N torsion (see Exercise 2.8).

(ii) Let $A_1 = \mathbb{Z}/\mathbb{Z}p \oplus 0$ and $A_2 = 0 \oplus \mathbb{Z}$. Then ${}_{\mathbb{Z}}A = A_1 \oplus A_2$. Let L denote the cyclic submodule $\mathbb{Z}(1 + \mathbb{Z}p, p)$ of A. Since, as an Abelian group, L is infinite cyclic, it follows that L is a uniform $\mathbb{Z}$-module. Suppose that K is a submodule of A and $L \leq_{\rm e} K$. Then K is uniform, and hence cyclic, because K is a finitely generated Abelian group (see Proposition 1.68). There exist elements $a, b \in \mathbb{Z}$ such that $K = \mathbb{Z}(a + \mathbb{Z}p, b)$. Now there exists $n \in \mathbb{Z}$ such that

$$(1 + \mathbb{Z}p, p) = n(a + \mathbb{Z}p, b),$$

and hence $1 - na \in \mathbb{Z}p$ and $p = nb$. It follows that $n = 1$ or -1, and hence $L = K$. Thus $L \leq_c A$.

We claim $L \notin \mathrm{Lift}_A(A)$. Suppose not. Define $\varphi \in \mathrm{Hom}_{\mathbb{Z}}(L, A)$ by $\varphi(1 + \mathbb{Z}p, p) = (0, 1)$. There exists $\theta \in \mathrm{Hom}_{\mathbb{Z}}(A, A)$ such that $\theta|_L = \varphi$. Suppose $\theta(1 + \mathbb{Z}p, 0) = (a_1 + \mathbb{Z}p, b_1)$, $\theta(0, 1) = (a_2 + \mathbb{Z}p, b_2)$ for some $a_1, a_2, b_1, b_2 \in \mathbb{Z}$. Then $p(a_1 + \mathbb{Z}p, b_1) = 0$ implies $b_1 = 0$. Hence

$$(0, 1) = \varphi(1 + \mathbb{Z}p, p) = \theta(1 + \mathbb{Z}p, p) = (a_1 + \mathbb{Z}p, 0) + p(a_2 + \mathbb{Z}p, b_2),$$

and this implies $1 = pb_2$, a contradiction. Thus $L \notin \mathrm{Lift}_A(A)$. □

We complete this section with the next proposition, which is essentially related to the Example 2.52 (ii).

Proposition 2.53. *Let ${}_{\mathbb{Z}}A = A_1 \oplus A_2$, where A_1 is torsion and A_2 is infinite cyclic. If every complement submodule in A is contained in* $\mathrm{Lift}_A(A)$, *then $A_1 = pA_1$ for each prime integer p.*

Proof. Let $A_2 = \mathbb{Z}a_2$ for some $0 \neq a_2 \in A_2$. Suppose $A_1 \neq pA_1$ for some prime p. Let $a_1 \in A_1$, $a_1 \notin pA_1$. Let $K = \mathbb{Z}(a_1, pa_2)$. Suppose $K \leq_e L$ for some $L \leq A$. Then for any $n \in \mathbb{Z}$, $n(a_1, pa_2) = (na_1, npa_2) = (0, 0)$ if and only if $na_1 = 0$, $npa_2 = 0$ if and only if $n = 0$ (A_2 is infinite cyclic). Therefore K is infinite cyclic, and hence K is a uniform $\mathbb{Z}$-module. Thus L is a uniform $\mathbb{Z}$-module. Let $x \in L$ and $a = (a_1, pa_2)$. Then $K + \mathbb{Z}x = \mathbb{Z}a + \mathbb{Z}x$ is finitely generated, so that $K + \mathbb{Z}x \leq L$, and is a direct sum of cyclic modules. But $K + \mathbb{Z}x$ is uniform, hence $K + \mathbb{Z}x$ is cyclic. Then $\mathbb{Z}a \subseteq K + \mathbb{Z}x = \mathbb{Z}y$ for some $y \in A$. Suppose $y = (a_1', ka_2)$ for some $a_1' \in A_1$ and $k \in \mathbb{Z}$. Then $a = sy$ for some $s \in \mathbb{Z}$. Hence $(a_1, pa_2) = s(a_1', ka_2)$, which gives $a_1 = sa_1'$, $pa_2 = ska_2$. Since A_2 is infinite cyclic, $s = \pm 1$ or $k = \pm 1$. If $k = \pm 1$, then $s = \pm p$, so that $a_1 = \pm pa_1' \in pA_1$, a contradiction. Thus $s = \pm 1$. Therefore $y \in \mathbb{Z}a$, and hence $x \in \mathbb{Z}y \subseteq \mathbb{Z}a$, i.e., $L \subseteq \mathbb{Z}a = K$. Thus $K = L$, so $K \leq_c A$.

Now define a homomorphism $\varphi : K \to A$ by $\varphi(a_1, pa_2) = (0, a_2)$. Suppose that φ can be lifted to $\theta : A \to A$. Then $\theta(a_1, 0) = (u, 0)$ for some $u \in A_1$ and $\theta(0, a_2) = (v, ta_2)$ for some $v \in A_1$, $t \in \mathbb{Z}$. Hence $(0, a_2) = \varphi(a_1, pa_2) = \theta(a_1, pa_2) = \theta(a_1, 0) + p\theta(0, a_2) = (u, 0) + p(v, ta_2)$. Then we obtain $0 = u + pv$, $a_2 = pta_2$, so that $1 = pt$, a contradiction. We see that φ cannot be lifted. It follows that $A_1 = pA_1$ for each prime p. □

Exercises

2.26. Determine $\mathrm{Lift}_{\mathbb{Z}}(\mathbb{Z})$, $\mathrm{Lift}_{\mathbb{Z}}(\mathbb{Q})$, $\mathrm{Lift}_{\mathbb{Z}}(\mathbb{Q}/\mathbb{Z})$.

2.27. Let p be any prime integer. Compute $\mathrm{Lift}_{\mathbb{Z}}(\mathbb{Z}(p^\infty))$, $\mathrm{Lift}_{\mathbb{Z}}((\mathbb{Z}/\mathbb{Z}p) \oplus (\mathbb{Z}/\mathbb{Z}p^2))$, and $\mathrm{Lift}_{\mathbb{Z}}((\mathbb{Z}/\mathbb{Z}p) \oplus (\mathbb{Z}/\mathbb{Z}p^3))$.

2.28. Provide an example such that B, $C \leq {}_RA$ with $B \cong C$ and $B \in \mathrm{Lift}_X(A)$, but $C \notin \mathrm{Lift}_X(A)$, i.e., $\mathrm{Lift}_X(A)$ is not a class.

2.29. Let ${}_RA$ be a nonsingular module and $N \leq_e A$. Assume that $N \in \mathrm{Lift}_X(N + Ra)$ for all $a \in A$. Prove that $N \in \mathrm{Lift}_X(A)$.

2.30. Show that we cannot remove nonsingularity in Exercise 2.29.

2.4 Ejectivity and elifting submodules

In this section we introduce a concept which generalizes the notion of A-injectivity of a module X. Also, we define elifting submodules in the same manner as the lifting submodules of a module and obtain a number of properties of elifting submodules which are analogous to results proved in the preceding section for lifting submodules.

Let A and X be left R-modules. We say that X is A-*ejective* if for each $K \leq A$ and each homomorphism $\varphi : K \to X$ there exist a homomorphism $\theta : A \to X$ and a $E \leq_e K$ such that $\theta(x) = \varphi(x)$ for all $x \in E$, i.e., $\theta|_E = \varphi|_E$. It is clear that if X is A-injective, then X is A-ejective. If X is A-ejective for all left R-modules A, then we say that X is *ejective*.

Let ρ be a *left exact radical* in the category of left R-modules (see [Ste75]). Amongst other properties this means:

(i) $N \leq A$ implies that $\rho(N) = \rho(A) \cap N$; (left exact)
(ii) $\rho(\rho(A)) = \rho(A)$; (idempotent)
(iii) $\rho(X \oplus Y) = \rho(X) \oplus \rho(Y)$ (for any preradical).

For example, the Goldie torsion theory $\rho = Z_2$ satisfies the aforementioned properties (see [Ste75, p. 148]). Given a preradical ρ, there is a smallest radical $\overline{\rho}$ larger than ρ ([Ste75, p. 137]). If ρ is idempotent or left exact, then so is $\overline{\rho}$. If ρ is left exact preradical, then $\rho(A)$ is an essential submodule of $\overline{\rho}(A)$. If s denotes the socle, then s is a left exact preradical, so $\overline{s}$ is a left exact radical and $\overline{s}(A)$ has $s(A)$ as an essential submodule. Since s and $\overline{s}$ have the same torsion-free class, $s(A) = 0$ if and only if $\overline{s}(A) = 0$.

Proposition 2.54. *Let ρ be a left exact preradical and $A = A_1 \oplus A_2$, where $\rho(A) \leq_e A_2$. Then A_1 is A_2-ejective.*

Proof. Let $K \leq A_2$, $\varphi : K \to A_1$ be a homomorphism, and $E = K \cap \rho(A)$. Then $E \leq_e K$. Let $\theta : A_2 \to A_1$ be the zero homomorphism. Then $\theta(x) = \varphi(x)$, for all $x \in E$, so A_1 is A_2-ejective. □

Lemma 2.55. *If A is R-ejective and $Z({}_RA) = 0$, then ${}_RA$ is injective.*

Proof. By adaptation of the proof of the Baer criterion. □

Compare the following result with Proposition 2.19.

Proposition 2.56. *Let A and X be R-modules. Then X is A-ejective if and only if there exists $E \leq_e A$ such that X is E-ejective and for any R-homomorphism $\varphi : E \to X$ there exists $K \leq_e E$ and an R-homomorphism $\theta : A \to X$ such that $\theta|_K = \varphi|_K$.*

Proof. ($\Rightarrow$) For this implication, take $E = A$.

($\Leftarrow$) Let $B \leq A$ and $\varphi : B \to X$ an R-homomorphism. Let $B_1 = B \cap E$. Since X is E-ejective, there exist $B_2 \leq_e B_1$ and $\varphi_1 : E \to X$ such that $\varphi|_{B_2} = \varphi_1|_{B_2}$. Now there exist $K \leq_e E$ and an R-homomorphism $\varphi_2 : A \to X$ such that $\varphi_2|_K = \varphi_1|_K$. Let $B_3 = B_2 \cap K$. Then $B_3 \leq_e B$ and $\varphi|_{B_3} = \varphi_2|_{B_3}$. Thus X is A-ejective. □

The next corollary gathers various examples for the ejectivity concept. First recall that a module A is called *(P-injective) mininjective* if every (principal) simple right ideal of R belongs to $\mathrm{Lift}_A(R)$. Recall also that a ring R is called *right simple injective* if for every right ideal K of R, every R-homomorphism $\gamma : K \to R$ with $\gamma(K)$ simple lifts to R [NY03].

Corollary 2.57. *Let X and A be R-modules.*

(i) *If X is R-ejective, then X is mininjective.*

(ii) *Assume that R_R is an essential extension of a minimal right ideal. Then X is R-ejective if and only if X is mininjective.*

(iii) *If X is P-injective and X is Rx-ejective for some $x \in R$, where $Rx \leq_e {}_RR$, then X is R-ejective.*

(iv) *If R_R is uniform with a minimal right ideal I such that $I = xR = Rx$ for some $x \in I$, then R is R-ejective. In particular, every subdirectly irreducible commutative ring R is R-ejective.*

(v) *If $\mathrm{soc}(R_R) \leq_e R$, $\mathrm{soc}(R_R)$ is finitely generated, and R is right simple injective, then R_R is R-ejective.*

(vi) *If $\mathrm{soc}(A) \leq_e A$, then X is A-ejective if and only if for each homomorphism $\varphi : \mathrm{soc}(A) \to X$ there exists a homomorphism $\theta : A \to X$ such that $\theta|_{\mathrm{soc}(A)} = \varphi$ (i.e., $\mathrm{soc}(A) \in \mathrm{Lift}_X(A)$).*

Proof. (i) is an immediate consequence of the definitions of X being R-ejective and X being mininjective.

(ii) follows from part (i) and Proposition 2.56.

(iii) follows from Proposition 2.56 by taking E in Proposition 2.56 to be Rx.

(iv) Since I is a minimal right ideal, R is I-ejective. Let $\varphi : I \to R$ be an R-homomorphism. Then $\varphi(I) \subseteq I$. So there exist $s, t \in R$ such that $\varphi(x) = xs = tx$. Define $\theta : R \to R$ by $\theta(r) = tr$ for all $r \in R$. Then $\theta(xr) = txr = \varphi(xr)$, for all $r \in R$. So $\theta|_I = \varphi|_I$. By Proposition 2.56, R is R-ejective.

(v) follows from Proposition 2.26, Proposition 2.56 and [NY03, Lem. 6.1].

(vi) follows from Proposition 2.26, Proposition 2.56. □

The following example illustrates Corollary 2.57, which also shows that relative ejectivity is different from relative injectivity.

Example 2.58.

(i) Let $R = \begin{bmatrix} D & D \\ 0 & D \end{bmatrix}$, where D is a division ring and $S = \begin{bmatrix} 0 & D \\ 0 & D \end{bmatrix}$. Then $S_R \leq_e R_R$. By Proposition 2.26, R_R is S-injective and hence R_R is S-ejective. But R_R is not R-ejective, since R_R is not mininjective (see [NY03, p. 37]).

(ii) There are subdirectly irreducible commutative rings which are not self-injective. These rings illustrate Corollary 2.57 (iv) (see [NY03]).

(iii) Examples illustrating Corollary 2.57 (v) are provided in [NY03, p. 133, 154].

The following result generalizes [DHSW94, Lem. 7.5] and will be very useful for obtaining structural results for certain type of modules in further sections of this book.

Theorem 2.59. *Let A_1 and A_2 be modules such that $A = A_1 \oplus A_2$. Then A_1 is A_2-ejective if and only if for every $K \leq A$ such that $K \cap A_1 = 0$, there exists $A_3 \leq A$ such that $A = A_1 \oplus A_3$ and $K \cap A_3 \leq_e K$.*

Proof. ($\Rightarrow$) Assume that A_1 is A_2-ejective. Let $\pi_i : A \to A_i$ ($i = 1, 2$) denote the canonical projections. Assume $K \leq A$ such that $K \cap A_1 = 0$. Observe that $\pi_2 : K \to A_2$ is an injection. Let $\overline{K} = \pi_2(K)$. Since A_1 is A_2-ejective, there exist $E \leq_e \overline{K}$ and a homomorphism $\theta : A_2 \to A_1$ such that $\theta|_E = \pi_1\pi_2^{-1}|_E$. Define $A_3 = \{\theta(y) + y : y \in A_2\}$. Then for $a \in A$ there exist $a_i \in A_i$ such that $a = a_1 + a_2 = (a_1 - \theta(a_2)) + (\theta(a_2) + a_2) \in A_1 + A_3$. Suppose $y \in A_1 \cap A_3$. Then there exists $y_i \in A_i$ such that $y = y_1 = \theta(y_2) + y_2$. Hence $y_1 - \varphi(y_2) = y_2 \in A_1 \cap A_2 = 0$. So $A_1 \cap A_3 = 0$. Thus $A = A_1 \oplus A_3$. Now let $0 \neq k \in K$. Then $k = \pi_1(k) + \pi_2(k)$. Recall that $\pi_2(k) \neq 0$ because $K \cap A_1 = 0$. So there is an $r \in R$ such that $0 \neq r\pi_2(k) = \pi_2(rk) \in E$. Hence $0 \neq rk = \pi_1(rk) + \pi_2(rk)$. But $\pi_1(rk) = \theta(\pi_2(rk))$, so $0 \neq rk = \theta(\pi_2(rk)) + \pi_2(rk) \in K \cap A_3$. Therefore $K \cap A_3 \leq_e K$.

($\Leftarrow$) Assume that for every $K \leq A$ such that $K \cap A_1 = 0$, there exists $A_3 \leq A$ such that $A = A_1 \oplus A_3$ and $K \cap A_3 \leq K$. Let $L \leq A_2$ and $\varphi : L \to A_1$ be a homomorphism. Put $H = \{-\varphi(x) + x : x \in L\}$. Then $H \leq A$ and $H \cap A_1 = 0$. By assumption, there exists $H' \leq A$ such that $A = A_1 \oplus H'$, and $H' \cap H \leq_e H$. Let $K = H' \cap H \cap L$. There exists $C \leq L$ such that $K \cap C = 0$ and $K \oplus C \leq_e L$. Let $B = \{b \in C : -\varphi(b) + b \in H'\}$. Observe that $B \leq C$. We claim that $B \leq_e C$. So let $0 \neq c \in C$. Then $-\varphi(c) + c \in H$. If $-\varphi(c) + c = 0$, then $c \in A_1 \cap L = 0$, a contradiction. Thus $-\varphi(c) + c \neq 0$. Then there exists $r \in R$ such that $0 \neq r(-\varphi(c) + c) = -\varphi(rc) + rc \in H' \cap H$. So $0 \neq rc \in B$. Hence, $B \leq_e C$.

Observe that $K \oplus B \leq_e L$. Now let $k + b \in K \oplus B$, where $k \in K$ and $b \in B$. Let $\pi : A \to A_1$ be the projection along H', i.e., $\ker \pi = H'$. Then $\pi(k + b) = \pi(b) = \pi(\varphi(b) - \varphi(b) + b) = \pi(\varphi(b)) + \pi(-\varphi(b) + b) = \pi(\varphi(b)) = \varphi(b)$. Recall that $k \in H \cap L$. Then there is a $y \in L$ such that $k = -\varphi(y) + y$. Hence

$\varphi(y) = y - k \in L \cap A_1 = 0$. So $y = k$ and $0 = \varphi(y) = \varphi(k)$. We conclude that $\pi(k+b) = \varphi(b) = \varphi(k+b)$, and so that A_1 is A_2-ejective. □

Corollary 2.60. *Let A_1 and A_2 be modules with $Z(A_1) = 0$ and $A = A_1 \oplus A_2$. Then A_1 is A_2-injective if and only if A_1 is A_2-ejective.*

Proof. ($\Rightarrow$) Clear.

($\Leftarrow$) Assume A_1 is A_2-ejective. Let $K \le A$ such that $K \cap A_1 = 0$. By Theorem 2.59, there exists $A_3 \le A$ such that $A = A_1 \oplus A_3$ and $K \cap A_3 \le_e K$. Let $0 \ne k \in K$. Then $0 \ne k = \pi_1(k) + \pi_3(k)$, where π_i are the canonical projections on A_1 and A_3. There exists $L \le_e R$ such that $Lk \subseteq K \cap A_3$ (Exercise 2.3). Hence $L\pi_1(k) = 0$. Since A_1 is nonsingular, $\pi_1(k) = 0$. Thus $K = K \cap A_3 \subseteq A_3$. By [DHSW94, Lem. 7.5], A_1 is A_2-injective. □

Proposition 2.61. *Let X and A be left R-modules. Then*

(i) *${}_RX$ is ${}_RA$-ejective for all ${}_RA \in R$-Mod if and only if ${}_RX$ is injective.*
(ii) *Assume $Z({}_RA) = 0$. Then ${}_RA$ is ${}_RR$-ejective if and only if ${}_RA$ is injective.*
(iii) *If ${}_RA$ is ${}_RR$-ejective and ${}_RA = {}_RD \oplus {}_RY$, then ${}_RD$ is ${}_RR$-ejective.*
(iv) *If $A/Z_2({}_RA)$ is ${}_RR$-ejective, then $A/Z_2({}_RA)$ is injective both as an R-module and $R/Z_2({}_RR)$-module. In particular, if ${}_RA$ is ${}_RR$-ejective and $A = Z_2({}_RA) \oplus B$, then $A/Z_2({}_RA)$ is injective both as an R-module and $R/Z_2({}_RR)$-module.*
(v) *Assume that ${}_RA = Z_2({}_RA)$ and $Z({}_RR)) = 0$. Then ${}_RA$ is ${}_RR$-ejective.*
(vi) *Assume that $\operatorname{soc}({}_RA) \le_e {}_RA$ and $\operatorname{soc}({}_RR) = 0$. Then ${}_RA$ is ${}_RR$-ejective.*

Proof. (i) Assume ${}_RX$ is ${}_RA$-ejective for all ${}_RA \in R$-Mod. Let $\varphi \in \operatorname{End}({}_RX)$. Then there exist ${}_RY \le_e {}_RX$ and a homomorphism $\theta : E(X) \to X$ such that $\theta(x) = \varphi(x)$ for all $x \in Y$. Since ${}_RY \le_e E(X)$, then θ is a monomorphism. Hence $\theta(E(X))$ is a direct summand of ${}_RX$. But $Y \subseteq \theta(E(X))$. Then $\theta(E(X)) = {}_RX$. Therefore ${}_RX$ is injective. The converse is obvious.

(ii) By part (i), if ${}_RA$ is injective, then ${}_RA$ is ${}_RR$-ejective. So assume ${}_RA$ is ${}_RR$-ejective. Let ${}_RI \le {}_RR$ and $\varphi : I \to A$ be an R-homomorphism. There exists ${}_RJ \le_e {}_RI$ and $\theta : R \to A$ such that $\theta(x) = \varphi(x)$ for all $x \in J$. Let $k \in I$. There exists ${}_RL \le_e {}_RR$ such that $Lk \subseteq J$ (Exercise 2.3). Then $L[\varphi(k) - \theta(k)] = 0$. Since $Z({}_RA) = 0$, then $\varphi(k) = \theta(k)$. By the Baer Criterion (Theorem 1.20), ${}_RA$ is injective.

(iii) Let ${}_RI \le {}_RR$, $\varphi : I \to D$ be an R-homomorphism, $i : D \to A$ the inclusion homomorphism, and $\pi : A \to D$ the projection. Then there is a ${}_RJ \le_e {}_RI$ and homomorphism $\theta : R \to A$ such that $\theta(x) = i(\varphi(x))$ for all $x \in J$ because ${}_RA$ is ${}_RR$-ejective. Then $\pi\theta : R \to D$ and $\varphi(x) = \pi(\theta(x))$. Therefore ${}_RD$ is ${}_RR$-ejective.

(iv) Note that $Z_2(R) \subseteq l_R(A/Z_2({}_RA))$. Thus $A/Z_2({}_RA)$ is an $R/Z_2({}_RR)$-module where multiplication by scalars is defined by $(r + Z_2({}_RR))(a + Z_2({}_RA)) = ra + Z_2({}_RA)$ for all $a \in A$ and $r \in R$. Let $\overline{A}$ and $\overline{R}$ denote $A/Z_2({}_RA)$ and $R/Z_2({}_RR)$, respectively. Thus $\overline{A}$ and $\overline{R}$ are both R and $\overline{R}$-modules. Now let ${}_{\overline{R}}\overline{K} \le {}_{\overline{R}}\overline{R}$ and $\varphi : \overline{K} \to \overline{A}$ be an $\overline{R}$-homomorphism. By (ii), $\overline{A}$ is an injective

R-module. Hence there exists an R-homomorphism $\theta : \overline{R} \to \overline{A}$. But θ is also an $\overline{R}$-homomorphism. Thus $\overline{A}$ is $\overline{R}$-injective. The particular case when ${}_RA$ is ${}_RR$-ejective and $A = Z_2({}_RA) \oplus Y$ follows from (iii) and the above argument. For (v) and (vi), let ${}_RI \leq {}_RR$ and $\varphi : I \to A$ an R-homomorphism. Suppose that there is a $0 \neq {}_RJ \leq {}_RI$ such that $J \cap \ker\varphi = 0$. Let $\ker\varphi = K$.

(v) There is a $y \in J$ such that $\varphi(y) \neq 0$. Since $Z({}_RA) \leq_e {}_RA$, there are $r \in R$ and ${}_RL \leq_e {}_RR$ such that $0 \neq r\varphi(y)$ and $\varphi(Lry) = 0$ (Exercise 2.3). Then $0 \neq ry$, but $Lry = 0$. This is contrary to $Z({}_RR) = 0$. Hence ${}_RK \leq_e {}_RI$. Let $\theta : {}_RR \to {}_RA$ be the zero homomorphism. Then $\varphi(k) = \theta(k)$, for all $k \in K$, and so ${}_RA$ is ${}_RR$-ejective.

(vi) Observe that $\varphi|_J : {}_RJ \to {}_RA$ is a monomorphism. Hence $\operatorname{soc}({}_RJ) \neq 0$, contrary to $\operatorname{soc}({}_RR) = 0$. So ${}_RK \leq_e {}_RI$. Therefore, as above, ${}_RA$ is ${}_RR$-ejective. □

The next proposition provides an equivalent condition for a module being relative ejective in terms of mappings into its injective hull. To this end compare the following result with Proposition 2.45.

Proposition 2.62. *Let A and X be modules. Then X is A-ejective if and only if for each $\varphi : A \to E(X)$ there exist $E \leq_e A$ and $\theta : A \to X$ such that $\theta|_E = \varphi|_E$.*

Proof. ($\Rightarrow$) Assume X is A-ejective and $\varphi : A \to E(X)$ is a homomorphism. Let $K = \varphi^{-1}(N)$. Then $K \leq_e A$ (Exercise 1.64 (v)). Moreover, there exist $E \leq_e K$ and $\theta : A \to X$ such that $\theta|_X = \varphi|_X$.

($\Leftarrow$) Assume that for each homomorphism $\varphi : A \to E(X)$ there exist $E \leq_e A$ and a homomorphism $\theta : A \to X$ such that $\theta|_E = \varphi|_E$. Assume $K \leq A$ and $\varphi : K \to X$ is a homomorphism. There exists $\overline{\varphi} : A \to E(X)$ such that $\overline{\varphi}|_K = \varphi$. Let $Y = \overline{\varphi}^{-1}(X)$. Then $K \leq Y \leq_e A$. Thus there exist $E \leq_e Y$ and a homomorphism $\theta : A \to X$ such that $\theta|_E = \overline{\varphi}|_E$. Let $\overline{E} = K \cap E \leq_e K$. Then $\theta|_E = \overline{\varphi}|_E = \varphi|_E$. Therefore, X is A-ejective. □

In the rest of this section we define elifting submodules of a module in a similar way to lifting submodules, and then we obtain analogues of results established in Section 2.3.

Let A and X be left R-modules. A submodule N of A is called an *elifting submodule for X in A* if for any $\varphi \in \operatorname{Hom}_R(N, X)$ there exist $\theta \in \operatorname{Hom}_R(A, X)$ and a $E \leq_e N$ such that $\theta|_E = \varphi|_E$, i.e., $\theta(x) = \varphi(x)$, $x \in E$. So we set

$$\operatorname{Elift}_X(A) = \{N : N \leq A \text{ and } N \text{ is an elifting submodule for } X \text{ in } A\}.$$

Lemma 2.63. $\operatorname{Lift}_X(A) \subseteq \operatorname{Elift}_X(A)$.

Proof. Let $N \in \operatorname{Lift}_X(A)$ and $\varphi : N \to X$ be a homomorphism. By hypothesis, there exists a homomorphism $\theta : A \to X$ such that $\theta|_N = \varphi$. Since $N \leq_e N$, we have that $N \in \operatorname{Elift}_X(A)$. □

Observe that, by Lemma 2.34 and Lemma 2.63, in particular direct summands of the module A are contained in $\mathrm{Elift}_X(A)$. However, the classes $\mathrm{Lift}_X(A)$ and $\mathrm{Elift}_X(A)$ are not identical (see, e.g., Example 2.58 (ii)).

Lemma 2.64. *The following statements are equivalent.*

(i) *X is A-ejective.*
(ii) *Every submodule of A is an elifting submodule for X in A.*
(iii) *Every essential submodule of A is an elifting submodule for X in A.*

Proof. (i) $\Rightarrow$ (ii) and (ii)$\Rightarrow$ (iii) are clear.

(iii) $\Rightarrow$ (i) Let $N \leq A$. Let N' be a complement of N in A. Then $N \oplus N' \leq_e A$. Let $\varphi \in \mathrm{Hom}_R(N, X)$. By Lemmas 2.34 and 2.63, there exist $\theta \in \mathrm{Hom}_R(N \oplus N', X)$ and $K \leq_e N$ such that $\theta|_K = \varphi|_K$. By (iii), there exist $\chi \in \mathrm{Hom}_R(A, X)$ and $V \leq_e N \oplus N'$ such that $\chi|_V = \theta|_V$. Since $K \cap V \leq_e N$ (Proposition 2.1) and $\chi|_{K\cap V} = \varphi|_{K \cap V}$, then X is A-ejective. □

Compare the following lemma with Lemma 2.36.

Lemma 2.65. *Let K, N be submodules of A such that $K \leq N$. Then*

(i) *$K \in \mathrm{Elift}_X(N)$, $N \in \mathrm{Elift}_X(A)$ implies that $K \in \mathrm{Elift}_X(A)$,*
(ii) *$K \in \mathrm{Elift}_X(A)$ implies that $K \in \mathrm{Elift}_X(N)$,*
(iii) *Assume that $K \leq_c N$ and $K \leq \mathrm{soc}\, N$. Then $N \in \mathrm{Elift}_X(A)$ implies that $N/K \in \mathrm{Elift}_X(A/K)$,*
(iv) *Assume that $K = \mathrm{soc} K \leq_e N$. Then $K \in \mathrm{Elift}_X(A)$, and $N/K \in \mathrm{Elift}_X(A/K)$ implies that $N \in \mathrm{Elift}_X(A)$.*

Proof. (i) Let $\varphi \in \mathrm{Hom}_R(K, X)$. Then there exist $\theta \in \mathrm{Hom}_R(N, X)$ and a $E_1 \leq_e K$ such that $\theta|_{E_1} = \varphi|_{E_1}$. Now, there exist $\theta_1 \in \mathrm{Hom}_R(A, X)$ and a $E_2 \leq_e N$ such that $\theta_1|_{E_2} = \theta|_{E_2}$. Since $E_1 \cap E_2 \leq_e K$ and $\theta_1|_{E_1 \cap E_2} = \varphi|_{E_1 \cap E_2}$ we conclude that $K \in \mathrm{Elift}_X(A)$.

(ii) Let $\varphi \in \mathrm{Hom}_R(K, X)$. There exist $\theta \in \mathrm{Hom}_R(A, X)$ and a $E \leq_e K$ such that $\theta|_E = \varphi|_E$. Now $f = \theta|_N : N \to X$ and $f|_E = \varphi|_E$. It follows that $K \in \mathrm{Elift}_X(N)$.

(iii) Let $\varphi \in \mathrm{Hom}_R(N/K, X)$. Let $\pi : N \to N/K$ be the canonical epimorphism. Then $\varphi\pi : N \to X$ is a homomorphism. Since $N \in \mathrm{Elift}_X(A)$, there exist $\theta \in \mathrm{Hom}_R(A, X)$ and a $E \leq_e N$ such that $\theta|_E = \varphi\pi|_E$. Note that $K \leq E$. Thus $E/K \leq_e N/K$ (Exercise 1.64). Define $\overline{\theta} : A/K \to X$ by $\overline{\theta}(a + K) = \theta(a)$ $(a \in A)$. Then $\overline{\theta}$ is well defined and $\overline{\theta} \in \mathrm{Hom}_R(A/K, X)$. For any $x \in E$, $\overline{\theta}(x + K) = \theta(x) = \varphi(x + K)$. It follows that $N/K \in \mathrm{Elift}_X(A/K)$.

(iv) Let $\varphi \in \mathrm{Hom}_R(N, X)$. Then $\varphi|_K \in \mathrm{Hom}_R(K, X)$. There exist $\theta \in \mathrm{Hom}_R(A, X)$ and a $E_1 \leq_e K$ such that $\varphi|_{E_1} = \theta|_{E_1}$. By hypothesis, $K = E_1$. Define $\chi : N/K \to X$ by $\chi(n+K) = \varphi(n) - \theta(n)$ $(n \in N)$. Note that χ is well defined and a homomorphism. There exists $\psi \in \mathrm{Hom}_R(A/K, X)$ and a $E_2/K \leq_e A/K$ such that $\psi|_{E_2/K} = \chi|_{E_2/K}$. Let $\pi : A \to A/K$ be the canonical epimorphism.

Let $\alpha = \psi\pi + \theta \in \mathrm{Hom}_R(A, X)$. Now $E_1 \cap E_2 \leq_{\mathrm{e}} K$ and $K \leq_{\mathrm{e}} N$ yield that $E_1 \cap E_2 \leq_{\mathrm{e}} N$. So $\alpha|_{E_1 \cap E_2} = \varphi|_{E_1 \cap E_2}$. Hence $N \in \mathrm{Elift}_X(A)$. □

Corollary 2.66. *For any* $N \in \mathrm{Elift}_X(A)$, $\mathrm{Elift}_X(N) = \{K \leq N : K \in \mathrm{Elift}_X(A)\}$.

Proof. Suppose $K \in \mathrm{Elift}_X(N)$. Then $K \leq N$ and by Lemma 2.65 (i), $K \in \mathrm{Elift}_X(A)$. Therefore $\mathrm{Elift}_X(N) \subseteq \{K \leq N : K \in \mathrm{Elift}_X(A)\}$. Conversely, suppose $K \leq N$ and $K \in \mathrm{Elift}_X(A)$. By Lemma 2.65 (ii), $K \in \mathrm{Elift}_X(N)$. □

Similarly to Example 2.38, one can construct examples such that if K, N are submodules of the module A with $K \leq N$, then $K \in \mathrm{Elift}_X(A)$ does not imply that $N \in \mathrm{Elift}_X(A)$.

Proposition 2.67. *Let* $N, K \leq A$ *such that* $N + K$ *and* $N \cap K = soc\,(N \cap K)$ *belong to* $\mathrm{Elift}_X(A)$. *Then* $N, K \in \mathrm{Elift}_X(A)$.

Proof. Let $\varphi \in \mathrm{Hom}_R(N, X)$. Then $\varphi|_{N \cap K} \in \mathrm{Hom}_R(N \cap K, X)$. There exist $\theta_1 \in \mathrm{Hom}_R(A, X)$ and a $E_1 \leq_{\mathrm{e}} N \cap K$ such that $\theta_1|_{E_1} = \varphi|_{E_1}$. By hypothesis, $E_1 = N \cap K$. Define $\chi : N + K \to X$ by $\chi(n + k) = \varphi(n) + \theta_1(k)$ $(n \in N, k \in K)$. It can be checked that χ is well defined. Clearly, $\chi \in \mathrm{Hom}_R(N + K, X)$, so by hypothesis, there exist $\theta \in \mathrm{Hom}_R(A, X)$ and a $E_2 \leq_{\mathrm{e}} N + K$ such that $\theta|_{E_2} = \chi|_{E_2}$. By Proposition 2.1, $E_1 \cap E_2 \leq_{\mathrm{e}} N \cap K$, and for any $x \in E_1 \cap E_2$, $\theta(x) = \chi(x) = \varphi(x)$. Thus, $\theta|_{E_1 \cap E_2} = \varphi|_{E_1 \cap E_2}$. It follows that $N \in \mathrm{Elift}_X(A)$. Similarly, $K \in \mathrm{Elift}_X(A)$. □

Corollary 2.68. *Let* N, K *be submodules of* A.

(i) *If* $N \cap K = 0$ *and* $N \oplus K \in \mathrm{Elift}_X(A)$ *then* $N, K \in \mathrm{Elift}_X(A)$.
(ii) *If* $A = N + K$ *and* $N \cap K = soc\, N \cap K \in \mathrm{Elift}_X(A)$ *then* $N, K \in \mathrm{Elift}_X(A)$.

Proof. Immediate by Proposition 2.67. □

Theorem 2.69. *Let* $X = \prod_{\lambda \in \Lambda} X_\lambda$. *Then* $\mathrm{Elift}_X(A) = \bigcap_{\lambda \in \Lambda} \mathrm{Elift}_{X_\lambda}(A)$, *for any module* A.

Proof. Let $\lambda \in \Lambda$ and $Y = X_\lambda$. Let $N \in \mathrm{Elift}_X(A)$. Let $\varphi \in \mathrm{Hom}_R(N, Y)$. Let $i : Y \to X$ denote the inclusion mapping and $\pi : X \to Y$ the canonical projection. Then $i\varphi \in \mathrm{Hom}_R(N, X)$. By hypothesis, there exist $\theta \in \mathrm{Hom}_R(A, X)$ and a $E \leq_{\mathrm{e}} N$ such that $\theta|_E = i\varphi|_E$.

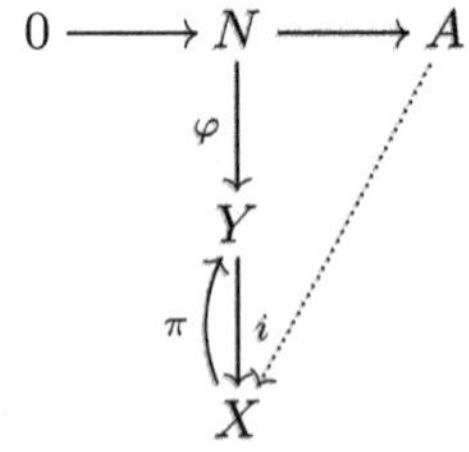

Now $\pi\theta \in \mathrm{Hom}_R(A, Y)$, and for any $x \in E$, $(\pi\theta)(x) = \pi i\varphi(x) = \varphi(x)$. Thus $\varphi|_E = \pi\theta|_E$. It follows that $N \in \mathrm{Elift}_Y(A)$. Hence $\mathrm{Elift}_X(A) \subseteq \mathrm{Elift}_Y(A)$. Therefore $\mathrm{Elift}_X(A) \subseteq \bigcap_{\lambda\in\Lambda} \mathrm{Elift}_{X_\lambda}(A)$.

Conversely, let $K \in \bigcap_{\lambda\in\Lambda} \mathrm{Elift}_{X_\lambda}(A)$. Let $\alpha \in \mathrm{Hom}_R(K, X)$. For each $\lambda \in \Lambda$, let $\pi_\lambda : X \to X_\lambda$ denote the canonical projection. Then $\pi_\lambda\alpha \in \mathrm{Hom}_R(K, X_\lambda)$, $\lambda \in \Lambda$. By hypothesis, for each $\lambda \in \Lambda$ there exist $\beta_\lambda \in \mathrm{Hom}_R(A, X_\lambda)$ and a $E_\lambda \leq_{\mathrm{e}} K$ such that $\beta_\lambda(x) = \pi_\lambda\alpha(x)$, $x \in E_\lambda$. Define $\beta : A \to X$ by $\beta(a) = \{\beta_\lambda(a)\}_{\lambda\in\Lambda}$ $(a \in A)$ and apply Zorn's Lemma to the family $\underline{S} = \{(E_\lambda, \beta_\lambda) : \beta_\lambda|_{E_\lambda} = \pi_\lambda\alpha|_{E_\lambda}\}$, which has a maximal member, say (E_w, β_w). Now, for each $x \in E_w$, $\beta(x) = \alpha(x)$. Hence $\beta|_{E_w} = \alpha|_{E_w}$, and so $K \in \mathrm{Elift}_X(A)$. □

Corollary 2.70. *Let $X = \prod_{\lambda\in\Lambda} X_\lambda$. Then X is A-ejective if and only if X_λ is A-ejective for all $\lambda \in \Lambda$.*

Proof. By Lemma 2.64 and Theorem 2.69. □

Exercises

2.31. Determine $\mathrm{Elift}_{\mathbb{Z}}(\mathbb{Z})$, $\mathrm{Elift}_{\mathbb{Z}}(\mathbb{Q})$, $\mathrm{Elift}_{\mathbb{Z}}(\mathbb{Q}/\mathbb{Z})$.

2.32. Let p be any prime integer. Calculate

$$\mathrm{Elift}_{\mathbb{Z}}(\mathbb{Z}(p^\infty)),\quad \mathrm{Elift}_{\mathbb{Z}}((\mathbb{Z}/\mathbb{Z}p) \oplus (\mathbb{Z}/\mathbb{Z}p^2)),\quad \text{and}\quad \mathrm{Elift}_{\mathbb{Z}}((\mathbb{Z}/\mathbb{Z}p) \oplus (\mathbb{Z}/\mathbb{Z}p^3)).$$

2.33. Provide an example such that $B, C \leq {}_RA$ with $B \cong C$ and $B \in \mathrm{Elift}_X(A)$ but $C \notin \mathrm{Elift}_X(A)$, i.e., $\mathrm{Elift}_X(A)$ is not a class.

2.34. Let F be a field and let $R_R = \begin{bmatrix} F & F \\ 0 & F \end{bmatrix}$. Prove that R is not R-ejective.

2.35. Let A be the Specker group $\prod \mathbb{Z}$. Prove that $\mathrm{Elift}_{\mathbb{Z}}(A/(\bigoplus \mathbb{Z})) = A/(\bigoplus \mathbb{Z})$.

2.36. Let V be a vector space over a field F with $\dim(V) \geq 2$ and let $R_R = \begin{bmatrix} F & & V \\ & \diagdown & \\ 0 & & F \end{bmatrix} = \{\begin{bmatrix} a & v \\ 0 & a \end{bmatrix} : a \in F, v \in V\}$. Prove R_R is not R_R-ejective.

2.5 Conditional direct summands

Since direct summands of a module are of special interest in Ring and Module Theory and play an important role throughout our work, we focus on this type of submodules in this section. To this end, we proceed as often done in algebra starting with direct summand(s) of a module in hand, and adding a condition which is based on them to produce a new direct summand.

Notice that the condition in § 2.3 which asserts that "complement submodules are direct summands" in a module (see Corollary 2.50 (i)) does not fit in the aforementioned general procedure.

By contrast, the following properties (see also Exercise 2.45) work well and are well behaved as companion conditions in certain classes of modules studied in further chapters.

Let A be a left R-module. We introduce the following properties of A:

(i) *property* C_2: if $X \leq A$ is isomorphic to a direct summand of A, then X is a direct summand of A; in other words, for each direct summand N of A and each monomorphism $\varphi : N \to A$, the submodule $\varphi(N)$ is also a direct summand of A;

(ii) *property* C_3: if A_1 and A_2 are direct summands of A such that $A_1 \cap A_2 = 0$, then $A_1 \oplus A_2$ is a direct summand of A;

(iii) the *summand intersection property*, SIP: if A_1 and A_2 are direct summands of A, then $A_1 \cap A_2$ is a direct summand of A.

Recall that there is a correspondence between direct summands and idempotent endomorphisms of a module. Suppose that ${}_RA = K \oplus K'$. Then the canonical projection $\pi : A \to K$ is an idempotent endomorphism of A, $\pi^2 = \pi \in \mathrm{End}({}_RA)$ and $K = A\pi$ (note that π is a right operator on A). Thus each direct summand of A is the image of an idempotent endomorphism of A. Conversely, if $e^2 = e \in \mathrm{End}({}_RA)$ then $1 - e$ is an idempotent in $\mathrm{End}({}_RA)$ and ${}_RA = Ae \oplus A(1-e)$.

Before looking at the relationships between C_2, C_3 and SIP properties, we prove the following easy lemma which will be used several places in the sequel.

Lemma 2.71. *Assume* $A = N \oplus N'$. *Let* $K \leq A$ *with* $N \cap K = 0$. *Then* $N \oplus K = N \oplus \pi(K)$, *where* $\pi : A \to N'$ *is the canonical projection.*

Proof. Let $x \in N \oplus K$. Then $x = n + k$, $k = y + y'$ for some $y \in N$ and $y' \in N'$ such that $y' = \pi(k)$. So $x = n + y + \pi(k) \in N + \pi(K)$. It follows that $N \oplus K \leq N + \pi(K) = N \oplus \pi(K)$.
Now, let $a \in N \oplus \pi(K)$. Then $a = b + c$, $b \in N$, $c \in \pi(K)$, and $c = \pi(d)$ for some $d \in K$ where $d = e + c$ for some $e \in N$. So $a = b + (d - e) = (b - e) + d \in N \oplus K$. Thus $N \oplus K = N \oplus \pi(K)$ as required. □

Lemma 2.72. *If a module* A *satisfies property* C_2, *then it satisfies property* C_3.

Proof. Let K, L be direct summands of A with $K \cap L = 0$. Then $A = K \oplus K'$ for some $K' \leq A$. Let $\pi : A \to K'$ denote the canonical projection. Since $K \cap L = 0$, we have that $\pi(L) \cong L$ and $\pi(L) \leq K'$. But $\pi(L)$ is a direct summand of A by C_2, i.e., $A = \pi(L) \oplus L'$ for some $L' \leq A$. Thus $K' = \pi(L) \oplus (K' \cap L')$ and $A = K \oplus \pi(L) \oplus (K' \cap L')$. Hence, $K \oplus \pi(L)$ is a direct summand of A. By Lemma 2.71, $K \oplus L = K \oplus \pi(L)$. Thus A has property C_3. □

The next example shows that none of the implications $C_3 \Rightarrow C_2$, SIP $\Rightarrow C_3$, $C_3 \Rightarrow$ SIP is true in general.

Example 2.73.

(i) Let $\mathbb{Z}$ be the $\mathbb{Z}$-module. Then $\mathbb{Z}$ satisfies C_3 but does not satisfy C_2.

(ii) Let A be any free $\mathbb{Z}$-module of non-zero finite rank k. Then A satisfies C_3 if and only if $k = 1$. So, if $A = \bigoplus_{i=1}^{k} \mathbb{Z}$ $(k \geq 2)$ then A has SIP but does not satisfy C_3.

(iii) Let p be a prime integer. Let A be the $\mathbb{Z}$-module $\mathbb{Z} \oplus (\mathbb{Z}/\mathbb{Z}p)$. Then A satisfies C_3, but not SIP.

Proof. (i) Since $\mathbb{Z}$ is an indecomposable module it clearly satisfies C_3. Let N denote the submodule $\mathbb{Z}2$ of $\mathbb{Z}$. Then $\varphi : N \to \mathbb{Z}$, $\varphi(n2) = n$ $(n \in \mathbb{Z})$ is an isomorphism. So $N \cong \mathbb{Z}$, but $\mathbb{Z}2$ is not a direct summand of $\mathbb{Z}$.

(ii) Suppose that $k \geq 2$. Let $f_1, f_2, \ldots, f_k$ be a basis of A. Let $K_1 = \mathbb{Z}f_1$ and $K_2 = \mathbb{Z}(f_1 + 2f_2)$. Clearly, $A = K_1 \oplus L = K_2 \oplus L$, where $L = \mathbb{Z}f_2 + \mathbb{Z}f_3 + \cdots + \mathbb{Z}f_k$. Also, $K_1 \cap K_2 = 0$, but $K_1 \oplus K_2 = \mathbb{Z}f_1 \oplus \mathbb{Z}2f_2$, which is not a direct summand of A. Thus A does not satisfy C_3. Now the $\mathbb{Z}$-module $\bigoplus_{i=1}^{k} \mathbb{Z}$ $(k \geq 2)$ does not satisfy C_3, but is has SIP by Exercise 2.37.

(iii) By Exercise 2.38, A satisfies C_3 and $B = \mathbb{Z}(1, 0 + \mathbb{Z}p)$, $C = \mathbb{Z}(1, 1 + \mathbb{Z}p) \leq_d A$ but $B \cap C$ is not a direct summand of A. □

The fact that the C_2, C_3 and SIP conditions are inherited by direct summands will be used in several results. In fact, we give the following lemma. Its proof is a direct applications of the modular law (Exercise 1.19) and it is included for the sake of completeness.

Lemma 2.74. *Let A be a module and $N \leq_{\mathrm{d}} A$. Then*

(i) *If A satisfies C_2, then so does N.*

(ii) *If A satisfies C_3, then so does N.*

(iii) *If A satisfies* SIP, *then so does N.*

Proof. (i) Let $X, K \leq N$ such that $X \cong K \leq_{\mathrm{d}} N$. So $K \leq_{\mathrm{d}} A$. Then condition C_2 yields that $X \leq_{\mathrm{d}} A$. Hence $A = X \oplus X'$ for some $X' \leq A$. By the modular law, $N = N \cap A = N \cap (X \oplus X') = X \oplus (N \cap X')$. Thus $X \leq_{\mathrm{d}} N$.

(ii) Let $K_1, K_2 \leq_{\mathrm{d}} N$ such that $K_1 \cap K_2 = 0$. Since $K_1, K_2 \leq_{\mathrm{d}} A$, by condition C_3, $K_1 \oplus K_2 \leq_{\mathrm{d}} A$. Hence $A = (K_1 \oplus K_2) \oplus K_3$ for some $K_3 \leq A$. By the modular law, $N = (K_1 \oplus K_2) \oplus (N \cap K_3)$, i.e., $K_1 \oplus K_2 \leq_{\mathrm{d}} N$.

(iii) Let $K, L \leq_{\mathrm{d}} N$. So $K, L \leq_{\mathrm{d}} A$. By hypothesis, $K \cap L \leq_{\mathrm{d}} A$. Hence $A = (K \cap L) \oplus X$ for some $X \leq A$. By the modular law, $N = (K \cap L) \oplus (N \cap X)$, i.e., $K \cap L \leq_{\mathrm{d}} N$. □

Lemma 2.75. *Let A be a left R-module where $R = ReR$ for some idempotent e in R and $S = eRe$. Let $K, K' \leq {}_RA$ and $N, N' \leq {}_S(eA)$. Then,*

(i) *$K = ReK$, and $N = eRN$,*

(ii) *$K \cap K' = 0$ if and only if $eK \cap eK' = 0$,*

(iii) *$N \cap N' = 0$ if and only if $RN \cap RN' = 0$.*

Proof. (i) Since $K \leq A$ we have that $K = RK = ReRK = ReK$. Similarly, $N = SN = eReN = eRN$.

(ii) If $K \cap K' = 0$, then $eK \cap eK' \leq K \cap K'$ gives $eK \cap eK' = 0$. Conversely, suppose $eK \cap eK' = 0$. Let $x \in K \cap K'$. Then $eRx \leq eK \cap eK' = 0$, and hence $ReRx = 0$. Thus $Rx = 0$ so that $x = 0$. It follows that $K \cap K' = 0$.

(iii) By (i) and (ii). □

Lemma 2.76. *Let $R = ReR$ and $S = eRe$ for some idempotent e in R and A is a left R-module. Let $L, N \leq {}_S(eA)$. Then L is a complement of N in ${}_S(eA)$ if and only if RL is a complement of RN in ${}_RA$.*

Proof. Suppose L is a complement of N in eA. Then $L \cap N = 0$ and hence $RL \cap RN = 0$. Suppose $RL \leq K \leq {}_RA$ and $K \cap RN = 0$. Then Lemma 2.75 (i) gives $L = eRL \leq eK \leq eA$ and $eK \cap N \leq K \cap RN = 0$. It follows that $L = eK$ and hence $RL = ReK = K$. Hence RL is a complement of RN in A.

Conversely, suppose that RL is a complement of RN in A. Then $L \cap N = 0$. Suppose $L \leq H \leq {}_S(eA)$ and $H \cap N = 0$. By Lemma 2.75, $RL \leq RH$ and $RH \cap RN = 0$. Thus $RL = RH$ and hence $L = eRL = eRH = H$, again by Lemma 2.75. Thus L is a complement of N in eA. □

Proposition 2.77. *Let A be a left R-module and let L be a submodule of A, where $R = ReR$ for some idempotent e in R and $S = eRe$. Then*

(i) *$L \leq_e {}_RA$ if and only if $eL \leq_e {}_S(eA)$,*
(ii) *$L \leq_c {}_RA$ if and only if $eL \leq_c {}_S(eA)$,*
(iii) *$L \leq_d {}_RA$ if and only if $eL \leq_d {}_S(eA)$.*

Proof. (i) Suppose $L \leq_e {}_RA$. Let $0 \neq N \leq {}_S(eA)$. Then $L \cap RN \neq 0$ and hence $eL \cap N \neq 0$, by Lemma 2.75. Thus, $eL \leq_e {}_S(eA)$. Conversely, suppose $eL \leq_e {}_S(eA)$. Let $0 \neq K \leq {}_RA$. By Lemma 2.75, $K = ReK$, so that $0 \neq eK \leq {}_S(eA)$. Hence $eK \cap eL \neq 0$. But $eK \cap eL \leq K \cap L$. Thus $K \cap L \neq 0$. It follows that $L \leq_e {}_RA$.

(ii) By Lemma 2.76.

(iii) Suppose $L \leq_d {}_RA$. Then $A = L \oplus L'$ for some $L' \leq {}_RA$. Thus $eA = eL + eL'$. But $eL \cap eL' \leq L \cap L' = 0$. Therefore $eA = eL \oplus eL'$. Conversely, suppose that $eA = eL \oplus K$ for some $K \leq {}_S(eA)$. By Lemma 2.75, $L \cap RK = 0$, and $A = ReA = R(eL + K) = ReL + RK = L + RK$. Thus ${}_RA = L \oplus RK$, i.e., $L \leq_d {}_RA$. □

Combining Proposition 2.77 together with Lemma 2.75, we have the following fact.

Proposition 2.78. *Let A be a left R-module where $R = ReR$ for some idempotent e in R and $S = eRe$. Then*

(i) $e\,\mathrm{soc}({}_RA) = \mathrm{soc}({}_S(eA))$. *In particular, ${}_RA$ is semisimple if and only if ${}_S(eA)$ is semisimple,*

(ii) $eZ({}_RA) = Z({}_S(eA))$. *In particular,* ${}_RA$ *is nonsingular if and only if* ${}_S(eA)$ *is nonsingular.*

Proof. (i) Follows from Lemma 2.75 and Proposition 2.77.

(ii) Let $ea \in eZ({}_RA)$. Then $a \in Z({}_RA)$. There exists an essential left ideal F of R such that $Fea = 0$. By Proposition 2.77 (i), $Re \cap F$ is essential in Re and hence $e(Re \cap F)$ is essential in ${}_S(eRe) = S$. But $ea \in eA$ and $e(Re \cap F) \leq eF \leq F$. Thus $ea[e(Re \cap F)] = 0$, and so $ea \in Z({}_S(eA))$. Now, let $ea \in Z({}_S(eA))$. Then $Gea = 0$ for some essential left ideal G of S. By Proposition 2.77, RG is essential in Re. Thus $RG \oplus R(1-e)$ is essential in ${}_RR$. Since $ea[RG \oplus R(1-e)] = 0$, we have $ea \in Z({}_RA)$ and hence $ea \in eZ({}_RA)$. The second part is clear. □

Theorem 2.79. *Let A be a left R-module, where $R = ReR$ for some idempotent e in R and $S = eRe$. Then*

(i) *the left R-module A satisfies C_2 if and only if the left S-module eA satisfies C_2,*

(ii) *the left R-module A satisfies C_3 if and only if the left S-module eA satisfies C_3,*

(iii) *the left R-module A has the* SIP *if and only if the left S-module eA has the* SIP.

Proof. (i) Let $K, L \leq A$ such that $K \cong L$, where L is a direct summand of ${}_RA$. Let $f : K \to L$ be an R-isomorphism. Then eL is a direct summand of ${}_S(eA)$. Now $\varphi = f|_{eK} : eK \to eL$ is an isomorphism. Therefore, eK is a direct summand of eA and hence K is a direct summand of ${}_RA$. For the converse, let $B, C \leq eA$ such that $B \cong C$ and C is a direct summand of ${}_S(eA)$. Thus RC is a direct summand of ${}_RA$. Suppose that $\varphi : B \to C$ is an isomorphism. Define $\theta : RB \to RC$ and $\theta' : RC \to RB$ by $\theta(\sum_{i=1}^n r_ib_i) = \sum_{i=1}^n r_i\varphi(b_i)$, $\theta'(\sum_{i=1}^n r_ic_i) = \sum_{i=1}^n r_i\varphi^{-1}(c_i)$ for all $n \geq 1$, $b_i \in B$, $c_i \in C$, $r_i \in R$ $(1 \leq i \leq n)$. Now suppose $\sum_{i=1}^n r_ib_i = 0$. Then $\sum_{i=1}^n ser_ib_i = 0$ for all $s \in R$. Therefore, $\sum_{i=1}^n ser_ieb_i = 0$ and hence $\sum_{i=1}^n ser_ie\varphi(b_i) = 0$. Thus $\sum_{i=1}^n ser_i\varphi(b_i) = 0$. It follows that $Re(\sum_{i=1}^n r_i\varphi(b_i)) = 0$, so that $ReR(\sum_{i=1}^n r_i\varphi(b_i)) = 0$, i.e., $\sum_{i=1}^n r_i\varphi(b_i) = 0$. We see that θ is a well-defined mapping. It is easy to check that θ is an R-homomorphism. Similarly, θ' is an R-homomorphism. Clearly, $\theta'\theta = 1|_{RB}$ and $\theta\theta' = 1|_{RC}$. Hence θ is an isomorphism. By hypothesis, RB is a direct summand of ${}_RA$ and then by Proposition 2.77, B is a direct summand of ${}_RA$.

(ii) Let B, C be direct summands of ${}_S(eA)$ with $B \cap C = 0$. Then $B = eRB$, $C = eRC$. Hence RB and RC are direct summands of ${}_RA$. Since $eRB \cap eRC = B \cap C = 0$, we have that $RB \cap RC = 0$ in ${}_RA$. Thus $RB \oplus RC$ is a direct summand of ${}_RA$. By Proposition 2.77, $B \oplus C = e(RB \oplus RC)$ is a direct summand of ${}_S(eA)$. Conversely, let K, L be direct summands of ${}_RA$ with $K \cap L = 0$. Therefore, eK and eL are direct summands of ${}_S(eA)$. Then $eK \cap eL \leq K \cap L = 0$ gives that $R(eK \oplus eL) = ReK \oplus ReL = K \oplus L$ is a direct summand of ${}_RA$.

(iii) Clear by Proposition 2.77. □

Corollary 2.80. *Let R a ring such that $R = ReR$ for some idempotent e in R. Then R_R satisfies C_2 (resp., C_3, or* SIP*) if and only if the right eRe-module Re satisfies C_2 (resp., C_3 or* SIP*).*

Proof. Immediate by Theorem 2.79. □

A module ${}_RA$ has the *summand sum property,* SSP, if the sum of every pair of direct summands of ${}_RA$ is a direct summand of ${}_RA$. Observe that the SSP versions of all of the results proven for the SIP can hold with easily modified proofs in which SIP is replaced with SSP. To demonstrate this, we collect more shared results. First we give the following notations which will be used in the rest of this section as well as in the sequel.

Let R be any ring. Then we will use $M_m(R)$, $T_m(R)$, and R^m to denote the full m-by-m matrix ring over R, the upper triangular m-by-m matrix ring over R, and the direct sum of m copies of R for any positive integer m, respectively.

Theorem 2.81. *$M_m(R)$ satisfies C_2 (resp., C_3,* SIP *or* SSP*) if and only if the free right R-module R^m satisfies C_2 (resp., C_3,* SIP *or* SSP*).*

Proof. Observe that $M_m(R) = M_m(R)eM_m(R)$, where e is the matrix unit with 1 in the $(1,1)$th position and zero elsewhere. The result now follows from Theorem 2.79 and Exercise 2.47. □

Theorem 2.81, leads naturally to the following general problem:
If R_R has SIP (resp., SSP), find conditions on R to ensure that every free right R-module with a basis of cardinality κ has SIP (resp., SSP).

It is easy to see that Kaplansky's result (Exercise 2.37) implies that the condition that R be a principal ideal domain is sufficient for the SIP version of the above problem. Moreover, since the class of (von Neumann) regular rings is closed with respect to forming the full m-by-m matrix ring for any positive integer m, Theorem 2.81 shows that regularity is a sufficient condition for the problem (see Exercise 2.20).

Notice that Corollary 2.80 together with the problem formulated above leads us to ask whether being SIP (resp., SSP) is Morita invariant. More precisely, let P be a ring theoretic property. Then P is said to be *Morita invariant* if the following holds: whenever a ring R has P, so do $M_n(R)$ for every $n \geq 2$ and eRe for all $e^2 = e \in R$ such that $R = ReR$. For example, being Artinian is Morita invariant (Exercise 2.51).

The next example shows that being SSP is not Morita invariant.

Example 2.82. Let $R = \mathbb{Z}$. Then by Exercise 2.40 the right R-module $R \oplus R$ does not have SSP. Thus, by Theorem 2.81, $M_2(R)$ does not have SSP.

Note that Example 4.33 shows that commutativity, no nonzero divisors, acc on ideals, and Krull dimension 2 do not ensure a solution for the SIP version of the problem when $\kappa \geq 3$. So, we conclude that being SIP is not Morita invariant either. Surprisingly, our next result shows that the right Ore condition and no nonzero divisors do ensure a solution for the SIP version of the problem when $\kappa = 2$. To understand this, we refer readers who are not familiar with the rings of quotients to the appendix for the construction of rings of quotients in terms of Gabriel topologies. Further recall that a nonzero idempotent e of R is called *primitive* if it cannot be written as a sum of two nonzero orthogonal idempotents.

Proposition 2.83. *If R is a right Ore domain, then the right R-module $R \oplus R$ has* SIP.

Proof. Let D be the classical right ring of quotients of R; D is a division ring. Routine calculations show that any nontrivial idempotent of $M_2(D)$ is primitive and has one of the following forms, where $a, d, f \in D$ with $a - a^2 \neq 0$ and $d \neq 0$:

$$\begin{bmatrix} 1 & f \\ 0 & 0 \end{bmatrix}, \begin{bmatrix} 0 & 0 \\ f & 1 \end{bmatrix}, \begin{bmatrix} 0 & f \\ 0 & 1 \end{bmatrix}, \begin{bmatrix} 1 & 0 \\ f & 0 \end{bmatrix}, \begin{bmatrix} a & (1-a)d \\ d^{-1}a & d^{-1}(1-a)d \end{bmatrix}.$$

One can then verify that if c and e are nontrivial idempotents of $M_2(D)$, then either $cM_2(D) \cap eM_2(D) = 0$, or $ce = e$ and $ec = c$. So $M_2(R)$ has SIP as a right module over itself. By Theorem 2.81, $(R \oplus R)_R$ has SIP. □

We complete the above discussion by the following example, the proof of which is left to the reader (Exercise 2.44).

Example 2.84. Let F be any field and let $R = T_2(F)$. Then R_R does not satisfy C_3 (and hence does not satisfy C_2 either). Furthermore, R_R has the SIP, but does not have SSP.

Using the idea in Section § 2.3 (see Theorem 2.48 and Corollaries 2.49, 2.50) we now connect conditions C_2 and C_3 with a collection of certain submodules of a module in terms of lifting submodules. To this aim, we first build up some special collections of submodules.

Let A be a module and n be a positive integer. Let us define the following classes in conjunction with the conditions C_2 and C_3, respectively:

$\underline{A}' = \{N \leq A : \text{there exists } K \leq_{\mathrm{d}} A \text{ such that } K \cong N\}$,

$\underline{A}^{(n)} = \{L_1 + L_2 + \cdots + L_n : L_i \leq_d A$ for $1 \leq i \leq n$ and $L_1 + L_2 + \cdots + L_n$ is a direct sum$\}$, and

$\underline{C}^{(n)} = \{C_1 + C_2 + \cdots + C_n : C_i \leq_{\mathrm{c}} A$ for $1 \leq i \leq n$ and $C_1 + C_2 + \cdots + C_n$ is a direct sum$\}$.

Then we have:

Proposition 2.85. *The following statements are equivalent for a module A.*

(i) *A has C_2.*

(ii) $\underline{A}' \subseteq \mathrm{Lift}_X(A)$ *for all left R-modules X.*

(iii) $\underline{A}' \subseteq \mathrm{Lift}_X(A)$ *for all $X \in \underline{A}'$.*

(iv) $\underline{A}' \subseteq \mathrm{Lift}_A(A)$.

Proof. (i) $\Leftrightarrow$ (ii) $\Leftrightarrow$ (iii) By Theorem 2.48.

(iii) $\Rightarrow$ (iv) Obvious.

(iv) $\Rightarrow$ (i) Let $N' \in \underline{A}'$. Then there exist $N \leq_{\rm d} A$ and an isomorphism $\varphi : N' \to N$. By Theorem 2.43, $\text{Lift}_A(A) \subseteq \text{Lift}_N(A)$. Thus, by (iv), $N' \in \text{Lift}_N(A)$, and there exists $\theta \in \text{Hom}_R(A, N)$ such that $\theta|_{N'} = \varphi$. For any $a \in A$, $\theta(a) \in N$, and hence $\theta(a) = \varphi(n')$ for some $n' \in N'$. Thus $\theta(a) = \theta(n')$ and so $a - n' \in \ker\theta$. It follows that $A = N' + (\ker\theta)$. But $N' \cap (\ker\theta) = \ker\varphi = 0$. Thus $A = N' \oplus (\ker\theta)$. Therefore A has C_2. □

Proposition 2.86. *The following statements are equivalent for a module A.*

(i) *A has C_3.*

(ii) *$\underline{A}^{(2)} \subseteq \text{Lift}_X(A)$ for all left R-modules X.*

(iii) *$\underline{A}^{(2)} \subseteq \text{Lift}_X(A)$ for all $X \in \underline{A}^{(2)}$.*

(iv) *$\underline{A}^{(2)} \subseteq \text{Lift}_A(A)$.*

Proof. (i) $\Leftrightarrow$ (ii) $\Leftrightarrow$ (iii) By Theorem 2.48.

(iii) $\Rightarrow$ (iv) Obvious.

(iv) $\Rightarrow$ (i) Let $K, L \leq_{\rm d} A$ with $K \cap L = 0$. Let $\pi : K \oplus L \to K$ be the canonical projection. By (iv) and Theorem 2.43, $K \oplus L \in \text{Lift}_A(A) \subseteq \text{Lift}_K(A)$, and hence there exists $\theta \in \text{Hom}_R(A, K)$ such that $\theta|_{K\oplus L} = \pi$. It follows that $A = K \oplus (\ker\theta)$. Now $\theta(L) = \pi(L) = 0$ implies $L \subseteq \ker\theta$. But $A = L \oplus L'$ for some submodule L' of A. Thus $\ker\theta = L \oplus (\ker\theta \cap L')$, and hence $A = K \oplus L \oplus (\ker\theta \cap L')$. It follows that A has C_3. □

Corollary 2.87. *Let A be a module. If A has C_3, then for any integer $n \geq 3$, every element of $\underline{A}^{(n)}$ is a direct summand of A.*

Proof. Let $L \in \underline{A}^{(n)}$. Then $L = L_1 \oplus L_2 \oplus \cdots \oplus L_n$ where $L_i \leq_{\rm d} A$ $(1 \leq i \leq n)$. By induction, $L_1 \oplus \cdots \oplus L_{n-1} \leq_{\rm d} A$ and hence $L_1 \oplus L_2 \oplus \cdots \oplus L_n \leq_{\rm d} A$, by C_3. □

Now, we prove the following equivalence for a direct sum of many complements in a module.

Proposition 2.88. *Let X be any left R-module. Then the following statements are equivalent for a module A.*

(i) *$\underline{C}^{(2)} \subseteq \text{Lift}_X(A)$.*

(ii) *$\underline{C}^{(n)} \subseteq \text{Lift}_X(A)$ for all $n \geq 2$.*

Proof. (ii) $\Rightarrow$ (i) Obvious.

(i) $\Rightarrow$ (ii) Suppose (i) holds. Let $k \geq 3$ and $N_i \leq_{\rm c} A$ $(1 \leq i \leq k)$ such that $N_1 + N_2 + \cdots + N_k$ is a direct sum. Let $N = N_1 + N_2 + \cdots + N_k$ and let $\varphi \in \text{Hom}_R(N, X)$. There exists $N' \leq_{\rm c} A$ such that $N_2 + \cdots + N_k \leq_{\rm e} N'$. By induction, $N_2 + \cdots + N_k \in \text{Lift}_X(A)$ and hence there exists $\alpha \in \text{Hom}_R(A, X)$ such that $\alpha(a) = \varphi(a)$ $(a \in N_2 + \cdots + N_k)$. Now $N_1 \cap N' = 0$, because $N_1 \cap (N_2 + \cdots + N_k) = 0$, so we can define $\beta \in \text{Hom}_R(N_1 \oplus N', X)$ by $\beta(n + n') = \varphi(n) + \alpha(n')$ $(n \in N_1,$

$n' \in N$). Then, by (i), there exists $\delta \in \mathrm{Hom}_R(A, X)$ such that $\delta|_{N_1 \oplus N'} = \beta$. For any $n_i \in N$ $(1 \le i \le k)$, $\delta(n_1 + \cdots + n_k) = \beta(n_1 + \cdots + n_k) = \varphi(n_1) + \alpha(n_2 + \cdots + n_k) = \varphi(n_1) + \varphi(n_2 + \cdots + n_k) = \varphi(n_1 + \cdots + n_k)$. Thus $\delta|_N = \varphi$. It follows that $N \in \mathrm{Lift}_X(A)$. Hence, $\underline{C}^{(k)} \subseteq \mathrm{Lift}_X(A)$. □

Exercises

2.37. Prove that a free module over a principal ideal domain has SIP.

2.38. Let p be a prime integer. Let A be the $\mathbb{Z}$-module $\mathbb{Z} \oplus (\mathbb{Z}/\mathbb{Z}p)$. Show that

(i) If $K \le_{\mathrm{d}} A$, then $K = 0$, A, $\mathbb{Z} \oplus 0$ or $\mathbb{Z}(1, a + \mathbb{Z}p)$ $(0 \ne a \in \mathbb{Z})$.
(ii) A has C_3.

2.39. Let p be a prime integer. Let A be the $\mathbb{Z}$-module $(\mathbb{Z}/\mathbb{Z}p) \oplus (\mathbb{Z}/\mathbb{Z}p^3)$. Show that A does not satisfy condition C_3.

2.40. Let A be the $\mathbb{Z}$-module $\mathbb{Z} \oplus \mathbb{Z}$. Let $N \le A$. Show that

(i) $N \le_{\mathrm{d}} {}_{\mathbb{Z}}A$ if and only if there are coprime integers a and b such that $N = \mathbb{Z}(a, b)$.
(ii) A does not satisfy the SSP.

2.41. Let A be a left R-module. Prove that A has the SIP (resp., SSP) if and only if for every pair of direct summands S and T with $\pi : A \to S$, the canonical projection map, $\ker(\pi|_T)$ (resp., $\mathrm{im}(\pi|_T)$) is a direct summand of A (resp., S) (This is a characterization of the SIP and SSP conditions in terms of mappings.)

2.42. Let A be a left R-module. If A has the SIP (resp., SSP) and $S \oplus T \le_{\mathrm{d}} A$, then for any homomorphism $\alpha : S \to T$, $\ker\alpha \le_{\mathrm{d}} S$ (resp., $\mathrm{im}\,\alpha \le_{\mathrm{d}} T$).

2.43. Let p be a prime integer. Let A be the $\mathbb{Z}$-module $(\mathbb{Z}/\mathbb{Z}p) \oplus (\mathbb{Z}/\mathbb{Z}p^2)$. Show that A has neither the SIP, nor the SSP.

2.44. Let F be any field and let $R = T_2(F)$. Prove that R_R does not satisfy C_3, has the SIP, but does not have the SSP.

2.45. A left R-module A is said to have *Absolute direct summand* property (or to be an *Ads-module*) if for any decomposition $A = B \oplus C$, whenever D is any complement of C in A, then $A = D \oplus C$. Prove that ${}_RA$ is an Ads-module if and only if for any decomposition $A = B \oplus C$, B is C-injective.

2.46. Provide examples which show that there are no implications between the Ads condition and the SIP or SSP conditions.

2.47. Replace SIP with SSP in the statements of Theorem 2.79 (iii) and Corollary 2.80. Then prove the corresponding new results.

2.48. A ring R is called *Abelian* if every idempotent element of R is central. Prove that if R is Abelian, then R_R is an Ads-module.

2.49. Let A be a left R-module. Let $S = \text{End}\,({}_RA)$. Prove that S is Abelian if and only if $A = K \oplus N = L \oplus N$ implies that $K = L$.

2.50. Obtain corresponding SIP and SSP versions of the Proposition 2.85 (or Proposition 2.86).

2.51. Prove that being Artinian (hereditary or Noetherian, respectively) is Morita invariant.

Remarks. More information about complement submodules and relative injective modules can be found in [AF74], [AMV75], [Fai76], [SHD90], [Wis91], [Wis96], and [DHSW94].

Lifting submodules were introduced in [Ter92]. The ejectivity concept was defined in[ABT09]. This new generalization of relative injectivity is extremely useful in analyzing the structure of certain type of modules. In a similar way to lifting submodules, elifting submodules were introduced. Most results of the last section are taken from [Ter92], [Ter03], [BKT06], [KT03], [Gar89], [Hau89], and [BR93].

Chapter 3

Extending Property and Related Concepts

This chapter is concerned with extending (or CS) property and some related concepts which are interesting in their own right. In the first section we deal with the direct sums of CS-modules and submodules or some distinguished class of submodules of a CS-module. The direct sum of two CS-modules is not a CS-module. One of the most interesting questions concerning CS-modules is when a (finite or infinite) direct sum of CS-modules is also CS. After we briefly provide a new characterization the of CS property, we focus on the problem when a direct sum of an arbitrary number of CS-modules is again a CS-module. We provide several results related to this problem. It is well known that the CS condition is not closed under essential extensions. However, we show that if a module is CS, then so is its rational hull. Quite often it is useful to determine whether a certain module property is inherited by submodules or some distinguished class of submodules, especially if the module at hand, is the ring itself. We give conditions which ensure that a submodule of a CS-module will be CS.

In the second section we study extending modules over commutative domains, and also somewhat generalized versions of the main result in the noncommutative setting. To this end we give the detailed proof of a fundamental decomposition theorem which states that a torsion-free. CS-module over a commutative domain is a finite direct sum of injective modules and uniform modules. We apply this result to get more information about CS-modules over Dedekind domains or principal ideal domains. Moreover, we provide two generalizations of the aforementioned decomposition theorem. Thus it is shown that a nonsingular reduced CS-module over a commutative ring with only a finite number of minimal prime ideals has a finite uniform dimension and that a nonsingular left CS-module over a semiprime left Goldie ring is a finite direct sum of injective modules and uniform modules.

Section 3 provides a counterexample which shows that being CS is not Morita invariant. In view of this fact, we explore matrix CS-rings. After we prove general results on generalized upper triangular matrix CS-rings, we obtain necessary and sufficient conditions under which a right nonsingular generalized upper triangular matrix ring and a split null extension are right CS-rings. We provide examples which illustrate our results.

Section 4 is based on some related concepts to the extending property. We consider the CS property together with a conditional direct summand condition C_2 (C_3). Modules of this kind are known as continuous (quasi-continuous) modules in literature. We show that continuous and quasi-continuous modules can be characterized in terms of lifting homomorphisms from certain submodules to the module itself. We provide several examples which demonstrate our results.

3.1 Extending modules

In this section we introduce CS-modules and obtain a new characterization in terms of idempotent endomorphisms of the injective hulls. One of the most interesting questions concerning CS-modules is when a (finite or infinite) direct sum of CS-modules is also CS. We provide several results in this direction. It is well known that the CS property is not closed under essential extensions. However, we prove that if a module is CS, then so is its rational hull. It is useful to determine whether the CS property is inherited by submodules or some distinguished class of submodules especially if the module under question is the ring itself. We provide conditions which ensure that a submodule of a CS-module will be CS.

Let R be any ring. A left R-module A is called a CS-*module* (or *extending*) if every submodule of A is essential in a direct summand of A. In view of Proposition 2.5, it is clear that a module A is CS if and only if every closed (or complement) submodule is of A direct summand, whence the reason for the name "CS-module". Proposition 2.10 has the immediate consequence that any direct summand of a CS-module is a CS-module (Exercise 3.2). The origins of these concepts are twofold:

(i) The work of von Neumann in the 1930's concerning continuous geometries and their realization as lattices of principal left ideals of (von Neumann) regular rings; his work was further developed by Utumi and others in the setting of rings and modules.
(ii) The theory of injective modules.

In order to understand the relationship between CS-modules and injective modules, we recall some definitions.

We call A *quasi-injective* (or *self-injective*) if A is A-injective. It is not difficult to prove that A is quasi-injective if and only if $\theta(A) \subseteq A$ for every endomorphism θ of $E(A)$, the injective hull of A (Exercise 1.95). For any ring R, a left R-module A is called *quasi-continuous* if $\varphi(A) \subseteq A$ for every idempotent endomorphism φ of $E(A)$ (the name derives from von Neumann's work). It can be shown that A is quasi-continuous if and only if for every finite collection N_i $(1 \leq i \leq k)$ of submodules of A such that $\sum_i N_i$ is direct, there exist submodules L_i $(1 \leq i \leq k+1)$ of A such that $A = L_1 \oplus L_2 \oplus \cdots \oplus L_{k+1}$ and $N_i \leq_e L_i$ $(1 \leq i \leq k)$. Thus, for any ring R and left R-module A, we have:

A is injective $\Rightarrow$ A is quasi-injective $\Rightarrow$ A is quasi-continuous $\Rightarrow$ A is CS.

For example, the $\mathbb{Z}$-module $\mathbb{Z}$ is quasi-continuous but not quasi-injective. Also, the $\mathbb{Z}$-module $\mathbb{Z} \oplus \mathbb{Z}$ is CS but not quasi-continuous (see Exercise 3.1).

Our first result gives various characterizations of quasi-continuous modules.

Proposition 3.1. *The following statements are equivalent for a module A with injective hull E.*

(i) *A is quasi-continuous.*

(ii) *If $E = E_1 \oplus \cdots \oplus E_n$ is a finite direct sum of submodules E_i $(1 \le i \le n)$, then $A = (E_1 \cap A) \oplus \cdots \oplus (E_n \cap A)$.*

(iii) *If $E = E_1 \oplus E_2$ is a direct sum of submodules E_1 and E_2, then $A = (E_1 \cap A) \oplus (E_2 \cap A)$.*

(iv) (a) *A is* CS.

(b) *For any $K, L \le_{\rm d} A$ with $K \cap L = 0$, the submodule $K \oplus L$ is also a direct summand of A (i.e., A satisfies C_3).*

(v) *If $L_i \le A$ $(1 \le i \le n)$ with $L_1 \oplus \cdots \oplus L_n \le A$, where n is a positive integer, then there exist $A_i \le A$ $(1 \le i \le n+1)$ such that $A = A_1 \oplus \cdots \oplus A_{n+1}$ and $L_i \le_{\rm e} A_i$ $(1 \le i \le n)$.*

(vi) *If L_1, $L_2 \le A$ with $L_1 \cap L_2 = 0$, then there exist $A_i \le A$ $(1 \le i \le 3)$ such that $A = A_1 \oplus A_2 \oplus A_3$ and $L_i \le_{\rm e} A_i$ $(i = 1, 2)$.*

(vii) *If L_1, $L_2 \le A$ with $L_1 \cap L_2 = 0$, then there exist A_1,$A_2 \le A$ such that $A = A_1 \oplus A_2$ and $L_i \le A_i$ $(i = 1, 2)$.*

Proof. (i) $\Rightarrow$ (ii) Let $\pi_i : E \to E_i$ $(1 \le i \le n)$ denote the canonical projections. Then π_i is an idempotent endomorphism of E $(1 \le i \le n)$. Thus $\pi_i(A) \le A$ $(1 \le i \le n)$ and hence $A \le \pi_1(A) \oplus \cdots \oplus \pi_n(A) \le (E_1 \cap A) \oplus \cdots \oplus (E_n \cap A) \le A$, i.e., $A = (E_1 \cap A) \oplus \cdots \oplus (E_n \cap A)$.

(ii) $\Rightarrow$ (iii) Clear.

(iii) $\Rightarrow$ (i) Let φ be an idempotent endomorphism of E. Then $E = \varphi(E) \oplus (1-\varphi)(E)$ and hence $A = [\varphi(E) \cap A] \oplus [(1-\varphi)(E) \cap A]$. It follows that $\varphi(A) = \varphi[\varphi(E) \cap A] \oplus \varphi[(1-\varphi)(E) \cap A] \le \varphi(E) \cap A \le A$. Thus A is quasi-continuous.

(i) $\Rightarrow$ (iv) Let $N \le A$. Then $E = E(N) \oplus F$ for some $F \le E$. By (iii), $A = [E(N) \cap A] \oplus [F \cap A]$. Clearly, $N \le_{\rm e} E(N) \cap A$. Thus A is CS.

Let $K, L \le_{\rm d} A$ with $K \cap L = 0$. Then $E = E(K) \oplus E(L) \oplus G$ for some $G \le E$. By (ii), $A = [E(K) \cap A] \oplus [E(L) \cap A] \oplus [G \cap A] = K \oplus L \oplus (G \cap A)$, i.e., $K \oplus L \le_{\rm d} A$.

(iv) $\Rightarrow$ (v) Let n be a positive integer and let $L_i \le A$ $(1 \le i \le n)$ such that $L_1 + \cdots + L_n$ is direct. By (iv) (a), for each $1 \le i \le n$, there exists $A_i \le_{\rm d} A$ such that $L_i \le_{\rm e} A_i$. Then $A_1 + \cdots + A_n$ is direct and, by (iv)(b), $A_1 \oplus \cdots \oplus A_n \le_{\rm d} A$.

(v) $\Rightarrow$ (vi) $\Rightarrow$ (vii) Clear.

(vii) $\Rightarrow$ (iii) Suppose that there exist $E_i \le E$ $(i = 1, 2)$ such that $E = E_1 \oplus E_2$. Let $L_i = E_i \cap A$ $(i = 1, 2)$. Then $A = A_1 \oplus A_2$ for some $A_i \le A$ $(i = 1, 2)$ such that $L_i \le A_i$ $(i = 1, 2)$. Since $L_i \le_{\rm e} E_i$ $(i = 1, 2)$ it follows that $L_i \le_{\rm e} A_i$ $(i = 1, 2)$. Let $x \in A_1$. There exist $y \in E_1$, $z \in E_2$ such that $x = y + z$. Suppose $z \ne 0$. There exists $r \in R$ such that $0 \ne rz \in A$. Then $rz = rx - ry \in A_1 \cap A_2 = 0$, a contradiction. Thus $z = 0$. Hence $x = y \in E_1 \cap A = L_1$. Thus $L_1 = A_1$. Similarly, $L_2 = A_2$. We conclude that $A = (E_1 \cap A) \oplus (E_2 \cap A)$. □

Recall that R-modules A_i $(i \in I)$ are called *relatively injective* if A_i is A_j-injective for all distinct i, j in I. We give another characterization of quasi-continuous modules.

Corollary 3.2. *A module A is quasi-continuous if and only if*

(i) *A is* CS, *and*

(ii) *Whenever $A = A_1 \oplus A_2$ is a direct sum of submodules A_1, A_2, then A_1 and A_2 are relatively injective.*

Proof. If A is quasi-continuous then (i) and (ii) follow by Proposition 3.1 and Exercise 3.4. Conversely, suppose that (i) and (ii) hold. Let L_1 and L_2 be submodules of A such that $L_1 \cap L_2 = 0$. By (i), there exist $A_1, A_2 \leq A$ such that $A = A_1 \oplus A_2$ and L_1 is essential in A_1. Clearly $A_1 \cap L_2 = 0$. By (ii), there exists a submodule A' of A such that $A = A \oplus A'$ and $L_2 \leq A'$ (Exercise 3.4). By Proposition 3.1, A is quasi-continuous. □

For a module A, consider the following relations on the set of submodules of A:

(i) $X \alpha Y$ if there exists $N \leq A$ such that $X \leq_e N$ and $Y \leq_e N$;

(ii) $X \beta Y$ if $X \cap Y \leq_e X$ and $X \cap Y \leq_e Y$. (Equivalently, $X \beta Y$ if and only if $X \cap N = 0$ implies $Y \cap N = 0$ and $Y \cap K = 0$ implies $X \cap K = 0$ for all $N, K \leq A$.)

Observe that if $X, Y \leq A$ are such that $X \alpha Y$, then $X \beta Y$.

A module A is called a UC-*module* if every submodule has a unique closure. (see Proposition 2.5).

For example, semisimple modules, uniform modules, and nonsingular modules are all examples of UC-modules (see Exercise 2.9). The $\mathbb{Z}$-module $(\mathbb{Z}/\mathbb{Z}p) \oplus (\mathbb{Z}/\mathbb{Z}p^3)$ is not UC (Exercise 3.5).

Lemma 3.3. *Let A be a module. Then*

(i) *α is reflexive and symmetric.*

(ii) *α is transitive if and only if A is a* UC-*module.*

(iii) *β is an equivalence relation.*

Proof. The proof is routine. □

Our next objective is to characterize CS-modules in terms of the β relation. First we give the following example, which shows that being isomorphic and, the β equivalence relation are different on the set of submodules of a module.

Example 3.4. (i) Let F be any field. Let $R_R = \begin{bmatrix} F & F \\ 0 & F \end{bmatrix}$. Take the right ideals $X = \begin{bmatrix} 0 & F \\ 0 & 0 \end{bmatrix}$ and $Y = \begin{bmatrix} 0 & 0 \\ 0 & F \end{bmatrix}$ of R. Then X and Y are not β-related. However it is clear that $X \cong Y$.

(ii) Take for R a noncommutative nonprincipal ideal domain. Let X be any right ideal of R. Then $X\beta R$, but X is not isomorphic to R_R.

Proposition 3.5. *Let A be a module. The following statements are equivalent.*

(i) *A is CS.*
(ii) *for each $X \leq A$, there exists a direct summand D of A such that $X\alpha D$.*
(iii) *for each $c = c^2 \in \text{End}\,(E(A))$ there is an $e = e^2 \in \text{End}\,(E(A))$ such that $Ae \leq A$, $E(A)e\beta E(A)c$, and there exists a homomorphism $h : E(A)c \to E(A)e$ such that $h|_{A\cap E(A)c}$ is the inclusion homomorphism.*

Proof. (i) $\Leftrightarrow$ (ii) Clear.

(i) $\Rightarrow$ (iii) Let $c = c^2 \in \text{End}(E(A))$ and $X = A \cap E(A)c$. Then there is $f = f^2 \in \text{End}(A)$ such that $X \leq_e Af$. Let $e \in \text{End}(E(A))$ be the projection $e : E(A) \to E(Af)$ (i.e., for $x \in E(A)$, $x = x_1 + x_2$, where $x_1 \in E(Af)$ and $x_2 \in E(A(1-f))$ and $e(x) = x_1$). So $X \leq_e E(A)c$ and $X \leq_e E(A)e$. Hence $E(A)e\beta E(A)c$. Now $Ae = (Af \oplus A(1-f))e = (Af)e \oplus A(1-f)e = Af$ gives that $Ae \leq A$. Since $E(A)e$ is injective, there is a monomorphism $h : E(A)c \to E(A)e$ that extends the inclusion $i : X \to E(A)e$.

(iii) $\Rightarrow$ (i) Let $Y \leq A$. Then $Y \leq_e E(Y) = E(A)c$ for some $c = c^2 \in \text{End}(E(A))$. So there is $e = e^2 \in \text{End}(E(A))$ such that $Ae \leq A$, $E(Ae)\beta E(A)c$, and there exists $h : E(A)c \to E(A)e$ such that $h|_{A\cap E(A)c}$ is the inclusion homomorphism. Hence, for $y \in Y \subseteq A \cap E(A)c$, $h(y) = y \in A \cap E(A)e = Ae$. Then $Y \leq Ae \leq_d A$ and since $E(A)e\beta E(A)c$, $Y \leq_e Ae$. Consequently, A is CS. □

It is clear that every module has at least one CS essential extension, namely its injective hull. However, since the CS condition is a generalization of injectivity, one may expect to find CS essential extensions intermediate between a module and its injective hull. It is well known that the CS condition is not closed under essential extensions (see Exercise 3.6). However, we show that if a module is CS, then so is its rational hull. Recall that for a module A, the rational hull of A is defined as the following submodule of $E(A)$: $\widetilde{E}(A) = \{x \in E(A) : h(A) = 0 \Rightarrow h(x) = 0$ for all $h \in \text{End}(E(A))\}$ (see Exercise 3.8).

Theorem 3.6. *If A is* CS, *then $\widetilde{E}(A)$ is* CS.

Proof. Let $K \leq \widetilde{E}(A)$ and $X = K \cap A$. So there exists $e = e^2 \in \text{End}(A)$ such that $X \leq_e Ae$. By Exercises 3.10 and 3.11, there exists $f \in \text{End}(\widetilde{E}(A))$ such that $f|_A = e$. Since $E(A)$ is injective, there exists $\bar{e} \in \text{End}(E(A))$ such that $\bar{e}\,|_{\widetilde{E}(A)}= f$. Let $a \in A$. Then $[\bar{e} - \bar{e}^2](a) = [e - e^2](a) = 0$. By the definition of $\widetilde{E}(A)$, $[\bar{e} - \bar{e}^2](y) = 0$ for all $y \in \widetilde{E}(A)$. Hence $f = f^2$. Assume $k \in K$ is such that

$fk-k \neq 0$. There exists $r \in R$ such that $0 \neq r(fk-k)$ and $rk \in A$. Then $rk \in X$. So $r(fk-k) = rfk - rk = rek - rk = 0$, a contradiction. Hence, $K \leq \widetilde{E}(A)f$. Let $0 \neq vf \in \widetilde{E}(A)f$. There exists $s \in R$ such that $0 \neq svf$ and $sv \in A$. Then $0 \neq svf = sve \in Ae$. So there exists $t \in R$ such that $0 \neq tsvf \in X \leq K$. Therefore $K \leq_e \widetilde{E}(A)f$, so $\widetilde{E}(A)$ is CS. □

We have already seen that every quasi-continuous module, and in particular any injective module, is a CS-module. Clearly, other examples of CS-modules include all uniform modules and all semisimple modules. However, the direct sum of two CS-modules is not a CS-module (Exercises 3.5, 3.7). One of the most interesting questions concerning CS-modules is when a (finite or infinite) direct sum of CS-modules is also CS. Even for the ring $\mathbb{Z}$ of integers, the situation is very interesting. For example, for any prime p, the $\mathbb{Z}$-module $(\mathbb{Z}/\mathbb{Z}p) \oplus (\mathbb{Z}/\mathbb{Z}p^2)$ is CS, but the module $(\mathbb{Z}/\mathbb{Z}p) \oplus (\mathbb{Z}/\mathbb{Z}p^3)$ is not (Exercise 3.5). Observe that $\mathbb{Z}/\mathbb{Z}p$ is not $\mathbb{Z}/\mathbb{Z}p^3$-injective and also not $\mathbb{Z}/\mathbb{Z}p^2$-injective. For the readers convenience we propose Exercise 3.12 which contains a sufficient condition for a finite direct sum of CS-modules to be a CS-module and a fundamental characterization of the CS property in terms of second singular submodule.

In an effort to understand the aforementioned examples better we study when arbitrary direct sum of CS-modules is a CS-module. Let R be any ring. Let A be an R-module such that $A = \bigoplus_{i \in I} A_i$ is the direct sum of CS-modules A_i $(i \in I)$, for some given index set I. We are interested in when A itself is a CS-module. For any set I, $|I|$ will denote its cardinality.

Theorem 3.7. *Let R be a ring and let $A = \bigoplus_{i \in I} A_i$ be the direct sum of R-modules A_i $(i \in I)$, for some index set I with $|I| \geq 2$. Then the following statements are equivalent.*

(i) *A is* CS.

(ii) *There exist $i, j \in I$, $i \neq j$, such that every closed submodule K of A with $K \cap A_i = 0$ or $K \cap A_j = 0$ is a direct summand.*

(iii) *There exist $i, j \in I$, $i \neq j$, such that every complement of A_i or of A_j in A is a* CS-*module and a direct summand of A.*

Proof. (i) ⇒ (ii) Clear.

(ii) ⇒ (iii) Let K be a complement of A_i in A. By (ii), K is a direct summand of A. Let L be a closed submodule of K. By Proposition 2.10, L is a closed submodule of A and clearly $L \cap A_i = 0$. By (ii), L is a direct summand of A, and hence also of K. Thus K is CS.

(iii) ⇒ (i) Let N be a closed submodule of A. There exists a closed submodule H of N such that $N \cap A_i$ is essential in H. Clearly $H \cap A_j = 0$. By Zorn's Lemma, there exists a complement P of A_j in A such that $H \leq P$. Now Proposition 2.10 shows that H is closed in A, and hence H is closed in P. Applying (iii) we see that H is a direct summand of the CS-module P and P is a direct summand of A. Hence H is a direct summand of A.

There exists a submodule H' of A such that $A = H \oplus H'$. The modular law gives $N = H \oplus (N \cap H')$. By Proposition 2.10, $N \cap H'$ is a closed submodule of A, and clearly $(N \cap H') \cap A_i = 0$. By the above argument, (iii) gives that $N \cap H'$ is a direct summand of A, and hence also of H'. It follows that N is a direct summand of A. Thus A is CS. □

Lemma 3.8. *Let $A = A_1 \oplus A_2$ be a module and let K be a submodule of A. Then K is a complement of A_2 in A if and only if there exists a homomorphism $\varphi : A_1 \to E(A_2)$ such that $K = \{x + \varphi(x) : x \in \varphi^{-1}(A_2)\}$.*

Proof. Suppose that K is a complement of A_2 in A. Let $\pi_i : A \to A_i$ $(i = 1, 2)$ denote the canonical projections. Note that $\pi_1|_K : K \to A_1$ is a monomorphism. If $i : A_2 \to E(A_2)$ is the inclusion mapping, then there exists a homomorphism $\varphi : A_1 \to E(A_2)$ such that $\varphi(\pi_1|_K) = i(\pi_2|_K)$. For any $x \in K$, $\varphi\pi_1(x) = \pi_2(x) \in A_2$, so that $\pi_1(x) \in \varphi^{-1}(A_2)$, and $x = \pi_1(x) \oplus \pi_2(x) = \pi_1(x) \oplus \varphi(\pi_1(x))$. Thus $K \subseteq \{y + \varphi(y) : y \in \varphi^{-1}(A_2)\} = K_1$ (say). But K_1 is a submodule of A and $K_1 \cap A_2 = 0$, so that $K = K_1$, as required.

Conversely, suppose that $\theta : A_1 \to E(A_2)$ is a homomorphism and $K = \{x + \theta(x) : x \in \theta^{-1}(A_2)\}$. Clearly, K is a submodule of A and $K \cap A_2 = 0$. Suppose that L is a submodule of A such that $K \subseteq L$ and $L \cap A_2 = 0$. Now suppose there exists $u \in L$ such that $\pi_2(u) \neq \theta\pi_1(u)$. Since $0 \neq \pi_2(u) - \theta\pi_1(u) \in E(A_2)$, there exists $r \in R$ such that $0 \neq r(\pi_2(u) - \theta\pi_1(u)) \in A_2$. But, in this case, $r\theta\pi_1(u) \in A_2$ and $r(\pi_2(u) - \theta\pi_1(u)) = \pi_2(ru) - \theta\pi_1(ru) = ru - (\pi_1(ru) + \theta\pi_1(ru)) \in (L + K) \cap A_2 = L \cap A_2 = 0$, contradiction.

Let $v \in L$. Then $\theta\pi_1(v) = \pi_2(v) \in A_2$, so that $\pi_1(v) \in \theta^{-1}(A_2)$ and $v = \pi_1(v) + \pi_2(v) = \pi_1(v) + \theta(\pi_1(v)) \in K$. It follows that $L = K$. Thus, K is a complement of A_2 in A. □

Now we wish to apply Lemma 3.8 to the module $A = \bigoplus_{i \in I} A_i$. For each $i \in I$ let A_{-i} denote $\bigoplus_{j \in I \setminus \{i\}} A_j$. Combining Theorem 3.7 and Lemma 3.8 we now have at once:

Theorem 3.9. *Let R be a ring and let $A = \bigoplus_{i \in I} A_i$ be the direct sum of R-modules A_i $(i \in I)$, for some index set I with $|I| \geq 2$. Then the following statements are equivalent.*

(i) *A is* CS.
(ii) *For each $i \in I$ and each homomorphism $\varphi : A_{-i} \to E(A_i)$, the submodule $\{x + \varphi(x) : x \in \varphi^{-1}(A_i)\}$ is a* CS*-module and a direct summand of A.*
(iii) *There exist $i \neq j$ in I such that for each $k \in \{i, j\}$ and each homomorphism $\varphi : A_{-k} \to E(A_k)$, the submodule $\{x + \varphi(x) : x \in \varphi^{-1}(A_k)\}$ is a* CS*-module and a direct summand of A.*

By Exercise 3.12, any finite direct sum of relatively injective CS-modules is CS. Let $A = \bigoplus_{i \in I} A_i$, where the modules A_i $(i \in I)$ are relatively injective. Let $i \in I$ and let $\varphi : A_{-i} \to E(A_i)$ be a homomorphism. By Proposition 2.20, A_i is

A_{-i}-injective and hence $\varphi(A_{-i}) \leq A_i$ (Proposition 2.22). Let $K = \{x + \varphi(x) : x \in \varphi^{-1}(A_i)\} = \{x + \varphi(x) : x \in A_{-i}\}$. Then $A = K \oplus A_i$. Thus in Theorem 3.9, if the modules A_i $(i \in I)$ are relatively injective, the condition (ii) becomes (ii′) For each $i \in I$, A_{-i} is CS, and (iii) becomes (iii′) There exist $i \neq j$ in I such that A_{-i} and A_{-j} are CS.

We also note that if $A = \bigoplus_{i \in I} A_i$ is a CS-module, for any family of modules A_i $(i \in I)$, and if $i \in I$, then $\varphi^{-1}(A_i)$ is a CS-module for any homomorphism $\varphi : A_{-i} \to E(A_i)$. This is because in Theorem 3.9, $\varphi^{-1}(A_i) \cong \{x + \varphi(x) : x \in \varphi^{-1}(A_i)\}$.

Let p be any prime integer and let A denote the $\mathbb{Z}$-module $(\mathbb{Z}/\mathbb{Z}p) \oplus (\mathbb{Z}/\mathbb{Z}p^3)$. Let $A_2 = 0 \oplus (\mathbb{Z}/\mathbb{Z}p^3)$. Then A_2 is a CS-module, being uniform. Let K be a closed submodule of A with $K \cap A_2 = 0$. By Zorn's Lemma, there exists a complement L of A_2 in A such that $K \leq L$. Now $L \oplus A_2$ is essential in A (Proposition 2.3), so that $(\mathbb{Z}/\mathbb{Z}p) \oplus 0 \subseteq L \oplus A_2$, and it follows that $A = L \oplus A_2$. Now $L \cong A/A_2$, which is simple, so that L is a CS-module and K is a direct summand of A: indeed $K = 0$, or L. Thus Theorem 3.7 fails if in (ii) or (iii) we take only one element i in I rather than distinct elements i, j. However, in a particular case, Theorem 3.7 does hold when we take only one element i in I. Recall that the $\mathbb{Z}$-module $(\mathbb{Z}/\mathbb{Z}p) \oplus (\mathbb{Z}/\mathbb{Z}p^3)$ is not UC (Exercise 3.5).

Theorem 3.10. *Let R be a ring and let A be a* UC*-module such that $A = \bigoplus_{i \in I} A_i$ is the direct sum of R-modules A_i $(i \in I)$, for some non-empty index set I. Then the following statements are equivalent.*

(i) *A is* CS.
(ii) *There exists $i \in I$ such that A_i is* CS *and every closed submodule K of A with $K \cap A_i = 0$ is a direct summand.*
(iii) *There exists $i \in I$ such that A_i is* CS *and every complement of A_i in A is a* CS*-module and a direct summand of A.*
(iv) *The module A_i is* CS *for each $i \in I$ and every closed submodule L of A with $L \cap A_i = 0$ $(i \in I)$ is a direct summand of A.*

Proof. (i) ⇒ (ii) By Exercise 3.2.

(ii) ⇒ (iii) Let L be a complement of A_i in A. By (ii), L is a direct summand of A. Let N be a closed submodule of L. By Proposition 2.10 and (ii), N is a direct summand of A, and hence also of L. Thus L is CS.

(iii) ⇒ (i) Let H be a closed submodule of A. By Exercise 3.13, $H \cap A_i$ is a closed submodule of A_i and hence, by (iii), $H \cap A_i$ is a direct summand of A. Thus $A = (H \cap A_i) \oplus H'$ for some submodule H' of A. Now $H = (H \cap A_i) \oplus (H \cap H')$ and $H \cap H'$ is a closed submodule of A (Proposition 2.10). Moreover, $(H \cap H') \cap A_i = 0$. From the proof of Theorem 3.7 (iii) ⇒ (i), it follows that $H \cap H'$ is a direct summand of A, and hence H is a direct summand of A.

(i) ⇒ (iv) By Exercise 3.2.

(iv) ⇒ (i) Let P be a closed submodule of A. For each $i \in I$, $P \cap A_i$ is closed in A_i (Exercise 3.13) and hence $A_i = (P \cap A_i) \oplus A_i'$ for some submodule A_i' of A.

Let $A' = \bigoplus_{i \in I} A_i'$, $P' = \bigoplus_{i \in I}(P \cap A_i)$. Then $A = P' \oplus A'$ and $P' \leq P$. It follows that $P = P' \oplus (P \cap A')$. By Proposition 2.10, $P \cap A'$ is closed in A and clearly $(P \cap A') \cap A_i = 0$ $(i \in I)$. By (iv), $P \cap A'$ is a direct summand of A. Thus P is a direct summand of A. We conclude that A is CS. □

Note that the $\mathbb{Z}$-module $(\mathbb{Z}/\mathbb{Z}p) \oplus (\mathbb{Z}/\mathbb{Z}p^2)$ is CS but not UC, and the $\mathbb{Z}$-module $A = (\mathbb{Z}/\mathbb{Z}p) \oplus (\mathbb{Z}/\mathbb{Z}p^3)$ is neither UC nor CS for any prime p (Exercise 3.5). Moreover, A is the direct sum of its submodules $A_1 = (\mathbb{Z}/\mathbb{Z}p) \oplus 0$ and $A_2 = 0 \oplus (\mathbb{Z}/\mathbb{Z}p^3)$, both of which are CS. In addition, if L is any non-zero submodule of A such that $L \cap A_1 = 0$ and $L \cap A_2 = 0$, then $L = \mathbb{Z}(1 + \mathbb{Z}p, a + \mathbb{Z}p^3)$ for some $a \in \mathbb{Z}p^2 \backslash \mathbb{Z}p^3$ and thus L is a direct summand of A, in fact $A = L \oplus A_2$. Thus Theorem 3.10 (iv) $\Rightarrow$ (i) fails if A is not UC.

Theorem 3.9 has the following companion result for UC-modules.

Theorem 3.11. *Let R be a ring and let A be a* UC*-module such that $A = \bigoplus_{i \in I} A_i$ is the direct sum of R-modules A_i $(i \in I)$, for some non-empty index set I. Then the following statements are equivalent.*

(i) *A is* CS.
(ii) *There exists $i \in I$ such that A_i is* CS *and for each homomorphism $\varphi : A_{-i} \to E(A_i)$ the submodule $\{x + \varphi(x) : x \in \varphi^{-1}(A_i)\}$ is a* CS*-module and a direct summand of A.*

Proof. Follows from Lemma 3.8 and Theorem 3.10. □

As we remarked above, there are CS-modules as well as not CS-modules which contain semisimple direct summands (Exercise 3.5). It is natural to ask for which modules A_1 and semisimple modules A_2 is the module $A_1 \oplus A_2$ CS. First we prove:

Proposition 3.12. *Let R be a ring and let A be a* UC *R-module such that $A = A_1 \oplus A_2$ is the direct sum of an R-module A_1 and a semisimple R-module A_2. Then A is* CS *if and only if A_1 is* CS.

Proof. The necessity is clear by Exercise 3.2. Conversely, suppose that A_1 is CS. Let K be a complement of A_1 in A. Then $A_1 \oplus K$ is essential in A (Proposition 2.22), and hence $A_2 \leq \operatorname{soc} A \leq A_1 \oplus K$. Thus $A = A_1 \oplus K$. It follows that $K \cong A/A_1 \cong A_2$, so that K is CS. By Theorem 3.10, A is CS. □

Proposition 3.13. *Let R be a ring, A_1 be an R-module with zero socle, and A_2 be a semisimple R-module. Then the R-module $A = A_1 \oplus A_2$ is* CS *if and only if A_1 is* CS *and A_2 is A_1-injective.*

Proof. Assume A is CS. By Exercise 3.2, A_1 is CS. Clearly, $A_2 = \operatorname{soc} A$. Let N be any submodule of A_1 and let $\varphi : N \to A_2$ be a homomorphism. Let $L = \{x - \varphi(x) : x \in N\}$. Then L is a submodule of A and $L \cap A_2 = 0$. There exist submodules K, K' of A such that $A = K \oplus K'$ and L is an essential submodule of K. Note that $\operatorname{soc} K = K \cap A_2 = 0$, so that $A_2 = \operatorname{soc} A \subseteq K'$. Thus, $K' = A_2 \oplus (K' \cap A_1)$ and $A = K \oplus A_2 \oplus (K' \cap A_1)$. Let $\pi : A \to A_2$ denote the projection with kernel $K \oplus (K' \cap A_1)$. Let $\theta = \pi|_{A_1}$. Then $\theta : A_1 \to A_2$ and $\theta(x) = \varphi(x)$ for all $x \in N$. It follows that A_2 is A_1-injective. Conversely, suppose that A_1 is CS and A_2 is A_1-injective. Clearly A_1 is A_2-injective. By Exercise 3.12, A is CS. □

Combining Theorem 3.7 with the proof of Proposition 3.12 we obtain the following general result.

Lemma 3.14. *Let R be a ring and let A_1 and A_2 be R-modules with A_2 semisimple. Then the R-module $A_1 \oplus A_2$ is* CS *if and only if every complement K of A_2 in A is a* CS*-module and direct summand of A.*

Theorem 3.15. *Let R be a ring, A_1 be a* CS *R-module, and A_2 be a semisimple R-module such that A_2 is (A_1/N)-injective for every non-zero submodule N of A_1. Then the R-module $A = A_1 \oplus A_2$ is* CS.

Proof. Let K be a complement of A_2 in A. There exists a homomorphism $\varphi : A_1 \to E(A_2)$ such that $K = \{x + \varphi(x) : x \in \varphi^{-1}(A_2)\}$ (Lemma 3.8). Let $Q = \varphi^{-1}(A_2)$ and let $P = \ker \varphi$. Then $P \leq Q$ are submodule of A_1.

Suppose that $P = 0$. Then $K \cap A_1 = 0$, and hence $A_1 \oplus K = A_1 \oplus \varphi(Q)$, which is a direct summand of A, because $\varphi(Q)$ is a direct summand of A_2. Thus K is a direct summand of A and, because K embeds in $A/A_1 \cong A_2$, K is semisimple and thus CS.

Now suppose that $P \neq 0$. By hypothesis, A_2 is (A_1/P)-injective. Now $Q/P \cong \varphi(Q)$, which is a direct summand of A_2. Thus Q/P is (A_1/P)-injective. There exists a submodule Q' of A_1 such that $P \subseteq Q'$ and $(A_1/P) = (Q/P) \oplus (Q'/P)$. Define $\theta : A_1 \to E(A_2)$ by $\theta(q+q') = \varphi(q)$ $(q \in Q, q' \in Q')$. It is easily checked that θ is well defined and a homomorphism. Moreover, $\theta|_Q = \varphi$. Let $K' = \{x + \theta(x) : x \in \theta^{-1}(A_2)\} = \{x + \theta(x) : x \in A_1\}$, noting that $\theta(A_1) = \varphi(Q) \leq A_2$. Lemma 3.8 gives that K' is a complement of A_2 in A. But $K \leq K'$, so that $K = K'$. Clearly $A = K \oplus A_2$. Thus K is a CS-module and a direct summand of A. By Lemma 3.14 A is CS. □

Now we aim to prove a partial converse to Theorem 3.15.

Lemma 3.16. *Let R be a ring, A_1 be a uniform R-module of finite composition length, and A_2 be a semisimple module such that the R-module $A = A_1 \oplus A_2$ is* CS. *Let $\varphi : A_1 \to E(A_2)$ be a homomorphism such that $\varphi(A_1) \nleq A_2$. Then $\varphi^{-1}(A_2) = 0$ or $\varphi^{-1}(A_2)$ is isomorphic to a simple submodule of A_2.*

Proof. Let $U = \varphi^{-1}(A_2)$. Let $K = \{x + \varphi(x) : x \in U\}$. By Lemma 3.8, K is a closed submodule and hence K is a direct summand. Note that $K \cong U \cong A_1$. Thus $K = 0$ or is uniform. Suppose that $K \neq 0$. By the Krull–Schmidt Theorem, $K \cong A_1$ or K is isomorphic to a simple submodule of A_2. Suppose that $K \cong A_1$. Comparing composition lengths, $U = A_1$ and hence $\varphi(A_1) \leq A_2$, a contradiction. Thus $U = 0$ or U is isomorphic to a simple submodule of A_2. □

Theorem 3.17. *Let R be a ring, A_1 be a uniform R-module of finite composition length and A_2 be a semisimple R-module. Then the R-module $A = A_1 \oplus A_2$ is* CS *if and only if A_2 is (A_1/N)-injective for every non-zero submodule N of A_1.*

Proof. The sufficiency was proved in Theorem 3.15. Conversely, suppose that A is CS. Suppose that N is a non-zero submodule of A_1, L is a submodule containing N, and there exists a monomorphism $\alpha : L/N \to A_2$. Note that $\alpha(L/N)$ is a direct summand of A_2 and hence $A_1 \oplus \alpha(L/N)$ is CS (Exercise 3.2). Thus without loss of generality, $\alpha : L/N \to A_2$ is an isomorphism.

Let $\pi : L \to L/N$ denote the canonical epimorphism. Let $\theta = \alpha\pi : L \to A_2$. Then θ can be lifted to a homomorphism $\varphi : A_1 \to E(A_2)$. Let $Q = \varphi^{-1}(A_2)$. Clearly $L \leq Q$. For any $q \in Q$ there exists $x \in L$ such that $\varphi(q) = \theta(x) = \varphi(x)$, so that $Q = L + \ker\varphi$. Moreover, $L \cap \ker\varphi = L \cap \ker\theta = N$. Thus $Q/N = (L/N) \oplus ((\ker\theta)/N)$. But $N \neq 0$ implies that the composition length of Q is at least 2. By Lemma 3.16, $\varphi(A_1) \leq A_2$, i.e., $Q = A_1$. Thus $A_1/N = (L/N) \oplus ((\ker\theta)/N)$. It follows that A_2 is (A_1/N)-injective. □

Theorem 3.17 has the following immediate consequence.

Corollary 3.18. *Let R be a ring and let A_1 be an R-module with unique composition series $A_1 > L > N > 0$. Then the R-module $A_1 \oplus (L/N)$ is not* CS.

Quite often it is useful to determine whether a certain module property is inherited by submodules or some distinguished class of submodules, especially if the module under question is the ring itself. For example, it is well known that if R is the $n \times n$ upper triangular matrix ring over a division ring, then every right ideal is CS. Clearly, if A is a uniform or semisimple (or CS) module, then every (closed) submodule is CS.

Next we provide conditions which ensure that a submodule of a CS-module will be CS.

Theorem 3.19.

(i) *Let A be a* CS*-module and X a submodule of A. If the intersection of X with any direct summand of A is a direct summand of X, then X is a* CS*-module.*
(ii) *Let A be a module, X a* CS *submodule of A, and D a direct summand of A. If $D + X$ is nonsingular, then $D \cap X$ is a direct summand of X.*
(iii) *If A is nonsingular and X a* CS *submodule, then the intersection of X with any direct summand of A is a direct summand of X.*

Proof. (i) Let N be a submodule of X. Then there exists a direct summand D of A such that N is essential in D. Then N is essential in $D \cap X$ and $D \cap X$ is a direct summand of X. Consequently, X is a CS-module.

(ii) Let D be a direct summand of A and $Y = D \cap X$. Then there exists a submodule C of X such that C is a direct summand of X and Y is essential in C. Assume $Y \neq C$. Then $D \neq D + C$. Let $d + c \in D + C$ such that $d + c \notin D$, where

$d \in D$ and $c \in C$. So $c \neq 0$. There exists an essential left ideal L of R such that $0 \neq Lc \subseteq Y$ (Exercise 2.3). Since D is nonsingular, $0 \neq L(d+c) \subseteq D$. Thus D is essential in $D + C$, which is a contradiction. We conclude that $Y = C$.

(iii) This part follows immediately by the same proof as in part (ii). □

A module is called *distributive* if its lattice of submodules is a distributive lattice. A submodule X of a module A is called *fully invariant* if $f(X) \subseteq X$ for all $f \in \operatorname{End} A$ (see Exercise 3.14).

Corollary 3.20. *Let A be a* CS-*module.*

(i) *If A is a distributive module, then every submodule is* CS.
(ii) *If X is a submodule of A such that $e(X) \subseteq X$ for all $e = e^2 \in \operatorname{End} A$, then X is a* CS-*module. In particular every fully invariant submodule of A is* CS.

Proof. (i) is a consequence of Theorem 3.19 (i).

(ii) Let D be a direct summand of A and $\pi : A \to D$ the projection map. Then $\pi(X) = X \cap D$. By Theorem 3.19 (i), X is CS. □

Recall that a ring (or right ideal) is *reduced* if it contains no nonzero nilpotent elements.

Corollary 3.21. *Let R be a right* CS-*ring.*

(i) *If R is Abelian, then every right ideal of R is right* CS.
(ii) *Every ideal of R is a* CS *submodule of R_R.*
(iii) *If X is a reduced right ideal of R, then X is* CS.

Proof. (i) and (ii) follows from Corollary 3.20 (ii).

(iii) There exists $e = e^2 \in R$ such that X is right essential in eR. Hence eR is reduced. By Exercise 3.15, eR is a reduced ring with unity. Also eR is right CS. By part (i), X is CS. □

Corollary 3.22. *Every nonsingular* CS-*module has* SIP.

Proof. This follows from Theorem 3.19 (iii). □

The following example shows that the nonsingularity condition is not necessary in order for the CS property to be inherited by right ideals of a noncommutative ring. However, it does show that the nonsingularity condition is not superfluous in Theorem 3.19 (iii) and Corollary 3.22.

Example 3.23. Let F be a field and let

$$R_R = \left\{ \begin{bmatrix} a & x & 0 & 0 \\ 0 & b & 0 & 0 \\ 0 & 0 & b & y \\ 0 & 0 & 0 & a \end{bmatrix} : a, b, x, y \in F \right\}.$$

Then every right ideal of R is CS and $Z(R) \leq_{\mathrm{e}} R_R$.

To see that the nonsingularity condition in Corollary 3.22 is not superfluous, let

$$e = e^2 = \begin{bmatrix} 0 & 1 & 0 & 0 \\ 0 & 1 & 0 & 0 \\ 0 & 0 & 1 & 0 \\ 0 & 0 & 0 & 0 \end{bmatrix} \quad \text{and} \quad c = c^2 = \begin{bmatrix} 0 & 0 & 0 & 0 \\ 0 & 1 & 0 & 0 \\ 0 & 0 & 1 & 0 \\ 0 & 0 & 0 & 0 \end{bmatrix}.$$

Then $eR \cap cR$ is nilpotent. Hence $eR \cap cR$ is not a direct summand of R. So R does not have SIP.

Since every module embeds in a CS-module (namely, its injective hull), it is natural to ask: how can the injective hull of a module be utilized to determine the CS property for the module? We now provide a few results that address this question.

Corollary 3.24. *Let X be a module such that $X \cap D$ is a direct summand of X, for all injective submodules D of $E(X)$. Then X is* CS.

Proof. Consider X a submodule of the injective (hence CS) module $E(X)$. Then use Theorem 3.19 (i). □

Corollary 3.25. *Let X be a nonsingular module. Then X is* CS *if and only if $X \cap D$ is a direct summand of X, for all injective submodules D of $E(X)$.*

Proof. Use Corollary 3.24 and Theorem 3.19 (ii). □

Proposition 3.26. *Let R be a right nonsingular right* CS*-ring and I a right ideal of R. Then the following statements are equivalent:*

(i) *I is* CS.
(ii) *For any $c = c^2 \in R$ there exists $e = e^2 \in Q$ such that $cR \cap I = eR \cap I = eI$, where Q denotes the maximal right quotient ring of R.*

Proof. (i) $\Rightarrow$ (ii) Let $c = c^2 \in R$. By Theorem 3.19 (iii), $cR \cap I$ is a direct summand of I. Let K be a complement of I in R and $\pi : I \oplus K \to cR \cap I$ the projection homomorphism. By the Baer criterion, there exists $e \in Q$ such that $\pi(x) = ex$ for all $x \in I \oplus K$. Then $(e - e^2)(I \oplus K) = 0$. Since Q is a right nonsingular R-module, it follows that $e = e^2$. Hence $eI = cR \cap I$. Then $eI = eR \cap I$.

(ii) $\Rightarrow$ (i) This implication follows from Theorem 3.19 (i). □

Note in Proposition 3.26 (ii), that eQ is a direct summand of cQ. Indeed by Theorem 3.19 (iii), $cQ \cap eQ$ is a direct summand of eQ. But $cR \cap I \subseteq cQ \cap eQ$, hence $cQ \cap eQ = eQ$. Thus by the Modular Law, $cQ = eQ \oplus (cQ \cap (1 - e)Q)$. Furthermore, $E(cR \cap I) = eQ$.

From Proposition 3.1 we have: a module A is quasi-continuous if and only if $E(A) = \bigoplus_{i=1}^n E_i$ implies $A = \bigoplus_{i=1}^n (A \cap E_i)$. This condition on the injective hull of A and the condition on the injective hull in Corollary 3.25 are distinct, as the following example shows. Moreover, this example shows condition (ii) of Proposition 3.26 cannot be weakened to "for any $c = c^2 \in R$, $cR \cap I = cI$ ".

Example 3.27. There exists a right nonsingular right CS-ring R with $c = c^2 \in R$ and a right ideal I such that $cR \cap I \neq cI$. Let $R = \begin{bmatrix} F & F \\ 0 & F \end{bmatrix}$, where F is a field, $c = \begin{bmatrix} 0 & 1 \\ 0 & 1 \end{bmatrix}$, and $I = \begin{bmatrix} 0 & 0 \\ 0 & F \end{bmatrix}$. For this example the e in Proposition 3.26 can be taken to be zero. Also from this example we have that $E(R_R) = E_1 \oplus E_2$, where $E_1 = \{\begin{bmatrix} x & y \\ x & y \end{bmatrix} : x, y \in F\}$ and $E_2 = \begin{bmatrix} 0 & 0 \\ F & F \end{bmatrix}$. But $R \neq (R \cap E_1) \oplus (R \cap E_2) = \begin{bmatrix} 0 & F \\ 0 & F \end{bmatrix}$.

We complete this section by the following two results. First recall that a ring R is *Baer* (*pp*) if the left annihilator of each non-empty subset (element) of R is generated by an idempotent (see Exercises 3.16 and 3.17).

Theorem 3.28. *Let R be a left* CS*-ring and X a cyclic left R-module. Then the following conditions are equivalent.*

(i) *X is nonsingular.*
(ii) *Every cyclic submodule of X is projective and* CS.
(iii) *Every cyclic submodule of X is projective.*

Proof. Since $Z(R)$ is left essential in $Z_2(R)$, we have $R = B \oplus Z_2(R)$ (left ideal decomposition). Because $Z_2(R)$ is an ideal of R, B is a left CS-ring with unity. Furthermore, B is left nonsingular as a ring and as a left R-module.

(i) $\Rightarrow$ (ii) Let Y be a left ideal of R such that X is isomorphic to R/Y. Since R/Y is nonsingular, $Z(_RR) \subseteq Y$. Let $c \in Z_2(R)$ such that $c \notin Y$. There exists an essential left ideal L of R such that $Lc \subseteq Z(_RR) \subseteq Y$. Hence that $c \in Z(R/Y)$, which is a contradiction. So $Z_2(R) \subseteq Y$. It follows that there exists a left ideal T of R such that $T \subseteq B$ and R/Y is R-isomorphic to B/T. But B/T is a nonsingular left B-module. Since B is left nonsingular left CS, by Exercise 3.17 B/T is a projective B-module. Hence there exists $e = e^2 \in B$ such that B/T is B-isomorphic to Be. Furthermore, this B-isomorphism is also an R-isomorphism and Be is R-isomorphic to Re. Therefore R/Y is R-isomorphic to Re. Hence X is projective and left CS. This argument can now be applied to any cyclic submodule of X.

(ii) $\Rightarrow$ (iii) Obvious.

(iii) $\Rightarrow$ (i) Assume that $0 \neq x \in Z(_RX)$. Then Rx is projective. Thus there exists $e = e^2 \in R$ such that Rx is R-isomorphic to Re. But there exists an essential left ideal K of R such that $Kx = 0 = Ke$, which is a contradiction. Hence $Z(_RX) = 0$. □

Corollary 3.29. *The following conditions are equivalent:*

(i) *R is left nonsingular and left* CS.
(ii) *R is left* CS *and a Baer ring.*
(iii) *R is left* CS *and* pp.
(iv) *R is left nonsingular and every principal left ideal of R is left* CS.

Proof. (i) $\Rightarrow$ (ii) This follows from Exercise 3.17
(ii) $\Rightarrow$ (iii) It is well known that every Baer ring is a left pp-ring.
(iii) $\Rightarrow$ (iv) This follows from Theorem 3.28.
(iv) $\Rightarrow$ (i) Obvious. □

Exercises

3.1. Show that any free Abelian group (i.e., $\mathbb{Z}$-module) is a CS-module if and only if it has finite rank.

3.2. Show that any direct summand of a CS-module is also a CS-module.

3.3. Let A be an indecomposable module. Prove that A is CS if and only if A is uniform.

3.4. (i) Let A_1 and A_2 be R-modules and let $A = A_1 \oplus A_2$. Prove that A_1 is A_2-injective if and only if for every submodule N of A such that $N \cap A_1 = 0$ there exists a submodule A' of A such that $A = A_1 \oplus A'$ and $N \subseteq A'$.
(ii) Let R be a ring such that $u - dim(R_R) < \infty$. Prove that R is a right CS-ring if and only if every maximal uniform right ideal is a direct summand.

3.5. Let p be any prime integer. Let ${}_{\mathbb{Z}}A = (\mathbb{Z}/\mathbb{Z}p) \oplus (\mathbb{Z}/\mathbb{Z}p^3)$ and ${}_{\mathbb{Z}}B = (\mathbb{Z}/\mathbb{Z}p) \oplus (\mathbb{Z}/\mathbb{Z}p^2)$. Show that
(i) A is neither CS, nor UC,
(ii) B is CS, but not UC,
(iii) $\mathbb{Z}/\mathbb{Z}p$ is not $\mathbb{Z}/\mathbb{Z}p^2$-injective.

3.6. Let F be any field and let $R = \begin{bmatrix} F & F & F \\ 0 & F & 0 \\ 0 & 0 & F \end{bmatrix}$. Show that
(i) $\operatorname{soc}(R_R) \leq_e R_R$,
(ii) R_R is not CS.

3.7. Let $R = \mathbb{Z}[x]$ and ${}_RA = R \oplus R$. Prove that ${}_RA$ is not a CS-module.

3.8. Let A be a left R-module. Show that
(i) $\widetilde{E}(A)$ is a submodule of $E(A)$,
(ii) If A is nonsingular, then $\widetilde{E}(A) = E(A)$.

3.9. Let p be any prime integer. Prove that $\widetilde{E}(\mathbb{Z} \oplus \mathbb{Z}/\mathbb{Z}p) = \mathbb{Z}_{(p)} \oplus (\mathbb{Z}/\mathbb{Z}p)$, where $\mathbb{Z}_{(p)}$ denotes the localization of $\mathbb{Z}$ at the prime ideal (p).

3.10. Let A be a left R-module and $N \leq A$. Prove that the following statements are equivalent:

(i) $N \leq_{\rm d} A$.
(ii) $\mathrm{Hom}_R(A/N, E(A)) = 0$.
(iii) For any submodule P such that $N \leq P \leq A$, $\mathrm{Hom}_R(P/N, A) = 0$.

3.11. Let ${}_RA$ be a module. Prove that the following statements are equivalent:

(i) $\widetilde{E}(A) = A$.
(ii) For any left R-modules $K \leq N$ such that $\mathrm{Hom}_R(N/K, E(A)) = 0$, any R-homomorphism $f : K \to A$ can be extended to N (necessarily uniquely). (Such modules are called *rationally complete*.)

3.12. Let ${}_RA$ be a module. Prove that

(i) A is CS if and only if $A = Z_2(A) \oplus A'$, for some submodule A' of A, such that A' and $Z_2(A)$ are both CS and $Z_2(A)$ is A'-injective.
(ii) If $A = A_1 \oplus A_2 \oplus \cdots \oplus A_n$ be a finite direct sum of relatively injective modules A_i, then A is CS if and only if all A_i are CS.

3.13. Let A be a module. Prove that A is UC if and only if $K \leq_{\rm c} A$ implies that $K \cap N \leq_{\rm c} N$, for any submodule N of A.

3.14. Let A be a left R-module. Show that

(i) When R is considered as right R-module, the fully invariant submodules are just two-sided ideals.
(ii) Any sum or intersection of fully invariant submodules of A is again a fully invariant submodule of A.
(iii) If $A = \bigoplus_{i \in I} X_i$ and B is a fully invariant submodule of A, then $B = \bigoplus_{i \in I} \pi_i(B) = \bigoplus_{i \in I}(X_i \cap B)$, where π_i is the ith canonical projection homomorphism of A.

3.15. Let R be a ring and let $I \leq_{\rm d} R_R$. Prove that if I contains an essential reduced right ideal, then I is a reduced ring with unity.

3.16. Prove the following statements:

(i) A ring is left pp if and only if every principal left ideal is projective.
(ii) Baer (or pp) rings are nonsingular.

3.17. Let R be a ring. Prove the following statements:

(i) If R is left nonsingular, then it is left CS if and only if each cyclic nonsingular left R-module is projective.
(ii) R is left nonsingular left CS if and only if R is a Baer ring and each left ideal of R which has zero right annihilator is essential.

3.18. Let ${}_RA$ be a CS-module. Prove the following statements.

(i) If R has acc on left annihilators for A then A is a direct sum of uniform modules.

(ii) If $A/\operatorname{soc} A$ has finite uniform dimension, then $A = A_1 \oplus A_2$ for some semisimple submodule A_1 of A and submodule A_2 with finite uniform dimension.

(iii) A has acc (respectively, dcc) on essential submodules if and only if $A = A_1 \oplus A_2$ for some semisimple A_1 and Noetherian (respectively, Artinian) submodule A_2.

3.19. Let A_1 and A_2 be uniform modules with local endomorphism rings such that $A = A_1 \oplus A_2$ is CS. Let C be a submodule of A_1 and let $f : C \to A_2$ be a homomorphism. Prove the following statements:

(i) If f cannot be extended to a homomorphism from A_1 to A_2, then f is a monomorphism and A_2 is embedded in A_1.

(ii) If any monomorphism $A_2 \to A_1$ is an isomorphism, then A_2 is A_1-injective.

(iii) If A_2 is not embedded in A_1, then A_2 is A_1-injective.

3.20. Let A be a module such that $A = A_1 \oplus A_2$, where A_1 and A_2 are CS-modules. Prove that if A_1 is nonsingular and A_2 is A_1-injective, then A is a CS-module.

3.2 Extending modules over commutative domains

In this section we deal with extending modules over commutative domains. To this end we give the detailed proof of a fundamental decomposition theorem which states that a torsion-free reduced CS module is a finite direct sum of uniform submodules. Moreover, we apply this result to get more information about CS-modules over Dedekind domains or principal ideal domains. Furthermore, we provide two generalized forms of the aforementioned decomposition theorem. Thus it is shown that a nonsingular reduced CS-module over a commutative ring with only a finite number of minimal prime ideals has a finite uniform dimension, and that a nonsingular right CS-module over a semiprime right Goldie ring is a direct sum of a finite number of injective modules and uniform modules.

Let R be a commutative domain with identity. We investigate when an R-module is CS. The first result, which is a special case of Exercise 3.12 (i), splits the problem into the torsion-free case and the torsion case.

Lemma 3.30. *Let R be a commutative domain. An R-module A is a* CS*-module if and only if A is a torsion* CS*-module or there exist $A_1, A_2 \leq A$ such that $A = A_1 \oplus A_2$, A_1 is a torsion injective module, and A_2 a torsion-free* CS*-module.*

An R-module A will be called *reduced* if it contains no non-zero injective submodules. Any torsion-free R-module is a direct sum of an injective module and a reduced module (Exercise 3.21).

Lemma 3.31. *Let R be a commutative domain. A torsion-free R-module A is* CS *if and only if there exist $A_1, A_2 \leq A$ such that $A = A_1 \oplus A_2$, A_1 is injective, and A_2 is reduced and* CS.

Proof. By Exercises 3.2 and 3.20. □

Let A be a torsion-free CS-module over R. Let $0 \neq a \in A$. Then $l(a) = \{r \in R : ar = 0\} = 0$. Thus A is a direct sum of uniform submodules (Exercise 3.18 (i)). If, in addition, A is reduced we can say more. First we prove:

Lemma 3.32. *Let R be a commutative domain with field of fractions Q. Let U be a torsion-free uniform R-module. Then there exists a monomorphism $\varphi : N \to Q$ such that $R \leq \varphi(U)$.*

Proof. Let $0 \neq u \in U$. Define $\theta : Ru \to Q$ by $\theta(ru) = r$ $(r \in R)$. Then θ is a monomorphism. Since Q is an injective R-module, θ can be lifted to a homomorphism $\varphi : U \to Q$. Now $\ker \varphi \cap Ru = \ker \theta = 0$ implies φ is a monomorphism, because U is uniform. Moreover $R = \theta(Ru) \leq \varphi(U)$. □

Theorem 3.33. *Let R be a commutative domain and let A be a torsion-free reduced* CS *R-module. Then A is a finite direct sum of uniform submodules.*

Proof. As we remarked above, A is a direct sum of uniform submodules. Suppose A is not a finite direct sum of uniform submodules. Then Exercise 3.2 and Lemma 3.32 allow us to assume without loss of generality that

$$A = A_0 \oplus A_1 \oplus A_2 \oplus \cdots,$$

and $R \leq A_i \leq Q$ $(i \geq 0)$, where Q is the field of fractions of R. Let $0 \neq c \in R$. Define $\alpha : A \to Q$ by

$$\alpha(a_0, a_1, a_2, \dots) = a_0 + a_1c^{-1} + a_2c^{-2} + \cdots$$

for all $(a_0, a_1, a_2, \dots) \in A$. Clearly, α is a homomorphism. Now $A/(\ker \alpha) \cong \operatorname{im} \alpha$ so that $A/(\ker \alpha)$ is torsion-free and hence $\ker \alpha \leq_c A$. By hypothesis, $A = (\ker \alpha) \oplus A'$ for some $A' \leq A$. Moreover, $A' \cong A/\ker \alpha \cong \operatorname{im} \alpha \leq Q$, so that A' is uniform. Let $0 \neq x \in A'$. Then $x \in A_0 \oplus \cdots \oplus A_t$ for some positive integer t. It is easy to check that $A' \leq A_0 \oplus \cdots \oplus A_t$.
Let $y = (0, \dots, 0, 1, 0, 0, \dots) \in A$, where $1 \in A_{t+1}$. Since $y \in \ker \alpha + A_0 + \cdots + A_t$, there exist $u_i \in A_i$ $(0 \leq i \leq t)$ such that

$$(u_0, \dots, u_t, -1, 0, 0, \dots) \in \ker \alpha.$$

Thus

$$u_0 + u_1c^{-1} + u_2c^{-2} + \cdots + u_tc^{-t} - c^{-t-1} = 0,$$

and hence

$$c^{-1} = u_0c^t + u_1c^{t-1} + \cdots + u_t \in A_0 + \cdots + A_t.$$

Let $V = A_0 + A_1 + A_2 + \cdots \leq Q$. Then $c^{-1} \in V$ for all $0 \neq c \in R$. It follows that $Q \leq V$ and hence $Q = V$. Define $\beta : A \to Q$ by

$$\beta(v_0, v_1, v_2, \cdots) = v_0 + v_1 + v_2 + \cdots$$

for all $(v_0, v_1, v_2, \dots) \in A$. By the above proof, β is an epimorphism.

Now $A/(\ker\beta) \cong Q$ gives $A/(\ker\beta)$ torsion-free and hence $\ker\beta \leq_c A$. Since A is CS, $A = (\ker\beta) \oplus A''$, for some $A'' \leq A$. But $A'' \cong A/(\ker\beta) \cong Q$, so that A'' is a non-zero injective submodule of A, contradiction. Thus A is a finite direct sum of uniform submodules. □

Combining Lemma 3.31 and Theorem 3.33 we see that if a torsion-free R-module A is CS then

$$A = A_1 \oplus A_2 \oplus \cdots \oplus A_n$$

for some positive integer n and $A_i \leq A$ $(1 \leq i \leq n)$ where A_1 is a Q-vector space and A_i is a uniform R-module $(2 \leq i \leq n)$. Now we suppose that $A = A_1 \oplus A_2$ is a CS-module over the ring R where $(R \leq)A_i \leq Q$ $(i = 1, 2)$ and obtain some information about A_1, A_2. Then any non-zero, non-essential submodule of A is uniform.

Lemma 3.34. *Let R be a commutative domain with field of fractions Q, let $A_i \leq {}_RQ$ $(i = 1, 2)$ and let $A = A_1 \oplus A_2$. Then $K \leq_c A$ if and only if $K = 0$, $K = A$ or there exist $q_1, q_2 \in Q$, not both zero, such that $K = QK \cap A = Q(q_1, q_2) \cap A$.*

Proof. Suppose that $K \leq_c A$. Suppose that $K \neq 0, A$. Then there exist $q_1, q_2 \in Q$, not both zero, such that $(q_1, q_2) \in K$. Now $K \leq_e Q(q_1, q_2) \cap A \leq A$ gives $K = Q(q_1, q_2) \cap A$.

Conversely, suppose that $K = Q(q_1, q_2) \cap A$, for some $q_1, q_2 \in Q$, not both zero. Then $QK = Q(q_1, q_2)$, a one-dimensional subspace of the Q-vector space $Q \oplus Q$, so that K is uniform. There exists $L \leq_c A$ such that $K \leq_e L$. Then

$$QK = E(K) = E(L) = QL,$$

and hence

$$L = QL \cap A = QK \cap A = Q(q_1, q_2) \cap A = K.$$

Thus $K \leq_c A$. □

Corollary 3.35. *Let R be a commutative domain with field of fractions Q, let $0 \neq A_i \leq {}_RQ$ $(i = 1, 2)$, and let $A = A_1 \oplus A_2$. Then A is a CS-module if and only if for all nonzero $q_1, q_2 \in Q$, there exist $x_1, x_2 \in Q$ such that*

$$A = [Q(q_1, q_2) \cap A] \oplus [Q(x_1, x_2) \cap A].$$

Proof. Suppose first that A is CS. Let $q_1, q_2 \in Q$, not both zero, and let $K = Q(q_1, q_2) \cap A$. By Lemma 3.34, $K \leq_c A$, and hence $A = K \oplus K'$ for some $K' \leq A$.

Clearly $K' \neq 0, A$, so that, again by Lemma 3.34, $K' = Q(x_1, x_2) \cap A$ for some $x_1, x_2 \in Q$.

Conversely, suppose that A satisfies the above condition. By Lemma 3.34, A is CS. □

Let R be a commutative domain with field of fractions Q. Let X and Y be R-submodules of Q and let S be a subring of Q with $R \leq S$. We define

$$(X\ {}_S\colon Y) = \{s \in S : Ys \subseteq X\}.$$

If $q \in Q$, we define

$$(X\ {}_S\colon q) = (X\ {}_S\colon qR) = \{s \in S : qs \in X\}.$$

Note that $(X\ {}_S\colon Y)$ is an R-submodule of Q for all such X, Y, S.

Lemma 3.36. *Let R be a commutative domain with field of fractions Q, let $0 \neq A_i \leq {}_R Q$ $(i = 1, 2)$, and let $A = A_1 \oplus A_2$. Let $0 \neq q_1, q_2 \in Q$. Then $Q(q_1, q_2) \cap A \leq_{\mathrm{d}} A$ if and only if*

$$R \leq [(A_1\ {}_Q : A_1) \cap (A_2 q_1\ {}_Q : A_1 q_2)] + [(A_2\ {}_Q : A_2) \cap (A_1 q_2\ {}_Q : A_2 q_1)]$$

Proof. Suppose first that $Q(q_1, q_2) \cap A \leq_{\mathrm{d}} A$. By Corollary 3.35,

$$A = [Q(q_1, q_2) \cap A] \oplus [Q(x_1, x_2) \cap A]$$

for some $x_1, x_2 \in Q$. Let $a_1 \in A_1$. Then

$$(a_1, 0) = u(q_1, q_2) + v(x_1, x_2)$$

for some $u \in (A_1\ {}_Q : q_1) \cap (A_2\ {}_Q : q_2)$ and $v \in (A_1\ {}_Q : x_1) \cap (A_2\ {}_Q : x_2)$. We deduce that

$$a_1 = uq_1 + vx_1 \quad \text{and} \quad 0 = uq_2 + vx_2,$$

and thus

$$\begin{aligned} a_1 q_2 &= (-vx_2)q_1 + vx_1 q_2 = v(x_1 q_2 - x_2 q_1), \\ a_1 x_2 &= ux_2 q_1 + (-uq_2)x_1 = u(x_2 q_1 - q_2 x_1). \end{aligned}$$

Consequently,

$$A_1 q_2 \leq (x_1 q_2 - x_2 q_1)[(A_1\ {}_Q : x_1) \cap (A_2\ {}_Q : x_2)] \tag{3.1}$$

and

$$A_1 x_2 \leq (x_1 q_2 - x_2 q_1)[(A_1\ {}_Q : q_1) \cap (A_2\ {}_Q : q_2)]. \tag{3.2}$$

Similarly,

$$A_2 q_1 \leq (x_1 q_2 - x_2 q_1)[(A_1\ {}_Q : x_1) \cap (A_2\ {}_Q : x_2)] \tag{3.3}$$

and

$$A_2x_1 \leq (x_1q_2 - x_2q_1)[(A_1 \ _Q : q_1) \cap (A_2 \ _Q : q_2)]. \tag{3.4}$$

Clearly, $x_1q_2 - x_2q_1 \neq 0$, for otherwise $q_1 = q_2 = 0$. Let

$$y_1 = -x_2q_1/(x_1q_2 - x_2q_1) \in Q \quad \text{and} \quad y_2 = x_1q_2/(x_1q_2 - x_2q_1) \in Q.$$

Then $1 = y_1 + y_2$. Moreover,

$$A_1q_1y_1 = A_1x_2q_1(x_1q_2 - x_2q_1)^{-1}(-q_1) \leq A_1q_1,$$

by (3.2), and

$$A_1q_2y_1 = A_1q_2x_2(x_1q_2 - x_2q_1)^{-1}(-q_1) \leq A_2q_1,$$

by (3.1). Similarly, (3.3) and (3.4) give $A_2q_2y_2 \leq A_2q_2$ and $A_2q_1y_2 \leq A_1q_2$. This proves the necessity.

Conversely, suppose that $1 = z_1 + z_2$ for some $z_1 \in (A_1 \ _Q : A_1) \cap (A_2q_1 \ _Q : A_1q_2)$ and $z_2 \in (A_2 \ _Q : A_2) \cap (A_1q_2 \ _Q : A_2q_1)$. Let

$$x_1 = -z_2/q_2, \quad x_2 = z_1/q_1.$$

Let $a_1 \in A_1$. Then

$$(a_1, 0) = a_1x_2(q_1, q_2) + (-a_1q_2)(x_1, x_2) \in [Q(q_1, q_2) \cap A] + [Q(x_1, x_2) \cap A].$$

Similarly,

$$(0, a_2) \in [Q(q_1, q_2) \cap A] + [Q(x_1, x_2) \cap A],$$

for all $a_2 \in A_2$. Thus

$$A = [Q(q_1, q_2) \cap A] + [Q(x_1, x_2) \cap A].$$

Also $x_2q_1 - x_1q_2 = 1$ easily gives

$$[Q(q_1, q_2) \cap A] \cap [Q(x_1, x_2) \cap A] = 0.$$

Thus

$$A = [Q(q_1, q_2) \cap A] \oplus [Q(x_1, x_2) \cap A]. \qquad \square$$

Corollary 3.37. *Let R be a commutative domain with field of fractions Q, let $0 \neq A_i \leq {}_RQ$ $(i = 1, 2)$ and let $A = A_1 \oplus A_2$. Then A is a CS-module if and only if, for all non-zero $q \in Q$,*

$$R \leq [(A_1 \ _Q\colon A_1) \cap (A_2 \ _Q\colon A_1q)] + [(A_2 \ _Q\colon A_2) \cap (A_1q \ _Q\colon A_2)].$$

Proof. ($\Rightarrow$) By Lemma 3.36 with $q_1 = 1$, $q_2 = q$.

($\Leftarrow$) Let q_1, q_2 be non-zero elements of Q. Then

$$Q(q_1, q_2) \cap A = Qq_1(1, q) \cap A = Q(1, q) \cap A,$$

where $q = q_2/q_1$. Now Lemma 3.36 shows that $Q(1, q) \cap A \leq_{\mathrm{d}} A$. It follows that A is CS. $\square$

We shall require Corollary 3.37 a little later. Now we prove:

Theorem 3.38. *Let R be a commutative domain and let $A = A_1 \oplus A_2$ be a torsion-free CS-module over R such that A_1 and A_2 are both uniform modules. Then A_1 embeds in A_2 or A_2 embeds in A_1. If, in addition, A is reduced, then A_1 embeds in A_2 and A_2 embeds in A_1.*

Proof. Let Q denote the field of fractions of R and let $0 \neq q \in Q$. By Corollary 3.37, we have

$$1 = z_1 + z_2$$

for some $z_1, z_2 \in Q$ such that

$$A_1 z_1 \leq A_1, \quad A_1 q z_1 \leq A_2, \quad A_2 z_2 \leq A_2 z_2, \quad A_2 z_2 \leq A_1 q.$$

If $z_1 \neq 0$, then

$$A_1 \cong A_1 q z_1 \leq A_2,$$

so A_1 embeds in A_2. If A_1 does not embed in A_2, then for all choices of q, $z_1 = 0$, and hence $z_2 = 1$ and

$$A_2 q^{-1} \leq A_1 \quad (0 \neq q \in Q).$$

Thus $Q = A_2 Q \leq A_1$ and hence $A_1 = Q$. The result follows. □

Recall that an ideal I of a ring R is said to be *invertible* if the following conditions are satisfied. There is an overring S of R such that if $A = \{s \in S : sI \subseteq R\} = (R :_S I)$ and $B = \{s \in S : Is \leq R\} = (R\ _S\!: I)$, then $AI = IB = R$. In this situation we have $A = B$ and we write $I^{-1} = A = B$. If R is a commutative domain, then every nonzero principal ideal of R is invertible (in the field of fractions of R) [Gil72].

Lemma 3.39. *Let R be a commutative domain and let I, J be ideals of R such that $I + J$ is invertible. Then $(I\ _R : J) + (J\ _R : I) = R$.*

Proof. Clearly, $(I\ _R : J) + (J\ _R : I) \leq R$. Suppose that $(I\ _R : J) + (J\ _S : I) \neq R$. Then there exists a maximal ideal P of R such that

$$(I\ _R : J) + (J\ _R : I) \leq P.$$

Further, there exist ideals A, B of R such that

$$I = (I + J)A \quad \text{and} \quad J = (I + J)B.$$

Then $A \leq (I\ _R : J) \leq P$ and $B \leq (J\ _R : I) \leq P$. Therefore,

$$I + J = (I + J)P$$

and hence $R = P$, a contradiction. Thus $(I\ _R : J) + (J\ _R : I) = R$. □

Corollary 3.40. *Let R be a commutative domain and let I be an ideal of R. Let a, b be non-zero elements of R such that $aR + bI$ is invertible. If $q = b/a$, then*

$$(Iq\ _R : R) + (R\ _R : Iq) = R.$$

Proof. $(Iq\ _R : R) = (Iqa\ _R : Ra) = (Ib : Ra)$, and $(R\ _R : Iq) = (Ra : Ib)$. The result follows by Lemma 3.39. □

Theorem 3.41. *Let R be a commutative domain and let I be an ideal of R such that $aR + bI$ is invertible for all $0 \neq a, b \in R$. Let A be a torsion-free uniform R-module. Then the R-module $A \oplus AI$ is* CS.

Proof. Let Q denote the field of fractions of R. Without loss of generality, $0 \neq A \leq Q$. Let $0 \neq q \in Q$. Then $q = b/a$ for some $0 \neq a, b \in R$. By Corollary 3.40, $1 = z_1 + z_2$ for some $z_1, z_2 \in R$ such that $z_1 \in Iq$, $Iqz_2 \leq R$. Thus $Az_1 \leq A$, $Az_1 \leq AIq$, $AIz_2 \leq AI$, $AIqz_2 \leq A$. By Corollary 3.37, $Q(q, 1) \cap (A \oplus AI) \leq_d A \oplus AI$. It follows that $A \oplus AI$ in CS. □

A commutative ring R is called *semihereditary* if every finitely generated ideal of R is projective. (see Exercise 1.93). Recall that a commutative hereditary (semihereditary) domain is called a *Dedekind* (Prüfer) domain.

Corollary 3.42. *Let R be a Prüfer domain, A be a torsion-free uniform R-module, and I be a finitely generated ideal of R. Then $A \oplus AI$ is* CS.

Proof. Let $0 \neq a, b \in R$. Then $aR + bI$ is finitely generated, hence invertible. The result follows from the Theorem 3.41. □

Theorem 3.43. *The following statements are equivalent for a commutative domain R.*

(i) *R is Prüfer.*
(ii) *$A \oplus AI$ is* CS *for any torsion-free uniform R-module A and finitely generated ideal I of R.*
(iii) *The R-module $R \oplus R$ is* CS.

Proof. (i) $\Rightarrow$ (ii) By Corollary 3.42.

(ii) $\Rightarrow$ (iii) Clear.

(iii) $\Rightarrow$ (i) Suppose that $F = R \oplus R$ is a CS R-module. Let $0 \neq a, b \in R$. By Lemma 3.36,

$$\begin{aligned} R &\leq [(R\ _Q : R) \cap (Ra\ _Q : Rb)] + [(R\ _Q : R) \cap (Rb\ _Q : Ra)] \\ &= (Ra\ _Q : Rb) + (Rb\ _Q : Ra), \end{aligned}$$

and hence $aR + bR$ is an invertible ideal of R (Lemma 3.39). It follows that R is a Prüfer domain. □

Lemma 3.44. *Let R be a commutative domain with field of fractions Q. Let U_i $(1 \le i \le 3)$ be R-submodules of Q such that $U_i \oplus U_j$ is* CS *for all $1 \le i \le j \le 3$. Then $U_3 \oplus (U_1 \cap U_2)$ is* CS.

Proof. Let $0 \ne q \in Q$. Since $U_1 \oplus U_2$ is CS, there exist $x_1, x_2 \in Q$ such that

$$1 = x_1 + x_2, \quad U_1x_1 \le U_1 \cap U_2, \quad U_2x_1 \le U_2, \quad U_2x_2 \le U_1 \cap U_2, \quad U_1x_2 \le U_1.$$

Similarly, since $U_1 \oplus U_3$ and $U_2 \oplus U_3$ are CS, there exist $y_1, y_3, z_2, z_3 \in Q$ such that

$$1 = y_1 + y_3,\ U_1y_1 \le U_1 \cap U_3q,\ U_3y_1 \le U_3,\ U_3y_3 \le U_3 \cap U_1q^{-1},\ U_1y_3 \le U_1,$$

and

$$1 = z_2 + z_3,\ U_2z_2 \le U_2 \cap U_3q,\ U_3z_2 \le U_3,\ U_3z_3 \le U_3 \cap U_2q^{-1},\ U_2z_3 \le U_2.$$

Now let

$$w_1 = y_3x_1 + z_3x_2 \in Q \quad \text{and} \quad w_2 = y_1x_1 + z_2x_2 \in Q.$$

Then $1 = w_1 + w_2$.

Now

$$\begin{aligned} w_2 = y_1x_1 + z_2x_2 &= (z_2 + z_3)y_1x_1 + z_2(y_1 + y_3)x_2 \\ &= z_2y_1 + z_3y_1x_1 + z_2y_3x_2 \\ &= z_2y_1 + z_3y_1(1 - x_1) + z_2y_3(1 - x_1) \\ &= z_2y_1 + z_3y_1 - z_3y_1x_1 + z_2y_3 - z_2y_3x_1. \end{aligned}$$

Note that

$$\begin{aligned} U_3z_2y_1 &\le U_3, \\ U_3z_3y_1 &\le U_3, \\ U_3z_3y_1x_2 &\le (U_2q^{-1})x_2y_1 \le U_1y_1q^{-1} \le U_3qq^{-1} = U_3, \\ U_3z_3y_3 &\le U_3, \quad \text{and} \\ U_3z_2y_3x_1 &\le (U_1q^{-1})x_1z_2 \le U_2z_2q^{-1} \le U_3qq^{-1} = U_3. \end{aligned}$$

Hence $U_3w_2 \le U_3$, and it follows that $U_3w_1 \le U_3$.

Next, observe that

$$w_1 = y_3x_1 + z_3x_2 = y_3x_1 + z_3(y_1 + y_3)x_2 = y_3x_1 + z_3y_1x_2 + z_3y_3x_2,$$

and hence

$$(U_1 \cap U_2)w_1 \le U_1y_3x_1 + U_2z_3y_1x_2 + U_2z_3y_3x_2.$$

Now

$$\begin{aligned} U_1y_3x_1 &\le U_1, \\ U_2z_3y_1x_2 &= U_2z_3x_2y_1 \le U_2x_2y_1 \le U_1y_1 \le U_1, \end{aligned}$$

and

$$U_2z_3y_3x_2 = U_2z_3x_2y_3 \leq U_2x_2y_3 \leq U_1y_3 \leq U_1.$$

Thus $(U_1 \cap U_2)w_1 \leq U_1$, and hence $(U_1 \cap U_2)w_2 \leq U_1$.

Further,

$$\begin{aligned}(U_1 \cap U_2)w_1 &\leq (U_1 \cap U_2)y_3x_1 + U_2z_3x_2 \\ &= (U_1 \cap U_2)(1 - y_1)x_1 + U_2z_3x_2 \\ &\leq (U_1 \cap U_2)x_1 + U_1y_1x_1 + U_2z_3x_2 \\ &\leq U_2 + U_2 + U_2 = U_2.\end{aligned}$$

Hence, $(U_1 \cap U_2)w_1 \leq (U_1 \cap U_2)$ and it follows that $(U_1 \cap U_2)w_2 \leq (U_1 \cap U_2)$. Moreover,

$$(U_3q)w_1 \leq (U_3q)y_3x_1 + (U_3q)z_3x_2 \leq U_1x_1 + U_2x_2 \leq U_1 \cap U_2,$$

and

$$\begin{aligned}(U_1 \cap U_2)w_2 &\leq (U_1 \cap U_2)y_1x_1 + (U_1 \cap U_2)z_2x_2 \\ &\leq (U_1 \cap U_2)y_1 + (U_1 \cap U_2)z_2 \leq U_3q.\end{aligned}$$

By Corollary 3.37, $U_3 \oplus (U_1 \cap U_2)$ is CS. □

Corollary 3.45. *Let R be a commutative domain with field of fractions Q. Let n be a positive integer, let U_i $(1 \leq i \leq n)$ be R-submodules of Q, and let $q_i \in Q$ $(1 \leq i \leq n-1)$. Suppose that $U_i \oplus U_j$ is* CS *for all $1 \leq i < j \leq n$. Then $U_n \oplus [U_1q_1 \cap \cdots \cap U_{n-1}q_{n-1}]$ is* CS.

Proof. By induction on n. Without loss of generality, $q_i \neq 0$ $(1 \leq i \leq n-1)$. We suppose that $U_n \oplus [U_1q_1 \cap \cdots \cap U_{n-2}q_{n-2}]$ and $U_{n-1} \oplus [U_1q_1 \cap \cdots \cap U_{n-2}q_{n-2}]$ are CS. Since

$$U_{n-1}q_{n-1} \oplus [U_1q_1 \cap \cdots \cap U_{n-2}q_{n-2}] \cong U_{n-1} \oplus [U_1q_1 \cap \cdots \cap U_{n-2}q_{n-2}],$$

it follows that $U_{n-1}q_{n-1} \oplus [U_1q_1 \cap \cdots \cap U_{n-2}q_{n-2}]$ is CS. Also, $U_{n-1}q_{n-1} \oplus U_n \cong U_{n-1} \oplus U_n$, so that $U_{n-1}q_{n-1} \oplus U_n$ is CS. By Lemma 3.44, $U_n \oplus [U_1q_1 \cap \cdots \cap U_{n-1}q_{n-1}]$ is CS. □

Theorem 3.46. *Let R be a commutative domain and let A be a torsion-free R-module such that $A = A_1 \oplus \cdots \oplus A_n$ for some positive integer n and uniform submodules A_i $(1 \leq i \leq n)$. Then A is* CS *if and only if $A_i \oplus A_j$ is* CS *for all $1 \leq i < j \leq n$.*

Proof. The necessity is clear. Conversely, suppose that $A_i + A_j$ is CS for all $1 \leq i < j \leq n$. Without loss of generality, we can assume that $A_i \leq Q$ $(1 \leq i \leq n)$, where Q is the field of fractions of R. We shall prove that A is CS by induction on

n. Let K be a maximal uniform submodule of A. There exist $q_i \in Q$ $(1 \le i \le n)$ such that $K = Q(q_1, \dots, q_n) \cap A$, i.e.,

$$K = \left[\bigcap_{i=1}^{n} (A_i \ _Q : q_i) \right] (q_1, \dots, q_n).$$

If $q_1 = 0$ then $K \le A_2 \oplus \cdots \oplus A_n$ and hence $K \le_{\text{d}} A$ by induction. Suppose that $q_i \neq 0$.

Let $\pi : A \to A_1 \oplus \cdots \oplus A_{n-1}$ denote the canonical projection. Then

$$\pi(K) \le Q(q_1, \dots, q_{n-1}, 0) \cap A = \left[\bigcap_{i=1}^{n-1} (A_i \ _Q : q_i) \right] (q_1, \dots, q_{n-1}, 0) \cong \bigcap_{i=1}^{n-1} A_i {q_i}^{-1}.$$

Let $L = Q(q_1, \dots, q_{n-1}, 0) \cap A$. Then $L \le_d A_1 \oplus \cdots \oplus A_{n-1}$, by induction on n. Moreover

$$K \le \pi(K) \oplus A_n \le L \oplus A_n.$$

But $L \oplus A_n \cong A_n \oplus \bigcap_{i=1}^{n-1} A_i {q_i}^{-1}$ which is CS, by Corollary 3.45. Thus $K \le_{\text{d}} L \oplus A_n$. Hence $K \le_{\text{d}} A$. By Exercise 3.4 (ii), A is CS. □

Corollary 3.47. *Let R be a commutative domain and let A be a torsion-free R-module. Then A is* CS *if and only if $A = A_0 \oplus A_1 \oplus \cdots \oplus A_n$ where n is a positive integer, A_0 is injective, A_i $(1 \le i \le n)$ is uniform, and $A_i \oplus A_j$ is* CS *for all $1 \le i < j \le n$.*

Proof. The necessity follows by Lemma 3.31, Theorem 3.33, and Exercise 3.2. The sufficiency follows by Theorem 3.46 and Exercise 3.20. □

Now we consider the special case of Dedekind domains and, more particularly, PIDs. Recall that a *discrete valuation ring* (or DVR) is a principal ideal domain that has exactly one non-zero prime ideal.

Lemma 3.48. *Let R be a Dedekind domain with field of fractions Q and let $N \le A \le Q$ such that $(N \ _R : A) \neq 0$. Then $N = A(N \ _R : A)$.*

Proof. If $N = A$, there is nothing to prove. Suppose $N \neq A$. Let $I = (N \ _R : A)$, a proper ideal of R. There exist a positive integer n and maximal ideals P_i $(1 \le i \le n)$ such that $I = P_1 \cdots P_n$. We prove the result by induction on n. Suppose first that $n = 1$, i.e., I is a maximal ideal of R. Suppose that $AI \lneqq N$. Then

$$(A/AI) = (N/AI) \oplus (K/AI)$$

for some $AI \lneqq K \lneqq A$. Let $x \in N \backslash AI$, $y \in K \backslash AI$. The ring R_I of R localised at I is a DVR. Thus $x \in yR_I$ or $y \in xR_I$. Suppose that $x \in yR_I$. Then there exist $r \in R$, $a \in I$ such that $(1 - a)x = ry \in N \cap K = AI$. Hence $x \in AI$, a contradiction. We conclude that $N = AI$.

Let $n \geq 2$. Suppose that $N \nleq AP_1$. By the above argument, AP_1 is maximal and hence

$$A = N + AP_1.$$

Then $AP_2 \cdots P_n \leq N$ and hence $P_2 \cdots P_n \leq I = P_1 \cdots P_n$, a contradiction. Thus $N \leq AP_1$ and hence

$$N_1 = NP_1^{-1} \leq A.$$

Now $(N_1 : A) = IP_1^{-1} = P_2 \cdots P_n$. By induction,

$$NP_1^{-1} = A(N_1 : A)$$

and hence

$$N = A(N_1 : A)P_1 = A(IP_1^{-1})P_1 = AI,$$

as required. □

Theorem 3.49. *Let R be a Dedekind domain and let $A = A_1 \oplus A_2$ be a torsion-free reduced R-module, where A_1 and A_2 are both uniform. Then the following statements are equivalent.*

(i) *A is* CS.
(ii) *A_1 embeds in A_2 and A_2 embeds in A_1.*
(iii) *$A_2 \cong A_1 I$ for some non-zero ideal I of R.*

Proof. (i) ⇒ (ii) By Theorem 3.38.
(ii) ⇒ (iii) By Lemma 3.48.
(iii) ⇒ (i) By Theorem 3.43. □

Corollary 3.50. *Let R be a Dedekind domain. Then a torsion-free R-module A is* CS *if and only if $A \cong A_0 \oplus A_1 I_1 \oplus \cdots \oplus A_1 I_n$ for some positive integer n, injective module A_0, ideals I_j $(1 \leq j \leq n)$ of R, and non-zero submodule A_1 of Q, the field of fractions of R.*

Proof. By Theorems 3.46 and 3.49. □

In case R is a PID we can say more:

Corollary 3.51. *Let R be a* PID *with field of fractions Q. Then a torsion-free R-module A is* CS *if and only if $A \cong A_0 \bigoplus A_1^n$ for some non-negative integer n, injective module A_0, and non-zero submodule A_1 of Q.*

Proof. By Corollary 3.50. □

As we promised at the beginning of this section, we now intend to obtain generalizations of Theorems 3.33. We shall use the following facts to prove our first main result. The next lemma is based on torsion theory. Recall that a *torsion theory* for R-Mod is a pair $(\underline{T}, \underline{F})$ of classes of left R-modules such that

(i) $\mathrm{Hom}(\underline{T}, \underline{F}) = 0$ for all $T \in \underline{T}$, $F \in \underline{F}$,

(ii) $\underline{T}$ and $\underline{F}$ are maximal classes having property (i).

The modules in $\underline{T}$ are called *torsion* modules and the modules in $\underline{F}$ are *torsion-free.* Any given non-empty class $\underline{G}$ of modules generates a torsion theory in the following way;

$$\underline{F} = \{F : \operatorname{Hom}(G, F) = 0 \text{ for all } G \in \underline{G}\},$$
$$\underline{T} = \{T : \operatorname{Hom}(T, F) = 0 \text{ for all } F \in \underline{F}\}.$$

In particular, the torsion theory which is generated by the class $\underline{G} = \{A/L : L \leq_{\mathrm{e}} A\}$ is called the *Goldie torsion theory.* Note that for the Goldie torsion theory $(\underline{T}, \underline{F})$, $\underline{F}$ consists precisely of the nonsingular left R-modules. For any module A, let $\tau(A)$ denote the torsion submodule of A. We have:

Lemma 3.52. *Let R be a ring and $(\underline{T}, \underline{F})$ a torsion theory for R-Mod. Let A be a torsion-free reduced left R-module. Let I be an ideal of R such that $I \leq \tau({}_RR)$. Then $IA = 0$ and A is a reduced left (R/I)-module.*

Proof. Let $\underline{S} = \{E : E$ is a left ideal of R and $R/I \in \underline{T}\}$. Then for any left R-module X, $\tau(X) = \{x \in X : Ex = 0$ for some $E \in \underline{S}\}$. Let $a \in A$, $r \in I$. There exists $G \in \underline{S}$ such that $Gr = 0$ and hence $G(ra) = (Gr)a = 0$. Thus $ra = 0$. It follows that $IA = 0$. Define $(r + I)a = ra$ $(a \in A,\ r \in R)$. With this definition the Abelian group A becomes a left (R/I)-module. Let N be an injective submodule of the left (R/I)-module A. Let J be a left ideal of R and $\theta : J \to N$ an R-homomorphism. Define $\varphi : (J + I)/I \to N$ by $\varphi(x + I) = \theta(x)$ $(x \in J)$. We show that φ is well defined. For, suppose $x, y \in I$ and $x + I = y + I$. Then $x - y \in I \leq \tau({}_RR)$. Thus there exists $F \in \underline{S}$ such that $F(x - y) = 0$. We have $F[\theta(x) - \theta(y)] = F\theta(x - y) = \theta(F(x - y)) = \theta(0) = 0$. But A is torsion-free. Thus $\theta(x) = \theta(y)$, i.e., θ is well defined. Clearly, φ is an (R/I)-homomorphism. Since N is (R/I)-injective it follows that there exists $n \in N$ such that $\varphi(x + I) = (x + I)n$ $(x \in J)$. Hence for all $x \in J$, $\theta(x) = \varphi(x + I) = (x + I)n = xn$. It follows that N is an injective submodule of the reduced R-module A. Thus $N = 0$. Hence A is a reduced (R/I)-module. □

Lemma 3.53. *Let R be a ring, I an ideal of R, and A a left R-module such that $IA = 0$. Then A is a left R/I-module. Moreover the R-module A is a* CS*-module if and only if the (R/I)-module A is a* CS*-module.*

Proof. Define $(r + I)a = ra$ $(a \in A, r \in R)$. Then the Abelian group A is a left (R/I)-module. Let N be a subgroup of A. Let $n \in N$, $r \in R$. Clearly $rn = (r+I)n$ implies $rn \in N$ if and only if $(r + I)n \in N$. Thus N is an R-submodule of A if and only if N is an (R/I)-submodule of A. Suppose that the R-module A is CS. Let $N \leq_{\mathrm{c}\ R/I} A$. Let K be an R-submodule of A such that $N \leq_{\mathrm{e}} K$. Let L be an (R/I)-submodule of K such that $N \cap L = 0$. Then L is an R-submodule of K and hence $L = 0$. Thus $N \leq_{\mathrm{e}\ R/I} K$, and hence $N = K$. Thus $N \leq_{\mathrm{c}\ R} A$. Since A is a CS-module, there exists an R-submodule N' of A such that $A = N \oplus N'$. So N'

is an (R/I)-submodule of A. It follows that every complement in ${}_{R/I}A$ is a direct summand of ${}_{R/I}A$. Thus the (R/I)-module A is a CS-module.

Similarly, if the (R/I)-module A is a CS-module, then the R-module A is a CS-module. □

Lemma 3.54. *Let I be an ideal of a ring R and E a left ideal of R such that $I \leq E$ and $E/I \leq_e {}_{R/I}(R/I)$. Then $E \leq_e {}_RR$.*

Proof. Let $0 \neq r \in R$. If $r \in I$ then $r \in E$ and hence $0 \neq Rr \leq Rr \cap E$. If $r \notin I$ then $(R/I)(r+I) \cap (E/I) \neq 0$. Thus $[(Rr \cap E) + I]/I = [(Rr + I)/I] \cap (E/I) = (R/I)(r+I) \cap (E/I) \neq 0$ and hence $Rr \cap E \neq 0$. It follows that $E \leq_e {}_RR$. □

Corollary 3.55. *Let I be an ideal of a ring R and A a nonsingular left R-module such that $IA = 0$. Then A is a nonsingular left (R/I)-module.*

Proof. Again make A into a left (R/I)-module by defining $(r+I)a = ra$ $(a \in A, r \in R)$. Suppose $a \in A$ and $\overline{E}a = 0$ for some $\overline{E} \leq_e {}_{\overline{R}}\overline{R}$, where $\overline{R} = R/I$. There exists a left ideal E of R containing I such that $\overline{E} = E/I$. By Lemma 3.54, $E \leq_e {}_RR$. Moreover, $Ea = 0$. Hence $a = 0$, because A is nonsingular. It follows that A is a nonsingular left (R/I)-module. □

Proposition 3.56. *Let R be a ring and A a left R-module such that A is a nonsingular reduced* CS*-module. Let $I \leq Z_2({}_RR)$. Then $IA = 0$ and the left (R/I)-module A is a nonsingular reduced* CS*-module.*

Proof. By Lemmas 3.52, 3.53 and Corollary 3.55. □

A ring R is called *left Goldie* if $u\text{-}\dim({}_RR) < \infty$ and R has acc on left annihilator ideals. (Right Goldie rings are defined similarly). An element c in a ring R is called *regular* if $cr \neq 0$ and $rc \neq 0$ for all $0 \neq r \in R$. Let R be any ring and A left R-module. Then A is called *divisible* if $A = cA$ for all regular $c \in R$.

The next lemma is stated for the semiprime left Goldie ring case. However, we remark that the first part of the proof works for any ring.

Lemma 3.57. *Let R be a semiprime left Goldie ring and A a nonsingular left R-module. Then A is injective if and only if A is divisible.*

Proof. Suppose A is injective. Let $c \in R$ be regular. Let $y \in A$. Define $\theta : Rc \to A$ by $\theta(rc) = ry$ $(r \in R)$. Then θ is well defined (because c is regular) and an R-homomorphism. Thus there exists $x \in A$ such that $\theta(rc) = rcx$ $(r \in R)$. In particular, $y = \theta(c) = cx \in cA$. It follows that $A = cA$. Hence A is divisible.

Conversely, suppose A is divisible. Let $E \leq_e {}_RR$ and $\varphi : E \to A$ be an R-homomorphism. Since R is semiprime left Goldie, E contains a regular element d (Exercise 3.26). Now $\varphi(d) \in A = dA$ and hence $\varphi(d) = da$ for some $a \in A$. Let $e \in E$. Then $rd = d'e$ for some $r, d' \in R$, with d' regular (Exercise 3.27). Thus $d'\varphi(e) = \varphi(d'e) = \varphi(rd) = r\varphi(d) = rda = d'ea$ and so $d'(\varphi(e) - ea) = 0$. Since

Rd' is essential in $_RR$ (Exercise 3.25) and A is nonsingular, $\varphi(e) = ea$. It follows that A is injective. $\square$

Lemma 3.58. *Let R be a commutative ring with a finite collection of prime ideals $P_1, P_2, \ldots, P_n$ such that $P_1 \cap P_2 \cap \cdots \cap P_n = 0$. Then R is semiprime Goldie.*

Proof. Without loss of generality we can ssume that P_i is not contained in P_j $(1 \leq i \neq j \leq n)$. Thus $P_1 \cap P_2 \cap \cdots P_{i-1} \cap P_{i+1} \cap \cdots \cap P_n$ is not contained in P_i $(1 \leq i \leq n)$. Now we apply Exercise 3.25. For, let $c \in R$ be regular. Suppose $Rc \cap I = 0$ for some ideal I of R. Then $Ic \leq Rc \cap I$, so $Ic = 0$ and $I = 0$. Thus Rc is essential in $_RR$. To complete the proof, let E be an essential ideal of R. Then E is not contained in P_i $(1 \leq i \leq n)$. Thus $E \cap P_1 \cap P_2 \cap \cdots P_{i-1} \cap P_{i+1} \cap \cdots \cap P_n$ is not contained in P_i $(1 \leq i \leq n)$. Let $d_i \in E \cap P_1 \cap P_2 \cap \cdots P_{i-1} \cap P_{i+1} \cap \cdots \cap P_n$ such that $d_i \notin P_i$ $(1 \leq i \leq n)$. Let $d = d_1 + d_2 + \cdots + d_n \in E$. Let $r \in R$ and suppose $dr = 0$. Let $1 \leq i \leq n$. If $d \in P_i$, then $d_i \in P_i$, a contradiction. Thus $d \notin P_i$. Therefore $dr = 0 \in P_i$ implies $r \in P_i$. Hence $r \in P_1 \cap P_2 \cap \cdots \cap P_n = 0$. It follows that d is regular. Thus R is semiprime Goldie. $\square$

Now we are in a position to give our first main result which extends Theorems 3.33.

Theorem 3.59. *Let R be a commutative ring with only finitely many minimal prime ideals $P_1, P_2, \ldots, P_n$. Let $_RA$ be a nonsingular reduced* CS*-module. Then A is a finite direct sum of uniform submodules.*

Proof. Let $N = \{r \in R : r^k = 0 \text{ for some } k \geq 1\}$. Then $N = P_1 \cap P_2 \cap \cdots \cap P_n$. Let $r \in N$ and $E = \{s \in R : sr = 0\}$. Let $0 \neq a \in R$. Then there exists $k \geq 1$ such that $r^{k-1}a \neq 0$, but $r^k a = 0$. Thus $0 \neq r^{k-1}a \in Ra \cap E$. It follows that E is essential in R. Hence $r \in Z(R)$. Thus $N \leq Z_2(R)$. By Proposition 3.56, $NA = 0$ and the (R/N)-module A is a nonsingular reduced CS-module. Thus, without loss of generality, $N = 0$, i.e., $P_1 \cap P_2 \cap \cdots \cap P_n = 0$. Then $P_1 P_2 \cdots P_n = 0$, so that $P_1 P_2 \cdots P_n A = 0$. Let $K = \{b \in A : P_1 b = 0\}$. Then $K \leq A$. Let $L \leq A$ such that K is essential in L. Let $x \in L$. There exists an essential ideal I of R such that $Ix \leq K$, and hence $P_1 I x = 0$. Thus $(P_1 x)I = 0$, so that $P_1 x = 0$. It follows that $x \in K$. Hence $L = K$, and K is a complement in A. By hypothesis, $A = K \oplus K'$ for some $K' \leq A$. Note that $K' \cong A/K$ and $P_1(P_2 \cap P_3 \cap \cdots \cap P_n) \leq P_1 \cap P_2 \cap \cdots \cap P_n = 0$ gives $P_1(P_2 \cap P_3 \cap \cdots \cap P_n)A = 0$, so that $(P_2 \cap \cdots \cap P_n)A \leq K$. Hence $(P_2 \cap \cdots \cap P_n)K' = 0$. By Lemma 3.53 and Corollary 3.55, the (R/P_1)-module K is a nonsingular CS-module and the $(R/(P_2 \cap \cdots \cap P_n))$-module K' is a nonsingular CS-module. Let X be an injective submodule of the (R/P_1)-module K. Let c be a regular element of R. Then $c \notin P_1$, and hence $X = cX$ by Lemma 3.57. Thus $_RX$ is divisible. By Lemmas 3.57 and 3.58, X is an injective R-module. But A is reduced, so that $X = 0$. Hence the (R/P_1)-module K is reduced. Since (R/P_1) is a domain, Theorem 3.33 implies that K has finite uniform dimension. By the same argument, the $(R/(P_2 \cap \cdots \cap P_n))$-module K' is reduced and K' has finite uniform dimension by induction on n. Thus $A = K \oplus K'$ has finite uniform dimension. Since A is a CS-module, the result follows. □

Corollary 3.60. *Let R be a commutative Noetherian ring and let ${}_RA$ be a nonsingular reduced* CS*-module. Then A is a finite direct sum of uniform submodules.*

Proof. Immediate by using Theorems 3.59. □

We give an example which shows that in Theorem 3.59 and Corollary 3.60 commutativity is not superfluous. First let us prove an easy fact to demonstrate the argument of the example.

Lemma 3.61. *Let R be a commutative ring. Then any nonsingular simple R-module is injective.*

Proof. Let U be a nonsingular simple R-module. Let $0 \neq u \in U$ and let $L = \{r \in R : ru = 0\}$. Then L is a maximal ideal of R and L is not essential. Thus $R = L \oplus V$ for some ideal V of R. Clearly, $V \cong R/L \cong U$. Thus without loss of generality, $U = V$, i.e., $R = L \oplus U$. Therefore, $U = Re$ for some $e^2 = e \in R$. Let I be an essential ideal of R and $\varphi : I \to U$ be a homomorphism. Note that $U \leq I$. Thus $\varphi(e) = xe$ for some $x \in R$. Let $a \in I$. Then $a = ae + a(1-e)$, so that $\varphi(a) = a\varphi(e) + \varphi(a)(1-e)$. But $\varphi(a) \in U = Re$ gives $\varphi(a)(1-e) = 0$. Thus $\varphi(a) = a\varphi(e) = axe = aexe = [ae + a(1-e)](xe) = axe$. Hence $\varphi(a) = a(xe)$ $(a \in I)$, and consequently U is injective. □

Example 3.62. There exist a (left) nonsingular (left) Artinian ring R and a nonsingular reduced CS left R-module A such that A has infinite uniform dimension.

Proof. Let K be any field and let $R = \begin{bmatrix} K & 0 \\ K & K \end{bmatrix}$. Let $U = \begin{bmatrix} 0 & 0 \\ 0 & K \end{bmatrix}$. Clearly U is a simple nonsingular left R-module. Let $e = \begin{bmatrix} 0 & 0 \\ 0 & 1 \end{bmatrix}$. Then $U = Re$. Define $\varphi : \begin{bmatrix} 0 & 0 \\ K & 0 \end{bmatrix} \to U$ by $\varphi\left(\begin{bmatrix} 0 & 0 \\ x & 0 \end{bmatrix}\right) = \begin{bmatrix} 0 & 0 \\ 0 & x \end{bmatrix}$ $(x \in K)$. Thus φ is an R-homomorphism. If U is injective, then there exists $y \in K$ such that $\begin{bmatrix} 0 & 0 \\ 0 & 1 \end{bmatrix} = \varphi\left(\begin{bmatrix} 0 & 0 \\ 1 & 0 \end{bmatrix}\right) = \begin{bmatrix} 0 & 0 \\ 1 & 0 \end{bmatrix}\begin{bmatrix} 0 & 0 \\ 0 & y \end{bmatrix} = \begin{bmatrix} 0 & 0 \\ 0 & 0 \end{bmatrix}$. Hence U is not injective. Let A be a direct sum of an infinite number of copies of U. Then A is CS, being semisimple. If N is a non-zero injective submodule of A, then N is isomorphic to a direct sum of copies of U, so U is injective, a contradiction. Thus A is reduced. □

Combining Lemmas 3.30, 3.31 and Theorem 3.33 we deduce that torsion-free extending modules over commutative domains are finite direct sums of injective modules and uniform modules. The purpose of the remaining part of this section is to generalize this fact.

Let R be any ring and A be a left R-module. Recall that the module A is CS (or extending) if and only if every submodule is essential in a direct summand of A. It follows that if I is an ideal of R and A is an injective (R/I)-module, then A is an extending R-module (see Lemma 3.53). We shall say that an R-module A has an *indecomposable decomposition* if A is a direct sum of indecomposable submodules.

Lemma 3.63. *The following statements are equivalent for a nonsingular extending R-module A.*

(i) *A has an indecomposable decomposition.*
(ii) *Every finitely generated submodule of A has finite uniform dimension.*
(iii) *Every cyclic submodule of A has finite uniform dimension.*
(iv) *R satisfies acc on left ideals of the form $\ell(a)$, where $a \in A$.*

Proof. (i) $\Rightarrow$ (ii) There exist an index set I and indecomposable submodules A_i $(i \in I)$ such that $A = \bigoplus_I A_i$. By Exercise 3.3, A_i is a uniform module for each $i \in I$. If L is a finitely generated submodule of A, then $L \subseteq \bigoplus_J A_i$, for some finite subset J of I, and hence L has finite uniform dimension.

(ii) $\Rightarrow$ (iii) Clear.

(ii) $\Rightarrow$ (iv) Let $a \in A$. Suppose that $\ell(a)$ is essential in a left ideal I of R. Let $x \in I$. There exists an essential left ideal E of R such that $Ex \subseteq \ell(a)$. It follows that $Exa = 0$ and hence $xa = 0$, i.e., $x \in \ell(a)$. Thus $\ell(a) = I$. We have proved that $\ell(a)$ is a complement in the R-module R, for each $a \in A$, (see Exercise 2.14). Moreover, $R/\ell(a) \cong Ra$ gives that the R-module $R/\ell(a)$ has finite uniform dimension. Now (iv) follows by Proposition 2.11.

(iv) $\Rightarrow$ (i) By Exercise 3.18. □

Corollary 3.64. *Let R be a ring with finite left uniform dimension and let A be a nonsingular extending R-module. Then A has an indecomposable decomposition.*

Proof. As we saw in the proof of Lemma 3.63, for each $a \in A$ the left ideal $\ell(a)$ is a complement in the R-module R. Hence, R satisfies the ascending chain condition on left ideals of the form $\ell(a)$, where $a \in A$ (see Proposition 2.11). Apply Lemma 3.63. □

Lemma 3.65. *Let U and V be nonsingular uniform R-modules such that the module $U \oplus V$ is extending. Then U embeds in V or U is V-injective.*

Proof. Let $A = U \oplus V$, $A_1 = U \oplus 0$ and $A_2 = 0 \oplus V$, so that $A = A_1 \oplus A_2$. Suppose that U does not embed in V, i.e., A_1 does not embed in A_2. Let N be any non-zero submodule of A such that $N \cap A_1 = 0$. There exists a direct summand K of A such that N is essential in K. Let K' be a submodule of A such that $A = K \oplus K'$. Note that K and K' are both uniform submodules of A.

Since $K \cap A_1 = 0$, it follows that A_1 embeds in K'. If $K' \cap A_1 = 0$, then K' embeds in A_2, so that A_1 embeds in A_2, a contradiction. Thus $K' \cap A_1 \neq 0$.

Hence $K' \cap A_1$ is an essential submodule of both K' and A_1. Now $(K'+A_1)/A_1 \cong K'/(K' \cap A_1)$ is a singular submodule of the nonsingular module A/A_1, so that $K' \subseteq A_1$. Similarly, $A_1 \subseteq K'$. Thus, $K' = A_1$ and hence $A = K \oplus A_1$. By Exercise 3.4, A_1 is A_2-injective, i.e., U is V-injective. □

Lemma 3.66. *Let A_i $(i \in I)$ be nonsingular uniform R-modules such that for each $i \in I$ there exists a monomorphism $\phi_i : A_i \to A_{i+1}$, which is not an isomorphism. Then the R-module $A = \bigoplus_{i \in I} A_i$ is not extending.*

Proof. The monomorphism $\phi_1^{-1} : \phi_1(A_1) \to E(A_1)$ lifts to a monomorphism $\theta_1 : A_2 \to E(A_1)$ and clearly $A_1 = \theta_1(\phi_1(A_1)) \subsetneqq \theta_1(A_2) \subseteq E(A_1)$. Since $A_2 \cong \theta_1(A_2)$, we can suppose without loss of generality that $A_1 \subsetneqq A_2 \subseteq E(A_1)$. Note that $E(A_1) = E(A_2)$. Repeating this argument for the submodules A_2, and A_3, we can suppose without loss of generality that $A_2 \subsetneqq A_3 \subseteq E(A_1)$. Proceeding in this way, we can suppose that

$$A_1 \subsetneqq A_2 \subsetneqq A_3 \subsetneqq \cdots \subseteq E(A_1).$$

Define

$$\chi : A \to E(A_1) \text{ by } \chi(a_1, a_2, a_3, \dots) = a_1 + a_2 + a_3 + \cdots$$

for all $(a_1, a_2, a_3, \dots) \in A$. Let $K = \ker \chi$. Then A/K is isomorphic to a submodule of the nonsingular module $E(A_1)$, so that K is a complement in A. Suppose that K is a direct summand of A, i.e., $A = K \oplus K'$ for some submodule K' of A. Then $K' \cong A/K$, so that K' is uniform. Let $0 \neq x \in K'$. Then $x \in N = A_1 \oplus \cdots \oplus A_n$ for some positive integer n. Since K'/Rx is singular and A/N is nonsingular, it follows that $K' \subseteq N$. Let $y \in A_{n+1}$. Then $y = y_1 + y_2$ for some $y_1 \in K$ and $y_2 \in K' \subseteq N$. Thus $y = \chi(y) = \chi(y_1) + \chi(y_2) = \chi(y_2) \in A_1 + \cdots + A_n = A_n$. It follows that $A_{n+1} = A_n$, a contradiction. Thus K is not a direct summand of A and A is not an extending module. □

Lemma 3.67. *Let R be a prime left Goldie ring. Then an R-module A is injective if and only if there exists a non-zero nonsingular R-module P such that A is P-injective.*

Proof. The necessity is clear. Conversely, suppose that there exists a non-zero nonsingular module P such that A is P-injective. Let $0 \neq p \in P$. Then $\ell(p)$ is not an essential left ideal of R. There exists a uniform left ideal U of R such that $U \cap \ell(p) = 0$. Then U embeds in the R-module $R/\ell(p) \cong Rp$. Thus A is U-injective (Proposition 2.19).

There exist a positive integer n and independent uniform left ideals U_i $(1 \leq i \leq n)$ of R such that $U_1 \oplus \cdots \oplus U_n$ is an essential left ideal of R (See Proposition 2.17). Let $1 \leq i \leq n$. Since R is prime, it follows that $U_i U \neq 0$. Let $u \in U$ such that $U_i u \neq 0$ and define $\phi_i : U_i \to U$ by $\phi_i(x) = xu$ $(x \in U_i)$. It is not difficult to check that because R is left nonsingular, the mapping ϕ_i is an R-monomorphism. Thus A is U_i-injective for all $1 \leq i \leq n$ and hence A is $(U_1 \oplus \cdots \oplus U_n)$-injective. By Exercise 3.26 there exists an element $c \in U_1 \oplus \cdots \oplus U_n$ with $\ell(c) = 0$. Then $R \cong Rc$ and A is R-injective. By Baer's Lemma, A is an injective R-module. □

Lemma 3.68. *Let R be a prime left Goldie ring and let A be a nonsingular reduced extending R-module. Then A has finite uniform dimension.*

Proof. Suppose that $A \neq 0$. By Corollary 3.64, there exist an index set I and uniform submodules A_i $(i \in I)$ such that $A = \bigoplus_{i \in I} A_i$. Suppose that I is infinite. For all distinct i, j in I, $A_i \oplus A_j$ is extending (see Exercise 3.2) and hence A_i embeds in A_j by Lemmas 3.65 and 3.67. By Lemma 3.66, there exists an infinite subset J of I such that $A_i \cong A_j$ for all $i, j \in J$.

Let $N = A_i$ for any $i \in J$. Suppose that $\phi : N \to N$ is a non-zero homomorphism. Then ϕ is a monomorphism because A is nonsingular. Suppose that ϕ is not an isomorphism. Then for each $i, j \in J$, ϕ induces a monomorphism $\theta_{i,j} : A_i \to A_j$ which is not an isomorphism, contradicting Lemma 3.66. Thus ϕ is an isomorphism. It follows that the ring End A of endomorphism of A is a division ring and by Exercise 3.19, N quasi-injective. By Lemma 3.67, N is injective, a contradiction. Therefore, A has finite uniform dimension. □

For any ring R, an R-module A is called *torsion-free* if $ca \neq 0$ for all $0 \neq a \in A$ and regular $c \in R$. Given an ideal I of R we set $C(I) = \{c \in R : c + I$ is a regular element of the ring $R/I\}$.

Lemma 3.69. *Let R be a semiprime left Goldie ring.*

(i) *An R-module A is nonsingular if and only if A is torsion-free.*
(ii) *For any ideal I of R with $C(0) \subseteq C(I)$, any divisible (R/I)-module is a divisible R-module.*

Proof. (i) By Exercise 3.26. (ii) Clear. □

Lemma 3.70. *Let R be a prime left Goldie ring and let A be a nonsingular extending R-module. Then $A = A_1 \oplus \cdots \oplus A_n$ for some positive integer n and R-modules A_i $(1 \leq i \leq n)$ such that for each $1 \leq i \leq n$, A_i is injective or uniform.*

Proof. By Zorn's Lemma, A contains a maximal divisible submodule N and, by Lemma 3.69, N is injective. Thus $A = N \oplus N'$ for some reduced submodule N'. By Exercise 3.2, N' is an extending module. Apply Lemma 3.68. □

Theorem 3.71. *Let R be a semiprime left Goldie ring and let A be a nonsingular extending R-module. Then $A = A_1 \oplus \cdots \oplus A_n$ for some positive integer n and R-modules A_i $(1 \leq i \leq n)$ such that for each $1 \leq i \leq n$, A_i is injective or uniform.*

Proof. Let P_i $(1 \leq i \leq n)$ be the minimal prime ideals of R for some positive integer n (Exercise 3.27). Let $K = \{a \in A : P_1 a = 0\}$. Clearly, K is a submodule of A. Moreover, the (R/P_1)-module K is nonsingular, for if $a \in K$ and $(E/P_1)a = 0$ for some essential left ideal E/P_1 of R/P_1, where E is a left ideal containing P_1, then E is an essential left ideal of R (Lemma 3.54) and $Ea = 0$ so that $a = 0$.

Next we show that K is a closed submodule of the R-module A. Let L be an essential extension of K in A. Let $x \in L$. There exists an essential left ideal F of R such that $Fx \subseteq K$. By Exercise 3.26, there exists a regular element c of R

such that $c \in F$ and hence $P_1cx = 0$. For any p in P_1 there exists $r \in R$, $c' \in C(0)$ such that $rc = c'p$ (Exercise 3.27). Since $c \in C(P_1)$ (Exercise 3.27), it follows that $r \in P_1$ and hence $rcxc'px = 0$. But Rc' is essential in R, so that $px = 0$. Hence $P_1x = 0$ and $x \in K$. Thus $K = L$.

By hypothesis, there exists a submodule K' of A such that $A = K \oplus K'$. Since $P_1(P_2 \cap \cdots \cap P_n) = 0$, it follows that $(P_2 \cap \cdots \cap P_n)A \subseteq K$ and hence $(P_2 \cap \cdots \cap P_n)K' = 0$. Let $S = P_2 \cap \cdots \cap P_n$. By the above argument, K' is a nonsingular (R/S)-module. Clearly, K and K' are extending modules over the rings R/P_1 and R/S, respectively. By Lemma 3.70 and induction on n, K is a finite direct sum of nonsingular injective (R/P_1)-modules and uniform (R/P_1)-modules, and K' is a finite direct sum of nonsingular injective (R/S)-modules and uniform (R/S)-modules. The result now follows from Lemmas 3.69 and 3.57. □

One might expect that whether we can remove the ring R is being left Goldie in previous theorem. However corollary of the following result eliminates this possibility.

Theorem 3.72. *Let R_i $(i \in I)$ be any non-empty collection of commutative domains and let $R = \prod_{i \in I} R_i$ be their direct product. Then the free R-module R is* CS.

Proof. Let K be any non-zero closed submodule of ${}_RR$. Let $0 \neq x = \{x_i\} \in K$, where $x_i \in R$ $(i \in I)$. Choose $j \in I$ such that $x_j \neq 0$. Let $e^{(j)} = \{e_i{}^{(j)}\}$ denote the element of R given by $e_i{}^{(j)} = 1$ for $i = j$, $e_i{}^{(j)} = 0$ for $i \neq j$. Then $0 \neq e^{(j)}x \in K$. Now let $E = R_jX_j \oplus (\prod_{i \neq j} R_i)$. It is clear that E is an essential ideal of R and $Ee^{(j)} = Re^{(j)}x \subseteq K$. Because K is closed in the nonsingular R-module R, $e^{(j)} \in K$. Hence $R_j = Re^{(j)} \subseteq K$.

Let $\underline{S} = \{J \subseteq I : \bigoplus_j R_j \subseteq K\}$. By the above argument, $\underline{S}$ is non-empty. By Zorn's Lemma, $\underline{S}$ has a maximal member P (say). Let $F = \bigoplus_I R_i$. Clearly F is an essential ideal of R and $F(\prod_P R_i) = \bigoplus_I R_i \subseteq K$. Since K is closed in ${}_RR$, $\prod_P R_i \subseteq K$.

Now suppose that $K \neq \prod_P R_i$. There exists $y = \{y_i\} \in K$ such that $y_j \neq 0$ for some $j \in I \backslash P$. By the above argument, $R_j \subseteq K$ and hence $\bigoplus_{P \cup \{j\}} R_i = (\bigoplus_P R_i) \bigoplus R_j \subseteq K$, contradicting the choice of P. Thus $K = \prod_P R_i$. Let $K' = \prod_{I \backslash P} R_i$. Then ${}_RR = K \oplus K'$. It follows that ${}_RR$ is CS. □

Corollary 3.73. *Let $R = \prod_{i \geq 1} R_i$ be the direct product of an infinite number of copies R_i of the ring $\mathbb{Z}$ of integers. Then R is a semiprime ring and the R-module R is a nonsingular reduced* CS*-module which does not have finite uniform dimension.*

Proof. Let $F =_R R$. Clearly, R is a semiprime, and hence nonsingular ring and F is a nonsingular R-module. By Theorem 3.72, F is a CS R-module. Note next that F contains the infinite direct sum $\bigoplus_{i \geq 1} R_i$. Let G be an injective submodule of F. Suppose that $G \neq 0$. Let $0 \neq g = \{g_i\} \in G$. Choose any j such that $g_i \neq 0$ and let $H = \{r \in R_j :$ there exists $h = \{h_i\} \in G$ such that $h_j = r\}$. Then H is a non-zero ideal of the ring $R_j (= \mathbb{Z})$. Let p be any prime in $\mathbb{Z}$. Let $c = \{c_i\} \in R$, where $c_i = p$ for all i. Clearly c is a regular element of R. Because G is injective, $G = cG$ (Lemma 3.57) and hence $H = pH$.

Thus $pH = H$ for all primes in $\mathbb{Z}$ and hence $H = 0$, a contradiction. Thus $G = 0$. It follows that the R-module F is reduced. □

Observe that Theorem 3.71 fails for semiprime rings R which are not left Goldie even if they are commutative (see Corollary 3.73). For commutative rings we give the following extension of Theorem 3.71.

Corollary 3.74. *Let R be a commutative ring with prime radical N such that R/N is a Goldie ring. Let A be a nonsingular extending R-module. Then A is a finite direct sum of injective (R/N)-modules and uniform R-modules.*

Proof. Let $a \in N$. For any non-zero r in R, there exists a positive integer k such that $a^{k-1}r \neq 0$ and $a^k r = 0$. Thus $0 \neq a^{k-1}r \in Rr \cap \ell(a)$. It follows that $\ell(a)$ is an essential ideal of R. For any $x \in A$, $\ell(a)ax = 0$ implies that $ax = 0$. Thus $NA = 0$ and A is a nonsingular extending (R/N)-module. The result follows by Theorem 3.71. □

Next we provide the final generalization of the theorem of Kamal and Müller (see Theorem 3.33).

Corollary 3.75. *Let R be a commutative ring which satisfies* acc *on annihilators and let A be a nonsingular extending R-module. Then A is a finite direct sum of injective (R/N)-modules and uniform R-modules, where N is the prime radical of R.*

Proof. By Exercise 3.30, the ring R/N is Goldie. Apply Corollary 3.73. □

Note that Theorem 3.71 and its corollaries have (partial) converses, because if A is a nonsingular module over an arbitrary ring R such that $A = A_1 \oplus A_2$ for some injective module A_1 and extending module A_2, then A is extending by Exercise 3.20. For example, Corollary 3.73 can be restated thus: Let R be a commutative ring with prime radical N such that R/N is a Goldie ring. Then a nonsingular R-module A is extending if and only if $A = A_1 \oplus A_2$ for some injective module A_1 and extending A_2 with A_2 having finite uniform dimension.

Exercises

3.21. Let R be a commutative domain and A a left R-module. Prove that

(i) $Z(_RA) = \{a \in A : \ell(a) \neq 0\}$ (i.e., singular and torsion submodules of A are the same) (see Exercise 1.44),

(ii) If $_RA$ is torsion-free, then it is the direct sum of an injective module and a reduced module.

3.22. Let R be commutative domain. Show that R is a Dedekind domain if and only if R is Noetherian and for every nonzero prime ideal P of R, the localization R_P of R at P is a DVR.

3.23. Show that every nonzero ideal of a Dedekind domain is invertible.

3.24. Prove that a commutative ring is semiprime if and only if it is nonsingular.

3.25. Let R be a ring with $u\text{--}\dim({}_RR) < \infty$ and let c be a regular element in R. Show that $Rc \leq_e {}_RR$.

3.26. Let R be a ring. Prove that R is semiprime left Goldie if and only if for any left ideal I in R, $I \leq_e {}_RR$ if and only if I contains a regular element.

3.27. Let R be a semiprime left Goldie ring. Show that

(i) R has only a finite number of minimal prime ideals P_i $(1 \leq i \leq k)$.
(ii) $C(0) = C(P_1) \cap \cdots \cap C(P_k)$.
(iii) Let $a, c \in R$ such that c is regular. Show that $da = rc$ for some $d, r \in R$ such that d is regular.

3.28. Let R be a semiprime left Goldie ring and let A a left R-module. Let $T(A) = \{a \in A : xa = 0 \text{ for some regular } x \in R\}$. Prove that

(i) $T(A)$ is submodule of A,
(ii) $Z(A) = T(A)$ (i.e., over a semiprime left Goldie ring R, the singular submodule of any left R-module A is a direct analog of the torsion submodule of a module over a commutative domain. Because of this analogy, we may use torsion terminology in this situation).
(iii) Provide an example which shows that (ii) is not true for all rings.

3.29. Let R be any ring. Prove that R is semiprime left Goldie if and only if $T(A) = Z(A)$ for all left R-modules A.

3.30. Let R be a commutative ring which satisfies the ascending chain condition on annihilators and let N be the prime radical of R. Prove that R/N is a Goldie ring.

3.3 Extending matrix rings

In this section we mainly focus on matrix CS-rings. We give an example which shows that being CS is not Morita invariant. After we prove general results on generalized triangular matrix (also called formal triangular matrix) CS-rings, we obtain necessary and sufficient conditions under which a right nonsingular generalized triangular matrix ring is a right CS-ring. Moreover, we restrict our considerations to the split null extensions (also called trivial extensions) and we provide necessary and sufficient conditions under which such matrix rings are right CS-rings. We give several examples which illustrate our results.

Let R be a ring and $M_m(R)$, $T_m(R)$ be the full matrix ring and upper triangular matrix ring over R, respectively. By Proposition 2.77, we have the following theorem.

Theorem 3.76. *Let A be a left R-module, where $R = ReR$ for some idempotent e in R, and let $S = eRe$. Then*

(i) *the left R-module A is* CS *if and only if the left S-module eA is* CS.
(ii) *R_R is* CS *if and only if the eRe-module Re is* CS.

Corollary 3.77. *$M_m(R)$ is a right CS-ring if and only if the free right R-module R^m is* CS.

Proof. It is clear that $M_m(R) = M_m(R)eM_m(R)$, where e is the matrix unit with 1 in the $(1,1)$th position and zero elsewhere. The result now follows from Theorem 3.76. □

Example 3.78. Let R be the polynomial ring $\mathbb{Z}[x]$. It is clear that $M_2(R) = M_2(\mathbb{Z}[x]) \cong M_2(\mathbb{Z})[x]$. Note that $M_2(\mathbb{Z})$ is a right CS-ring by Corollary 3.77. However, Corollary 3.77 together with Exercise 3.7 show that $M_2(\mathbb{Z}[x])$ is not a right CS-ring.

Observe that Example 3.78 shows that over a right CS-ring neither the ring of polynomials, nor the full matrix ring need to be a right CS-ring in general. Thus being CS is not Morita invariant. In conjuction with Example 3.78, we next deal with matrix CS-rings over domains. A domain S is called *right Ore* if $xS \cap yS \neq 0$ for all nonzero $x, y \in S$. *Left Ore domains* are defined similarly. Note that S is a right Ore domain if and only if S_S is uniform.

Lemma 3.79. *Let S be a domain. If $M_2(S)$ is a right* CS*-ring then S is left and right Ore.*

Proof. Let $R = M_2(S)$. Let $0 \neq x, y \in S$ and suppose $Sx \cap Sy = 0$. Set $u = \begin{bmatrix} x & 0 \\ y & 0 \end{bmatrix}$. Then $\begin{bmatrix} x & 0 \\ y & 0 \end{bmatrix}\begin{bmatrix} r & s \\ w & v \end{bmatrix} = \begin{bmatrix} xr & xs \\ yr & ys \end{bmatrix}$, $\begin{bmatrix} r & s \\ w & v \end{bmatrix} \in R$. Hence uR is essential in eR for some $e^2 = e \in R$. Let say $e = \begin{bmatrix} a & b \\ c & d \end{bmatrix}$, where $a, b, c, d \in S$. Thus

$$\begin{bmatrix} x & 0 \\ y & 0 \end{bmatrix} = \begin{bmatrix} a & b \\ c & d \end{bmatrix}\begin{bmatrix} x & 0 \\ y & 0 \end{bmatrix}$$

and hence $\begin{bmatrix} x & 0 \\ y & 0 \end{bmatrix} = \begin{bmatrix} ax+by & 0 \\ cx+dy & 0 \end{bmatrix}$. Then $x = ax + by$ and $y = cx + dy$. Now $(1-a)x = by$. Therefore $(1-a)x = 0$ and hence $a = 1$, $b = 0$. Also $(1-d)y = cx$ implies that $d = 1$, $c = 0$. Thus $e = \begin{bmatrix} 1 & 0 \\ 0 & 1 \end{bmatrix}$. We conclude that uR is essential in R.

On the other hand,

$$0 \neq \begin{bmatrix} 1 & 0 \\ 0 & 0 \end{bmatrix} \begin{bmatrix} a' & b' \\ c' & d' \end{bmatrix} = \begin{bmatrix} xr & xs \\ yr & ys \end{bmatrix} \quad (a', b', c', d' \in S),$$

gives that $\begin{bmatrix} a' & b' \\ 0 & 0 \end{bmatrix} = \begin{bmatrix} xr & xs \\ yr & ys \end{bmatrix}$. Hence $r = s = 0$. Therefore, $a' = b' = 0$, a contradiction. It follows that S is left Ore.

Since R is a right CS-ring, S is a right CS-ring too (Corollary 3.77). Let $0 \neq z \in S$. Then zS is essential in fS_S for some $f^2 = f \in S$. Thus $f = 1$, so that zS_S is essential in S_S. Hence S is right Ore. □

Recall that a right nonsingular right CS-ring is a right pp-ring. However, the converse is not true in general, as the following example illustrates.

Example 3.80. Let K be a field of characteristic $p > 0$. Let $G = \langle x : x^p = 1 \rangle$, the cyclic group of order p. Let R denote the group algebra $K[G]$. Then R is a commutative Artinian CS-ring which is not pp.

Proof. Note that R is a (quasi-) Frobenius algebra and hence a self-injective Artinian ring. In particular, R is a CS-ring. Since $J(R) = Z(R) \neq 0$, R is not a pp-ring by Exercise 3.16. □

We have the following generalization.

Lemma 3.81. *Suppose R is a semiprime right and left Goldie ring. Then the following statements are equivalent.*

(i) *R is a right CS-ring.*
(ii) *R is a right and left pp-ring.*
(iii) *R is a left CS-ring.*

Proof. (i) $\Rightarrow$ (ii) Since R is a right (left) nonsingular, every right (left) annihilator is a complement in R, so is a direct summand. Therefore, R is a right (left) pp-ring.

(ii) $\Rightarrow$ (i) Let Q denote the semiprime Artinian classical ring of quotients of R. Note that a right ideal of Q is minimal if and only if it is uniform. Let $u \in R$. Suppose uR is uniform. Then uQ is a minimal right ideal of Q. We have $Qu = Qe$ for some idempotent e, so that $r(u) = (1 - e)Q$. Thus

$$eQ \cong Q/(1 - e)Q = Q/r(u) \cong uQ,$$

so that eQ is a minimal right ideal of the semisimple Artinian ring Q. Hence Qe is a minimal left ideal of Q. Therefore, Qu is minimal, i.e., Ru is uniform.

Now, let U be a maximal uniform right ideal of R and let $0 \neq u \in U$. Then $\ell(u) = Rf$ for some $f^2 = f \in R$. We have $uR \leq U$, so that uR and Ru are uniform. Also $Ru \cong R(1 - f)$, so that $R(1 - f)$ is a uniform left ideal. Hence $(1 - f)R$ is a uniform right ideal. But $uR \leq (1 - f)R$, so that $U \cap (1 - f)R \neq 0$. Hence

$U+(1-f)R$ is also uniform (R is nonsingular). Since $(1-f)R \leq_{\rm d} (U+(1-f)R)$, we must have $U+(1-f)R = (1-f)R$. Thus $U \leq (1-f)R$, and the maximality of U gives $U = (1-f)R$. By Exercise 3.4, R is a right CS-ring.

(ii) $\Leftrightarrow$ (iii) By symmetry. □

Now we prove the following result on matrix rings over domains. We should mention that the next theorem holds true if we replace 2 by n with the proof requiring only slight modifications.

Theorem 3.82. *Let S be a domain. Then the following statements are equivalent.*

(i) *$M_2(S)$ is a right* CS*-ring.*

(ii) *S is right and left Ore and every 2-generator right or left ideal is projective.*

(iii) *$M_2(S)$ is a left* CS*-ring.*

Proof. (i) $\Rightarrow$ (ii) By Lemma 3.79, S is a right and left Ore domain. Also, $M_2(S)$ is a right nonsingular, and hence a right and left pp-ring. By Exercise 3.35, every 2-generator right or left ideal is projective.

(ii) $\Rightarrow$ (i) Now $M_2(S)$ is a right and left pp-ring by Exercise 3.35. Since $M_2(S)$ is a prime Goldie ring, then by Lemma 3.81, $M_2(S)$ is a right CS-ring.

(i) $\Leftrightarrow$ (iii) By symmetry. □

In contrast to the case of full matrix rings or polynomial rings, the CS property transfers to essential overrings. Recall that T is a *right essential overring* of a ring R if T is an overring of R such that R_R is essential in T_R (Exercise 3.32). So we have

Corollary 3.83. *If $T_m(R)$ is a right* CS*-ring, then R^m is a right* CS *R-module.*

Proof. By Exercise 3.31, $M_m(R)$ is a right essential overring of $T_m(R)$. Hence the result follows from Exercise 3.32 and Corollary 3.77. □

Example 3.84. Let $R = T_2(\mathbb{Z})$. Then R is not a right CS-ring.

Proof. Let $u = \begin{bmatrix} 0 & 1 \\ 0 & 2 \end{bmatrix}$. Then uR is a uniform right ideal of R. It is easy to see that the identity element of R is the only idempotent element e of R such that $eu = u$, i.e., such that $uR \subseteq eR$. It follows that R is not a right CS-ring. □

Example 3.84 shows that the converse of Corollary 3.83 does not hold; this leads to the question of, when a generalized triangular matrix ring is a right CS-ring.

Recall that a 2-by-2 generalized triangular matrix ring T has the form

$$T = \begin{bmatrix} R & A \\ 0 & S \end{bmatrix},$$

where R and S are associative rings with unity and A is an (R, S)-bimodule. Such rings T are useful in many areas of Algebra (e.g., finite-dimensional algebras and

Morita contexts) and Functional Analysis (e.g., operator algebras). Observe that when $R = \text{End}(A_S)$, then T provides a convenient way of encoding information about A. Since A is an (R, S)-bimodule, there is always a natural ring homomorphism from R into $\text{End}(A_S)$. This ring homomorphism is injective if and only if ${}_RA$ is faithful. We will use $\ell_R(A)$ to denote the annihilator of A in R, i.e., $\ell_R(A) = \{r \in R : rA = 0\}$.

Lemma 3.85. *I is essential in T if and only if there exist essential submodules B of $\ell_R(A)$, N of A_S, and E of S_S such that* $\begin{bmatrix} B & N \\ 0 & E \end{bmatrix} \leq I$.

Proof. Simple to check. □

Lemma 3.86. *If T is a right CS-ring, then*

(i) *for every complement K in A_S there exists $r^2 = r \in R$ such that $K = rA$.*
(ii) *S is a right CS-ring.*

Proof. (i) Let K be a complement in A_S. Then $\begin{bmatrix} 0 & K \\ 0 & 0 \end{bmatrix} \leq T_T$. By hypothesis, there exists an idempotent $e = \begin{bmatrix} r & a \\ 0 & s \end{bmatrix} \in T$ such that $\begin{bmatrix} 0 & K \\ 0 & 0 \end{bmatrix}$ is essential in eT. Note that r is idempotent in R. Let $n \in K$. Then

$$\begin{bmatrix} 0 & n \\ 0 & 0 \end{bmatrix} = \begin{bmatrix} r & a \\ 0 & s \end{bmatrix} \begin{bmatrix} 0 & n \\ 0 & 0 \end{bmatrix} = \begin{bmatrix} 0 & rn \\ 0 & 0 \end{bmatrix}.$$

Thus $n = rn \in rA \leq A_S$, so that $K \leq rA$. Let $x \in A$ with $rx \neq 0$. Hence

$$0 \neq \begin{bmatrix} r & a \\ 0 & s \end{bmatrix} \begin{bmatrix} 0 & x \\ 0 & 0 \end{bmatrix} = \begin{bmatrix} 0 & rx \\ 0 & 0 \end{bmatrix}.$$

Since $\begin{bmatrix} 0 & K \\ 0 & 0 \end{bmatrix}$ is essential in eT, there exists $\begin{bmatrix} r' & a' \\ 0 & s' \end{bmatrix} \in T$ such that

$$0 \neq \begin{bmatrix} 0 & rx \\ 0 & 0 \end{bmatrix} \begin{bmatrix} r' & a' \\ 0 & s' \end{bmatrix} \in \begin{bmatrix} 0 & K \\ 0 & 0 \end{bmatrix}.$$

Therefore $\begin{bmatrix} 0 & rxs' \\ 0 & 0 \end{bmatrix} \in \begin{bmatrix} 0 & K \\ 0 & 0 \end{bmatrix}$. It follows that $0 \neq rxs' \in rxT \cap K$, i.e., K is essential in $(rA)_S$. Consequently, $K = rA$.

(ii) Let G be a right ideal of S. Let $B = \begin{bmatrix} 0 & 0 \\ 0 & G \end{bmatrix}$. Clearly, B is a right ideal of T. By hypothesis, there exists $\alpha \in T$ such that $\alpha^2 = \alpha$ and B is essential in αT. Now $\alpha = \begin{bmatrix} r & a \\ 0 & s \end{bmatrix}$ for some $r \in R$, $s \in S$, $a \in A$. Note that $s^2 = s$. Let $g \in G$. Then

$$\begin{bmatrix} 0 & 0 \\ 0 & g \end{bmatrix} = \alpha \begin{bmatrix} 0 & 0 \\ 0 & g \end{bmatrix} = \begin{bmatrix} r & a \\ 0 & s \end{bmatrix} \begin{bmatrix} 0 & 0 \\ 0 & g \end{bmatrix} = \begin{bmatrix} 0 & ag \\ 0 & sg \end{bmatrix}.$$

Therefore, $g = sg \in sS$. Hence $G \leq sS$. Now let $s_1 \in S$ with $ss_1 \neq 0$. Thus $\begin{bmatrix} r & a \\ 0 & s \end{bmatrix}\begin{bmatrix} 0 & 0 \\ 0 & s_1 \end{bmatrix} = \begin{bmatrix} 0 & as_1 \\ 0 & ss_1 \end{bmatrix} \neq 0$. It follows that $0 \neq \begin{bmatrix} 0 & as_1 \\ 0 & ss_1 \end{bmatrix}\begin{bmatrix} r_2 & a_2 \\ 0 & s_2 \end{bmatrix} \in B$, where $\begin{bmatrix} r_2 & a_2 \\ 0 & s_2 \end{bmatrix} \in T$. Thus $\begin{bmatrix} 0 & as_1s_2 \\ 0 & ss_1s_2 \end{bmatrix} = \begin{bmatrix} 0 & 0 \\ 0 & h \end{bmatrix}$ for some $h \in G$, and so $0 \neq ss_1s_2 = h \in G$. It follows that G is essential in $(sS)_S$, so that S_S is a CS-module. □

Lemma 3.87. *If T is a right CS-ring, then A_S is an injective module.*

Proof. Let $X \leq S_S$ and $\varphi : X \to A_S$ be a homomorphism. Let

$$F = \left\{ \begin{bmatrix} 0 & \varphi(x) \\ 0 & x \end{bmatrix} : x \in X \right\}.$$

Then F is a right ideal of T. By hypothesis, there exists $e = \begin{bmatrix} r & a \\ 0 & s \end{bmatrix} \in T$ such that $e^2 = e$ and F is essential in eT. Suppose $r \neq 0$. Then

$$\begin{bmatrix} r & a \\ 0 & s \end{bmatrix}\begin{bmatrix} 1 & 0 \\ 0 & 0 \end{bmatrix} = \begin{bmatrix} r & 0 \\ 0 & 0 \end{bmatrix} \in eT.$$

Therefore there exist $r' \in R$, $s' \in S$, $a' \in A$ such that

$$0 \neq \begin{bmatrix} r & 0 \\ 0 & 0 \end{bmatrix}\begin{bmatrix} r' & a' \\ 0 & s' \end{bmatrix} = \begin{bmatrix} rr' & ra' \\ 0 & 0 \end{bmatrix} \in F,$$

and so

$$0 \neq \begin{bmatrix} rr' & ra' \\ 0 & 0 \end{bmatrix} = \begin{bmatrix} 0 & \varphi(x) \\ 0 & x \end{bmatrix}$$

for some $x \in X$. It follows that $rr' = 0$ and $x = 0$. Then $ra' = \varphi(x) = 0$, a contradiction. Hence $r = 0$. Now for each $x \in X$,

$$\begin{bmatrix} 0 & \varphi(x) \\ 0 & x \end{bmatrix} = \begin{bmatrix} 0 & a \\ 0 & s \end{bmatrix}\begin{bmatrix} 0 & \varphi(x) \\ 0 & x \end{bmatrix} = \begin{bmatrix} 0 & ax \\ 0 & sx \end{bmatrix},$$

so that $\varphi(x) = ax$. It follows A_S is injective. □

We arrive at the following general result.

Theorem 3.88. *If T is a right CS-ring, then*

(i) *for every complement K in A_S there exists $r^2 = r \in R$ such that $K = rA$.*
(ii) *S is a right CS-ring.*
(iii) *A_S is an injective module.*
(iv) *$\ell_R(A)$ (without 1) is a CS-module.*

Proof. (i), (ii) by Lemma 3.86.

(iii) By Lemma 3.87.

(iv) Let $E \leq \ell_R(A)$. Then

$$\begin{bmatrix} E & 0 \\ 0 & 0 \end{bmatrix} T = \begin{bmatrix} ER & EA \\ 0 & 0 \end{bmatrix} = \begin{bmatrix} E & 0 \\ 0 & 0 \end{bmatrix} \leq T.$$

By hypothesis, there exists $\alpha^2 = \alpha \in T$ such that $\begin{bmatrix} E & 0 \\ 0 & 0 \end{bmatrix}$ is essential in αT, where $\alpha = \begin{bmatrix} r & a \\ 0 & s \end{bmatrix}$. Note that $r^2 = r$. Suppose $r \notin \ell_R(A)$. Then there exists $n \in A$ such that $rn \neq 0$. Therefore

$$0 \neq \begin{bmatrix} r & a \\ 0 & s \end{bmatrix} \begin{bmatrix} 0 & n \\ 0 & 0 \end{bmatrix} = \begin{bmatrix} 0 & rn \\ 0 & 0 \end{bmatrix} \in \alpha T.$$

So, $0 \neq \begin{bmatrix} 0 & rn \\ 0 & 0 \end{bmatrix} \begin{bmatrix} r' & a' \\ 0 & s' \end{bmatrix} \in \begin{bmatrix} E & 0 \\ 0 & 0 \end{bmatrix}$ for some $\begin{bmatrix} r' & a' \\ 0 & s' \end{bmatrix} \in T$, which is a contradiction. It follows that $r \in \ell_R(A)$. Now let $e \in E$. Then

$$\begin{bmatrix} e & 0 \\ 0 & 0 \end{bmatrix} = \begin{bmatrix} r & a \\ 0 & s \end{bmatrix} \begin{bmatrix} e & 0 \\ 0 & 0 \end{bmatrix} = \begin{bmatrix} re & 0 \\ 0 & 0 \end{bmatrix}.$$

Hence $e = re \in rE$ so that $E \leq rE \leq rA$, i.e., $E \leq rA$. Let $c \in \ell_R(A)$ with $cr \neq 0$. Then $\begin{bmatrix} c & 0 \\ 0 & 0 \end{bmatrix} \begin{bmatrix} r & a \\ 0 & s \end{bmatrix} = \begin{bmatrix} cr & 0 \\ 0 & 0 \end{bmatrix} \neq 0$. Thus

$$0 \neq \begin{bmatrix} cr & 0 \\ 0 & 0 \end{bmatrix} \begin{bmatrix} r' & a' \\ 0 & s' \end{bmatrix} = \begin{bmatrix} crr' & 0 \\ 0 & 0 \end{bmatrix} \in \begin{bmatrix} E & 0 \\ 0 & 0 \end{bmatrix}.$$

It follows that $\begin{bmatrix} crr' & 0 \\ 0 & 0 \end{bmatrix} = \begin{bmatrix} b & 0 \\ 0 & 0 \end{bmatrix}$ for some $0 \neq b \in E$, so that $0 \neq crr' = b \in E$. Hence E is essential in rA, i.e., $\ell_R(A)$ is a (without 1) CS-module. □

We remark that the converse of the Theorem 3.88 is not true in general. We will give an example to clarify the case later in this section. Now we aim to determine when the converse of the former result holds true. For our purpose, in the rest of this section we will assume that ${}_RA$ is faithful, i.e., $\ell_R(A) = 0$ for a generalized triangular matrix ring T. Recall that in this case we can think of R as a subring of $\operatorname{End}(A_S)$, because the mapping $\varphi : R \to \operatorname{End}(A_S)$ given by

$$\varphi(r)(a) = ra \quad (r \in R, \quad a \in A),$$

is a ring monomorphism. Let us prove the following basic lemma on generalized triangular matrix rings.

Lemma 3.89.

(i) *Suppose A_S is nonsingular and $rN = 0$ for some $r \in R$ where $N \leq A_S$. Then $N \leq_e A_S$ implies that $r = 0$.*

(ii) *T is right nonsingular if and only if A_S and S_S are nonsingular.*

Proof. (i) Let $0 \neq a \in A$. Then $aE \leq N$ for some $E \leq_e S$. Therefore, $raE = 0$, and hence $ra = 0$, i.e., $rA = 0$. Thus $r = 0$.

(ii) ($\Leftarrow$) Let $\begin{bmatrix} r & a \\ 0 & s \end{bmatrix} \in Z(T_T)$. By Lemma 3.85, there exist essential submodules N of A_S and E of S_S such that $rN = 0$, $aE = 0$, $sE = 0$. Thus $a = s = 0$ and since N is essential in A_S, we have $r = 0$, by part (i).

($\Rightarrow$) Let $a \in Z(A_S)$. Let E denote the right annihilator of a in S. Then E is essential in S_S. Note that $\begin{bmatrix} 0 & a \\ 0 & 0 \end{bmatrix} \begin{bmatrix} R & A \\ 0 & S \end{bmatrix} = 0$. Hence

$$\begin{bmatrix} R & A \\ 0 & E \end{bmatrix} \leq r\left(\begin{bmatrix} 0 & a \\ 0 & 0 \end{bmatrix}\right),$$

so that $\begin{bmatrix} 0 & a \\ 0 & 0 \end{bmatrix} \in Z(T_T) = 0$. It follows that $a = 0$, i.e., A_S is nonsingular.

Let $x \in Z(S_S)$ and let F denote the right annihilator of x in S. Hence F is essential in S_S. Note that $\begin{bmatrix} 0 & 0 \\ 0 & x \end{bmatrix} \begin{bmatrix} R & A \\ 0 & E \end{bmatrix} = 0$. Thus $\begin{bmatrix} R & A \\ 0 & E \end{bmatrix} \leq r\left(\begin{bmatrix} 0 & 0 \\ 0 & x \end{bmatrix}\right)$, so that $\begin{bmatrix} 0 & 0 \\ 0 & x \end{bmatrix} \in Z(T_T) = 0$, and hence $x = 0$. □

Lemma 3.90. *Suppose S is a right nonsingular right CS-ring and A_S is nonsingular such that for every complement K in A_S there exists $e^2 = e \in R$ such that $K = eA$. Let C be a complement in T_T. Then there exists $c^2 = c \in C$ such that $C = cT \oplus B$, $B \leq \begin{bmatrix} 0 & A \\ 0 & S \end{bmatrix}$, and $B \cap \begin{bmatrix} 0 & A \\ 0 & 0 \end{bmatrix} = 0$, where $B = (1-c)T \cap C$.*

Proof. Let C be a complement in T_T. Define

$$K = \left\{a \in A : \begin{bmatrix} 0 & a \\ 0 & 0 \end{bmatrix} \in C\right\}.$$

Then $K \leq A_S$. Also define

$$X = \{r \in R : rA \leq K\}.$$

Then $X \leq R_R$. We first prove that

$$\begin{bmatrix} X & K \\ 0 & 0 \end{bmatrix} \leq C \leq \begin{bmatrix} X & A \\ 0 & S \end{bmatrix}. \tag{3.5}$$

Let $r \in X$. Then

$$\begin{bmatrix} r & 0 \\ 0 & 0 \end{bmatrix} \begin{bmatrix} 0 & A \\ 0 & S \end{bmatrix} = \begin{bmatrix} 0 & rA \\ 0 & 0 \end{bmatrix} \leq \begin{bmatrix} 0 & K \\ 0 & 0 \end{bmatrix} \leq C$$

implies $\begin{bmatrix} r & 0 \\ 0 & 0 \end{bmatrix} \in C$ (Exercise 2.16). Thus $\begin{bmatrix} X & K \\ 0 & 0 \end{bmatrix} \leq C$. Next, let $\alpha = \begin{bmatrix} r & a \\ 0 & s \end{bmatrix} \in C$. Then $\begin{bmatrix} 0 & rA \\ 0 & 0 \end{bmatrix} = \alpha \begin{bmatrix} 0 & A \\ 0 & 0 \end{bmatrix} \leq C$ implies $rA \leq K$, so $r \in X$. This proves (3.5). Next we prove that K is a complement in A_S. Let $a \in A$ such that $aE \leq K$ for some essential right ideal E of S. Then

$$\begin{bmatrix} 0 & a \\ 0 & 0 \end{bmatrix} \begin{bmatrix} 0 & A \\ 0 & E \end{bmatrix} = \begin{bmatrix} 0 & aE \\ 0 & 0 \end{bmatrix} \leq C$$

implies $\begin{bmatrix} 0 & a \\ 0 & 0 \end{bmatrix} \in C$ (Lemma 3.85 and Exercise 2.16), so $x \in K$. By Exercise 2.16 (iii), K is a complement in A_S. Let us show that there exists $e^2 = e \in R$ such that

$$X = eR, \quad K = eA. \tag{3.6}$$

By hypothesis, there exists $e^2 = e \in R$ such that $K = eA$. Thus $e \in X$. Let $r \in X$. Then $rA \leq K$. Let $a \in A$. Then $ra = ea'$ for some $a' \in A$ and hence

$$ra = ea' = e^2a' = e(ea') = era,$$

so that $(r - er)a = 0$. Thus $(r - er)A = 0$, and hence $r = er$. It follows that $X = eX \leq eR \leq X$, so that $X = eR$. This proves (3.6).

Let $c = \begin{bmatrix} e & 0 \\ 0 & 0 \end{bmatrix} \in C$, by (3.5). Note that $c^2 = c$. Thus $T = cT \oplus (1-c)T$ and hence $C = cT \oplus B$, where $B = C \cap (1-c)T \leq T_T$. Let $\beta \in B$. Then

$$\beta = \begin{bmatrix} 1-e & 0 \\ 0 & 1 \end{bmatrix} \begin{bmatrix} r & a \\ 0 & s \end{bmatrix} = \begin{bmatrix} (1-e)r & (1-e)a \\ 0 & s \end{bmatrix}$$

for some $r \in R$, $s \in S$, $a \in A$. Since $\beta \in C$, (3.5) gives $(1-e)r \in X$, so that $(1-e)r = e(1-e)r = 0$ by (3.6). Thus $\beta \in \begin{bmatrix} 0 & A \\ 0 & S \end{bmatrix}$. Suppose $\beta \in \begin{bmatrix} 0 & A \\ 0 & 0 \end{bmatrix}$. Then $s = 0$, and $\beta \in C$ implies $(1-e)a \in K$, so $(1-e)a = e(1-e)a = 0$ by (3.6). We conclude that $\beta = 0$, whence $\beta \leq \begin{bmatrix} 0 & A \\ 0 & S \end{bmatrix}$ and $B \cap \begin{bmatrix} 0 & A \\ 0 & 0 \end{bmatrix} = 0$, which completes the proof of Lemma 3.90. □

Now we prove the following main result which was pointed out at the beginning of this section.

Theorem 3.91. *T is a right nonsingular right* CS*-ring if and only if*

(i) *for every complement K in A_S there exists an idempotent e in R such that $K = eA$,*

(ii) *S is a right nonsingular right* CS*-ring, and*

(iii) *the right S-module A is nonsingular injective.*

Proof. The necessity follows by Lemmas 3.86, 3.87 and 3.89. For the converse, suppose and (i), (ii), and (iii) hold. In view of Lemma 3.90, it is sufficient to prove that, if B is a right ideal of T such that $B \leq \begin{bmatrix} 0 & A \\ 0 & S \end{bmatrix}$ and $B \cap \begin{bmatrix} 0 & A \\ 0 & 0 \end{bmatrix} = 0$, then B is essential in a direct summand of T_T. Let

$$X = \left\{ (a, s) \in A \oplus S : \begin{bmatrix} 0 & a \\ 0 & s \end{bmatrix} \in B \right\}.$$

Then X is a S-submodule of $A \oplus S$ with $X \cap A = 0$. Let $\pi_1 : A \oplus S \to A$, $\pi_2 : A \oplus S \to S$ denote the canonical projections. Note that $\pi_2|_B : B \to S$ is a monomorphism. Consider the diagram

$$\begin{array}{ccccc} 0 & \longrightarrow & B & \xrightarrow{\pi_2|_B} & S \\ & & {\scriptstyle \pi_1|_B}\big\downarrow & & \\ & & A & & \end{array}$$

Since A_S is injective, there exists a mapping $\varphi : S \to A$ such that $\varphi\pi_2|_B = \pi_1|_B$. Let

$$E = \left\{ \begin{bmatrix} 0 & \varphi(s) \\ 0 & s \end{bmatrix} : s \in S \right\}.$$

Then E is a right ideal of T. Also, for any $b \in B$,

$$b = \begin{bmatrix} 0 & \pi_1(b) \\ 0 & \pi_2(b) \end{bmatrix} = \begin{bmatrix} 0 & \varphi\pi_2(b) \\ 0 & \pi_2(b) \end{bmatrix} \in E.$$

Because S is a right CS-ring, there exists an idempotent e in S such that $C = \pi_2(B)$ is essential in eS. Let $f = \begin{bmatrix} 0 & \varphi(e) \\ 0 & e \end{bmatrix}$. Then f is an idempotent in T. For any $c \in C$, $c = ec$, and hence

$$\begin{bmatrix} 0 & \varphi(c) \\ 0 & c \end{bmatrix} = \begin{bmatrix} 0 & \varphi(e) \\ 0 & e \end{bmatrix} \begin{bmatrix} 0 & \varphi(c) \\ 0 & c \end{bmatrix}.$$

Thus $B = fB$. There exists an essential right ideal D of S such that $eD \leq C$. Let $d \in D$. Then

$$\begin{bmatrix} 0 & \varphi(e) \\ 0 & e \end{bmatrix} \begin{bmatrix} 0 & A \\ 0 & d \end{bmatrix} = \begin{bmatrix} 0 & \varphi(ed) \\ 0 & ed \end{bmatrix} \in B.$$

It follows that B is essential in fT. Thus T is a right CS-ring. □

Corollary 3.92. *Let K be a field and V a non-zero vector space over K. Let $R = \mathrm{End}_K(V)$, the ring of K-endomorphisms of V. Let $T = [\begin{smallmatrix} R & V \\ 0 & K \end{smallmatrix}]$. Then T is a right nonsingular right* CS*-ring with right uniform dimension $1 + \dim_K(V)$.*

Proof. Let U be a complement in V_K. Then $V = U \oplus U'$ for some $U' \leq V$. Let $\pi : V \to U$ denote the canonical projection. Then $\pi^2 = \pi \in R$ and $U = \pi(V)$. The other conditions of Theorem 3.91 are obviously satisfied. Thus T is a right nonsingular right CS-ring.

Let $I = \begin{bmatrix} 0 & V \\ 0 & K \end{bmatrix}$. Then I is essential in T_T, $I = \begin{bmatrix} 0 & V \\ 0 & 0 \end{bmatrix} \oplus \begin{bmatrix} 0 & 0 \\ 0 & K \end{bmatrix}$ and $u\text{-}\mathrm{dim\,I} = \dim_K(V) + 1$. If $A_1 \oplus A_2 \oplus \cdots$, is a direct sum of non-zero right ideals of T, then $A_i \cap I \neq 0$ $(1 \leq i)$ and hence $(A_1 \cap I) \oplus (A_2 \cap I) \oplus \cdots$ is a direct sum of non-zero submodules of I_T. It follows that $u\text{-}\mathrm{dim\,T} = 1 + \dim_K(V) = u\text{-}\mathrm{dim}\, I$. $\square$

Next we give an example which shows that neither is the converse of Theorem 3.88 true in general, nor is the right nonsingularity condition in Theorem 3.91 superfluous.

Example 3.93. Let p be any prime integer and

$$T = \begin{bmatrix} \mathbb{Z}/\mathbb{Z}p^2 & \mathbb{Z}/\mathbb{Z}p^2 \\ 0 & \mathbb{Z}/\mathbb{Z}p^2 \end{bmatrix}.$$

Then $T = \begin{bmatrix} R & A \\ 0 & S \end{bmatrix}$, ${}_RA$ is faithful, and A_S is injective uniform, S is a CS-ring, but T is not a right CS-ring and $Z_2(T) = T$ (in fact T is finite).

Proof. $\mathbb{Z}/\mathbb{Z}p^2$ has a unique composition series:

$$\mathbb{Z}/\mathbb{Z}p^2 \geq \mathbb{Z}p/\mathbb{Z}p^2 \geq \mathbb{Z}p^2/\mathbb{Z}p^2 = 0.$$

Note that $\mathbb{Z}/\mathbb{Z}p^2$ is a uniform $\mathbb{Z}/\mathbb{Z}p^2$-module and its singular ideal is $\mathbb{Z}p/\mathbb{Z}p^2$. Now

$$(1 + \mathbb{Z}p^2)(\mathbb{Z}p/\mathbb{Z}p^2) \leq \mathbb{Z}p/\mathbb{Z}p^2,$$

so that $1 + \mathbb{Z}p^2 \in Z_2(\mathbb{Z}/\mathbb{Z}p^2)$ and hence $Z_2(\mathbb{Z}/\mathbb{Z}p^2) = \mathbb{Z}/\mathbb{Z}p^2$. Thus $T = \begin{bmatrix} R & R \\ 0 & R \end{bmatrix}$ and ${}_RR$ is faithful, so by Lemma 3.85, I is essential in T_T if and only if there exist essential submodules X, Y of R_R such that $\begin{bmatrix} 0 & X \\ 0 & Y \end{bmatrix} \leq I$, if and only if $\begin{bmatrix} 0 & \mathbb{Z}p/\mathbb{Z}p^2 \\ 0 & \mathbb{Z}p/\mathbb{Z}p^2 \end{bmatrix} \leq I$. Now

$$\begin{bmatrix} \mathbb{Z}p/\mathbb{Z}p^2 & \mathbb{Z}p/\mathbb{Z}p^2 \\ 0 & \mathbb{Z}p/\mathbb{Z}p^2 \end{bmatrix} \begin{bmatrix} 0 & \mathbb{Z}p/\mathbb{Z}p^2 \\ 0 & \mathbb{Z}p/\mathbb{Z}p^2 \end{bmatrix} = 0,$$

gives $Z(T) = \begin{bmatrix} \mathbb{Z}p/\mathbb{Z}p^2 & \mathbb{Z}p/\mathbb{Z}p^2 \\ 0 & \mathbb{Z}p/\mathbb{Z}p^2 \end{bmatrix}$. Also $\begin{bmatrix} \mathbb{Z}/\mathbb{Z}p^2 & \mathbb{Z}/\mathbb{Z}p^2 \\ 0 & \mathbb{Z}/\mathbb{Z}p^2 \end{bmatrix} \begin{bmatrix} 0 & \mathbb{Z}p/\mathbb{Z}p^2 \\ 0 & \mathbb{Z}p/\mathbb{Z}p^2 \end{bmatrix} \leq Z(T)$. Thus $T = Z_2(T)$. Let

$$B = \begin{bmatrix} p + \mathbb{Z}p^2 & 0 \\ 0 & 0 \end{bmatrix} T + \begin{bmatrix} 0 & 1 + \mathbb{Z}p^2 \\ 0 & p \end{bmatrix} T = \left\{ \begin{bmatrix} pa + \mathbb{Z}p^2 & pb + c + \mathbb{Z}p^2 \\ 0 & pc + \mathbb{Z}p^2 \end{bmatrix} : a, b, c \in \mathbb{Z} \right\}.$$

Let $\overline{\mathbb{Z}} = \mathbb{Z}/\mathbb{Z}p^2$ and $\overline{a} = a + \mathbb{Z}p^2$ $(a \in \mathbb{Z})$. Then

$$B = \left\{ \begin{bmatrix} \overline{pa} & \overline{pb + c} \\ 0 & \overline{pc} \end{bmatrix} : a, b, c \in \mathbb{Z} \right\}.$$

Clearly, B is a right ideal of T. Suppose B is essential in a right ideal C of T. Note that $\operatorname{soc}(B) = \begin{bmatrix} 0 & \overline{\mathbb{Z}p} \\ 0 & 0 \end{bmatrix}$, so B is uniform (T is a right and left Artinian ring). Thus C is uniform. Let $r = \begin{bmatrix} \overline{x} & \overline{y} \\ 0 & \overline{z} \end{bmatrix} \in C$, $r \neq 0$. Therefore $\begin{bmatrix} \overline{x} & 0 \\ 0 & 0 \end{bmatrix} = \begin{bmatrix} \overline{x} & \overline{y} \\ 0 & \overline{z} \end{bmatrix} \begin{bmatrix} \overline{1} & 0 \\ 0 & 0 \end{bmatrix} \in C$ and $\begin{bmatrix} 0 & \overline{x} \\ 0 & 0 \end{bmatrix} = \begin{bmatrix} \overline{x} & 0 \\ 0 & 0 \end{bmatrix} \begin{bmatrix} 0 & \overline{1} \\ 0 & 0 \end{bmatrix} \in C$.

If $p \nmid x$, then $\begin{bmatrix} 0 & \overline{1} \\ 0 & 0 \end{bmatrix} \in C$, and since $\begin{bmatrix} 0 & \overline{1} \\ 0 & \overline{p} \end{bmatrix} \in B \subseteq C$, we have $\begin{bmatrix} 0 & 0 \\ 0 & \overline{p} \end{bmatrix} \in C$. Thus C is not uniform, because $\begin{bmatrix} 0 & \overline{1} \\ 0 & 0 \end{bmatrix} T \cap \begin{bmatrix} 0 & 0 \\ 0 & \overline{p} \end{bmatrix} T = 0$. Thus $p|x$, $x = px'$ and $\begin{bmatrix} \overline{x} & 0 \\ 0 & 0 \end{bmatrix} = \begin{bmatrix} \overline{p} & 0 \\ 0 & 0 \end{bmatrix} \begin{bmatrix} \overline{x'} & 0 \\ 0 & 0 \end{bmatrix} \in B$. Hence $\begin{bmatrix} 0 & \overline{y} \\ 0 & \overline{z} \end{bmatrix} \in C$.

Suppose $\overline{y} \neq 0$ or $\overline{z} \neq 0$ (otherwise $r \in B$). Then $0 \neq \begin{bmatrix} 0 & \overline{y} \\ 0 & \overline{z} \end{bmatrix} \begin{bmatrix} \overline{u} & \overline{v} \\ 0 & \overline{w} \end{bmatrix} \in B$, i.e., $0 \neq \begin{bmatrix} 0 & \overline{yw} \\ 0 & \overline{zw} \end{bmatrix} = \begin{bmatrix} 0 & \overline{pb + c} \\ 0 & \overline{pc} \end{bmatrix}$ for some $b, c \in \mathbb{Z}$. Therefore $\overline{yw} = \overline{pb + c}$ and $\overline{zw} = \overline{pc}$. Hence $\overline{z}\,\overline{w} = \overline{p}\,\overline{c} = \overline{p}(\overline{y}\,\overline{w} - \overline{p}\,\overline{b}) = \overline{p}\,\overline{y}\,\overline{w}$. If $p \nmid w$, then $\overline{w}$ is a unit so $\overline{z} = \overline{p}\,\overline{y}$ and $\begin{bmatrix} 0 & \overline{y} \\ 0 & \overline{z} \end{bmatrix} = \begin{bmatrix} 0 & \overline{y} \\ 0 & \overline{py} \end{bmatrix} \in B$. Thus $r \in B$. Suppose $p|w$. Then $\overline{z}\,\overline{w} = \overline{p}\,\overline{y}\,\overline{w} = 0$. Hence $\overline{y}\,\overline{w} = 0$, so $p \nmid y$. It follows that $\overline{y}$ is a unit and $\begin{bmatrix} 0 & \overline{1} \\ 0 & \overline{zy}^{-1} \end{bmatrix} = \begin{bmatrix} 0 & \overline{y} \\ 0 & \overline{z} \end{bmatrix} \begin{bmatrix} 0 & 0 \\ 0 & \overline{y}^{-1} \end{bmatrix} \in C$. But $\begin{bmatrix} 0 & \overline{1} \\ 0 & \overline{p} \end{bmatrix} \in B \subseteq C$, so $\begin{bmatrix} 0 & 0 \\ 0 & \overline{p} - \overline{zy}^{-1} \end{bmatrix} \in C$. If $\overline{p} - \overline{z}\,\overline{y}^{-1} \neq 0$, then $\begin{bmatrix} 0 & 0 \\ 0 & \overline{p} \end{bmatrix} \in C$, so $\begin{bmatrix} 0 & \overline{1} \\ 0 & 0 \end{bmatrix} \in C$ and C is not uniform. Therefore $\overline{p} - \overline{z}\,\overline{y} = 0$, $\overline{z} = \overline{p}\,\overline{y}$ and $\begin{bmatrix} 0 & \overline{y} \\ 0 & \overline{z} \end{bmatrix} = \begin{bmatrix} 0 & \overline{y} \\ 0 & \overline{py} \end{bmatrix} = \begin{bmatrix} 0 & \overline{1} \\ 0 & \overline{p} \end{bmatrix} \begin{bmatrix} 0 & 0 \\ 0 & \overline{y} \end{bmatrix} \in B$. We see that $r \in B$ and hence $B = C$, i.e., B is a complement in T_T. Next we show that B does not contain non-zero idempotents. For, let $e = e^2 \in B$. Then $e = \begin{bmatrix} \overline{pa} & \overline{p}\overline{b} + \overline{c} \\ 0 & \overline{pc} \end{bmatrix}$ for some $a, b, c \in \mathbb{Z}$. Now $e^2 = \begin{bmatrix} 0 & \overline{p}\overline{c}(\overline{a}+\overline{c}) \\ 0 & 0 \end{bmatrix}$, so $\overline{p}\,\overline{a} = 0, \overline{p}\,\overline{c} = 0$ and $e = \begin{bmatrix} 0 & \overline{p}\overline{b}+\overline{c} \\ 0 & 0 \end{bmatrix}$ gives $e^2 = 0$. Thus $e = 0$. We conclude that T is not a right CS-ring. □

Theorem 3.91 can be applied to construct CS or non-CS special matrix rings. To this end, we collect some useful matrix rings in the following example.

Example 3.94. Let K be a field. Then

(i) $T = \begin{bmatrix} K & K & K \\ 0 & K & K \\ 0 & 0 & K \end{bmatrix}$ is a right CS-ring. Moreover, T_T does not satisfy C_2.

(ii) $T = \begin{bmatrix} K & 0 & K \\ 0 & K & K \\ 0 & 0 & K \end{bmatrix}$ is not a right CS-ring. Also T_T does not satisfy C_2.

(iii) $T = \begin{bmatrix} K & K & K \\ 0 & K & 0 \\ 0 & 0 & K \end{bmatrix}$ is not a right CS-ring.

Proof. (i) Take $S = \begin{bmatrix} K & K \\ 0 & K \end{bmatrix}$, $A = \begin{bmatrix} K & K \end{bmatrix}$ and $R = K$. Then S is a right nonsingular right CS-ring, A is a right nonsingular injective uniform S-module. By Theorem 3.91, T is a right CS-ring.

For the second part, let $I = \begin{bmatrix} 0 & 0 & 0 \\ 0 & 0 & 0 \\ 0 & 0 & K \end{bmatrix}$, $J = \begin{bmatrix} 0 & 0 & K \\ 0 & 0 & 0 \\ 0 & 0 & 0 \end{bmatrix}$. Clearly $I \cong J$, $I \leq_{\mathrm{d}} T_T$ but J is not a direct summand of T_T. Hence T_T does not satisfy C_2.

(ii) Take $R = K$, $S = \begin{bmatrix} K & K \\ 0 & K \end{bmatrix}$, $A = \begin{bmatrix} 0 & K \end{bmatrix}$. Then S is a right nonsingular right CS-ring and A_S is a simple non-injective module. By Theorem 3.91, T is not a right CS-ring. Now, let $E = \begin{bmatrix} 0 & 0 & 0 \\ 0 & 0 & 0 \\ 0 & 0 & K \end{bmatrix}$, $F = \begin{bmatrix} 0 & 0 & 0 \\ 0 & 0 & K \\ 0 & 0 & 0 \end{bmatrix}$. Then $E \cong F$, $E \leq_{\mathrm{d}} T_T$ but F is not a direct summand of T_T. If follows that T_T does not satisfy C_2.

(iii) Take $R = K$, $A = \begin{bmatrix} K & K \end{bmatrix}$ and $S = \begin{bmatrix} K & 0 \\ 0 & K \end{bmatrix}$. Then S is semisimple ring and A_S is a nonsingular injective module. Suppose T is a right CS-ring. Since $L = K \oplus 0 \leq_c A_S$, then $L = e(K \oplus K)$ for some $e = e^2 \in K$. But $e = 0$ or 1, as K is a field. Hence $L = 0$ or $K \oplus K$, a contradiction. □

We now restrict our attention to the special matrix rings. We we conclude this section with a result on split null (or trivial) extensions (see Exercise 1.7).

Theorem 3.95. *Let R be a ring, M be an ideal of R, and denote $S = S(R, M)$. If S_S is extending, then R_R is extending.*

Proof. First observe that for any (R, R)-bimodule M and $\begin{bmatrix} e & m \\ 0 & e \end{bmatrix} = \begin{bmatrix} e & m \\ 0 & e \end{bmatrix}^2 \in S$, then $e = e^2 \in R$, $em + me = m$ and $eme = 0$. So

$$\begin{bmatrix} e & me \\ 0 & e \end{bmatrix} = \begin{bmatrix} e & me \\ 0 & e \end{bmatrix}^2, \quad \begin{bmatrix} e & m \\ 0 & e \end{bmatrix}\begin{bmatrix} e & me \\ 0 & e \end{bmatrix} = \begin{bmatrix} e & me \\ 0 & e \end{bmatrix}$$

and

$$\begin{bmatrix} e & me \\ 0 & e \end{bmatrix}\begin{bmatrix} e & m \\ 0 & e \end{bmatrix} = \begin{bmatrix} e & m \\ 0 & e \end{bmatrix}.$$

Hence

$$\begin{bmatrix} e & m \\ 0 & e \end{bmatrix} S = \begin{bmatrix} e & me \\ 0 & e \end{bmatrix} S.$$

Assume that S_S is extending. Let $X_R \le R_R$. Then $\begin{bmatrix} X & & X\cap M \\ & \diagdown & \\ 0 & & X \end{bmatrix} \le S_S$.

There exists $\begin{bmatrix} e & m \\ 0 & e \end{bmatrix} = \begin{bmatrix} e & m \\ 0 & e \end{bmatrix}^2 \in S$ such that $\begin{bmatrix} X & & X\cap M \\ & \diagdown & \\ 0 & & X \end{bmatrix} \le_e \begin{bmatrix} e & m \\ 0 & e \end{bmatrix} S_S$.

Since $\begin{bmatrix} x & 0 \\ 0 & x \end{bmatrix} = \begin{bmatrix} e & m \\ 0 & e \end{bmatrix}\begin{bmatrix} x & 0 \\ 0 & x \end{bmatrix}$, for all $x \in X$, then $ex = x$ for all $x \in X$ and $mX = 0$. So $\begin{bmatrix} X & & X\cap M \\ & \diagdown & \\ 0 & & X \end{bmatrix} \le \begin{bmatrix} e & m \\ 0 & e \end{bmatrix} S_S$. Let $0 \ne er \in eR$. We need to find $t \in R$ such that $0 \ne ert \in X$. Since $\begin{bmatrix} X & & X\cap M \\ & \diagdown & \\ 0 & & X \end{bmatrix} \le_e \begin{bmatrix} e & m \\ 0 & e \end{bmatrix} S_S$, there exists $\begin{bmatrix} v & k \\ 0 & v \end{bmatrix} \in S$ such that

$$0 \ne \begin{bmatrix} e & me \\ 0 & e \end{bmatrix}\begin{bmatrix} r & 0 \\ 0 & r \end{bmatrix}\begin{bmatrix} v & k \\ 0 & v \end{bmatrix} = \begin{bmatrix} erv & erk + merv \\ 0 & erv \end{bmatrix} \in \begin{bmatrix} X & & X\cap M \\ & \diagdown & \\ 0 & & X \end{bmatrix}.$$

So

$$\begin{bmatrix} erv & erk + merv \\ 0 & erv \end{bmatrix} = \begin{bmatrix} e & 0 \\ 0 & e \end{bmatrix}\begin{bmatrix} erv & erk + merv \\ 0 & erv \end{bmatrix} = \begin{bmatrix} erv & erk \\ 0 & erv \end{bmatrix}.$$

Hence $0 \ne erv \in X$ or $0 \ne erk \in X \cap M \subseteq X$. Therefore, R_R is extending. □

Exercises

3.31. Let R be a ring and let $m \ge 1$ be an integer. Prove that $M_m(R)$ is an essential overring of $T_m(R)$.

3.32. Let S be an essential overring of a ring R. Prove that if R_R is CS, then so is S_R (i.e., the CS property transfers to essential overrings).

3.33. Let $R = F[x]$, where F is a field. Prove that $M_2(R)$ is a right CS-ring.

3.34. Let $T = \begin{bmatrix} \mathbb{R} & \mathbb{C} \\ 0 & \mathbb{C} \end{bmatrix}$ and $G = \begin{bmatrix} \mathbb{Z}/2\mathbb{Z} & \mathbb{Z}/2\mathbb{Z} \\ 0 & \mathbb{Z} \end{bmatrix}$. Then

(i) Show that T is a right CS-ring.
(ii) Show that G is not a right CS-ring.
(iii) Prove that every principal right ideal of G is essential in a direct summand of G_G.

3.35. Let R be any ring and let n be a positive integer. Prove that $M_n(R)$ is right pp-ring if and only if every n-generator right ideal is projective.

3.36. Let V be a vector space over a field F with $u\text{–}\dim(V) = n \geq 1$ and let $R_R = \begin{bmatrix} F & & V \\ & \diagdown & \\ 0 & & F \end{bmatrix}$. Prove that R_R is CS if and only if $n = 1$.

3.4 Continuous and quasi-continuous modules

This section is devoted on some concepts related to the CS property. To this end, we deal with modules satisfying the CS property together with a conditional direct summand condition $C_2(C_3)$. Modules of this kind are interesting in their own right and known in the literature as continuous (quasi-continuous) modules. We show that they can be characterized in terms of lifting homomorphisms from certain submodules to the module itself. We provide several examples which demonstrate our results. Moreover, we apply results in the previous section to get information on continuous and quasi-continuous full matrix rings and generalized triangular matrix rings.

Let R be a ring and A a left R-module. The module A is called *continuous* if it satisfies CS and the condition C_2 (see Section 2.5). Recall that any module satisfying C_2 also satisfies C_3 (Lemma 2.72). The module A is called *quasi-continuous* if it is a CS-module and satisfies the condition C_3. Thus, continuous modules are quasi-continuous. As promised, we show that continuous and quasi-continuous modules can be characterized by the lifting of homomorphisms from certain submodules of A to A itself. Note that there exists a different lifting condition to characterize continuous modules. Perhaps it is better to begin with this fact.

Theorem 3.96. *The following are equivalent for a module A.*

(i) *A is continuous.*
(ii) *If $B \oplus C \leq A$ and $f : B \oplus C \to A$ is a homomorphism with $\operatorname{im} f$ closed in A and $\ker f = C$, then there exists $g \in \operatorname{End}(A)$ extending f.*

(iii) *Every partial endomorphism f of A with both* $\ker f$ *and* $\operatorname{im} f$ *closed in A extends to an endomorphism of A.*

Proof. (i) $\Rightarrow$ (iii) Suppose $N \leq A$ and $f \in \operatorname{Hom}(N, A)$ with $\ker f$ and $\operatorname{im} f$ closed in A. By hypothesis, $\ker f$ and $\operatorname{im} f$ are direct summands of A. Thus, for some submodule B of A, $A = \ker f \oplus B$ and $N = \ker f \oplus (N \cap B)$. By C_2, $N \cap B \cong \operatorname{im} f$ is also a direct summand of A. Since a continuous module is quasi-continuous, $N = \ker f \oplus (N \cap B)$ is a direct summand of A, and clearly can be extended to an endomorphism of A.

(iii) $\Rightarrow$ (ii) Let f be given as in (ii). Let D be the closure of C in A (see Exercise 2.9). Then f can be extended to $\overline{f} : B \oplus D \to A$ with $\overline{f}|_D = 0$. Then (iii) applies, so $\overline{f}$, and hence f, extends to A.

(ii) $\Rightarrow$ (i) By Exercise 3.37, A is quasi-continuous. It remains to show that A satisfies C_2. Suppose that $B \leq_{\mathrm{d}} A$ and $\varphi : N \to B$ is an isomorphism. By hypothesis, there exists $D \leq_{\mathrm{d}} A$ such that $N \leq_{\mathrm{e}} D$; say $A = D \oplus C$ for some $C \leq A$. Now define $f : N \oplus C \to B$ by $f(n + c) = \varphi(n)$, where $n \in N$, $c \in C$. Thus $\ker f = C$ and $\operatorname{im} f = B$ are closed in A, so there exists $g \in \operatorname{End}(A)$ extending f. Since $g|_N = \varphi$ is one to one and $N \leq_{\mathrm{e}} D$, $g|_D$ is one to one. Thus $B = \varphi(N) = g(N) \leq_{\mathrm{e}} g(D)$, and it follows that $g(N) = g(D)$ since B is closed in A. Therefore $N = D \leq_{\mathrm{d}} A$; i.e., C_2 is satisfied and A is continuous. □

We return to our main aim in this section. The first lemma gives a flavor of what is to follow.

Lemma 3.97. *Let K be a complement in A. Then K is a direct summand of A if and only if there exists a complement L of K in A such that $K \oplus L \in \operatorname{Lift}_A(A)$.*

Proof. Suppose first that K is a direct summand of A. Then $A = K \oplus K'$ for some submodule K' of A. Clearly, $L = K'$ will do.

Conversely, suppose that there exists a complement L of K in A with the stated property. Let $\varphi : K \oplus L \to A$ be the homomorphism defined by

$$\varphi(x + y) = x \quad (x \in K, y \in L).$$

By hypothesis, there exists a homomorphism $\theta : A \to A$ such that $\theta(x + y) = x$ $(x \in K, y \in L)$. Note that $K \subseteq \operatorname{im} \theta$ and $L \subseteq \ker \theta$.

Let $0 \neq v \in \operatorname{im} \theta$. Then there exists $u \in A$ such that $v = \theta(u)$. Note that $u \notin L$. Thus $K \cap (L + Ru) \neq 0$. There exist $x \in K, y \in L$, and $r \in R$ such that $0 \neq x = y + ru$. Then $x = \theta(x) = \theta(y + ru) = rv$. It follows that $Rv \cap K \neq 0$ for all non-zero $v \in \operatorname{im} \theta$. Thus K is an essential submodule of $\operatorname{im} \theta$. But K is a complement in A. Hence $K = \operatorname{im} \theta$.

Now it is easy to check that $A = K \oplus (\ker \theta)$. Thus K is a direct summand of A. □

Corollary 3.98. *A module A is CS if and only if for every complement K in A there exists a complement L of K in A such that $K \oplus L \in \operatorname{Lift}_A(A)$.*

Proof. Immediate by using Lemma 3.97. □

Let n be a positive integer. We consider the following condition for a module A:

$$P_n : \qquad \underline{C}^{(n)} \subseteq \text{Lift}_A(A).$$

It is clear that if A satisfies P_n, then A satisfies P_{n-1}, for all $n \geq 2$. Modules satisfying P_1 will be considered in Chapter 5.

Next we establish the connection between P_n and the quasi-continuity of a module A, as was pointed out at the beginning of this section.

Theorem 3.99. *The following statements are equivalent for a module A.*

(i) *A is quasi-continuous.*
(ii) *A satisfies P_n for every positive integer n.*
(iii) *A satisfies P_n for some integer $n \geq 2$.*
(iv) *A satisfies P_2.*

Proof. (i)⇒ (ii) ⇒ (iii) ⇒ (iv) Obvious.
(iv) ⇒ (i) Follows by Proposition 2.86 and Corollary 3.98. □

Let n be a positive integer. Consider the following condition for a module A:

$$Q_n : \quad \text{For every } K \in \underline{A}^{(n)} \text{ such that } K = K_1 \oplus \cdots \oplus K_n \text{ and } K_i \in A' \, (1 \leq i \leq n),\ K \in \text{Lift}_A(A).$$

It is clear that if A satisfies Q_n, then A satisfies Q_{n-1}, for all $n \geq 2$. Moreover, if A satisfies Q_n, then A satisfies P_n, for all $n \geq 1$.

Theorem 3.100. *The following statements are equivalent for a module A.*

(i) *A is continuous.*
(ii) *A satisfies Q_n for every positive integer n.*
(iii) *A satisfies Q_n for some integer $n \geq 2$.*
(iv) *A satisfies Q_2.*
(v) *A is CS and satisfies Q_1.*

Proof. (i)⇒ (ii) ⇒ (iii) ⇒ (iv) Clear.
(iv) ⇒ (i) By Corollary 3.98 and Proposition 2.85.
(i) ⇒ (v) Clear.
(v) ⇒ (i) By Proposition 2.85. □

Theorems 3.99 and 3.100 show us that the following implications hold.

$$\begin{array}{ccccccc}
 & & \text{quasi-injective} \ \Rightarrow & \text{continuous} & \Rightarrow & \text{quasi-continuous} & & \\
 & & & \Updownarrow & & \Updownarrow & & \\
Q_n & \Leftrightarrow & & Q_2 & \Rightarrow & P_2 & \Leftrightarrow & P_n \\
 & & & \Downarrow & & \Downarrow & & \Downarrow \\
 & & & Q_1 & \Rightarrow & P_1 & \Leftarrow & CS,
\end{array}$$

for any integer $n \geq 2$. No other implications can be added to this table in general. To see why this is the case we shall give a number of examples. First of all, note that by Exercise 3.38, there is a continuous module which is not quasi-injective. The next example is easy.

Example 3.101. Let $\mathbb{Z}$ denote the ring of rational integers. Then the $\mathbb{Z}$-module $\mathbb{Z}$ satisfies P_2, but does not satisfy Q_1.

Proof. Let $A = {}_{\mathbb{Z}}\mathbb{Z}$. It is clear that A is CS and satisfies C_3, so that A satisfies P_2, by Theorem 3.99. Let N denote the submodule $2\mathbb{Z}$ of $\mathbb{Z}$. Then $N \cong A$, but the homomorphism $\varphi : N \to A$ given by $\varphi(2n) = n$ $(n \in \mathbb{Z})$ does not lift to A. Suppose that there exists a homomorphism $\theta : A \to A$ such that $\theta|_N = \varphi$. Then there exists $x \in A$ such that $\theta(a) = xa$ $(a \in A)$. Therefore, $2xa = \theta(2a) = \varphi(2a) = a$, so that $2x = 1$, a contradiction. Thus A does not satisfy Q_1. □

Example 3.101 shows that none of the implications

$$\text{quasi-continuous} \Rightarrow \text{continuous}, \quad P_2 \Rightarrow Q_2, \quad P_1 \Rightarrow Q_1,$$

is true in general.

Note that $P_n \not\Rightarrow C_3$. Let A be any free $\mathbb{Z}$-module of non-zero finite rank. Then A is CS (Exercise 3.1). However, by Example 2.73, A does not satisfy P_n. Hence $CS \Rightarrow P_n$ is not true in general. Observe that the module in Example 2.51 satisfies P_1, but does not satisfy P_2. Thus neither of the implications $P_1 \Rightarrow P_2$, $P_1 \Rightarrow CS$, is true in general. Moreover the module A in Example 2.51 does not satisfy Q_1 (Exercise 3.39). Now we turn to the condition Q_2. The next example shows that Q_1 does not imply Q_2.

Example 3.102. There exists a commutative local ring R such that the R-module R satisfies Q_1 but it is not CS.

Proof. By Exercise 3.40, there exists a commutative valuation domain S such that every homomorphic image of S is a self-injective ring, but S is not Noetherian. Suppose that every proper image of S has non-zero socle. Then every homomorphic image of S in finitely cogenerated (S is a valuation ring), and hence S is Artinian (Exercise 3.40). By Exercises 3.41 and 3.42 S is Noetherian, a contradiction. Thus there exists an ideal I of S such that the ring S/I has zero socle. Let $T = S/I$ and note that T is a local self-injective ring. Let J denote the unique maximal ideal of T. Let R denote the subring of the commutative ring $T \oplus T$ defined by

$$R = \{(t, t') : t - t' \in J\}.$$

Then R is nothing other than the pullback of

$$\begin{array}{ccccc} & & T & & \\ & & \downarrow{\scriptstyle \pi} & & \\ T & \xrightarrow{\pi} & T/J & \longrightarrow & 0 \end{array}$$

where $\pi : T \to T/J$ is the canonical epimorphism. The ring R is a commutative local ring with unique maximal ideal $J \oplus J$. Let A denote the R-module R. Then, being local, A is not CS and hence A does not have Q_2 (Theorem 3.100).

It remains to show that A satisfies Q_1. Let t and t' be any non-zero elements of T. If t has zero annihilator in T, then $T = Tt$, because T is self-injective (see Lemma 3.57), and hence $A = R(t, t')$. Similarly, $A = R(t, t')$ if t' has zero annihilator. Now suppose that both t and t' have non-zero annihilator. Then $R(t, t') \cap (J \oplus 0) \neq 0$ and $R(t, t') \cap (0 \oplus J) \neq 0$. Thus $R(t, t')$ is an essential submodule of A. It follows that the only complements in A are $0, A, J \oplus 0$, and $0 \oplus J$.

Let N be a submodule and K a complement submodule of A such that there exists an isomorphism $\alpha : K \to N$. Let $\varphi : N \to A$ be a homomorphism. If $K = 0$, then $N = 0$ and φ can be lifted to A. Now suppose that $K = A$. In this case, $N = R(t, t')$ for some elements t, t' in T such that both t and t' have zero annihilator. As we have just seen, this gives $N = R(t, t') = A$. Again φ lifts to A trivially.

Now suppose that $K = J \oplus 0$. For any $a \in J$, $\alpha(a, 0) = (b, c)$ for some $b, c \in T$. Now $(a, 0)(0 \oplus J) = 0$ implies $(b, c)(0 \oplus J) = 0$, and hence $cJ = 0$. Since T, has zero socle, we have $c = 0$. Thus $N = \alpha(K) \subseteq J \oplus 0$. It follows that $N = L \oplus 0$ for some proper ideal L of T. Now consider $\varphi : N \to A$. Since $N(0 \oplus J) = 0$, the same argument gives $\varphi(N) \subseteq J \oplus 0$. Thus φ induces a homomorphism $\varphi' : L \to T$. But T is self-injective, and hence φ' can be lifted to T and this allows us to lift φ to A. A similar proof shows that if $K = 0 \oplus J$, then φ can be lifted to A. It follows that A satisfies Q_1. □

In the rest of this section we deal with continuous and quasi-continuous matrix rings. First we have the following corollary.

Corollary 3.103. *Let A be a left R-module where $R = ReR$ for some idempotent e in R and $S = eRe$. Then the left R-module A is (quasi) continuous if and only if the left S-module eA is (quasi) continuous.*

Proof. By Theorems 2.79 and 3.76. □

Let R be a ring such that $M_2(R)$ is a right quasi-continuous ring. By Corollary 3.77 and Theorem 2.79, the right R-module R^2 is quasi-continuous and hence, by Corollary 3.2, R is right self-injective. Thus we have the following result.

Corollary 3.104. *The following statements are equivalent for a ring R.*

(i) *R is right self-injective.*
(ii) *$M_2(R)$ is right quasi-continuous.*
(iii) *$M_n(R)$ is right quasi-continuous for every positive integer n.*
(iv) *$M_n(R)$ is right self-injective for every positive integer n.*

Now we consider generalized triangular (quasi) continuous matrix rings and we apply results which were proved in the previous section. Recall that a generalized triangular matrix ring T has the form $T = \begin{bmatrix} R & A \\ 0 & S \end{bmatrix}$ (see § 3.2).

Proposition 3.105. *If T is right nonsingular right quasi-continuous, then $A = 0$ (so $R = 0$).*

Proof. Suppose T is right nonsingular and right quasi-continuous. Then by Lemmas 3.86, 3.89, S is a right nonsingular right CS-ring. Let $0 \neq a \in A$. Then $aS \cong S/r(a)$. Now $r(a)$ is a complement in S_S by Exercise 2.14, so $r(a)$ is a direct summand of S_S. Hence $S = r(a) \oplus E$ for some right ideal E of S. Thus $S/r(a) \cong aS \cong E$. Let $\alpha : aS \to S$ be an S-monomorphism. Note that $T = \begin{bmatrix} R & A \\ 0 & 0 \end{bmatrix} \oplus \begin{bmatrix} 0 & 0 \\ 0 & S \end{bmatrix}$. Therefore, $\begin{bmatrix} 0 & 0 \\ 0 & S \end{bmatrix}$ is $\begin{bmatrix} R & A \\ 0 & 0 \end{bmatrix}$-injective (Corollary 3.2).

Define

$$\beta : \begin{bmatrix} 0 & aS \\ 0 & 0 \end{bmatrix} \to \begin{bmatrix} 0 & 0 \\ 0 & S \end{bmatrix} \quad \text{by} \quad \beta\left(\begin{bmatrix} 0 & as \\ 0 & 0 \end{bmatrix}\right) = \begin{bmatrix} 0 & 0 \\ 0 & \alpha(as) \end{bmatrix} \quad (s \in S).$$

Then β is a T-homomorphism. Now β lifts to a T-homomorphism

$$\theta : \begin{bmatrix} R & A \\ 0 & 0 \end{bmatrix} \to \begin{bmatrix} 0 & 0 \\ 0 & S \end{bmatrix}.$$

Note that

$$\theta\left(\begin{bmatrix} 1 & 0 \\ 0 & 0 \end{bmatrix}\right) = \begin{bmatrix} 0 & 0 \\ 0 & z \end{bmatrix} = \begin{bmatrix} 0 & 0 \\ 0 & z \end{bmatrix}\begin{bmatrix} 0 & 0 \\ 0 & 1 \end{bmatrix} = \left(\theta\left(\begin{bmatrix} 1 & 0 \\ 0 & 0 \end{bmatrix}\right)\right)\begin{bmatrix} 0 & 0 \\ 0 & 1 \end{bmatrix}$$
$$= \theta\left(\begin{bmatrix} 1 & 0 \\ 0 & 0 \end{bmatrix}\begin{bmatrix} 0 & 0 \\ 0 & 1 \end{bmatrix}\right) = \theta\left(\begin{bmatrix} 0 & 0 \\ 0 & 0 \end{bmatrix}\right) = \begin{bmatrix} 0 & 0 \\ 0 & 0 \end{bmatrix}.$$

But

$$\begin{bmatrix} 0 & 0 \\ 0 & \alpha(a) \end{bmatrix} = \beta\left(\begin{bmatrix} 0 & a \\ 0 & 0 \end{bmatrix}\right) = \theta\left(\begin{bmatrix} 0 & a \\ 0 & 0 \end{bmatrix}\right) = \theta\left(\begin{bmatrix} 1 & 0 \\ 0 & 0 \end{bmatrix}\begin{bmatrix} 0 & a \\ 0 & 0 \end{bmatrix}\right)$$
$$= \left(\theta\left(\begin{bmatrix} 1 & 0 \\ 0 & 0 \end{bmatrix}\right)\right)\begin{bmatrix} 0 & a \\ 0 & 0 \end{bmatrix} = \begin{bmatrix} 0 & 0 \\ 0 & 0 \end{bmatrix}.$$

Hence $\alpha(a) = 0$ and $a = 0$, a contradiction. Thus $A = 0$. □

Corollary 3.106. *If T is right nonsingular right continuous, then $A = 0$ (so $R = 0$).*

Proof. Immediate by Proposition 3.105. □

By a *pp-ring* we mean a ring which is both a right pp-ring and a left pp-ring. Note that a right pp-ring which does not contain an infinite collection of orthogonal idempotents is also a left pp-ring (Exercise 3.43). A pp-ring is said to be *limited* if it does not contain an infinite collection of orthogonal idempotents. Therefore, limited pp-rings are Baer rings which satisfy the ascending and descending chain conditions on right and left annihilator ideals (see Exercise 3.43).

Lemma 3.107. *Let T be a limited pp-ring. Then there exist rings R and S and a left R-, right S-bimodule A such that S is semiprime, A is a faithful left R-module, and* $T \cong \begin{bmatrix} R & A \\ 0 & S \end{bmatrix}$.

Proof. By Exercise 3.43, T satisfies the ascending chain condition on right annihilator ideals. If P is a maximal in the collection of proper two-sided ideals which are right annihilator ideals, then $P = r\ell(P)$ and P is a prime ideal of T. Let U be the intersection of all such prime ideals P. Since $U = r\ell(U)$ and T is a Baer ring (Exercise 3.43) it follows that there exists an idempotent f in T such that $U = fT$. Then $(1-f)Tf = 0$ and $T \cong \begin{bmatrix} R & A \\ 0 & S \end{bmatrix}$, where $R = fTf$, $A = fT(1-f)$, and $S = (1-f)T(1-f) = (1-f)T \cong T/U$. Thus S is a semiprime ring and it can easily be checked that A is a faithful left R-module. □

Corollary 3.108. *Suppose T is a right continuous and right limited pp-ring. Then T is semiprime Artinian.*

Proof. By Lemma 3.107, $T \cong \begin{bmatrix} R & A \\ 0 & S \end{bmatrix}$, where ${}_RA$ is faithful and S is semiprime. Then by Proposition 3.105, $A = R = 0$, so $T \cong S$. Now the result follows by using Exercise 3.44. □

Exercises

3.37. Prove that a module A is quasi-continuous if and only if, whenever $N \oplus B \leq A$ and $\pi : N \oplus B \to N$ is the canonical projection, there exists $f \in \mathrm{End}\,({}_RA)$ such that $f|_{N\oplus B} = \pi$.

3.38. Let $\{D_\alpha\}$ be an infinite family of division rings, and F_α be a proper division subring of D_α for each α. Let S be the subring of the direct sum of D_α consisting of all elements x such that all but a finite number of α-components of x belong to F_α. Prove that S is a continuous regular ring which is not self-injective.

3.39. Show that the $\mathbb{Z}$-module $A = (\mathbb{Z}/\mathbb{Z}p) \oplus \mathbb{Q}$ does not satisfy Q_1.

3.40. Let F be a field and x an indeterminate; and let W be the family of all well-ordered sets $\{i\}$ of nonnegative real numbers, the order relation being the natural order of the real numbers. Then let R be the set of all "formal power" series $\sum_{i\in\{i\}} a_i x^i$ with $a_i \in F$ and $\{i\}$ in W. Show that R is not a Noetherian domain such that every homomorphic image of R is a self-injective ring.

3.41. An R-module A is said to *finitely embedded* if there exist finitely many simple modules $S_1, S_2, \ldots, S_k$ such that $E(A) \cong E(S_1) \oplus E(S_2) \oplus \cdots \oplus E(S_k)$. Prove that an R-module A is Artinian if and only if every factor module of A is finitely embedded.

3.42. Prove that every (right) Artinian ring is (right) Noetherian.

3.43. Show that limited pp-rings are Baer rings which satisfy acc and dcc on right and left annihilator ideals.

3.44. Prove that a right continuous ring with acc on essential right ideals is right Artinian.

3.45. Let ${}_RA$ be a module. Let $S = \text{End}\,({}_RA)$ and $J = J(S)$, $\triangle = \{f \in S : \ker f \leq_e A\}$. Prove the following statements.

(i) A is quasi-continuous if and only if whenever B and C are complements of each other in A, then $S = \ell_S(B) \oplus \ell_S(C)$.

(ii) If A is continuous, then $S/\triangle$ is a (von Neumann) regular ring and $\triangle = J$.

3.46. Let X be a topological space and let $C(X)$ denote the ring of real-valued continuous functions on X. Assume that X is a Hausdorff space such that whenever $V \subseteq X$ is closed and $x \notin V$, there exists an $f \in C(X)$ such that $f(x) = 1$ and $f(y) = 0$ for all $y \in V$. Prove that the following statements are equivalent.

(i) $C(X)$ is a continuous ring

(ii) $C(X)$ is a regular ring and every open set in X has an open closure.

Remarks. The results of this chapter are mainly taken from [HSTT96, BKP99, Smi94, Smi97, Ter92, Ter95], and [ST92].

Theorem 3.6 appeared in [ABT09, Thm. 5.3]. Theorem 3.96 is due to Li and Zelmanowitz [LZ88, Thm. 1.3]. Example 3.102, which we owe to B.L. Osofsky, is Example 11 in [ST92].

Good early sources for background about this chapter are [HS93, KaM88, Smi94, Lev66, CH77, CK80, Smi97, DT10, MM90], and [DHSW94].

Chapter 4

Inner Generalizations of Extending Modules

This chapter is devoted to generalizations of extending modules such that either for every submodule, or for a kind of special submodules N there exists a direct summand D with the property that $D \oplus N$ is essentially contained in the module (so-called inner generalization). We present basic results on inner generalizations as well as recent developments in the theory. We give several interesting results in this direction and prove implications between the conditions and provide counterexamples when implications are not valid.

4.1 Weak CS-modules

The main task of this section is to explore a "Weak" version of the extending property of modules, where weak means that the CS condition applies only to the semisimple submodules of the module. A module A is a weak CS-module if every semisimple submodule is essential in a direct summand of A. A direct sum of a semisimple submodule and an injective module is a weak CS-module, but in general is not a CS-module, even for Dedekind domains. Moreover, for any Dedekind domain, every module with finite uniform dimension is a weak CS-module. Weak CS-modules share some of the properties of CS-modules. For example, if a weak CS-module satisfies the ascending chain condition on essential submodules, then it is a direct sum of a semisimple module and a Noetherian module. For sources of references, see [Smi90], [Ter01], [Er99].

Definition 4.1. An R-module A is called a *weak* CS-*module* provided, for each semisimple submodule S of A, there exists a direct summand K of A such that $S \leq_{\mathrm{e}} K$.

Any module with zero socle is a weak CS-module. Thus, for the ring $\mathbb{Z}$ of rational integers, any free $\mathbb{Z}$-module of infinite rank is a weak CS-module which is not a CS-module (Exercise 3.1). One might expect that every weak CS-module with essential socle is a CS-module. However, the next example shows that this is not the case.

Example 4.2. Let p be a prime number and let A be the $\mathbb{Z}$-module $(\mathbb{Z}/\mathbb{Z}p) \oplus (\mathbb{Z}/\mathbb{Z}p^3)$. Then A is a weak CS-module with essential socle, but is not a CS-module.

Proof. It is clear that $\operatorname{soc} A \leq_e A$. Note that, by Exercise 3.5, A is not a CS-module. (In fact, the submodule $K = \mathbb{Z}(1 + \mathbb{Z}p, p + \mathbb{Z}p^3)$ is a complement submodule of A which is not a direct summand).

Now let us show that A is a weak CS-module. Note that A has uniform dimension 2. Let S be a semisimple submodule of A. If S is not simple, then $S \leq_e A$. Suppose that S is simple. Then $S = \mathbb{Z}(a + \mathbb{Z}p, p^2b + \mathbb{Z}p^3)$, for some integers a, b such that $0 \leq a, b \leq p - 1$. If $a = 0$, then $S \leq_e L = 0 \oplus (\mathbb{Z}/\mathbb{Z}p^3)$. If $a \neq 0$, then $A = S \oplus L$. Thus, in any case, S is essential in a direct summand of A. It follows that A is a weak CS-module. □

The above example leads us the following natural problem. Let A be a module such that $A = A_1 \oplus A_2$, where A_1 and A_2 are weak CS-modules. Is A a weak CS-module? A counterexample is constructed in [Zh99]. We next present this example.

Example 4.3. Let R be the trivial extension of $\mathbb{Z}_4$ with the $\mathbb{Z}_4$-module $2\mathbb{Z}_4$, i.e.,

$$R = \begin{bmatrix} \mathbb{Z}_4 & & 2\mathbb{Z}_4 \\ & \diagdown & \\ 0 & & \mathbb{Z}_4 \end{bmatrix} = \left\{ \begin{bmatrix} \overline{a} & 2\overline{b} \\ 0 & \overline{a} \end{bmatrix} : \overline{a}, \overline{b} \in \mathbb{Z}_4 \right\}.$$

Let $I = 2\mathbb{Z}_4 \oplus 0$ and $J = 0 \oplus 2\mathbb{Z}_4$. Let $A_1 = R/I$ and $A_2 = R/J$. Then $0 \subset \operatorname{soc} A_1 \subset A_1$ and $0 \subset \operatorname{soc} A_2 \subset A_2$ are composition series for A_1 and A_2, respectively. Therefore A_1 and A_2 are weak CS-modules (both A_1 and A_2 are uniform modules). Now, let $N = \{((\overline{0}, \overline{0}) + I, (\overline{0}, \overline{0}) + J), ((\overline{0}, \overline{2}) + I, (\overline{2}, \overline{0}) + J)\}$. Then N is a simple submodule of A since $N \subseteq \operatorname{soc} A_1 \oplus \operatorname{soc} A_2$. Suppose that A is a weak CS-module. Then there exists a direct summand P of A such that N is essential in P. By direct calculations, $P = N$, and so $A_1 \oplus A_2 = N \oplus X$ for some submodule X of A. For each i, A_i is an indecomposable module of finite length, and hence $\operatorname{End}(A_i)$ $(i = 1, 2)$ is a local ring. Then $N \cong A_1$ or $N \cong A_2$, which is a contradiction. It follows that A is not a weak CS-module.

It is known that if R is a ring and A is a CS R-module and A' a semisimple or injective module, then $A \oplus A'$ need not be a CS-module (even for Dedekind domains, see Example 2.51). The corollary of the next result shows that weak CS-modules are better behaved.

Lemma 4.4. *Let $A = A_1 \oplus A_2$ be a direct sum weak* CS*-modules A_1 and A_2 such that A_1 is A_2-injective. Then A is weak* CS.

Proof. Let S be a semisimple submodule of A. Then there exists a submodule B of A such that $S = (S \cap A_1) \oplus B$. Since A_1 is weak CS, there exists a direct summand K of A_1 such that $S \cap A_1$ is essential in K. By, [DHSW94, Lem. 7.5], there exists a submodule A' of A such that $A = A' \oplus A_1$ and $B \subseteq A'$. Then $A' \cong A_2$, thus A'

is weak CS, so there exists a direct summand L of A' such that B is essential in L. Therefore, $S = (S \cap A_1) \oplus B$ is essential in $K \oplus L$ which is a direct summand of A. It follows that A is weak CS. □

Theorem 4.5. *If $A = A_1 \oplus \cdots \oplus A_n$, where A_i are weak* CS *and for each i, A_i is A_k-injective, $k > i$, then A is weak* CS.

Proof. Immediate by induction and Lemma 4.4. □

Corollary 4.6. *Let A be a module such that $A = A_1 \oplus A_2$ for some weak* CS*-module A_1 and semisimple or injective module A_2. Then A is a weak* CS*-module.*

Proof. By Lemma 4.4. □

It is easy to see that a direct summand of a CS-module is also a CS-module (Exercise 3.2). However, the corresponding problem for the weak CS case is still open (see Open Problems and Questions 1). We give an affirmative answer for modules with the conditional direct summands property, namely C_3 and a UC, i.e., unique closure property (see [Er99], [Smi90]).

Proposition 4.7. *Any direct summand of a module which is both weak* CS *and* UC *is weak* CS.

Proof. Let A be a module which is both weak CS and UC. Let K be a direct summand of A and S a semisimple submodule of K. Now S is essential in a direct summand T of A, and S is essential in a complement L of K. So L and T are complements in A. Thus $L = T$ and T is also a direct summand of K. Hence K is weak CS. □

Corollary 4.8. *Any direct summand of a nonsingular module which is weak* CS *is also weak* CS.

Proof. By Exercise 2.9 (iii) and Proposition 4.7. □

Proposition 4.9. *If A is a weak* CS*-module satisfying C_3, then any direct summand is weak* CS.

Proof. Let K be a direct summand of A, that is, $A = K \oplus K'$ for some submodule K' of A. Let S be a semisimple submodule of K. By hypothesis, there exists a direct summand L of A such that S is essential in L. Then $L \cap K' = 0$. Now, let $\pi : A \to K$ be the canonical projection. By Lemma 2.71, $\pi(L) \oplus K' = L \oplus K'$. Also, since $L \oplus K'$ is a direct summand of A by the C_3 assumption, then so is $\pi(L)$, whence $\pi(L)$ is a direct summand of K, too. Since $S = \pi(S)$, S is in $\pi(L)$. Now since S is essential in L, $S \oplus K'$ is essential in $L \oplus K' = \pi(L) \oplus K'$. Thus $S = \pi(L) \cap (S \oplus K')$ is essential in $\pi(L) \cap (\pi(L) \oplus K') = \pi(L)$, and the result follows. □

Corollary 4.10. *If A is a weak* CS*-module satisfying C_2, then any direct summand is weak* CS.

Proof. By Lemma 2.72 and Proposition 4.9. □

Now we return to the case of Dedekind domains. Let us begin with the following lemma, which plays a key role in our main result in this context.

Lemma 4.11. *Let R be a Dedekind domain, P a maximal ideal of R, and A a non-zero finitely generated R-module such that $P^n A = 0$ for some positive integer n. Let S be a simple submodule of A. Then there exists a direct summand K of A such that S is an essential submodule of K.*

Proof. It is known that A is a finite direct sum of cyclic submodules, say $A = Ra_1 \oplus \cdots \oplus Ra_k$, for some positive integer k and elements $a_i \in A$ $(1 \leq i \leq k)$. For each $1 \leq i \leq k$, there exists $n(i) \geq 1$ such that $P^{n(i)} = \{r \in R : ra_i = 0\}$. Without loss of generality, we can suppose that $n(1) \leq n(2) \leq \cdots \leq n(k) = n$. Suppose that $n(1) = n$. Then A is a CS-module by [MM90], and thus S is essential in a direct summand of A.

Suppose that $n(1) \neq n$. There exist elements $r_i \in R$ $(1 \leq i \leq k)$ such that $S = R(r_1a_1 + \cdots + r_ka_k)$. Suppose that $r_1a_1 = 0$. Then $S \subseteq Ra_2 \oplus \cdots \oplus Ra_k$, and, by induction on k, there exists a direct summand K of $Ra_2 \oplus \cdots \oplus Ra_k$, and hence also of A, such that S is essential in K. Thus we suppose $r_1a_1 \neq 0$. Similarly, $r_ia_i \neq 0$ $(2 \leq i \leq k)$.

Note that $PS = 0$. In particular, for each $1 \leq i \leq k$, $Pr_ia_i = 0$, hence $Pr_i \subseteq P^{n(i)}$ and $r_i \in P^{n(i)-1}$, because P is invertible. There exists an ideal I_i of R such that $Rr_i = I_iP^{n(i)-1}$ and $R = I_i + P$, for each $1 \leq i \leq k$. It follows that $r_iA = P^{n(i)-1}$ $(1 \leq i \leq k)$, and hence $r_1A \supseteq r_2A \supseteq \cdots \supseteq r_kA$. In particular, for any $2 \leq i \leq k$, $r_ia_i = r_1(b_{i1}a_1 + \cdots + b_{ik}a_k)$, for some $b_{ij} \in R$ $(1 \leq j \leq k)$. Thus $r_ia_i = r_1b_{ii}a_i$ $(2 \leq i \leq k)$. Now $S = R(r_1a_1 + \cdots + r_ka_k) = R(r_1a_1 + r_1b_{22}a_2 + \cdots + r_1b_{kk}a_k) = r_1R(a_1 + b_{22}a_2 + \cdots + b_{kk}a_k)$. Let $K = R(a_1 + b_{22}a_2 + \cdots + b_{kk}a_k)$. Note that $A = K \oplus A_2 \oplus \cdots \oplus A_k$, where $A_i = Ra_i$ $(2 \leq i \leq k)$. Moreover, $S \subseteq K$, K is cyclic, and $P^nK = 0$, so that P is uniserial and hence S is essential in K. □

Theorem 4.12. *Let R be a Dedekind domain. Then any finitely generated R-module is a weak* CS-*module.*

Proof. Let A be a finitely generated R-module. Then $A = A_1 \oplus A_2$, for some torsion module A_1 and torsion-free module A_2. Let S be any semisimple submodule of A. Then $S \subseteq A_1$. Note that A_1 has finite composition length. Let U be a simple submodule of S. There exists a maximal ideal P of R such that $PU = 0$. Let $A' = \{a \in A : P^ka = 0$ for some positive integer $k\}$. Then A' is a direct summand of A_1, and $P^nA' = 0$ for some positive integer n. Note that $U \subseteq A'$, so that, by Lemma 4.11, there exists a direct summand K of A' such that U is essential in K. Now $A_1 = K \oplus K'$ for some submodule K' of A_1. Also, we have $S \subseteq \operatorname{soc} A_1 = (\operatorname{soc} K) \oplus (\operatorname{soc} K') = U \oplus (\operatorname{soc} K') \subseteq U \oplus K'$. Hence, $S = U \oplus (S \cap K')$. But $S \cap K'$ is a semisimple submodule of K' and K' has smaller length than A_1. Thus, by induction on the length of A_1, there exists a direct summand L of K'

such that $S \cap K'$ is essential in L. Then $S = U \oplus (S \cap K')$ is an essential submodule of the direct summand $K \oplus L$ of A. It follows that A is a weak CS-module. □

Corollary 4.13. *Let R be a Dedekind domain and A an R-module with finite uniform dimension. Then A is a weak* CS*-module.*

Proof. If A is torsion-free, then A has zero socle and, in this case, A is certainly a weak CS-module. Now suppose that A is not torsion-free. By [Kap52, Thm. 9], it follows that $A = A_1 \oplus A_2$, for some finitely generated module A_1 and injective submodule A_2. By Theorem 4.12 and Corollary 4.6, A is a weak CS-module. □

As already mentioned, weak CS-modules enjoy some of the properties of CS-modules. Our next objective is to give these properties. Let R be a ring and A an R-module. Following [CY91], we call A *eventually semisimple* if for every direct sum $A_1 \oplus A_2 \oplus \cdots$ of submodules A_i $(i \geq 1)$ of A, there exists a positive integer k such that A_i is semisimple for all $i \geq k$. Clearly, semisimple modules and modules with finite uniform dimension are examples of eventually semisimple modules. Moreover, any eventually semisimple module with zero socle has finite uniform dimension. It is known that if $A/(\operatorname{soc} A)$ has finite uniform dimension, then A is eventually semisimple (Exercise 4.4). The converse is false (see, e.g., [Jat98, Exa. 6.3.15]). Other examples of eventually semisimple modules are given by the following result.

Lemma 4.14. *Let $A = A_1 \oplus A_2$, where A_1 is a semisimple module and A_2 a module with finite uniform dimension. Then the module A is eventually semisimple.*

Proof. Let $N_1 \oplus N_2 \oplus \cdots$ be a direct sum of submodules of A. For each $1 \leq i \leq j$, let $K_i = N_i \oplus N_{i+1} \oplus \cdots$, and $K_{ij} = N_i \oplus \cdots \oplus N_j$. Suppose that $K_1 \cap A_2 \neq 0$. Then $K_{1s} \cap A_2 \neq 0$ for some $s \geq 1$. Suppose that $K_{s+1} \cap A_2 \neq 0$. Then $K_{s+1t} \cap A_2 \neq 0$ for some $t \geq s+1$. Repeating this argument we produce a direct sum

$$K_{1s} \oplus K_{s+1t} \oplus \cdots$$

of non-zero submodules of A_2. Since A_2 has finite uniform dimension, it follows that this process must stop. Thus there exists a positive integer q such that $K_q \cap A_2 = 0$. Now K_q embeds in A_1, and thus K_q is semisimple. It follows that A is eventually semisimple. □

Following [Smi90], we shall call a module A *almost semisimple* if A has essential socle and every finitely generated semisimple submodule of A is a complement in A. Clearly, semisimple modules are almost semisimple. On the other hand, let R be a commutative von Neumann regular ring which is not Noetherian. There exists a semisimple module S which is not injective (Theorem 1.54). Let A denote the injective hull of S. Note that A is not semisimple, but has essential socle. Moreover, A is almost semisimple because every finitely generated semisimple submodule of A is injective and hence a direct summand of A, by Proposition 1.30.

Besides, if R is a left Noetherian ring, then every almost semisimple left R-module is semisimple, as the following result shows. Recall that a module is locally Noetherian if every finitely generated submodule is Noetherian (see Section 2.2).

Lemma 4.15. *Any locally Noetherian almost semisimple module is semisimple.*

Proof. Let A be a locally Noetherian almost semisimple module. Let $a \in A$. Then Ra is Noetherian. Thus $\operatorname{soc}(Ra)$ is finitely generated, and hence is a complement in A, by hypothesis. It follows that $\operatorname{soc}(Ra)$ is a complement in Ra. But $\operatorname{soc}(Ra)$ is essential in Ra. Thus $Ra = \operatorname{soc}(Ra)$. It follows that A is semisimple. □

Lemma 4.16. *Let A be an eventually semisimple module. Then there exists an almost semisimple complement K in A such that A/K has finite uniform dimension.*

Proof. Let $S = \operatorname{soc} A$. If $S = 0$, then A has finite uniform dimension so we take $K = 0$. Suppose that $S \neq 0$. Let N_1 be a complement of S in A. Since $N_1 \cap S = 0$, it follows that N_1 has finite uniform dimension. Let K_1 be a complement of N_1 in A with $S \subseteq K_1$. Note that, by Proposition 2.1, S is essential in K_1. Suppose that K_1 is not almost semisimple. Then there exists a finitely generated submodule S_1 of K_1 such that S_1 is not a complement in K_1. Let N_2 be a maximal essential extension of S_1 in K_1. Then $S_1 \neq N_2$, $N_1 \cap N_2 = 0$, and N_2 has finite uniform dimension. Let K_2 be a complement of $N_1 \oplus N_2$ in A. Note that $\operatorname{soc} K_2$ is essential in K_2, by the choice of N_1 and Exercise 1.140.

If K_2 is not almost semisimple, then there exists a finitely generated semisimple submodule S_2 of K_2 such that S_2 is not a complement in K_2. Repeat this process. It produces a direct sum $N_1 \oplus N_2 \oplus N_3 \oplus \cdots$ of submodules N_i $(i \geq 1)$ of A such that N_i has finite uniform dimension for all $i \geq 1$ and N_i is not semisimple for all $i \geq 2$. Since A is eventually semisimple, this process must stop, and hence K_t is almost semisimple for some $t \geq 1$. Note that $L = N_1 \oplus \cdots \oplus N_t \oplus K_t$ is an essential submodule of A. Since K_t is a complement in A, it follows that L/K_t is an essential submodule of A/K_t. But $L/K_t \cong N_1 \oplus \cdots \oplus N_t$, which has finite uniform dimension. Thus A/K_t has finite uniform dimension. □

Corollary 4.17. *Let A be a locally Notherian eventually semisimple module. Then there exists a semisimple complement K in A such that A/K has finite uniform dimension.*

Proof. By Lemmas 4.15 and 4.16. □

Lemma 4.18. *Let A be a weak* CS*-module. Then every almost semisimple submodule of A is semisimple.*

Proof. Let K be an almost semisimple submodule of A. Let $0 \neq x \in K$. Suppose that $\operatorname{soc}(Rx)$ is not finitely generated. Then $\operatorname{soc}(Rx) = L_1 \oplus L_2 \oplus \cdots$ for some infinitely generated submodules L_i $(i \geq 1)$ of Rx. For each $i \geq 1$, there exists a direct summand N_i of A such that L_i is essential in N_i. Then the sum $N_1 + N_2 + \cdots$ is direct, and there exists a positive integer t such that N_t is semisimple. In this case, $N_t = L_t$ so that L_t is a direct summand of A, and hence also of Rx. It follows that L_t is cyclic, a contradiction. Thus $\operatorname{soc}(Rx)$ is finitely generated. By hypothesis, $\operatorname{soc}(Rx)$ is a direct summand of Rx. But $\operatorname{soc}(Rx)$ is essential in Rx, and hence $Rx = \operatorname{soc}(Rx)$. It follows that K is semisimple. □

Lemmas 4.16 and 4.18 enable us to obtain the following extension of [CY91, Lem. 2].

Theorem 4.19. *A module A is an eventually semisimple weak* CS*-module if and only if $A = A_1 \oplus A_2$ for some semisimple module A_1 and weak* CS*-module A_2 with finite uniform dimension.*

Proof. Suppose that A is an eventually semisimple weak CS-module. By Lemmas 4.16 and 4.18, $A = A_1 \oplus A_2$ for some semisimple module A_1 and module A_2 with finite uniform dimension. Let S be a semisimple submodule of A_2. Then $A_1 \oplus S$ is a semisimple submodule of A. There exists a direct summand K of A such that $A_1 \oplus S$ is essential in K. Note that $K = A_1 \oplus (K \cap A_2)$, from which we infer that $K \cap A_2$ is a direct summand of A, and hence also of A_2. Furthermore, $S \subseteq K \cap A_2$. By Proposition 2.1, S is an essential submodule of the direct summand $K \cap A_2$ of A_2. Thus A_2 is a weak CS-module. Conversely, suppose that $A = A_1 \oplus A_2$ for some semisimple module A_1 and weak CS-module A_2 with finite uniform dimension. By Lemma 4.14, A is eventually semisimple, and, by Corollary 4.6, a weak CS-module. □

Corollary 4.20. *Let A be a weak* CS*-module such that $A/(\operatorname{soc} A)$ has finite uniform dimension. Then $A = A_1 \oplus A_2$ for some semisimple module A_1 and module A_2 with finite uniform dimension.*

Proof. By Theorem 4.19 and [CY91, Lem. 1]. □

Using Corollary 4.20, we can now generalize [CY91, Prop. 5].

Theorem 4.21. *A weak* CS*-module A satisfies the ascending (respectively, descending) chain condition on essential submodules if and only if $A = A_1 \oplus A_2$ for some semisimple module A_1 and Noetherian (respectively, Artinian) module A_2.*

Proof. We give the proof in the Noetherian case; the proof in the Artinian case is similar. If A is a direct sum of a semisimple module and a Noetherian module, then A satisfies acc on essential submodules, by [CY91, Lem. 4].

Conversely, suppose that A is a weak CS-module which satisfies acc on essential submodules. By [CY91, Lem. 4], $A/(\operatorname{soc} A)$ is Noetherian. By Corollary 4.20, $A = A_1 \oplus A_2$ for some semisimple module A_1 and submodule A_2 with finite uniform dimension. From Proposition 2.17, there exist a positive integer k and uniform submodules U_i $(1 \leq i \leq k)$ of A_2 such that $E = U_1 \oplus U_2 \oplus \cdots \oplus U_k$ is essential in A_2. Now A having acc on essential submodules implies that U_i is Noetherian $(1 \leq i \leq k)$ and also that A_2/E is Noetherian. Thus A_2 is Noetherian. □

Exercises

4.1. Let p be any rational prime, $A_1 = \mathbb{Z}/\mathbb{Z}p$, and $A_2 = \mathbb{Z}(p^\infty)$, the Prüfer p-group. Show that the $\mathbb{Z}$-module $A = A_1 \oplus A_2$ is a weak CS-module but not a CS-module.

4.2. A module A is called a CESS-*module* if every complement in A with essential socle is a direct summand of A. Then

(i) Show that every CESS-module is weak CS-module.
(ii) Find a weak CS-module which is not CESS-module.
(iii) Prove that any direct summand of a CESS-module is also a CESS-module.

4.3. Let A be a CESS-module. Then prove that $A = A_1 \oplus A_2$ for some CS-module A, with essential socle and module A_2 with zero socle.

4.4. (i) Prove that if $A/(\operatorname{soc} A)$ has finite uniform dimension, then A is eventually semisimple module.

(ii) Find an example which shows that the condition $A/(\operatorname{soc} A)$ has finite uniform dimension in Corollary 4.20 is not superfluous.

4.2 Construction of C_{11}-Modules

We introduce C_{11}-modules as a generalization of CS-modules. The focus of our discussion in this section is mainly on the properties of C_{11}-modules. We shall exhibit similarities and contrasts between C_{11}-modules and CS-modules. Most of the material presented can be found in [ST93], [Ter92], and [ST04].

Definition 4.22. A module A *satisfies* C_{11} if every submodule of A has a complement which is a direct summand of A, i.e., for each submodule N of A there exists a direct summand K of A such that K is a complement of N in A.

For purposes of comparison, we first prove.

Proposition 4.23. *A module A is* CS *if and only if for all submodules N and L such that $N \cap L = 0$ there exists a direct summand K of A such that $L \leq K$ and $N \cap K = 0$. Moreover, in this case $N \oplus K \leq_e A$.*

Proof. Suppose first that A is CS. Let N and L be submodules of A such that $N \cap L = 0$. There exists a complement K of N in A such that $L \leq K$. By hypothesis, K is a direct summand of A.

Conversely, suppose that A satisfies the stated condition. Let L be a complement in A. There exists a submodule N of A such that L is a complement of N in A. By hypothesis, there exists a direct summand K of A such that $L \leq K$ and $N \cap K = 0$. Thus $L = K$. It follows that every complement in A is a direct summand. Therefore A is CS. The last part follows from Proposition 2.3 (see also, the proof of Lemma 4.24 below). □

Lemma 4.24. *Let N be a submodule of a module A and let K be a direct summand of A. Then K is a complement of N in A if and only if $K \cap N = 0$ and $K \oplus N \leq_e A$.*

Proof. Suppose K is a complement of N in A. Then $K \cap N = 0$. Let $0 \neq x \in A$. If $x \in K$, then $0 \neq Rx = Rx \cap K \subseteq Rx \cap (K \oplus N)$. If $x \notin K$, then $N \cap (Rx + K) \neq 0$ and hence $Rx \cap (K \oplus N) \neq 0$. Then $Rx \cap (K \oplus N) \neq 0$ for all $0 \neq x \in A$. Thus $K \oplus N \leq_e A$.

Conversely, suppose that K and N have the stated properties. There exists a submodule K' of A such that $A = K \oplus K'$. Suppose that there exists a submodule K_1 of A such that $K \subseteq K_1$ and $K_1 \cap N = 0$. Then $K_1 = K_1 \cap A = K_1 \cap (K \oplus K') = K \oplus (K_1 \cap K')$. Let $0 \neq y \in (K_1 \cap K')$. Therefore, $0 \neq ry = n + k$ for some $n \in N$, $k \in K$, $r \in R$. So $ry - k = n \in K_1 \cap N = 0$. Thus $ry = k \in K \cap K' = 0$, a contradiction. Hence $K_1 \cap K' = 0$ and $K = K_1$. That is, K is a complement of N in A. □

Compare the next result with Proposition 4.23.

Proposition 4.25. *The following statements are equivalent for a module A.*

(i) *A has C_{11}.*
(ii) *For any complement submodule L in A, there exists a direct summand K of A such that K is a complement of L in A.*
(iii) *For any submodule N of A, there exists a direct summand K of A such that $N \cap K = 0$ and $N \oplus K \leq_e A$.*
(iv) *For any complement submodule L in A, there exists a direct summand K of A such that $L \cap K = 0$ and $L \oplus K \leq_e A$.*

Proof. (i) ⇒ (ii), (iii) ⇒ (iv) Obvious.

(i) ⇔ (iii), (ii) ⇔ (iv) Clear by Lemma 4.24.

(iv) ⇒ (i) Let B be any submodule of A. Then there exists a complement submodule C in A such that $B \leq_e C$. By hypothesis, there exists a direct summand K of A such that $C \cap K = 0$ and $C \oplus K \leq_e A$. By Lemma 4.24, K is a complement of C in A. Note that $K \cap B = 0$. Suppose that K' is a submodule of A which properly contains K. Therefore $K' \cap C \neq 0$ and hence $K' \cap C \cap B \neq 0$, i.e., $K' \cap B \neq 0$. Thus K is a complement of B in A. □

It is clear from Proposition 4.23 that any CS-module A satisfies C_{11}, because any complement submodule of A is a direct summand (or, see Proposition 4.25). In particular, uniform modules, semisimple modules, and injective modules satisfy C_{11}. On the other hand, any indecomposable module with C_{11} is uniform. We shall show that every module which is a direct sum of uniform modules satisfies C_{11}. In particular, for any prime p, the $\mathbb{Z}$-module $A = (\mathbb{Z}/\mathbb{Z}p) \oplus (\mathbb{Z}/\mathbb{Z}p^3)$ satisfies C_{11}. However, A does not satisfy the extending property (Exercise 3.5). Thus in general a direct sum of uniform modules is not CS.

Theorem 4.26. *Any direct sum of modules with C_{11} satisfies C_{11}.*

Proof. Let A_λ $(\lambda \in \Lambda)$ be a non-empty collection of modules, each satisfying C_{11}. Let $A = \bigoplus_{\lambda\in\Lambda} A_\lambda$. Let N be any submodule of A. Let $\lambda \in \Lambda$. Note that $N \cap A_\lambda$ is a submodule of A_λ and A_λ satisfies C_{11}. By Proposition 4.25, there exists a direct summand K_λ of A_λ such that $(N \cap A_\lambda) \cap K_\lambda = 0$ and $(N \cap A_\lambda) \oplus K_\lambda \leq_e A_\lambda$. Note that $N \cap A_\lambda = 0$, $(N \oplus K_\lambda) \cap A_\lambda = (N \cap A_\lambda) \oplus K_\lambda$, and $(N \oplus K_\lambda) \cap A_\lambda \leq_e A_\lambda$. Let Λ' be a non-empty subset of Λ containing λ such that there exists a direct summand K' of $A' = \bigoplus_{\lambda\in\Lambda'} A_\lambda$ with $N \cap K' = 0$ and with $(N \oplus K') \cap A' \leq_e A'$. Suppose $\Lambda' \neq \Lambda$. Let $\mu \in \Lambda$, $\mu \notin \Lambda'$. Now $L = (N \oplus K') \cap A_\mu$ is a submodule of A_μ, so there exists a direct summand K_μ of A_μ such that $L \cap K_\mu = 0$ and $L \oplus K_\mu \leq_e A_\mu$. Let $\Lambda'' = \Lambda' \cup \{\mu\}$ and $A'' = \bigoplus_{\lambda\in\Lambda''} A_\lambda = A' \oplus A_\mu$. Note that $K' \cap K_\mu = 0$. Let $K'' = K' \oplus K_\mu$. Then K'' is a direct summand of A'' and moreover $N \cap K'' = 0$.

Consider the submodule $N \oplus K''$. Note that $(N \oplus K'') \cap A'$ contains $(N \oplus K') \cap A'$, so that $(N \oplus K'') \cap A' \leq_e A'$. Moreover, $(N \oplus K'') \cap A_\mu = (N \oplus K' \oplus K_\mu) \cap A_\mu = [(N \oplus K') \cap A_\mu] \oplus K_\mu = L \oplus K_\mu$, which is an essential submodule of A_μ. It follows that $(N \oplus K'') \cap A''$ is an essential submodule of A''. Repeating this argument, there exists a direct summand K of A such that $N \cap K = 0$ and $N \oplus K \leq_e A$. By Proposition 4.25, A satisfies C_{11}. □

Corollary 4.27. *Any direct sum of* CS*-modules satisfies* C_{11}.

Proof. Immediate by Theorem 4.26. □

Corollary 4.28. *Any direct sum of uniform modules satisfies* C_{11}.

Proof. Immediate by Corollary 4.27. □

The next result shows that the study of modules with the property C_{11} reduces to the case of Goldie torsion modules and nonsingular modules. It is analogue of [DHSW94, 7.11].

Theorem 4.29. *A module A satisfies C_{11} if and only if $A = Z_2(A) \oplus K$ for some (nonsingular) submodule K of A and both $Z_2(A)$ and K satisfy C_{11}.*

Proof. The sufficiency is an immediate consequence of Theorem 4.26. Conversely, suppose A satisfies C_{11}. We prove first that $Z_2(A)$ is a direct summand of A. Let $L = Z_2(A)$. There exist submodules K and K' of A such that $A = K \oplus K'$, $L \cap K = 0$, and $L \oplus K \leq_e A$ (Proposition 4.25). Now $L = Z_2(A) = Z_2(K \oplus K') = Z_2(K) \oplus Z_2(K')$. It is clear that $Z_2(K) = 0$. Thus $L = Z_2(K') \subseteq K'$. Since $L \oplus K$ is essential in A, L is essential in K', and hence K'/L is singular. Thus $L = K'$, and L is a direct summand of A.

We have proved that $A = L \oplus K$. Now we prove that L satisfies C_{11}. Let N be any submodule of L. Then $N \oplus K$ is a submodule of A. Since A satisfies C_{11}, there exist submodules P, P' of A such that $A = P \oplus P'$, $(N \oplus K) \cap P = 0$ and $N \oplus K \oplus P \leq_e A$. Note that $P \cap K = 0$ and hence P embeds in $A/K \cong L$. Thus $P = Z_2(P)$ and $P \leq L$. It follows that P is a direct summand of L (in fact, $L = P \oplus (L \cap P')$) and $N \oplus P \leq_e L$. By Proposition 4.25, L satisfies C_{11}.

Finally we prove that K satisfies C_{11}. Let $\pi : A \to K$ denote the canonical projection. Let H be any submodule of K. Then $L \cap H = 0$, and there exist submodules Q and Q' of A such that $A = Q \oplus Q'$, $(L \oplus H) \cap Q = 0$, and $L \oplus H \oplus Q \leq_e A$. Note that $L = Z_2(A) = Z_2(Q) \oplus Z_2(Q') = Z_2(Q')$, because $Q \cap L = 0$ gives $Z_2(Q) = 0$. Hence $L \leq Q'$, and $Q' = L \oplus (Q' \cap K)$. Now $A = Q \oplus Q' = Q \oplus L \oplus (Q' \cap K)$. Thus $L \oplus Q$ is a direct summand of A. By Lemma 2.71, $L \oplus Q = L \oplus \pi(Q)$. Therefore, the submodule $\pi(Q)$ of K is a direct summand of A, and hence a direct summand of K. But $H \oplus \pi(Q) \oplus L \leq_e A$, and so $H \oplus \pi(Q) \leq_e K$. By Proposition 4.25, K satisfies C_{11}. □

In a similar vein to Theorem 4.29 we show that the study of nonsingular modules satisfying C_{11} reduces to the case of modules with essential socle and modules with zero socle. First we prove.

Lemma 4.30. *Let A be a module which satisfies C_{11}. Then $A = A_1 \oplus A_2$ where A_1 is a submodule of A with essential socle and A_2 a submodule of A with zero socle.*

Proof. Let S denote the socle of A. There exist submodules K and K' of A such that $A = K \oplus K'$, $S \cap K = 0$ and $S \oplus K \leq_e A$. So $S = \operatorname{soc} A = (\operatorname{soc} K) \oplus (\operatorname{soc} K')$. Clearly $\operatorname{soc} K = 0$ so that $S \leq K'$. Now $S \oplus K \leq_e A$ implies $S \leq_e K'$, and the result is proved. □

The converse of Lemma 4.30 is not true in general. Recall that [Lev66] provides an example of a commutative local ring R with zero socle which is not a C_{11} R-module (see Example 3.102). Let I be the unique maximal ideal of R. Now, let $A_1 = {}_RR$ and $A_2 = R/I$. Note that $\operatorname{soc} A_1 = 0$ and $\operatorname{soc} A_2 = A_2$ which is essential in A_2. Let A be the direct sum $A_1 \oplus A_2$ of R-modules A_1 and A_2. Since A_1 is not a C_{11}-module and A_2 is simple, A is not a C_{11}-module.

Let A be a nonsingular module. It is well known that for any submodule N of A there exists a unique complement $c(N)$ in A such that $N \leq_e c(N)$, namely $c(N) = \{a \in A : Ea \leq N \text{ for some essential left ideal } E \text{ of } R\}$ (Exercise 2.9). The proof of the next result is very similar to that of Theorem 4.29.

Theorem 4.31. *A nonsingular module A satisfies C_{11} if and only if $A = A_1 \oplus A_2$, where A_1 is a module satisfying C_{11} and having essential socle and A_2 is a module satisfying C_{11} and having zero socle.*

Proof. The sufficiency is clear by Theorem 4.26. Conversely, suppose that A satisfies C_{11}. By Lemma 4.30, $A = A_1 \oplus A_2$, where A_1 has essential socle and A_2 zero socle. Let S denote the socle of A. Clearly $A_1 = c(S)$.

We prove next that A_1 satisfies C_{11}. Let N be any submodule of A_1. By Proposition 4.25, there exists a direct summand P of A such that $(N \oplus A_2) \cap P = 0$ and $N \oplus A_2 \oplus P$ is an essential submodule of A. Now P embeds in A_1, and hence P has essential socle $S \cap P$. Thus $P = c(S \cap P) \leq c(S) = A_1$. Hence P is a direct summand of A_1 and $N \oplus P$ is an essential submodule of A_1. By Proposition 4.25, A_1 satisfies C_{11}.

Now consider A_2. Let $\pi : A \to A_2$ denote the canonical projection. Let H be any submodule of A_2. By Proposition 4.25, A has submodules Q and Q' such that $A = Q \oplus Q'$, $(A_1 \oplus H) \cap Q = 0$, and $A_1 \oplus H \oplus Q$ is essential in A. Now $S \cap Q = 0$ implies $S \subseteq Q'$, by Exercise 1.140. Thus $A_1 = c(S) \subseteq Q'$. It follows that A_1 is a direct summand of Q' and hence $A_1 \oplus Q$ is a direct summand of A. This implies that $A_1 \oplus \pi(Q)$ is a direct summand of A, $\pi(Q)$ is a direct summand of A_2, and $H \oplus \pi(Q)$ is essential in A_2. By Proposition 4.25, A_2 satisfies C_{11}. □

Theorems 4.29 and 4.31 raise the following natural question. Let A be a module which satisfies C_{11}. Does any direct summand of A satisfy C_{11}? If A is a module with uniform dimension 2, then any direct summand of A is 0, A or is uniform. Thus if A satisfies C_{11}, then so too does any direct summand of A. We shall show that there exists a C_{11}-module with uniform dimension 3 which has a direct summand which is not a C_{11}-module. First we prove the following result.

Proposition 4.32. *Let R be a ring such that the left R-module R is a C_{11}-module and every direct summand of a C_{11}-module is a C_{11}-module. Then every indecomposable projective left R-module is uniform.*

Proof. Let P be an indecomposable projective R-module. Then there exists a free R-module F such that $F = P \oplus P'$ for some submodule P'. By Theorem 4.26, F satisfies C_{11} and, by hypothesis, so does P. Thus P is uniform. □

In view of Proposition 4.32, we are interested in when indecomposable projective modules are uniform. If R is a commutative Noetherian domain, then Bass [Bas63, Cor. 4.5] proved that every non-finitely generated projective R-module is free. In the special case when $R = F[x_1, x_2, \ldots, x_n]$ is the polynomial ring in the indeterminates $x_1, x_2, \ldots, x_n$ over a field F, the Quillen–Suslin Theorem states that every (finitely generated) projective R-module is free (see [Lam78] for a good account of the Quillen–Suslin Theorem and related results). If R is a commutative Noetherian ring of Krull dimension $n \geq 0$, then it follows from a celebrated theorem of Serre (see Theorem IV.2.5 in [Bas68]) that all finitely generated indecomposable projective R-modules have uniform dimension at most n. In particular, if R is a commutative Noetherian ring of Krull dimension 1, then every finitely generated indecomposable projective R-module is uniform, and hence, by Corollary 4.28, every finitely generated projective R-module satisfies C_{11}.

Example 4.33 ([Lam99, Exa. 17.36]). Let $\mathbb{R}$ be the real field and S the polynomial ring $\mathbb{R}[x, y, z]$. Then the ring $R = S/Ss$, where $s = x^2+y^2+z^2-1$, is a commutative Noetherian domain of Kull dimension 2. Moreover, the free R-module $A = R \oplus R \oplus R$ satisfies C_{11}, but A contains a direct summand K which does not satisfy C_{11}.

It is standard to check that R is a commutative Noetherian domain of Kull dimension 2. The free R-module A satisfies C_{11} by Corollary 4.28. Let $\phi : A \to R$ be the homomorphism defined by $\phi(a + Ss, b + Ss, c + Ss) = ax + by + cz$ for all $a, b, c \in S$. Clearly, ϕ is an epimorphism, and hence, its kernel K is a direct summand of A, i.e., $A = K \oplus K'$ for some submodule $K' \cong R$. Clearly, K is not uniform.

Note that K is the R-module of regular sections of the tangent bundle of the 2-sphere S^2. Since the Euler characteristic $\chi(S^2) = 2 \neq 0$, it follows that the 2-sphere cannot have a non-vanishing regular section of its tangent bundle (see Corollary VI.13.3 in [Bre93]). Thus, K is an indecomposable module. It follows that K does not satisfy C_{11}.

More generally, if $n \geq 3$ is any odd integer, S is the polynomial ring $\mathbb{R}[x_1, x_2, \ldots, x_n]$ in the indeterminates $x_1, x_2, \ldots, x_n$ over $\mathbb{R}$, $s = x_1^2 + x_2^2 + \cdots + x_n^2 - 1$, and R is the commutative Noetherian domain S/Ss, then the free R-module $A = R^{(n)}$ satisfies C_{11}, but contains an indecomposable direct summand K with uniform dimension $n-1$ and K does not satisfy C_{11} (see also Exercise 4.7).

Let R be a left Noetherian domain. Then the left R-module R is uniform because R embeds in every non-zero left ideal of R. By Corollary 4.28, any free left R-module satisfies C_{11}. In contrast to Serre's Theorem mentioned above, Hodges and Osterburg [HO87] gave an example of a (left and right) Noetherian domain R of Krull dimension 1 such that there exists a finitely generated indecomposable projective left R-module P of uniform dimension 2. Levy [Lev89] showed that for every positive integer n, there exist a Noetherian domain R that is a module-finite algebra over a discrete valuation ring (and hence has Krull dimension 1) and a finitely generated indecomposable projective left R-module P such that P has uniform dimension n. Hodges and Stafford [HO87] proved that if $\{n_i : i \in I\}$ is any set of positive integers, then there exists a Noetherian domain R which is integral over its centre and has Krull dimension 1 such that for each $i \in I$, there exists a finitely generated indecomposable projective left R-module P_i with uniform dimension n_i. To summarize, there are many examples of left Noetherian domains R of Krull dimension 1 and finitely generated indecomposable projective left R-modules P which are not uniform, and hence do not satisfy C_{11} despite the fact that they are direct summands of free left R-modules which do satisfy C_{11}.

The next results deal with special cases when a direct summand of a C_{11}-module is a C_{11}-module. First we prove

Lemma 4.34. *Let N be a direct summand of a module A and let K be an injective submodule of A such that $N \cap K = 0$. Then $N \oplus K$ is a direct summand of A.*

Proof. There exists a submodule N' of A such that $A = N \oplus N'$. Let $\pi : A \to N'$ be the canonical projection. Then $N \cap K = 0$ implies $K \cong \pi(K)$, so that $\pi(K)$ is injective. It follows that $\pi(K)$ is a direct summand of N'. But $N \oplus K = N \oplus \pi(K)$, by Lemma 2.71, and hence $N \oplus K$ is a direct summand of A. □

Proposition 4.35. *Let A be a module which satisfies C_{11}. Let N be a direct summand of A such that A/N is an injective module. Then N satisfies C_{11}.*

Proof. Let L be any submodule of N. There exists an injective submodule N' of A such that $A = N \oplus N'$. Consider the submodule $L \oplus N'$. There exists a direct summand K of A such that $(L \oplus N') \cap K = 0$ and $L \oplus N' \oplus K$ is an essential submodule of A (Proposition 4.25). By Lemma 4.34, $N' \oplus K$ is a direct summand of A. But $N' \oplus K = N' \oplus \pi(K)$, where $\pi : A \to N$ denotes the canonical projection (Lemma 2.71). Thus $\pi(K)$ is a direct summand of N. However, $L \oplus \pi(K) \oplus N'$ is essential in A. It follows that $L \oplus \pi(K)$ is an essential submodule of N. By Proposition 4.25, N satisfies C_{11}. □

The next few results concern a left exact preradical r in the category of left modules over a ring R (see Section 2.4). The main focus is on the structure of modules and, in particular, on when we can obtain a decomposition of a left R-module A with respect to a preradical r in the category of left R-modules. Obviously, not all preradicals are of relevance, since $r(A)$ will be zero in many cases. In fact, the only preradicals of interest are those which are subgenerated by some submodule K of A and the related class of radical modules is just the class $\sigma[K]$, in which case any $r(A)$ is of the form $Tr(\sigma[K], A)$. For more details see [Wis91].

First we have the following result which generalizes Lemma 4.30.

Lemma 4.36. *Let R be a ring, r a left exact preradical in the category of left R-modules, and A a left R-module which satisfies C_{11}. Then $A = A_1 \oplus A_2$, where $r(A_1)$ is essential in A_1 and $r(A_2) = 0$.*

Proof. By Proposition 4.25, there exist submodules A_1, A_2 of A such that $A = A_1 \oplus A_2$, $r(A) \cap A_2 = 0$, and $r(A) \oplus A_2$ is an essential submodule of A. Since r is left exact, it follows that $r(A_2) = A_2 \cap r(A) = 0$. Let $\pi : A \to A_1$ denote the canonical projection. Then $\pi(r(A)) \subseteq r(A_1)$. For any $0 \neq a \in A_1$, there exists $t \in R$ such that $0 \neq ta \in r(A) \oplus A_2$, and hence, $0 \neq ta = \pi(ta) \in \pi(r(A)) \subseteq r(A_1)$. It follows that $r(A_1)$ is an essential submodule of A_1. □

In Lemma 4.36, we have been unable to show that A_1 or A_2 satisfies C_{11}, except in special cases (Open Problems and Questions 26).

Lemma 4.37. *Let N be a submodule of a module A such that N has a unique closure K in A. Then K is the sum of all submodules L of A containing N and such that N is essential in L.*

Proof. Let H be the sum of submodules L of A such that N is an essential submodule of L. Since N is essential in its closure K, it follows that $K \subseteq H$. Conversely, let L be any submodule of A such that N is an essential submodule of L. Let L' be any closure of L in A. Clearly, L' is a closure of N in A, and so $L' = K$. Thus $L \subseteq K$. It follows that $H \subseteq K$ and hence $H = K$. □

This brings us to the following result.

Theorem 4.38. *Let R be a ring, r a left exact preradical for the category of left R-modules, and A a left R-module such that $r(A)$ has a unique closure in A. Then A is a C_{11}-module if and only if $A = A_1 \oplus A_2$ is a direct sum of C_{11}-modules A_1 and A_2 such that $r(A_1)$ is essential in A_1 and $r(A_2) = 0$.*

Proof. The sufficiency follows by Theorem 4.26. Conversely, suppose A is a C_{11}-module. By Lemma 4.36, $A = A_1 \oplus A_2$ with $r(A_1)$ essential in A_1 and $r(A_2) = 0$. Note that $r(A) = r(A_1) \oplus r(A_2) = r(A_1)$, so A_1 is the (unique) closure of $r(A)$ in A. Let $\pi_i : A \to A_i$ $(i = 1, 2)$ denote the canonical projections.

Let N be any submodule of A_1. By Proposition 4.25, there exist submodules K, K' of A such that $A = K \oplus K'$, $(N \oplus A_2) \cap K = 0$, and $N \oplus A_2 \oplus K$ is essential in A. Since $K \cap A_2 = 0$, it follows that $K \cong \pi_1(K)$. Note that because r is left exact, $r(\pi_1(K)) = \pi_1(K) \cap r(A_1)$ is essential in $\pi_1(K)$. Hence, $r(K)$ is essential in K and, in addition, $r(A) = r(K) \oplus r(K')$ is essential in $K \oplus r(K')$. By Lemma 4.37, $K \oplus r(K') \subseteq A_1$ and, in particular, $K \subseteq A_1$. Now $A_1 = K \oplus (A_1 \cap K')$, and $N \oplus K = (N \oplus A_2 \oplus K) \cap A_1$ is essential in A_1. By Proposition 4.25, A_1 is a C_{11}-module.

Next, let H be any submodule of A_2. By Proposition 4.25, there exist submodules L, L' of A such that $A = L \oplus L'$, $(H \oplus A_1) \cap L = 0$, and $H \oplus A_1 \oplus L$ is essential in A. Note that $r(A) \subseteq A_1$ gives that $r(L) = L \cap r(A) \subseteq L \cap A_1 = 0$, and hence, $r(A) = r(L) \oplus r(L') = r(L') \subseteq L'$. Let L'' be a closure of $r(A)$ in L'. Since L' is a direct summand of A, it follows that L'' is a closure of $r(A)$ in A (see Proposition 2.10), and hence, $A_1 = L'' \subseteq L'$. Now $L' = A_1 \oplus (L' \cap A_2)$ and $A = L \oplus L' = L \oplus A_1 \oplus (L' \cap A_2) = \pi_2(L) \oplus A_1 \oplus (L' \cap A_2)$. We deduce that $\pi_2(L)$ is a direct summand of A_2 and $\pi_2(L) \oplus H = (\pi_2(L) \oplus A_1 \oplus H) \cap A_2 = (L \oplus A_1 \oplus H) \cap A_2$, which is essential in A_2. By Proposition 4.25, A_2 satisfies C_{11}. □

Theorem 4.38 applies in the case where $r(A)$ is a complement submodule of A and, in particular, when $r(A)$ is a direct summand of A. Thus, Theorem 4.38 generalizes Theorem 4.31 (see Exercise 2.9).

Following [Ste75, p. 152], a hereditary torsion theory is called *stable* if the class of torsion modules is closed under injective envelopes. By Proposition 7.3 in [Ste75, p. 153], the Goldie torsion theory is stable. Thus, Theorem 4.29 is a special case of the following result.

Corollary 4.39. *Let R be a ring and r the left exact radical for a stable hereditary torsion theory for the category of left R-modules. Then a left R-module A satisfies C_{11} if and only if $A = r(A) \oplus A'$ for some submodule A' and both $r(A)$ and A' satisfy C_{11}.*

Proof. The sufficiency follows by Theorem 4.26. Conversely, suppose A satisfies C_{11}. By Lemma 4.36, $A = A_1 \oplus A_2$ such that $r(A_1)$ is essential in A_1 and $r(A_2) = 0$. By hypothesis, $r(A_1) = A_1$. Moreover, $r(A) = r(A_1) \oplus r(A_2) = A_1$, and hence $A = r(A) \oplus A_2$. Now the result follows from Theorem 4.38. □

Before giving another case when a direct summand of a C_{11}-module is a C_{11}-module, we first prove an easy lemma.

Lemma 4.40. *Let $A = A_1 \oplus A_2$. Then A_1 satisfies C_{11} if and only if for every submodule N of A_1, there exists a direct summand K of A such that $A_2 \subseteq K$, $K \cap N = 0$, and $K \oplus N$ is an essential submodule of A.*

Proof. Suppose A_1 satisfies C_{11}. Let N be any submodule of A_1. By Proposition 4.25, there exists a direct summand L of A_1 such that $N \cap L = 0$ and $N \oplus L$ is essential in A_1. Clearly, $(L \oplus A_2) \cap N = 0$ and $(L \oplus A_2) \oplus N$ is essential in A. Conversely, suppose A_1 has the stated property. Let H be a submodule of A_1. By hypothesis, there exists a direct summand K of A such that $A_2 \subseteq K$, $K \cap H = 0$, and $K \oplus H$ is an essential submodule of A. Now $K = K \cap (A_1 \oplus A_2) = (K \cap A_1) \oplus A_2$ so that $K \cap A_1$ is a direct summand of A, and hence also of A_1, $H \cap (K \cap A_1) = 0$, and $H \oplus (K \cap A_1) = A_1 \cap (H \oplus K)$, which is an essential submodule of A_1. By Proposition 4.25, A_1 satisfies C_{11}. □

The next result applies in the case that A is a C_{11}-module satisfying condition C_3 (see Corollary 4.47).

Theorem 4.41. *Let $A = A_1 \oplus A_2$ be a C_{11}-module such that for every direct summand K of A with $K \cap A_2 = 0$, $K \oplus A_2$ is a direct summand of A. Then A_1 is a C_{11}-module.*

Proof. Let N be any submodule of A_1. By hypothesis, there exists a direct summand K of A such that $(N \oplus A_2) \cap K = 0$ and $N \oplus A_2 \oplus K$ is an essential submodule of A by Proposition 4.25. Moreover, $A_2 \oplus K$ is a direct summand of A. Now result follows from Lemma 4.40. □

Corollary 4.42. *Let A be a C_{11}-module and K a direct summand of A such that A/K is K-injective. Then K satisfies C_{11}.*

Proof. There exists a submodule K' of A such that $A = K \oplus K'$ and, by hypothesis, K' is K-injective. Let L be a direct summand of A such that $L \cap K' = 0$. By Lemma 7.5 in [DHSW94], there exists a submodule H of A such that $H \cap K' = 0$, $A = H \oplus K'$, and $L \subseteq H$. Now L is a direct summand of H, and hence $L \oplus K'$ is a direct summand of $A = H \oplus K'$. By Theorem 4.41, K satisfies C_{11}. □

Corollary 4.43. *Let $A = A_1 \oplus A_2$ be a direct sum of a submodule A_1 and an injective submodule A_2. Then A satisfies C_{11} if and only if A_1 satisfies C_{11}.*

Proof. If A satisfies C_{11}, then A_1 satisfies C_{11} by Corollary 4.42. Conversely, if A_1 satisfies C_{11}, then A satisfies C_{11} by Theorem 4.26. □

We shall complete this section by results on Abelian groups (i.e., $\mathbb{Z}$-modules). First we should remark the following fact. Let U be a torsion uniform $\mathbb{Z}$-module. Then the injective hull $E(U)$ of U is a torsion indecomposable injective $\mathbb{Z}$-module, so that $E(U)$ is a quasicyclic group $\mathbb{Z}(p^\infty)$ for some prime p (see, e.g., [Fu70, Thm. 23.1]).

Lemma 4.44. *Let A be an Abelian p-group for some prime p. Suppose that the $\mathbb{Z}$-module A is a direct sum of uniform modules. Then every direct summand of A is a direct sum of uniform modules.*

Proof. Our above remark shows that $A = A_1 \oplus A_2$, where A_1 is an injective submodule of A and the submodule A_2 is a direct sum of cyclic groups. Let N be a direct summand of A. There exists a submodule N' of A such that $A = N \oplus N'$. Let $\pi : A \to N$ denote the canonical projection. Then $\pi(A_1)$ is injective ([Fu70, p. 98(D)]), and hence $N = \pi(A_1) \oplus L$ for some submodule L of N. Let $L' = \pi(A_1) \oplus N'$. Then $A = L \oplus L'$. Let $\pi_1 : A \to L$ denote the canonical projection. Then $\pi_1(A_1) = 0$, and hence $A_1 \leq \ker \pi_1 = L'$. Thus $L' = A_1 \oplus (L' \cap A_2)$. Now $L \oplus A_1 \oplus (L' \cap A_2) = A = A_1 \oplus A_2$, so that $L \oplus (L' \cap A_2) \cong A_2$. By a theorem of Kulikov ([Fu70, Thm. 18.1]) L is a direct sum of cyclic groups. It follows that L is a direct sum of uniform submodules. Moreover, $\pi(A_1)$ is a direct sum of uniform submodules, by [Fu70, Thm. 23.1]). Thus $N = \pi(A_1) \oplus L$ is a direct sum of uniform modules. □

For any Abelian group A, let $T(A)$ denote its torsion subgroup and, for any prime p, let $T_p(A)$ denote the p-component of $T(A)$ [Fu70, p. 43]. We now prove

Theorem 4.45. *Let A be a $\mathbb{Z}$-module such that A is a direct sum of uniform modules. Then any direct summand of A is a direct sum of uniform modules.*

Proof. Each uniform $\mathbb{Z}$-module is torsion or torsion-free. Thus $A = A_1 \oplus A_2$, where A_1 is a torsion-free module, A_2 is a torsion module, and both A_1 and A_2 are direct sums of uniform modules. Let N be a direct summand of A. Then $A = N \oplus N'$ for some submodule N' of A. Note that $A_2 = T(N) \oplus T(N')$, so that both $T(N)$ and $T(N')$ are direct summands of A. It follows that $N = T(N) \oplus K$ and $N' = T(N') \oplus K'$ for some submodules K of N and K' of N'. Now $A_1 \oplus A_2 = A = N \oplus N' = A_2 \oplus K \oplus K'$, and hence $A_1 \cong K \oplus K'$. By [Fu73, Thm. 86.7], K is a direct sum of uniform modules.

Next, every torsion uniform submodule of A_2 is a p-group for some prime p [Fu70, Thm. 8.4]. Thus $A_2 = \bigoplus T_p(A_2)$, where the direct sum is over all primes p, and $T_p(A_2)$ is a direct sum of uniform modules for each p, again by [Fu70, Thm. 8.4]. Moreover, it is clear that $T_p(A_2) = T_p(N) \oplus T_p(N')$, for each prime p. By Lemma 4.44, $T_p(N)$ is a direct sum of uniform modules for each prime p, and hence so too is $T(N) = \oplus_p T_p(N)$. Thus $N = T(N) \oplus K$ is a direct sum of uniform modules. □

Combining Corollary 4.28 and Theorem 4.45, we conclude that over the ring $\mathbb{Z}$, every direct summand of a module which is a direct sum of uniform modules, satisfies C_{11}.

Exercises

4.5. Find examples to show that the weak CS and C_{11} conditions are different.

4.6. Let A be an indecomposable module. Prove that A is a C_{11}-module if and only if A is uniform.

4.7. Let $\mathbb{C}$ be the complex field and n any odd integer with $n \geq 3$. Let S be the polynomial ring $\mathbb{C}[x_1, x_2, \ldots, x_n]$ in the indeterminates $x_1, x_2, \ldots, x_n$ over $\mathbb{C}$. Let R be the ring S/Ss, where $s = x_1^2 + x_2^2 + \cdots + x_n^2 - 1$. Prove that the free R-module $A = \bigoplus_{i=1}^{n} R$ satisfies C_{11}, but A contains a direct summand which does not satisfy C_{11}.

4.8. Show that being injective cannot be replaced with CS or quasi-continuous in Proposition 4.35.

4.3 C_{11}-modules with conditional direct summand properties

The central aim of this section is to obtain a decomposition into uniform submodules of a C_{11}-module which satisfy chain conditions like acc or dcc on left annihilators or essential submodules. Moreover, we also derive a decomposition of C_{11}-modules into a direct sum of a semisimple submodule and a module with finite uniform dimension, as in Corollary 4.20 and Theorem 4.21 for weak CS-modules (see Exercise 3.18).

We showed in §4.2 that direct summands of modules with C_{11} need not satisfy C_{11}. This is in contrast to the situation for CS-modules (Exercise 3.2). Let P be some property of modules. Then we say that a module A *satisfies* P^+ if every direct summand of A satisfies P. For example, an indecomposable module satisfies P^+ if and only if it satisfies P. A module A satisfies C_1 (i.e., CS) if and only if A satisfies C_1^+. We can abbreviate this fact to $C_1^+ = C_1$. We know that, in general, $C_{11}^+ \neq C_{11}$. Observe that if a module is a direct sum of two uniform modules, then it satisfies C_{11}^+.

Before proceeding any further we consider the following example.

Example 4.46. Let p be a prime number. Let A denote the $\mathbb{Z}$-module $(\mathbb{Z}/\mathbb{Z}p) \oplus \mathbb{Q}$. Then A satisfies C_{11}^+ and C_2, but not C_1.

Proof. By Corollary 4.28, A satisfies C_{11}. Since A has uniform dimension 2, it satisfies C_{11}^+. On the other hand, A does not satisfies C_1, because $K = R(1+\mathbb{Z}p, 1)$ is a complement submodule which is not a direct summand, where R is the local ring $\mathbb{Z}_{(p)}$ (see Example 2.51).

Now we show that A satisfies C_2. Let L be a non-zero direct summand of A. Suppose $L \neq A$. Then L is uniform because A has uniform dimension 2. By Example 2.51, $L = (\mathbb{Z}/\mathbb{Z}p) \oplus 0$, $L = 0 \oplus \mathbb{Q}$, or $L = R(1+\mathbb{Z}p, q)$ for some $0 \neq q \in \mathbb{Q}$.

Now $A = L \oplus L'$ for some submodule L' of A. Suppose $L = R(1 + \mathbb{Z}p, q)$ for some $0 \neq q \in \mathbb{Q}$. Then $pL \cap L' = 0$, so that $R(0, pq) \cap L' = 0$ and hence $L' \cap (0 \oplus \mathbb{Q}) = 0$. Thus L' embeds in $\mathbb{Z}/\mathbb{Z}p$, which is simple. It follows that $L' = (\mathbb{Z}/\mathbb{Z}p) \oplus 0$, and consequently $A = L \oplus L' = (\mathbb{Z}/\mathbb{Z}p) \oplus Rq$, a contradiction because $\mathbb{Q} \neq Rq$. Thus $L = (\mathbb{Z}/\mathbb{Z}p) \oplus 0$ or $0 \oplus \mathbb{Q}$.

Let $\varphi : L \to A$ be a monomorphism. If $L = (\mathbb{Z}/\mathbb{Z}p) \oplus 0$, then $\varphi(L)$ is simple, so that $\varphi(L) = L$. If $L = 0 \oplus \mathbb{Q}$, then $\varphi(L)$ is torsion-free injective, and hence $\varphi(L) = L$. If $L = A$, then $\varphi(L) = (\mathbb{Z}/\mathbb{Z}p) \oplus \mathbb{Q} = L$. Thus $\varphi(L) = L$ for every direct summand L of A and monomorphism $\varphi : L \to A$. Hence, A satisfies C_2. □

Recall that $C_2^+ = C_2$ and $C_3^+ = C_3$ (see Lemma 2.74). Now we have the following consequence of Theorem 4.41.

Corollary 4.47. *Let A be a module which satisfies C_{11} and C_3. Then A satisfies C_{11}^+.*

Proof. Immediate by Theorem 4.41. □

The following result and its corollary were motivated by similar results for the CS condition (Theorem 3.19 and Corollary 3.20).

Proposition 4.48. *Let A be a C_{11}-module and X a submodule. If the intersection of X with any direct summand of A is a direct summand of X, then X is a C_{11}-module.*

Proof. Let B be a submodule of X. Then there exists a direct summand N of A such that $B \cap N = 0$ and $B \oplus N$ is essential in A. Now $A = N \oplus K$ for some submodule K of A. Hence $X \cap (B \oplus N) = B \oplus (X \cap N)$ is essential in X. By hypothesis, $X \cap N$ is a direct summand of X. It follows that X is a C_{11}-module. □

Corollary 4.49. *Let A be a C_{11}-module.*

(i) *If A is distributive, then every submodule of A is* CS.
(ii) *If X is a submodule of A such that $Xe \subseteq X$ for all $e^2 = e \in \mathrm{End}(_RA)$, then X is a C_{11}-module. In particular, every fully invariant submodule of A is a C_{11}-module.*
(iii) *If A has* SIP, *then A has C_{11}^+.*

Proof. (i) Let N be a complement submodule of A. Then there exists $e^2 = e \in \mathrm{End}(_RA)$ such that Ae is a complement of N. Then $N = N \cap A = N \cap (Ae \oplus A(1-e)) = (N \cap Ae) \oplus (N \cap A(1-e)) = N \cap A(1-e) \leq A(1-e)$. Since N is a complement in A, then $N = A(1-e)$. Hence A is CS. By Corollary 3.20, every submodule of A is CS.

(ii) Let D be a direct summand of A and $e : A \to D$ be the canonical projection. Then $Xe = D \cap X$. By Proposition 4.48, X is a C_{11}-module.

(iii) is an immediate consequence of Proposition 4.48. □

We now consider conditions on a module A which imply that A is a direct sum of uniform modules. First we recall the following definitions.

Let R be a ring and A a left R-module. For any a in A, let $\ell(a)$ denote the annihilator of a in A, i.e., $\ell(a) = \{r \in R : ra = 0\}$ (see Exercise 2.8). We shall say that A *satisfies* (A) if every ascending chain of left ideals of the form $\ell(a_1) \subseteq \ell(a_2) \subseteq \cdots$, with $a_i \in A$ $(i \geq 1)$, terminates. For example, if R is a left Noetherian ring, then every left R-module satisfies (A).

Let $X = \bigoplus_{\lambda \in \Lambda} X_\lambda$ be a direct sum of submodules X_λ $(\lambda \in \Lambda)$ of a module A. Then X is called a *local summand* of A if $\bigoplus_{\lambda \in \Lambda'} X_\lambda$ is a direct summand of A for each finite subset Λ' of Λ (see [MM90, Def. 2.15]). The proof of the next result is taken from [MM90, Prop. 2.18].

Lemma 4.50. *Let A be a module which has (A). Then every local summand of A is a complement in A.*

Proof. Let $X = \bigoplus_{\lambda \in \Lambda} X_\lambda$ be a local summand of A. There exists a complement submodule X^* of A such that X is essential in X^*. Suppose $X \neq X^*$. Choose $a \in X^* \backslash X$ with $\ell(a)$ maximal. There exists $0 \neq r \in R$ such that $0 \neq ra \in X$. There exists a finite subset Λ' of Λ such that $ra \in \bigoplus_{\lambda \in \Lambda'} X_\lambda$. Now $A = (\bigoplus_{\lambda \in \Lambda'} X_\lambda) \oplus Y$ for some submodule Y of A. Hence $a = x + y$ for some $x \in \bigoplus_{\lambda \in \Lambda'} X_\lambda$ and $y \in Y$. Note that $y = a - x \in X^*$, $y \notin X$. Also, if $s \in \ell(a)$, then $0 = sa = sx + sy$, which gives $sy = 0$. Therefore, $\ell(y)$ contains $\ell(a)$. Hence $\ell(a) = \ell(y)$, by the choice of a. But $ry = ra - rx \in (\bigoplus_{\lambda \in \Lambda'} X_\lambda) \cap Y = 0$, and so $ra = 0$, a contradiction. It follows that $X = X^*$. □

Lemma 4.51. *Let A be a module which satisfies (A). Suppose further that A satisfies C_{11} and C_3. Then every local summand of A is a direct summand.*

Proof. Let $X = \bigoplus_{\lambda \in \Lambda} X_\lambda$ be a local summand of A. Since A satisfies C_{11}, there exists a direct summand K of A such that $X \cap K = 0$ and $X \oplus K$ is essential in A (Proposition 4.25). Consider $X \oplus K$. For any finite subset Λ' of Λ, $Y = \bigoplus_{\lambda \in \Lambda'} X_\lambda$ is a direct summand of A and hence $Y \oplus K$ is a direct summand of A, because A satisfies C_3. Thus $X \oplus K$ is a local summand. By Lemma 4.50, $X \oplus K$ is a complement in A. But $X \oplus K$ is an essential submodule of A. Thus $A = X \oplus K$ (Proposition 2.7). □

Theorem 4.52. *Let A be a module which satisfies (A), C_{11} and C_3. Then A is a direct sum of uniform submodules.*

Proof. By Lemma 4.51 and [MM90, Thm. 2.17], A is a direct sum of indecomposable submodules. But by Corollary 4.47, every indecomposable direct summand of A satisfies C_{11}, and hence is uniform (Exercise 4.6). The result follows. □

Corollary 4.53. *Let R be a left Noetherian ring and A a left R-module which satisfies C_3. Then A satisfies C_{11} if and only if A is direct sum of uniform modules.*

Proof. By Corollary 4.28 and Theorem 4.52. □

Corollary 4.54. *Let R be a ring with finite left uniform dimension and A a nonsingular left R-module which satisfies C_3. Then A satisfies C_{11} if and only if A is direct sum of uniform modules.*

Proof. The sufficiency follows by Corollary 4.28. Conversely, suppose A satisfies C_{11} and C_3. Consider an ascending chain of left ideals $\ell(a_1) \subseteq \ell(a_2) \subseteq \cdots$, where $a_i \in A$ ($i \geq 1$). Since R has finite left uniform dimension, there exists $k \geq 1$ such that $\ell(a_i)$ is essential in $\ell(a_{i+1})$ for all $i \geq k$. Let $i \geq k$ and let $r \in \ell(a_{i+1})$. Then $Er \leq \ell(a_i)$ for some essential left ideal E. Thus $Era_i = 0$, and so $ra_i = 0$. It follows that $\ell(a_i) = \ell(a_{i+1})$ for all $i \geq k$. Thus A satisfies (A). By Theorem 4.52, A is direct sum of uniform modules. □

It is proved in [Yu94, Thm. 2.11] that over a ring with acc on essential left ideals, the finite exchange property implies the full exchange property for every quasi-continuous module. Recall that a module A is said to have the $\aleph$-*exchange property* if whenever $A \oplus N = \bigoplus_{i\in I} A_i$ with $|I| < \aleph$, there exist submodules $A_i' \leq A_i$ such that $A \oplus N = A \oplus (\bigoplus_{i\in I} A_i')$. In particular, A is said to have the *finite exchange property* when the above condition is satisfied for any finite index set I. Next we consider the exchange property on C_{11}-modules. First, recall that if a module has an indecomposable decomposition, then the finite exchange property implies the full exchange property (see [ZH84, Cor. 6]). To this end, we only state the following result, which can be proven by direct modification of the proof of Lemma 3.63.

Proposition 4.55. *The following statements are equivalent for a nonsingular C_{11}-module A which satisfies C_3.*

(i) *A has an indecomposable decomposition.*
(ii) *Every finitely generated submodule of A has finite uniform dimension.*
(iii) *Every cyclic submodule of A has finite uniform dimension.*
(iv) *A satisfies (A).*

Now we prove the following result.

Theorem 4.56. *Let A be a nonsingular module such that Ra has finite uniform dimension for each $a \in A$. If A satisfies C_{11} and C_3, then the finite exchange property implies the full exchange property.*

Proof. By Proposition 4.55 and [ZH84, Cor. 6]. □

Corollary 4.57. *Let R be a ring with finite uniform dimension and A a nonsingular left R-module which satisfies C_{11} and C_3. Then the finite exchange property implies the full exchange property.*

Proof. Let R and A be as stated. Let $a \in A$. Then $\ell(a)$ is a complement in the left R-module R (Exercise 2.14). Hence the R-module $R/\ell(a)$ has finite uniform dimension (Proposition 2.18). Since $Ra \cong R/\ell(a)$, we can apply the Theorem 4.56. □

Theorem 4.58. *Let A be a module which satisfies (A). Suppose further that A satisfies C_{11} and C_3. Then the finite exchange property implies the full exchange property.*

Proof. By Lemma 4.51 and [MM90, Thm. 2.17]. □

Observe that one cannot expect that the assumptions of Theorem 4.58 imply quasi-continuity of the module (see Example 4.46). Applying Theorem 4.58 to locally Noetherian modules, we have

Corollary 4.59. *Let A be a locally Noetherian module with C_{11} and C_3. Then the finite exchange property implies the full exchange property.*

Proof. Let $a \in A$. Then $R/\ell(a) \cong Ra$ so that $R/\ell(a)$ is Noetherian left R-module. It follows that A has (A). Then Theorem 4.58 yields the result. □

For instance, by Lemma 2.72, the aforementioned result hold true if we replace C_3 by C_2. Incidentally, we next generalize a fundamental property of the endomorphism rings of continuous modules (Exercise 3.45 (i) or [MM90, Prop. 3.5]).

Let $S, \triangle$, and J be as in Exercise 3.45. Then we prove the following generalization.

Theorem 4.60. *Let A be a module which satisfies C_{11} and C_2. Then $S/\triangle$ is a (von Neumann) regular ring and $\triangle$ equals J.*

Proof. Let $\alpha \in S$ and $K = \ker \alpha$. By C_{11}, there exists a direct summand L of A such that L is a complement of K in A. Since $\alpha|_L$ is a monomorphism, $\alpha(L)$ is a direct summand of A, by C_2. Hence there exists $\beta \in S$ such that $\beta\alpha = i|_L$. Then $(\alpha - \alpha\beta\alpha)(K \oplus L) = (\alpha - \alpha\beta\alpha)(L) = 0$, and so $K \oplus L \leq \ker(\alpha - \alpha\beta\alpha)$. Since $K \oplus L$ is essential in A, $\alpha - \alpha\beta\alpha \in \triangle$. Therefore, $S/\triangle$ is a regular ring. This also proves that $J \leq \triangle$.

Let $a \in \triangle$. Since $\ker a \cap \ker(1-a) = 0$ and $\ker a$ is essential in A, $ker(1-a) = 0$. Hence $A(1-a)$ is a direct summand of A by C_2. However, $A(1-a)$ is essential in A since $\ker a \leq A(1-a)$. Thus $A(1-a) = A$, and therefore $1-a$ is a unit in S. It follows that $a \in J$, and hence $\triangle \leq J$. □

We can say a little bit more for nonsingular modules with the properties C_{11} and C_2. First we need to prove the following basic lemma [Sand68, Lem. 1.3].

Lemma 4.61. *Let A be a nonsingular left R-module. Then $\triangle = 0$.*

Proof. Let $f \in \triangle$ and $N = \ker f$. Then for any $x \in A$ there exists an essential left ideal L of R such that $0 \neq Lx \leq N$. Now $Lf(x) = 0$. Since A is nonsingular, $f(x) = 0$, and since x was arbitrary, $f = 0$. □

Corollary 4.62. *Let A be a nonsingular left R-module which satisfies C_{11} and C_2. Then S is a (von Neumann) regular ring.*

Proof. By Lemma 4.61, $\triangle = 0$ and so the result follows from Theorem 4.60. □

Recall that weak CS-modules share some of the decomposition properties of CS-modules (Corollary 4.20 and Theorem 4.21). Now we obtain similar decompositions results for modules with C_{11}^+.

Lemma 4.63. *Let A be a module such that A satisfies C_{11}^+ and $A/(\operatorname{soc} A)$ has finite uniform dimension. Suppose that $\operatorname{soc} A$ is contained in a finitely generated submodule of A. Then A has finite uniform dimension.*

Proof. Suppose A does not have finite uniform dimension. Then $\operatorname{soc} A$ is not finitely generated. There exist submodules S_1, S_2 of $\operatorname{soc} A$ such that S_i is not finitely generated for $i = 1, 2$ and $\operatorname{soc} A = S_1 \oplus S_2$. There exist submodules K, K' of A such that $A = K \oplus K'$, $S_1 \cap K = 0$, and $S_1 \oplus K$ is essential in A. Note that $S_1 \oplus S_2 = \operatorname{soc} A = \operatorname{soc}(S_1 \oplus K) = S_1 \oplus (\operatorname{soc} K)$. Thus $\operatorname{soc} K \cong S_2$ and hence $\operatorname{soc} K$ is not finitely generated. On the other hand, $\operatorname{soc} K \oplus \operatorname{soc} K' = \operatorname{soc} A = S_1 \oplus \operatorname{soc} K$ so that $\operatorname{soc} K' \cong S_1$, and hence $\operatorname{soc} K'$ is not finitely generated.

By hypothesis, there exists a finitely generated submodule N of A such that $\operatorname{soc} A \leq N$. Suppose that $K = \operatorname{soc} K$. Then $\operatorname{soc} K$ is a direct summand of A and hence also of N. It follows that $\operatorname{soc} K$ is finitely generated, a contradiction. Thus $K \neq \operatorname{soc} K$. Similarly $K' \neq \operatorname{soc} K'$. But $A/\operatorname{soc} A \cong [K/(\operatorname{soc} K)] \oplus [K'/(\operatorname{soc} K')]$. It follows that the modules $K/(\operatorname{soc} K)$ and $K'/(\operatorname{soc} K')$ each have smaller uniform dimension than $A/(\operatorname{soc} A)$. By induction on the uniform dimension of $A/(\operatorname{soc} A)$, we conclude that K and K' both have finite uniform dimension, and hence so does $A = K \oplus K'$, a contradiction. Thus A has finite uniform dimension. □

Theorem 4.64. *Let A be a module such that A satisfies C_{11}^+ and $A/(\operatorname{soc} A)$ has finite uniform dimension. Then A contains a semisimple submodule A_1 and a submodule A_2 with finite uniform dimension such that $A = A_1 \oplus A_2$.*

Proof. If $A = \operatorname{soc} A$ then there is nothing to prove. Suppose that $A \neq \operatorname{soc} A$. Let $a \in A$, $a \notin \operatorname{soc} A$. There exist submodules K, K' of A such that $A = K \oplus K'$, $Ra \cap K = 0$ and $Ra \oplus K$ is an essential submodule of A. Let $\pi' : A \to K'$ denote the canonical projection. Then clearly $Ra \oplus K = R\pi'(a) \oplus K$ (Lemma 2.71). It follows that $R\pi'(a)$ is an essential submodule of K' and hence $\operatorname{soc} K' \leq R\pi'(a)$. By Lemma 4.63, K' has finite uniform dimension.

Note that $R\pi'(a) \cong Ra$ and hence $\pi'(a) \in K'$, $\pi'(a) \notin \operatorname{soc} K'$. Thus $K \neq \operatorname{soc} K'$. Now $A/(\operatorname{soc} A) \cong [K/(\operatorname{soc} K)] \oplus [K'/(\operatorname{soc} K')]$ implies that the module $K/(\operatorname{soc} K)$ has smaller uniform dimension than $A/\operatorname{soc} A$. By induction on the uniform dimension of $A/(\operatorname{soc} \mathrm{A})$, there exist submodules K_1, K_2 of K such that $K = K_1 \oplus K_2$, K_1 is semisimple and K_2 has finite uniform dimension. Thus A is the direct sum of the semisimple submodule K_1 and the submodule $K_2 \oplus K'$ which has finite uniform dimension. □

Corollary 4.65. *Let A be a module which satisfies C_{11} and acc (respectively, dcc) on essential submodules. Then $A = A_1 \oplus A_2$ for some semisimple submodule A_1 and Noetherian (respectively, Artinian) submodule A_2.*

Proof. We prove the result in the acc case; the dcc case is similar. Suppose A satisfies acc on essential submodules. By Proposition 2.13, $A/(\operatorname{soc} A)$ is Noetherian. Hence, by Theorem 4.64, $A = A_1 \oplus A_2$ for some semisimple submodule A_1 and submodule A_2 with finite uniform dimension. Now $\operatorname{soc} A = A_1 \oplus (\operatorname{soc} A_2)$, and hence $A/(\operatorname{soc} A) \cong A_2/(\operatorname{soc} A_2)$. Thus $A_2/(\operatorname{soc} A_2)$ is Noetherian. But $\operatorname{soc} A_2$ is Noetherian, because A_2 has finite uniform dimension. Therefore A_2 is Noetherian. □

Exercises

4.9. Find a module with C_{11}^{+} that does not have SIP.

4.10. Let $R_R = T_2(\mathbb{Z})$. Show that

(i) R_R satisfies the assumptions of Theorem 4.58.

(ii) R_R is not quasi-continuous.

4.11. Show that C_2 cannot be replaced with C_3 in Theorem 4.60.

4.4 Structural properties of C_{11}-modules and related concepts

In this section, we continue the study of C_{11}-modules. To this end, we consider connections between condition C_{11} and various other generalizations of condition CS. Besides, we deal with decompositions of C_{11}-modules into direct summands which also satisfy C_{11}. Several results are obtained which generalize some earlier results in this direction.

Let R be a ring and A a left R-module. If X is a fully invariant submodule of A (see Section 3.1), we write $X \trianglelefteq A$. Recall that a module is FI-*extending* if every fully invariant submodule is essential in a direct summand (see [BCFG01] or [BMR02]). Let us begin with a characterization of FI-extending modules in terms of lifting homomorphisms for fully invariant submodules (see §3.4).

Lemma 4.66. *Let A be a module. The following conditions are equivalent.*

(i) *A is* FI-*extending.*

(ii) *Every fully invariant submodule of A has a complement which is a direct summand of A.*

(iii) *For each $X \trianglelefteq A$, there exist a complement submodule L of A and a complement K of L such that $X \leq_{\mathrm{e}} L$ and $L \oplus K \in \operatorname{Lift}_A(A)$.*

Proof. (i) $\Leftrightarrow$ (ii) Let X be a fully invariant submodule of A. First assume that A is FI-extending. There exists $e = e^2 \in \text{End}(A)$ such that X is essential in Ae. Hence $A(1-e)$ is the desired complement. Conversely, let Ac be a complement of X, where $c = c^2 \in \text{End}(A)$. Let $x \in X$. Then $x = xc + x(1-c)$. Since X is fully invariant, $xc \in X \cap cA = 0$. Thus $X \subseteq A(1-c)$, and so X is essential in $A(1-c)$.

(i) $\Leftrightarrow$ (iii) This equivalence is a direct consequence of Lemma 3.97. □

Now we have the following relationships between CS, C_{11}, and FI-extending conditions.

Proposition 4.67. *Let A be a module. Consider the following statements*

(i) *A is CS.*
(ii) *A is C_{11}.*
(iii) *A is FI-extending.*

Then (i) $\Rightarrow$ (ii) $\Rightarrow$ (iii). *In general, the converses to these implications do not hold.*

Proof. (i) $\Rightarrow$ (ii) This implication is clear.

(ii) $\Rightarrow$ (iii) This implication is a direct consequence of Lemma 4.66. Let R be the ring of 2-by-2 upper triangular matrices over the integers, i.e., $R = T_2(\mathbb{Z})$. Then R_R has C_{11}, by Theorem 4.26 (see also Exercise 4.12). However, R_R is not CS, by Example 3.84. Finally, observe that a domain D has C_{11} as a D-module if and only if D is a right Ore domain. Hence any domain which is not right Ore is FI-extending, but does not satisfy C_{11} as a module over itself. □

Observe that if every submodule of A is fully invariant, then in Proposition 4.67 (i) $\Leftrightarrow$ (ii) $\Leftrightarrow$ (iii). Since property C_{11} lies strictly between the CS and FI-extending properties, it is natural to seek conditions which ensure that a C_{11}-module has CS, or that a FI-extending module has C_{11}. Such conditions are illustrated in our next result. First recall the following definitions. Let A be a module. A is said to be *ADS* when for any decomposition $A = B \oplus C$, if D is any complement of C in A, then $A = D \oplus C$ [BR93] (see also Exercise 2.45). From [AB96], A is called *complement bounded* if every nonzero complement submodule contains a nonzero fully invariant submodule of A.

Proposition 4.68.

(i) *Let A be an R-module such that* $\text{End}\,({}_RA)$ *is Abelian and* $X \leq A$ *implies* $X = \sum_{i\in I} h_i(A)$, *where* $h_i \in \text{End}\,({}_RA)$. *Then A is quasi-continuous if and only if A has C_{11}.*
(ii) *A is quasi-continuous if and only if A has C_{11} and the ADS property.*
(iii) *Let A be complement bounded. Then A has C_{11} if and only if A is* FI-*extending.*

Proof. (i) Assume ${}_RA$ has C_{11} and $X \leq A$. Then $X = \sum_{i\in I} h_i(A)$, where each $h_i \in \text{End}\,({}_RA)$. There exists $e = e^2 \in \text{End}\,({}_RA)$ such that Ae is a complement of X. Let $0 \neq x \in X$. Then $x = xe + x(1-e)$. But $x = \sum_{i\in I} h_i(a_i)$, where $a_i \in A$. So

$xe = \sum_{i\in I} h_i(a_i)e = \sum_{i\in I} h_i(a_ie) \in X \cap Ae = 0$. Hence $X \leq_{\mathrm{e}} A(1-e)$. Therefore ${}_RA$ has CS. Since $\mathrm{End}\,({}_RA)$ is Abelian, ${}_RA$ is quasi-continuous (Exercise 2.49). The converse follows from Proposition 4.67.

(ii) By [BR93, p. 138] and Proposition 4.67, quasi-continuous implies C_{11} and ADS. So assume A has C_{11} and ADS. Let $Y \leq A$ and let K be a complement of Y in A such that $A = K \oplus X$, for some $X \leq A$. Let $\overline{Y}$ be the closure of Y in A. By the ADS property, $A = K \oplus \overline{Y}$. Thus A has CS. Now, let B and C be direct summands of A such that $B \cap C = 0$. There exists a complement D of B in A such that $C \leq D$. By the ADS property, $A = B \oplus D$, but $D = C \oplus E$ for some $E \leq A$. Hence $A = B \oplus C \oplus E$. So $B \oplus C$ is a direct summand of A. Therefore A is quasi-continuous.

(iii) Assume A is complement bounded and FI-extending. Let $Y \leq A$ and K a complement of Y. If $K = 0$, then K is a direct summand of A. Assume $K \neq 0$. Then there exists $0 \neq X \trianglelefteq A$ such that X is the sum of all fully invariant submodules of A contained in K. Then $X \leq_{\mathrm{e}} K$. So there exists $e = e^2 \in \mathrm{End}({}_RA)$ such that $X \leq_{\mathrm{e}} Ae$. Hence $Y \cap Ae = 0$ and $Y \oplus Ae \leq_{\mathrm{e}} A$. By Proposition 4.25, A has C_{11}. The converse follows from Proposition 4.67. □

Recall an R-module A is said to be a *multiplication module* if for each $X \leq A$ there exists ${}_RI \leq {}_RR$ such that $X = IA$. Now we have.

Corollary 4.69. *If A is an R-module satisfying any of the following conditions, then A has* CS *if and only if A has C_{11}.*

(i) *${}_RA = {}_RR$ and R is Abelian.*
(ii) *A is cyclic and R is commutative.*
(iii) *A is multiplication module and R is commutative.*

Proof. By Proposition 4.68, the result is true for condition (i). Now assume that A is cyclic and R is commutative. There exists ${}_RB \leq {}_RR$ such that ${}_RA$ is isomorphic to R/B. Let Y/B be an R-submodule of R/B. So $Y/B = (\sum_{i\in I} Ry_i) + B = R(\sum_{i\in I} y_i + B)$, where each $y_i \in Y$. Define $h_i : R/B \to R/B$ by $h_i(r+B) = ry_i + B$. Then $h_i \in \mathrm{End}({}_R(R/B))$. Hence $Y/B = \sum_{i\in I} h_i(R/B)$. Since R is commutative, $\mathrm{End}({}_R(R/B))$ is commutative. Thus Proposition 4.68 (i) yields the result for condition (ii).

Finally, assume that A is a multiplication module and R is commutative. Let $X = IA$, where ${}_RI \leq {}_RR$. For each $x \in I$, define $h_x : A \to A$ by $h_x(a) = xa$ for $a \in A$. Then $X = IA = \sum_{x\in I} h_x(A)$. Observe that every submodule of a multiplication module is fully invariant (Exercise 4.14). By [BMR02, Lem. 1.9], if $e = e^2 \in \mathrm{End}\,({}_RA)$, then e and $1-e$ are right semicentral idempotents. Hence e is central. We conclude that $\mathrm{End}\,({}_RA)$ is Abelian. Again, Proposition 4.68 yields the result. □

The next example, which illustrates Proposition 4.68 (iii), is taken from [BMR02, Exa. 4.12].

Example 4.70. There exists a ring R such that R is strongly right bounded (i.e., every nonzero right ideal contains a nonzero ideal), R is right FI-extending, but R is not right CS. Let $T = \mathbb{Z}_2 S$ be the semigroup ring with $\mathbb{Z}_2$, the integers mod 2, and S, the semigroup on the set $\{a, b\}$ satisfying the relation $xy = y$ for $x, y \in S$. Let R be the Dorroh extension of T (i.e., the ring with unity formed from $T \times \mathbb{Z}$ with componentwise addition and with multiplication given by $(x, k)(y, n) = (xy + nx + ky, kn)$). Observe that $R = (a, 1)R \oplus (T, 0)$ as right ideals. To see that R is not right CS, observe that the right ideal $\{(k(a + b), 2k) : k \in \mathbb{Z}\}$ is not right essential in an idempotent generated right ideal. By direct calculation R is right FI-extending and complement bounded.

Recall that a module with C_{11} and C_3 has C_{11}^+ (Corollary 4.47). However, C_{11} and C_3 do not imply CS (Example 4.46). We provided an example showing that C_{11} does not imply C_{11}^+ and that a direct sum of C_{11}^+-modules is not C_{11}^+, although a direct sum of C_{11}-modules is C_{11} (Example 4.33 and Theorem 4.26). In the rest of this section, we study decompositions of C_{11}-modules into direct summands which also satisfy C_{11}.

Lemma 4.71. *Let $A = A_1 \oplus A_2$, where $A_1 \trianglelefteq A$. Then A has C_{11} if and only if both A_1 and A_2 have C_{11}.*

Proof. Assume A has C_{11}. By Corollary 4.49, A_1 has C_{11}. There exists a right semicentral idempotent $e \in \text{End}(_R A)$ such that $A_1 = Ae$ ([BMR02, Lem. 1.9]). Let $Y \leq A_2$. There exists $c = c^2 \in \text{End}(_R A)$ such that Ac is a complement of $A_1 \oplus Y$. Since e is right semicentral idempotent, $b = e + c - ce$ is an idempotent in $\text{End}(_R A)$ and $Ab = A_1 \oplus Ac$ ([BKP01, Prop. 1.3 (ix)]). Then $Ab = Ab \cap (A_1 \oplus A_2) = A_1 \oplus (Ab \cap A_2)$. Hence $Ab \cap A_2$ is a direct summand of A_2. Moreover, $Y \oplus Ab \leq_e A$, so $A_2 \cap (Y \oplus Ab) = Y \oplus (A_2 \cap Ab) \leq_e A_2$. By Proposition 4.25, A_2 has C_{11}. The converse follows from Theorem 4.26. □

The last part of the proof of Lemma 4.71 showing that A_2 has C_{11} can also be proved by invoking Theorem 4.41, where the K of Theorem 4.41 corresponds to Ac in Lemma 4.71 and the roles of A_1 and A_2 are reversed in Lemma 4.71.

The next theorem generalizes Theorem 4.29 and Corollary 4.39.

Theorem 4.72. *Let A be a C_{11}-module and $X \trianglelefteq A$. Then $A = A_1 \oplus A_2$, where $X \leq_e A_1$. Moreover*

(i) *$Z_2(A_1)$ has C_{11} implies A_1 has C_{11}.*
(ii) *$Z_2(A_2)$ has C_{11} implies A_2 has C_{11}.*
(iii) *$A_1 \trianglelefteq A$ (e.g., $Z_2(A) \leq A_1$) implies both A_1 and A_2 have C_{11}.*
(iv) *$Z_2(A) \leq A_2$ implies both A_1 and A_2 have C_{11}.*

Proof. By Proposition 4.67, A is FI-extending. So $A = A_1 \oplus A_2$, where $X \leq_e A_1$. Now part (iii) follows from Lemma 4.71 and [BMR02, Prop. 1.10]. Since $Z_2(A) \trianglelefteq A$, $Z_2(A) = (Z_2(A) \cap A_1) \oplus (Z_2(A) \cap A_2) = Z_2(A_1) \oplus Z_2(A_2)$ (Exercise 3.14). Using the previous part of the proof with $X = Z(A)$ or using Theorem 4.29, $A =$

$Z_2(A) \oplus K$ for some $K \leq A$, and $Z_2(A)$ and K are both C_{11}-modules. Hence $A_1 = Z_2(A_1) \oplus K_1$ and $A_2 = Z_2(A_2) \oplus K_2$ for some $K_1 \leq A_1$ and $K_2 \leq A_2$. Then $A_1 + Z(A_2) = K_1 \oplus Z_2(A_1) \oplus Z_2(A_2)$. Since $X + Z_2(A) \leq_{\text{e}} A_1 + Z_2(A)$ and $X + Z_2(A) \trianglelefteq A$, there exists $e = e^2 \in \text{End}(_RA)$ such that $Ae \trianglelefteq A$ from [BMR02, Prop. 1.10]. By applying part (iii) to $A = Ae \oplus K_2$ we obtain that Ae and K_2 are C_{11}-modules. Now using Theorem 4.29 on Ae, we have that K_1 has C_{11}. Thus $A = Z_2(A_1) \oplus K_1 \oplus Z_2(A_2) \oplus K_2$. (i) and (ii) follow from Theorem 4.26. For (iv), $Z_2(A) \leq A_2$ implies that $Z_2(A) = Z_2(A_2)$. Hence $Z_2(A_1) = 0$ has C_{11} and $Z_2(A_2)$ has C_{11}. By (i) and (ii), A_1 and A_2 are C_{11}-modules. □

It can be shown that if r is any radical for a stable hereditary torsion theory for the category of left R-modules with $r \leq Z_2$, then Theorem 4.72 remains true when Z_2 is replaced by r.

Corollary 4.73. *Let A be a C_{11}-module with $X \trianglelefteq A$ and $Z(_RX) = 0$. Then $A = A_1 \oplus A_2$, where $X \leq_{\text{e}} A_1$ and A_1 and A_2 are C_{11}-modules.*

Proof. $Z(_RX) = 0$ implies that $Z_2(A_1) = 0$. Hence $Z_2(A) = Z_2(A_2) \leq A_2$. The result is now a consequence of Theorem 4.72. □

Corollary 4.73 generalizes Theorem 4.31. Recall from [Birk76] that the *minimal direct summand* (idempotent generated right ideal) containing the nilpotent elements, MDSN, is a semicompletely prime ideal (i.e., if $x^n \in$ MDSN, then $x \in$ MDSN) which is the intersection of all idempotent generated right ideals containing the set of nilpotent elements. The MDSN may not exist for a given ring. We use N_t to denote the ideal generated by the set of nilpotent elements of index two. A nonempty subset X of R is *reduced* if it contains no nonzero nilpotents. Also from [Birk76], a right ideal X of R is *densely nil*, DN, if for each $0 \neq x \in X$ there exists $r \in R$ such that $xr \neq 0$, but $(xr)^2 = 0$. A ring R is said to be *strongly regular* if R is regular and reduced (see Exercise 4.16).

Corollary 4.74. *Let R be a ring (respectively semiprime ring). Then R_R has C_{11} if and only if $R = R_1 \oplus R_2$ is a right ideal direct sum, where R_1 has C_{11} and is the* MDSN *and an essential extension of N_t and R_2 is a right ideal (respectively ideal) which has* CS *and is also a reduced right* CS*-ring.*

Proof. By Theorem 4.72, $R = R_1 \oplus R_2$ where $N_t \leq_e R_1$ and R_2 is reduced. Let $0 \neq x \in Z(R_2)$. Then there exists $L \leq_{\text{e}} R$ such that $xL = 0$. Since $xR \cap L \neq 0$, there exists $r \in R$ such that $xr \neq 0$, but $x^2r = 0$. Then $xrx \in N_t \cap R_2 = 0$. So $xr \in N_t \cap R_2 = 0$, a contradiction. Hence $Z(R_2) = 0$, so $Z_2(R_2) = 0$. Thus $Z_2(R) \leq R_1$. By Theorem 4.72, $R_1 \trianglelefteq R$ and R_1 and R_2 are C_{11}-modules. By Corollary 4.69 (i), R_2 has CS and is a right CS-ring. Since $R_1 \trianglelefteq R$, R_2 is a reduced right C_{11}-ring. By [Birk76, Prop. 1.2], R_1 is the MDSN of R. If R is semiprime, then R_1 and R_2 are ideals, hence $R_1 \oplus R_2$ is a ring decomposition ([Birk83, p. 569]). The converse follows from Theorem 4.26. □

The following example shows that the semiprime condition in Corollary 4.74 is not enough to ensure that R_1 is DN.

Example 4.75. Let $R = M_n(D)$, where $n > 1$, and D is a right but not a left Ore domain. Then R_R has C_{11}, R is semiprime and $R = \text{MDSN}$, but R is not DN (see [BT77], [Ter03]).

Corollary 4.76. *Let R be a regular ring. Then R is right C_{11} if and only if $R = R_1 \oplus R_2$ (ring direct sum), where R_2 is a strongly regular continuous ring and R_1 is a right C_{11}-ring in which each nonzero right ideal contains a nonzero nilpotent element (i.e., R_1 is* DN*).*

Proof. By Corollary 4.74, we only need to show that R_1 is DN. Assume there exists $x \in R_1$ such that xR_1 is reduced. There exists $e = e^2 \in R_1$ such that $xR_1 = eR_1$. By [Birk83, p. 569], e is a right semicentral idempotent of R_1. Since R_1 is semiprime, e is a central idempotent of R_1. By [Birk81, Lem. 2] and Corollary 4.74, $x = 0$. Hence, R_1 is DN. □

Example 4.77. The ring $M_2(R)$ in [Goo79, Exa. 13.8] is a von Neumann regular ring which is not a Baer ring. Hence, it satisfies neither the left nor the right CS conditions. However, since R is a continuous ring, $M_2(R)$ satisfies both the left and right C_{11} conditions by [Ter03, Cor. 3.8] or Corollary 4.82. Moreover, since $M_2(R)$ is regular, it has SIP. Thus, by Corollary 4.49 (iii), $M_2(R)$ satisfies both the left and right C_{11}^+ conditions.

Our next result generalizes a result of [Fai95, Thm. 7], but specializes [BMR02, Thm. 3.6 (iii)] to C_{11}-rings. We use $\mu(R)$ to denote the maximal regular ideal of R. An ideal B of R is an *ideal essential extension* of an ideal X of R if $X \leq B$ and $X \cap C \neq 0$ for all $0 \neq C \trianglelefteq R$ for which $C \cap B \neq 0$.

Proposition 4.78. *Let R be a right C_{11}-ring satisfying C_2. If R is left* FI*-extending, then $R = \mu(R) \oplus B$ (ring direct sum) and B is an ideal essential extension of $J(R)$.*

Proof. This result is a consequence of [BMR02, Thm. 3.6 (iii)] and Theorem 4.60. □

If we let $R = R_1 \oplus R_2$ (ring direct sum) where R_1 is the ring indicated in Example 4.77 and $R_2 = \mathbb{Z}_{p^n}$ for $n > 1$, then R illustrates Proposition 4.78, but R does not satisfy the hypothesis of [Fai95, Thm. 7]. Our next result shows that certain cyclic modules of a right C_{11}-ring also have C_{11}.

Proposition 4.79. *Let R be a ring, $X \trianglelefteq R$, and $\overline{R} = R/X$. If R_R has C_{11} and $Z(\overline{R}_R) = 0$, then $\overline{R}_R$ and $\overline{R}_{\overline{R}}$ have C_{11}.*

Proof. Let $\overline{A}_R \leq \overline{R}_R$. Then $\overline{A} = A/X$ for some $A_R \leq R_R$. First we show that $\overline{R}_R$ has C_{11}. Since R_R has C_{11}, there exists $e = e^2 \in R$ such that eR_R is a complement of A_R in R_R. Observe that $X \leq (1-e)R$, hence $(eR \oplus X/X)_R$ is a direct summand of $\overline{R}_R$ and $(A/X) \cap (eR \oplus X/X) = 0$. We claim that $(A \oplus eR/X)_R \leq_e \overline{R}_R$. Let

$\overline{Y}_R \leq \overline{R}_R$ such that $(A \oplus eR/X) \cap \overline{Y} = 0$. There exists $Y_R \leq R_R$, where $\overline{Y} = Y/X$. Let $y \in Y$. Since $(A \oplus eR)_R \leq_e R_R$, there exists $L \leq_e R_R$ such that $yL \subseteq A \oplus eR$ (Exercise 2.3). Thus $(y+X)L \subseteq (A \oplus eR/X) \cap Y/X = 0$. Since $Z(\overline{R}_R) = 0$, $y \in X$. We conclude that $\overline{Y} = 0$, and so $(A \oplus eR/X)_R \leq_e \overline{R}_R$. By Proposition 4.25, $\overline{R}_R$ has C_{11}.

Using [Go76, Prop. 1.28] and arguing in much the same way as in the above proof, we get that $\overline{R}_{\overline{R}}$ has C_{11}. □

Exercises

4.12. Find a ring that does not have a MDSN.

4.13. Show that if R has a MDSN, then so does $R[x]$.

4.14. Prove that every submodule of a multiplication module is fully invariant.

4.15. Find N_t for the rings $R = T_2(\mathbb{Z})$ and $S = T_2(F)$, where F is a field.

4.16. Prove that R is a strongly regular ring if and only if it is an Abelian regular ring.

4.17. Prove Theorem 4.72 in the case where r is any radical for a stable hereditary torsion theory for the category of left R-modules with $r \leq Z_2$.

4.5 Extensions of C_{11}-rings and modules

In this section, we study the transference of the C_{11} condition from a given ring or module to various ring or module extensions. In particular, we show that if R is a right C_{11}-ring (i.e., R_R is a C_{11}-module), then the ring of column finite matrices of size Γ over R, the ring of m-by-m upper triangular matrices over R, and any right essential overring T of R are all right C_{11}-rings. We also provide necessary and sufficient conditions under which the generalized triangular matrix ring is a right C_{11}-ring. For a module A, we prove that all essential extensions of A satisfying C_{11} are essential extensions of C_{11}-modules constructed from A and certain subsets of idempotent elements of the ring of endomorphisms of the injective hull of A. Finally, we prove that if A is a C_{11}-module, then so is its rational hull.

First, let us consider full matrix rings. We have the following theorem and its corollaries.

Theorem 4.80. *Let A be a left R-module, where $R = ReR$ for some idempotent e in R and $S = eRe$. Then:*

(i) *The left R-module A has C_{11} if and only if the left S-module eA has C_{11}.*
(ii) *R_R has C_{11} if and only if the eRe-module Re has C_{11}.*

Proof. Follows from Proposition 2.77. □

Corollary 4.81. *Let R be any ring. Then $M_m(R)$ is a right C_{11}-ring if and only if the free right R-module $R^{(m)}$ has C_{11}.*

Proof. By Theorem 4.80. □

Corollary 4.82. *If R is a right C_{11}-ring, then so does the full matrix ring $M_m(R)$.*

Proof. Since R is a right C_{11}-ring, $R^{(m)}$ is a right C_{11} R-module by Theorem 4.26. Now Corollary 4.81 gives the result. □

Observe that Corollary 4.82 and Corollary 4.69 ensure that over commutative rings being C_{11} is a Morita invariant property (Open Problems and Questions 3).

Now we consider the behavior of the C_{11} condition under various matrix extensions and essential extensions of both rings and modules. Let $CFM_\Gamma(R)$ and $CRFM_\Gamma(R)$ designate the column finite and row and column finite $\Gamma \times \Gamma$ matrices over R, respectively. Then we have the following theorem.

Theorem 4.83. *Let R be a right C_{11}-ring. Then*

(i) *$CFM_\Gamma(R)$ is a right C_{11}-ring, for all Γ.*
(ii) *$CRFM_\Gamma(R)$ is a right C_{11}-ring, for all Γ.*
(iii) *$\mathrm{End}(F_R)$ is a right C_{11}-ring, where F_R is a free right R-module.*

Proof. Let S be an arbitrary ring with unity. Observe that $CFM_\Gamma(S) \cong \mathrm{End}(F_S)$, where F_S is a right S-module with a basis of cardinality Γ. Hence (i) $\Leftrightarrow$ (iii). Let $M(-)$ denote either $CFM_\Gamma(-)$ or $CRFM_\Gamma(-)$. Let $X_{M(S)} \leq e_{11}M(S)_{M(S)}$. Then there exists $V_S \leq S_S$ such that $X \leq e_{11}M(V)$ and $e_{11}M(V)e_{kk} \leq X$, for all possible k where $e_{ii} \in M(S)$ with 1 in the (i,i)th position and zero elsewhere.
Claim. $e_{11}M(R)_{M(R)}$ is C_{11}.

Let $X_{M(R)} \leq e_{11}M(R)_{M(R)}$. From above, there exists $V_R \leq R_R$ such that $X \leq e_{11}M(R)$ and $e_{11}M(V)e_{kk} \leq X$, for all possible k. Since R_R is C_{11}, there exists $c = c^2 \in R$ such that cR_R is a complement of V_R in R_R. Observe that $X \cap K = 0$, where $K = ce_{11}M(R)$, and that $K_{M(R)}$ is a direct summand of $e_{11}M(R)_{M(R)}$. Let $0 \neq a \in e_{11}M(R)$. There exists an ordinal k such that $0 \neq ae_{kk}$. Let s be the nonzero entry of ae_{kk}. There exists $r \in R$ such that $0 \neq sr \in V \oplus cR$ (Lemma 4.24). So $0 \neq ae_{kk}r \in e_{11}M(V \oplus cR)e_{kk} = e_{11}M(V)e_{kk} \oplus e_{11}M(cR)e_{kk} \leq X \oplus K$. By Proposition 4.25, $e_{11}M(R)_{M(R)}$ has C_{11}.

Now each of $M(R)$ is isomorphic to $e_{11}M(R)_{M(R)}$. Hence all rows of $M(R)$ are C_{11}. $M(R)_{M(R)}$ is the direct sum of its rows. By Theorem 4.26, $M(R)_{M(R)}$ has C_{11}. □

Recall that in §3.3 we dealt with conditions under which a 2-by-2 generalized triangular matrix ring enjoys the CS property. Our next result will characterize when a generalized triangular matrix ring satisfies C_{11}.

Theorem 4.84. *For rings S and R, assume that ${}_SA_R$ is an (S,R)-bimodule. Let $T = \begin{bmatrix} S & A \\ 0 & R \end{bmatrix}$ be the corresponding generalized triangular matrix ring. Then T is a right C_{11}-ring if and only if the following conditions hold.*

(i) *R is a right C_{11}-ring;*

(ii) *For any $X_S \leq S_S$ and $N_R \leq A_R$ such that $XA \subseteq N$, there exists $e = e^2 \in S$ such that eA_R is a complement of N_R in A_R and $eS \cap X = 0$; moreover if Y_S is a complement of X_S in S_S such that $eS \subsetneqq Y$, then $YA \nsubseteq eA$.*

Proof. ($\Rightarrow$) By Theorem 4.72 (iii), R is a right C_{11}-ring and $H_T = \begin{bmatrix} S & A \\ 0 & 0 \end{bmatrix}$ is C_{11}. Let $X_S \leq S_S$ and $N_R \leq A_R$ such that $XA \subseteq N$. Then $B_T = \begin{bmatrix} X & N \\ 0 & 0 \end{bmatrix} \leq H_T$. So, there exists $\begin{bmatrix} e & y \\ 0 & 0 \end{bmatrix} \in T$ such that $e = e^2$, $ey = y$, and $\begin{bmatrix} e & y \\ 0 & 0 \end{bmatrix} T$ is a complement of B in H. Observe that

$$\begin{bmatrix} e & y \\ 0 & 0 \end{bmatrix} T = \begin{bmatrix} eS & eA + yR \\ 0 & 0 \end{bmatrix} = \begin{bmatrix} eS & eA \\ 0 & 0 \end{bmatrix},$$

since $y = ey$.

Claim. eA_R is a complement of N_R in A_R.

Assume that there exists $K_R \leq A_R$ such that $eA \subseteq K$ and $K \cap N = 0$. Then

$$\begin{bmatrix} eS & K \\ 0 & 0 \end{bmatrix} \cap B = 0 \quad \text{and} \quad \begin{bmatrix} e & y \\ 0 & 0 \end{bmatrix} T \subseteq \begin{bmatrix} eS & K \\ 0 & 0 \end{bmatrix}.$$

Since $\begin{bmatrix} e & y \\ 0 & 0 \end{bmatrix} T_T$ is a complement of B_T in H_T, we have $\begin{bmatrix} e & y \\ 0 & 0 \end{bmatrix} T = \begin{bmatrix} eS & K \\ 0 & 0 \end{bmatrix}$. Hence $eA = K$. Thus eA_R is a complement of N_R in A_R.

If Y_S is a complement of X_S in S_S such that $eS \subsetneqq Y$ and if $YA \subseteq eA$, then

$$\begin{bmatrix} Y & eA \\ 0 & 0 \end{bmatrix} \cap B = 0 \quad \text{and} \quad \begin{bmatrix} eS & eA \\ 0 & 0 \end{bmatrix} \subsetneqq \begin{bmatrix} Y & eA \\ 0 & 0 \end{bmatrix},$$

a contradiction to $\begin{bmatrix} e & y \\ 0 & 0 \end{bmatrix} T_T$ being a complement of B_T in H_T. Hence $YA \nsubseteq eA$.

($\Leftarrow$) By Theorem 4.72 (iii) we only need to show that H_T is C_{11}. Let $B_T \leq H_T$. Then there exist $X_S \leq S_S$ and $N_R \leq A_R$ such that $XA \subseteq N$ and $B = \begin{bmatrix} X & N \\ 0 & 0 \end{bmatrix}$. By assumption, there exists $e = e^2 \in S$ such that eA_R is a complement of N_R in A_R and $eS \cap X = 0$. Let $K = \begin{bmatrix} e & 0 \\ 0 & 0 \end{bmatrix} T = \begin{bmatrix} eS & eA \\ 0 & 0 \end{bmatrix}$. Then $K \cap B = 0$. Assume there exists $\overline{K}_T \leq H_T$ such that $K \subseteq \overline{K}$ and $\overline{K} \cap B = 0$. Then $\overline{K} = \begin{bmatrix} Y & V \\ 0 & 0 \end{bmatrix}$, where $YA \leq V$, $V_R \leq A_R$, and $Y_S \leq S_S$. Since $eA \leq V$ and $V \cap N = 0$, we see that $V = eA$. So, by assumption, $Y = eS$. Thus, H_T has C_{11}. □

Corollary 4.85. *Let R be a ring. Then R is a right C_{11}-ring if and only if $T_m(R)$ is a right C_{11}-ring.*

Proof. ($\Leftarrow$) By Theorem 4.84 (i), R is right C_{11}.

($\Rightarrow$) Clearly, the result is true for $m = 1$. So assume $m = 2$. Let $X_R \leq R_R$ and $N_R \leq R_R$ such that $XR = X \leq N$. Since R_R has C_{11}, there exists $e = e^2 \in R$ such that eR is a complement of N_R in R_R. Hence, $X \cap eR \subseteq N \cap eR = 0$. Assume Y_R is a complement of X_R such that $eR \subsetneq Y$. Then $Y \not\subseteq eR$. So, by Theorem 4.84, $T_2(R)$ is a right C_{11}-ring. Now assume that $T_k(R)$ is a right C_{11}-ring and consider $T_k(R)$ for $k > 2$. In Theorem 4.84, take $S = (1 - e_{kk})T(1 - e_{kk})$, $A = (1 - e_{kk})Te_{kk}$, and $R = e_{kk}Te_{kk}$ where e_{ij} is the matrix unit with 1 in the (i, i)th position and zero elsewhere. By assumption, $(1 - e_{kk})T_k(R)(1 - e_{kk})$ is a right C_{11}-ring. The remainder of the proof is a routine verification that Theorem 4.84(ii) applies and an induction argument. □

Although a full or upper triangular matrix ring over a right CS-ring is not necessarily a right CS-ring (Examples 3.78, 3.84), Theorem 4.83 and Corollary 4.85 show that such matrix rings satisfy at least some aspects of the CS condition, namely, the C_{11} condition.

Proposition 4.86. *Assume R is a ring, M is an ideal of R, and $S = S(R, M)$ is the trivial extension of M by R. If ${}_RM$ is faithful and R_R is a C_{11}-module, then S_S is a C_{11}-module.*

Proof. Let $0 \neq Y_S \leq S_S$ and $X = \{m \in M : (0, m) \in Y\}$. Since $X_R \leq R_R$, there exists $e = e^2 \in R$ such that eR is a complement of X. We claim that $(e, 0)S$ is a complement of $Y \cap (\{0\} \times M)$. To see this let $(e, 0)(r, m) \in (e, 0)S$. Then $(e, 0)(r, m) = (er, em)$. Assume $(er, em) \in Y \cap (\{0\} \times M)$. Then $er = 0$ and $em \in eR \cap X = 0$. Hence, $(e, 0)S \cap (Y \cap (\{0\} \times M)) = 0_S$. To complete the proof of our claim, we need to show that $(e, 0)S \oplus (Y \cap (\{0\} \times M)) \leq_e S_S$. Observe that $(e, 0)S \oplus (Y \cap (\{0\} \times M)) = (eR, eM \oplus X)$. Let $0_S \neq (a, k) \in S$. First assume $a \neq 0_R$. Since eR is a complement of X, then $eR \oplus X \leq_e R$. So there is a $v \in R$ such that $0 \neq av \in eR \oplus X$. Since ${}_RM$ is faithful, there exists $m \in M$ such that $0_R \neq avm \in eM \oplus X$. Hence, $0_S \neq (a, k)(0, vm) = (0, avm) \in (eR, eM \oplus X)$. Now assume $a = 0$. Then $0 \neq k$, so there exists $w \in R$ such that $0 \neq kw \in (eR \oplus X) \cap M = eM \oplus X$. Then $0_S \neq (a, k)(w, 0) = (0, k)(w, 0) = (0, kw) \in (eR, eM \oplus X)$. Hence $(e, 0)S \oplus (Y \cap (\{0\} \times M)) = (eR, eM \oplus X)_S \leq_e S_S$. Therefore $(e, 0)S$ is a complement of $eM \times X(Y \cap (\{0\} \times M))$. From [Birk90, Lem. 4 (iv)], $(\{0\} \times M)_S \leq_e S_S$. Hence, $(Y \cap (\{0\} \times M))_S \leq_e Y_S$. So $(e, 0)S$ is a complement of Y. Therefore, S_S is a C_{11}-module. □

Since the module K in Example 4.33 is indecomposable, but of uniform dimension 2, it is an essential extension of a direct sum of two uniform submodules which satisfy C_{11} by Corollary 4.28, but K does not satisfy C_{11}. Surprisingly, our next result shows that the C_{11} condition does transfer from a ring R to any right essential overring (see Exercises 3.32 and 3.6).

Theorem 4.87. *If R is a right C_{11}-ring and T is a right essential overring (i.e., $R_R \leq_e T_R$) of R, then T is a right C_{11}-ring and T has C_{11} as a right R-module.*

Proof. First we prove that T_R has property C_{11}. Let $X_R \leq T_R$ and take $\overline{X} = X \cap R$. Then there exists $e = e^2 \in R$ such that eR is a complement of $\overline{X}$ in R. We will show that eT_R is a complement of X in T_R.

Claim 1. $X \cap eT_R = 0$. Assume to the contrary that there exists $0 \neq x \in X \cap eT$. There exists $r \in R$ such that $0 \neq xr \in X \cap R = \overline{X}$. But $x = ex$, so $0 \neq xr = exr \in \overline{X} \cap eR = 0$, a contradiction. Therefore $X \cap eT_R = 0$.

Claim 2. eT_R is a complement of X. Assume $eT_R \leq Y_R$ and $Y_R \cap X_R = 0$. Then $eR \leq Y \cap R$. But $(Y \cap R) \cap (X \cap R) = 0$. Hence, $eR = Y \cap R \leq_e Y_R$, and so $eT_R \leq_e Y_R$. Since eT is direct summand of T_R, $eT = Y$. By Claims 1 and 2, T has C_{11} as an R-module. Now we prove that T has C_{11} as a T-module. Let $X_T \leq T_T$ and take $\overline{X} = X \cap R$. Then, as above, $X \cap eT_T = 0$ and eT_T is a complement of X as a T-module. □

The following corollary is an application of Theorem 4.87 which gives an alternate proof of Theorem 4.83 when Γ is finite, as well as an alternate proof of Corollary 4.82.

Corollary 4.88. *If R is a right C_{11}-ring, then $M_m(R)$ is a right C_{11}-ring, for each positive integer m.*

Proof. By Corollary 4.85, $T_m(R)$ is a C_{11}-ring. From Exercise 3.31, $M_m(R)$ is a right essential overring of $T_m(R)$. By Theorem 4.87, $M_m(R)$ is a right C_{11}-ring. □

In [Osof64, Osof67], Osofsky raised the following question: if $E(R_R)$ has a ring multiplication which extends its right R-module scalar multiplication, must $E(R_R)$ be right self-injective? In [CHN07], examples were constructed giving a negative answer to this question. Our next result shows that if R_R satisfies C_{11} and $E(R_R)$ has such a compatible ring structure, then $E(R_R)_{E(R_R)}$ at least satisfies C_{11}.

Corollary 4.89. *Let R is a right C_{11}-ring. If $E(R_R)$ has a ring multiplication which extends its right R-module scalar multiplication, then $E(R_R)$ is a right C_{11}-ring.*

The following example illustrates several of our previous results and, in particular, shows that the C_{11} condition is preserved or improved as one moves through essential overrings.

Example 4.90. Let $R = \begin{bmatrix} \mathbb{Z}_4 & 2\mathbb{Z}_4 \\ 0 & \mathbb{Z}_4 \end{bmatrix}$, $T = T_2(\mathbb{Z}_4)$, $S = \begin{bmatrix} \mathbb{Z}_4 & \mathbb{Z}_4 \\ 2\mathbb{Z}_4 & \mathbb{Z}_4 \end{bmatrix}$, $W = M_2(\mathbb{Z}_4)$.

(i) $R_R \leq_e T_R$, $T_T \leq_e S_T$, $T_T \leq_e W_T$, $S_S \leq_e W_S$, but R_R is not essential in S_R.

(ii) From Theorem 4.84, taking $X = A = 2\mathbb{Z}_4$ and $N = 0$, R is not a right C_{11}-ring.

(iii) By Corollary 4.85, T is a right C_{11}-ring, but $T_2(\mathbb{Z}_4)$ is a not a right CS-ring (Example 3.93). By [BPR06, Thm. 3.11], S is a right CS-ring. Since the QF property is preserved by passing to matrices (Exercise 4.19), W is self-injective.

(iv) Since R is a Kasch ring (i.e., every simple right R-module can be embedded in R_R), R is its own maximal right ring of quotients (see [Lam99, p. 371]).

(v) By [Osof67] or [Lam99, pp. 79–80], $E(R_R)$ does not have a ring multiplication which extends its right R-module scalar multiplication. Observe that although W_W is injective, $W \neq E(R_R)$ because R_R is not essential in W_R.

The next two propositions and the example below show that for any module one can construct an essential extension which satisfies C_{11} from A and a certain subset of the idempotents of $\mathrm{End}(E(_RA))$. Moreover, every essential extension of A which has C_{11} is an essential extension of a C_{11}-module which can be constructed in this way. Thus, in general, a module A has essential extensions which have C_{11} and are intermediate between A and $E(A)$. The fact that they are "closer" to A than $E(A)$ usually allows for a better transfer of information between them and A rather than between A and $E(A)$. This is somewhat reminiscent of considering various localizations intermediate between a commutative domain and its field of quotients.

For a ring R, and a $\emptyset \neq Y \subseteq R$, let us use $\mathbf{I}(R)$ and $\langle Y\rangle$ to denote the set of all idempotent elements of R and the subring of R generated by Y, respectively. We have:

Proposition 4.91. *Let $K \subseteq \mathbf{I}(\mathrm{End}(E(_RA)))$ such that for each $X \leq {_RA}$ there exists $k \in K$ such that $E(_RA)k$ is a complement of X in $E(_RA)$. Then $_RA\langle K\rangle$ is a C_{11}-submodule of $E(_RA)$ which contains $_RA$.*

Proof. Since $0 \leq {_RA}$ there exists $h \in K$ such that $(E(_RA))h = E(_RA)$. Since $h^2 = h$, then $E(_RA) = E(_RA)h \oplus E(_RA)(1-h) = E(_RA)h$. Hence $1 - h = 0$, so $1 \in \langle K\rangle$. Thus $_RA \leq {_RA\langle K\rangle}$. Let $_RX \leq {_RA\langle K\rangle}$. Then $X \cap {_RA} \leq_{\mathrm{e}} {_RX}$. There exists $k \in K$ such that $E(_RA)k$ is a complement of $_RX$ in $E(_RA)$. Then $A\langle K\rangle k$ is a direct summand of $A\langle K\rangle$ and $A\langle K\rangle k$ is a complement of X in $A\langle K\rangle$. Therefore $_RA\langle K\rangle$ is a C_{11}-module. □

Lemma 4.92. *Assume $_RA \leq {_RN} \leq E(_RA)$. Let $e \in \mathbf{I}(\mathrm{End}(_RN))$. There exists $c \in \mathbf{I}(End(E(_RA)))$ such that $c|_N = e$.*

Proof. $E(N) = E(Ne) \oplus E(N(1-e))$. Let c be the projection of $E(N)$ onto $E(Ne)$. Then $cx = c(xe) + c(x(1-e)) = c(xe) = xe$, for all $x \in N$. □

Proposition 4.93. *Assume that $_RA \leq {_RN} \leq E(_RA)$. If $_RN$ is a C_{11}-module, then there exists $K \subseteq \mathbf{I}(\mathrm{End}(E(_RA)))$ with the property that for each $X \leq {_RA}$ there* *exists $k \in K$ such that $E(_RA)k$ is a complement of X in $E(_RA)$ and $A\langle K\rangle$ is a C_{11} submodule of $_RN$ which contains $_RA$.*

Proof. Since $0 \leq {}_RA$ we can choose 1 (the unity) of $\mathrm{End}(E(_RA))$ to be in K. The result is now a consequence of Proposition 4.91 and Lemma 4.92. □

Example 4.94. Let $R = \mathbb{Z}$, p be a positive prime integer, and $\mathbb{Q}_p$ be the group of rational numbers with denominators prime to p. Let A be the direct product of the $\mathbb{Q}_p$ as p varies through the positive prime integers. By [BCFG01, Lem. 6.3] and Proposition 4.67, A is not a C_{11}-module. Let $A_k = S_k \oplus P_k$, where S_k is the direct sum of the $\mathbb{Q}_p$ for the first k primes and P_k is the direct product of $\mathbb{Q}_p$ for the remaining primes. As k assumes increasing positive integer values, the A_k form an infinite descending chain of C_{11}-modules that are essential extensions of A. Then from [MM90, p. 19] A_k is not a CS-module for $k > 1$. Note that if in the foregoing construction we replace each $\mathbb{Q}_p$ with $\mathbb{Z}$, then again A is not a C_{11}-module [ST93, Lem. 3.4] (see also §5.1), but each A_k is a CS-module.

Our next objective is to show that condition C_{11} does transfer from a module to its rational hull (see Theorem 3.6). First we prove the following result, which gives a new characterization of property C_{11} in terms of the injective hull.

Proposition 4.95. *Let A be a module. Then A has C_{11} if and only if for each $c \in \mathbf{I}(\mathrm{End}(_RE(A)))$ there exists $e \in \mathbf{I}(\mathrm{End}(_RE(A)))$ such that $Ae \leq A$, $E(A)c \cap E(A)e = 0$, and $E(A)c \oplus E(A)e = E(A)$.*

Proof. ($\Rightarrow$) Assume A has C_{11} and $c \in \mathbf{I}(\mathrm{End}(_RE(A)))$. Let $X = E(A)c \cap A$. Then there exists $f \in \mathbf{I}(\mathrm{End}(_RE(A)))$ such that Af is a complement of X. Let $e \in \mathbf{I}(\mathrm{End}(_RE(A)))$ such that $E(A)e = E(Af)$ is the canonical projection homomorphism. Then $Ae = Af \leq A$, so $E(A)e \cap E(A)c = 0$. Since $E(A)c \oplus E(A)e$ is injective and essential in $E(A)$, we have $E(A)c \oplus E(A)e = E(A)$.

($\Leftarrow$) Conversely, assume that for each $c \in \mathbf{I}(\mathrm{End}(_RE(A)))$ there exists $e \in \mathbf{I}(\mathrm{End}(_RE(A)))$ such that $Ae \leq A$, $E(A)c \cap E(A)e = 0$, and $E(A)c \oplus E(A)e = E(A)$. Let $_RX \leq {}_RA$. Then $E(X) = E(A)c$ for some $c \in \mathbf{I}(\mathrm{End}(_RE(A)))$. Hence, there exists $e \in \mathbf{I}(\mathrm{End}(_RE(A)))$ such that $Ae \leq A$, $E(A)c \cap E(A)e = 0$, and $E(A)c \oplus E(A)e = E(A)$. Let $f = e|_A$. Then $f \in \mathbf{I}(\mathrm{End}(_RA))$ and Af is a complement of X. □

Proposition 4.96. *Let $K \leq_{\mathrm{e}} A$ such that K satisfies C_{11} and*

for each $e \in \mathbf{I}(\mathrm{End}(_RK))$ there exists $\overline{e} \in \mathbf{I}(\mathrm{End}(_RA))$ such that $\overline{e}|_K = e$.

Then A has C_{11}.

Proof. Assume K has C_{11}. Let $X \leq A$ and $Y = X \cap K$. There exists $e \in \mathbf{I}(\mathrm{End}(_RK))$ such that Ke is a complement of Y in K. By hypothesis, there exists $\overline{e} \in \mathbf{I}(\mathrm{End}(_RA))$ such that $\overline{e}|_K = e$. Let $0 \neq a\overline{e} \in A\overline{e}$. There exists $r \in R$ such that $0 \neq ra\overline{e} \in K$. Then $0 \neq ra\overline{e} = (ra\overline{e})\overline{e} = (ra\overline{e})e \in Ke$, so $Y \oplus Ke \leq_{\mathrm{e}} X \oplus A\overline{e} \leq_{\mathrm{e}} A$. Since $A\overline{e}$ is closed in A, $A\overline{e}$ is a complement of X in A. Therefore, A has C_{11}. □

Corollary 4.97. *Let A be a left R-module. If A has C_{11}, then so does the rational hull $\widetilde{E}(A)$ of A.*

Proof. Let $e \in \mathbf{I}(\mathrm{End}(_RA))$. By [Lam99, Prop. 8.13], there exist homomorphisms $\overline{e} : \widetilde{E}(A) \to \widetilde{E}(A)$ and $\overline{\overline{e}} : E(A) \to E(A)$ such that $\overline{\overline{e}}|_{\widetilde{E}(A)} = \overline{e}$ and $\overline{\overline{e}}|_A = e$. Let $\pi : E(A) \to E(Ae)$ be the canonical projection. We claim that $\overline{e} \in \mathbf{I}(\mathrm{End}(\widetilde{E}(A))$. Since $\pi|_A = e$, we have $[\pi - \overline{\overline{e}}](A) = 0$. So $\pi|_{\widetilde{E}(A)} = \overline{\overline{e}}|_{\widetilde{E}(A)} = \overline{e}$. Hence $\overline{e} \in \mathbf{I}(\mathrm{End}(\widetilde{E}(A))$. Thus the result follows from Proposition 4.96. □

Exercises

4.18. Let $\mathbb{N} = \{1, 2, 3, \dots\}$ and K be a field. Let $R = \{f : \mathbb{N} \times \mathbb{N} \to K : \forall i \in \mathbb{N}\ \exists n_i \in \mathbb{N} \text{ such that } f(i,j) = 0 \quad (\forall j \geq n_i)\}$. Given $f, g \in R$ define $f + g$ and fg by

$$(f+g)(i,j) = f(i,j) + g(i,j)$$

and

$$(fg)(i,j) = \sum_{k=1}^{\infty} f(i,k)g(k,j), \quad \text{where} \quad (i,j) \in \mathbb{N} \times \mathbb{N}.$$

Then R is a ring with identity $1 : \mathbb{N} \times \mathbb{N} \to K$ defined by

$$1(i,j) = \delta_{ij} = \begin{cases} 1, & \text{if } i = j, \\ 0, & \text{if } i \neq j. \end{cases}$$

(In fact, R is the *ring of row-finite matrices.*) Then

(i) Show that R is a prime ring.
(ii) Show that $e_{11}R$ is a minimal right ideal of R.
(iii) Show that R is not right Artinian.
(iv) Prove that $\mathrm{soc}\, R_R$ is not finitely generated.
(v) Deduce from (iv) that $\mathrm{soc}\, R_R$ is not a direct summand of R.

4.19. If a ring R is right Noetherian and right self-injective then it is said to be a QF (*quasi-Frobenius*)-ring. Prove that if R is a QF-ring, then so is $M_m(R)$.

4.20. Find an example which shows that the assumption that $_RM$ is faithful in Proposition 4.86 is not superfluous.

4.6 The CS property via classes of submodules

In this section, we introduce a framework which encompasses most of the generalizations of the CS property and allows us to target specific sets of submodules of a module for application of the CS property.

Let $\emptyset \neq \underline{C} \subseteq L(A)$, where $L(A)$ denotes the set of all submodules of A. We say A is *$\underline{C}$-extending*, if for each $X \in \underline{C}$ there exists a direct summand D of A such that X is essential in D. This concept was introduced by Oshiro in [Oshi83] with slightly different terminology and notation. However, in the results of his article, Oshiro assumed that $\underline{C}$ is closed under isomorphisms and essential extensions. In [DS98], Dogruoz and Smith introduced type 1 χ-extending and type 2 χ-extending modules, where they assume that χ is a class of modules containing the zero submodule and closed under isomorphisms (see Section 2.2). In contrast to [DS98] and [Oshi83], we do not implicitly assume that $\underline{C}$ is closed under isomorphisms or essential extensions. In fact, we consider various sets $\underline{C}$ which are not closed under isomorphisms or essential extensions (e.g., $\underline{C}$ = {fully invariant submodules}). Thus for $\underline{C} = L(A)$, $\underline{C} = \{X \in L(A) : X \text{ is uniform}\}$, $\underline{C} = \{X \in L(A) : \operatorname{soc} X = X\}$, $\underline{C} = \{X \leq A : X \text{ is } \tau\text{-dense submodule of A}\}$ and $\underline{C} = \{X \in L(A) : X \text{ is fully invariant in } A\}$ we get that A is extending, A is uniform extending, A is weak CS, A is τ-CS [CC07], and A is FI-extending, respectively. In [LOR98], Lopez-Permouth, Oshiro and Rizvi defined A to be N-extending for a module N when A is $\underline{C}$-extending where $\underline{C} = \{B \leq A : \text{there exists } X \leq N \text{ and } f \in \operatorname{Hom}(X, A) \text{ such that } f(X) \leq_{\mathrm{e}} B\}$.

Observe that the invariance of certain submodules with respect to some subset of endomorphisms of a module is useful and often related to the extending property. For example, if A is injective, then all fully invariant submodules are quasi-injective and all projection invariant submodules are quasi-continuous (Exercise 4.21). Recall from [Fu70] that a submodule X of A is *projection invariant* if for each $e = e^2 \in \operatorname{End}(_RA)$, $Xe \subseteq X$. This motivates us to define $\underline{H}\,\underline{I}(A) = \{X \leq A : X\underline{H} \subseteq X, \text{ with } \emptyset \neq \underline{H} \subseteq \operatorname{End}(_RA)\}$, where $X\underline{H}$ denotes the submodule of A generated by $\{\alpha(x) : x \in X \text{ and } \alpha \in \underline{H}\}$. Then $X \in \underline{P}\,\underline{I}(A)$ (resp., $\underline{E}_A\,\underline{I}(A)$) is the same as saying X is a projection (resp., fully) invariant submodule of A, where $\underline{P} = \{e : e = e^2 \in \operatorname{End}(_RA)\}$ and $\underline{E}_A = \operatorname{End}(_RA)$. We say that A is *$\underline{H}\,\underline{I}$-extending* if A is $\underline{C}$-extending for $\underline{C} = \underline{H}\,\underline{I}(A)$. Note that if $\underline{H} = \{1\}$, $\underline{H} = \underline{P}$, or $\underline{H} = \operatorname{End}(_RA)$, then A being $\underline{H}\,\underline{I}$-extending means that A is extending, $\underline{P}\,\underline{I}$-extending, or FI-extending, respectively.

Our investigation focuses on the behavior of $\underline{C}$-extending modules with respect to direct sums and direct summands. We obtain various well-known results about extending modules and generalizations as corollaries of our results thereby demonstrating the unifying nature of our framework. Also, by applying our results to certain sets of submodules of A such as $\underline{P}\,\underline{I}(A)$, we obtain new results, such as a characterization of the $\underline{P}\,\underline{I}$-extending Abelian groups.

Let $\underline{A}_A$, $\underline{E}_A$, and $\underline{P}$ denote the set of automorphisms of A, the ring of endomorphisms of A, and the set of idempotent elements of the ring of endomorphisms, respectively. (Subscripts may be omitted when the context is clear.) So $\underline{P}_A = \mathbf{I}\,(\underline{E}_A)$. (see §4.5).

A module A is called *polyform* if every essential submodule of A is dense (or rational) in A [Wis96]. Recall a submodule X is *dense* (or *rational*) in A if

$a, b \in A$ with $0 \neq a$ implies that there exists $r \in R$ such that $0 \neq ra$ and $rb \in X$. A submodule X of ${}_RA$ is *pure* in A if $LA \cap X = LX$ for each right ideal L of R [AF74]. Recall that $S_l(R) = \{e = e^2 \in R : xe = exe \text{ for all } x \in R\}$ is the set of left semicentral idempotents of R and $S_r(R) = \{c = c^2 \in R : cx = cxc \text{ for all } x \in R\}$ is the set of right semicentral idempotents of R [Birk83].

Let us provide several fundamental results and examples of $\underline{H}$-invariant submodules of A (i.e., of $\underline{H}\,\underline{I}(A)$). These results and examples support results throughout this section and the rest of this Chapter, especially applications to those involving the $\underline{P}\,\underline{I}$-extending property.

Proposition 4.98.

(i) *If A is indecomposable, then $L(A) = \underline{P}\,\underline{I}(A)$.*
(ii) *If A is a distributive module, then $L(A) = \underline{P}\,\underline{I}(A)$.*
(iii) *Let H be an indecomposable submodule of A. Then $H \in \underline{P}\,\underline{I}(A)$ if and only if $A = L \oplus N \Rightarrow H \subseteq L$ or $H \subseteq N$.*
(iv) *If $A = R$ and R is Abelian, then $L(A) = \underline{P}\,\underline{I}(A)$.*

Proof. (i) Since $\{0, 1\} = \underline{P} \subseteq \underline{E}_A$, $L(A) = \underline{P}\,\underline{I}(A)$.

(ii) Let $X \leq A$ and $e \in \underline{P}$. Then $X = X \cap A = X \cap (Ae \oplus A(1-e)) = X \cap Ae \oplus X \cap A(1-e) = Xe \oplus X(1-e)$. Hence $Xe \subseteq X$. Therefore, $L(A) = \underline{P}\,\underline{I}(A)$.

(iii) Since $H \in \underline{P}\,\underline{I}(A)$, $H = (H \cap L) \oplus (H \cap N)$. Since H is indecomposable, we have either $H = H \cap L$ or $H = H \cap N$. Hence $H \subseteq L$ or $H \subseteq N$. Now let $f^2 = f \in \underline{P}_A$. Then $A = Af \oplus A(1-f)$. By hypothesis, $H \subseteq Af$ or $H \subseteq A(1-f)$. If $H \subseteq Af$, then $Hf \subseteq H$, so we are finished. So assume that $H \subseteq A(1-f)$. Then $Hf = \{0\} \subseteq H$. Therefore $H \in \underline{P}\,\underline{I}(A)$.

(iv) is straightforward. □

Our next example exhibits modules with projection invariant submodules which are not fully invariant.

Example 4.99.

(i) Let $A = R$ be a simple domain (e.g., Weyl algebras, $A_n(K)$, over a field K of characteristic zero see [Lam99, p. 45]) which is not a division ring. Then $\{\{0\}, R\} = \underline{E}_A\underline{I}(A) \subsetneqq \underline{P}\,\underline{I} = L(A)$.
(ii) Let B be a ring such that B_B is indecomposable and B has a $0 \neq X \leq B$ such that X is not an ideal of B (e.g., B is the free ring in two indeterminates, or B is a simple domain that is not a division ring). Let I be the ideal of B generated by X. Take $A = R = T_2(B)$. Then $\begin{bmatrix} X & X \\ 0 & 0 \end{bmatrix}$ and $\begin{bmatrix} X & I \\ 0 & X \end{bmatrix} \in \underline{P}\,\underline{I}(A)$, since $\underline{P} = \{0, \begin{bmatrix} 1 & x \\ 0 & 0 \end{bmatrix}, \begin{bmatrix} 0 & y \\ 0 & 1 \end{bmatrix}, \begin{bmatrix} 1 & 0 \\ 0 & 1 \end{bmatrix} : x, y \in B\}$. However, neither $\begin{bmatrix} X & X \\ 0 & 0 \end{bmatrix}$ nor $\begin{bmatrix} X & I \\ 0 & X \end{bmatrix}$ are fully invariant, since they are not ideals of R.

(iii) Let B and X be as in (ii), and take $A = R = B \oplus B$. Then $X \oplus 0$, $0 \oplus X \in \underline{P}\,\underline{I}(A)$. However, neither are fully invariant. This example illustrates Proposition 4.98 (iii).

(iv) Take $A = R$ with R an Abelian ring. Then any right ideal X that is not an ideal satisfies $X \in \underline{P}\,\underline{I}(A)$, but $X \notin \underline{E}_A\underline{I}(A)$ (i.e., $X \ntrianglelefteq A$).

The following lemma is well known (see, e.g., [Wis96, p. 88]).

Lemma 4.100. *If a module A is polyform (e.g., $Z(A) = 0$), then A is a* UC*-module.*

The following Proposition corrects an error in the proof of [BTY14, Prop. 2.4].

Proposition 4.101. *Let A be a polyform module and $\emptyset \neq \underline{H} \subseteq \underline{C}$. If $X \in \underline{H}\,\underline{I}(A)$ and K is an essential closure of X, then $K \in \underline{H}\,\underline{I}(A)$.*

Proof. Assume to contrary that $K \notin \underline{H}\,\underline{I}(A)$. Then there exist $\alpha \in \underline{H}$ and $k \in K$ such that $\alpha(k) \notin K$. We claim that $X \leq_{\rm e} X + R\alpha(k)$. To see this, let $x + s\alpha(k) \in X + R\alpha(k)$, where $x \in X$, $s \in R$, and $x + s\alpha(k) \notin X$. Let $L = \{r \in R : r(x+sk) \in X\}$. Since A is polyform, $X \oplus L$ is dense in A. So there exists $t \in R$ such that $t(x + s\alpha(k)) \neq 0$ and $t(sk) \in X \oplus L$. By [BPR02, Lem. 1.3 (i)] $t \in \{v \in R : vsk \in X\}$. Hence $tsk \in X$. So we have $0 \neq t(x + s\alpha(k))$ and $tsk \in X$. Then $0 \neq t(x + s\alpha(k)) \in X$. Thus the claim is true. By Lemma 4.100, $X + R\alpha(k) \subseteq K$, a contradiction. Therefore, $K \in \underline{H}\,\underline{I}(A)$. □

Next we obtain characterizations of the $\underline{C}$-extending condition for an arbitrary module, as well as for a polyform module. We apply these results to obtain characterizations for $\underline{P}\,\underline{I}$-extending modules.

Theorem 4.102. *Let $\underline{C} \subseteq L(A)$. Then the following statements are equivalent.*

(i) *A is $\underline{C}$-extending.*

(ii) *For each $X \in \underline{C}$, there exists $e = e^2 \in \underline{E}_{E(A)}$ such that $X \leq_{\rm e} E(A)e$ and $Ae \leq A$.*

(iii) *For each $X \in \underline{C}$, there exist a complement submodule K of A and a complement L of K in A such that $X \leq_{\rm e} K$ and $K \oplus L \in \mathrm{Lift}_A(A)$.*

Proof. (i) $\Rightarrow$ (ii) Assume that A is $\underline{C}$-extending and $X \in \underline{C}$. There exists a direct summand D of A such that $X \leq_{\rm e} D$ and $A = B \oplus D$ for some complement B of D. Then $E(A) = E(B) \oplus E(D)$. Let $e : E(A) \to E(D)$ be the projection. For $a \in A$, if $a = b + d$ with $b \in B$ and $d \in D$, then $e(a) = e(d) = d$. Therefore, $X \leq_{\rm e} E(D) = E(A)e$ and $Ae \subseteq D \subseteq A$.

(ii) $\Rightarrow$ (i) Let $X \in \underline{C}$. Then $X \leq_{\rm e} E(A)e$ and $Ae \subseteq A$ for some $e = e^2 \in \underline{E}_{E(A)}$. Then $X \leq_{\rm e} A \cap E(A)e = Ae$. Since $A(1-e) \subseteq A$, Ae is a direct summand of A. Therefore, A is $\underline{C}$-extending.

(i) $\Leftrightarrow$ (iii) This equivalence follows from Lemma 3.97. □

Corollary 4.103. *Let A be a module. The following conditions are equivalent.*

(i) *A is $\underline{P}\,\underline{I}$-extending (resp., FI-extending).*

(ii) *Each $X \in \underline{P}\,\underline{I}(A)$ (resp., $\underline{E}_A\underline{I}(A)$) has a complement which is a direct summand of A.*

(iii) *For each $X \in \underline{P}\,\underline{I}(A)$ (resp., $\underline{E}_A\underline{I}(A)$), there exists $e = e^2 \in \underline{E}_{E(A)}$ such that $X \leq_{\mathrm{e}} e(E(A))$ and $e(A) \leq A$.*

(iv) *For each $X \in \underline{P}\,\underline{I}(A)$ (resp., $\underline{E}_A\underline{I}(A)$), there exist a complement submodule K of A and a complement L of K in A such that $X \leq_{\mathrm{e}} K$ and $K \oplus L \in \mathrm{Lift}_A(A)$.*

Proof. The equivalences (i) $\Leftrightarrow$ (iii) $\Leftrightarrow$ (iv), follow from Theorem 4.102, where $\underline{C} = \underline{P}\,\underline{I}(A)$ (resp., $\underline{E}_A\underline{I}(A)$).

(i) $\Leftrightarrow$ (ii) First assume that A is $\underline{P}\,\underline{I}$-extending and $X \in \underline{P}\,\underline{I}(A)$. Then there exists $e = e^2 \in \underline{E}_A$ such that $X \leq_{\mathrm{e}} Ae$, and hence $A(1-e)$ is the desired complement. Conversely, let Ac be a complement of X, where $c = c^2 \in \underline{E}_A$. Let $x \in X$. Then $x = xc + x(1-c)$. Since $X \in \underline{P}\,\underline{I}(A)$, $xc \in X \cap Ac = 0$. Thus $X \subseteq A(1-c)$, whence $X \leq_{\mathrm{e}} A(1-c)$. The proof for the FI-extending case is similar. □

Note that the FI-extending version of Corollary 4.103 (i), (ii), and (iv) appears in [BT07, Lem. 1.1], and Corollary 4.103 (iii) appears in [BMR02, Prop. 1.7].

Proposition 4.104. *Let $\underline{C} \subseteq L(A)$ such that $\underline{C}$ is closed under essential closures. Then A is $\underline{C}$-extending if and only if every complement submodule of A which is contained in $\underline{C}$ is a direct summand of A.*

Proof. The proof is straightforward. □

The following corollary is well known.

Corollary 4.105. *A module A is extending (resp., uniform extending) if and only if every complement submodule (resp., uniform submodule) is a direct summand.*

Proof. Take $\underline{C} = L(A)$ (resp., $\underline{C} = \{X \in L(A) : X \text{ is uniform}\}$) in Proposition 4.104. □

Corollary 4.106. *Let A be a polyform module and $\emptyset \neq \underline{H} \subseteq \underline{E}_A$. Then A is $\underline{H}\,\underline{I}$-extending if and only if every complement submodule of A which is contained in $\underline{H}\ \underline{I}(A)$ is a direct summand. In particular, A is $\underline{P}\,\underline{I}$-extending (resp.,* FI-*extending) if and only if every complement submodule of A which is contained in $\underline{P}\,\underline{I}(A)$ (resp., $\underline{E}_A\underline{I}(A)$) is a direct summand.*

Proof. The proof follows from Propositions 4.101 and 4.104. □

The FI-extending version of Corollary 4.106 is [BPR02, Prop. 1.5].

Proposition 4.107. *Let A be a module. Then A is extending $\Rightarrow$ A is $\underline{H}\,\underline{I}$-extending for each $\emptyset \neq \underline{H} \subseteq \underline{E}_A$ $\Rightarrow$ A is* FI-*extending.*

Proof. The proof is straightforward by recalling that for $\underline{H} = \{1_A\}, \underline{HI}(A) = L(A)$, and that $\underline{HI}(A) = \{X \trianglelefteq A\}$ when $\underline{H} = \underline{E_A}$. □

Proposition 4.108. *Let A be a module. Consider the following statements.*

(i) *A is extending;*
(ii) *A is a C_{11}-module;*
(iii) *A is $\underline{PI}$-extending;*
(iv) *A is FI-extending.*

Then (i) $\Rightarrow$ (ii) $\Rightarrow$(iii) $\Rightarrow$ (iv). *Moreover,* (ii) $\not\Rightarrow$ (i) *and* (iv) $\not\Rightarrow$ (iii).

Proof. (i) $\Rightarrow$ (ii) and (iii) $\Rightarrow$ (iv) are clear. Corollary 4.103 shows that (ii) $\Rightarrow$ (iii). Now, we give examples which show that certain converses of these implications, in general, do not hold.

(ii) $\not\Rightarrow$ (i) For example, let $R = T_2(\mathbb{Z})$. It is well known that R_R is not a CS-module (Example 3.84). However, R_R is a C_{11}-module by Corollary 4.28.

(iv)$\not\Rightarrow$ (iii) For example, let R be any domain which is not right Ore. Since R is a prime ring, every nonzero ideal is essential in R. Hence R_R is FI-extending. Since R_R is indecomposable but not uniform, we conclude that R_R is not $\underline{PI}$-extending (see Proposition 4.109 (i)). □

We conjecture that (iii) $\not\Rightarrow$ (ii) in Proposition 4.108; however at this time we have no example to support this conjecture (see Open Problems and Questions 8).

Proposition 4.109.

(i) *Let A be an indecomposable module. Then the following conditions are equivalent.*
  (a) *A is uniform.*
  (b) *A is* CS.
  (c) *A is $\underline{PI}$-extending.*
(ii) *Let A be a module such that $\underline{E_A}$ is Abelian and $X \leq A$ implies that $X = \sum_{i\in I} h_i(A)$, where each $h_i \in \underline{E_A}$. Then A is* CS *if and only if A is $\underline{PI}$-extending. In particular, if $A = R$ is Abelian, then R_R is* CS *if and only if R_R is $\underline{PI}$-extending.*
(iii) *Let A be a distributive module. Then A is* CS *if and only if A is $\underline{PI}$-extending.*

Proof. (i) Clearly, (a) $\Rightarrow$ (b) and (b) $\Rightarrow$ (c). For (c) $\Rightarrow$ (a), assume $0 \neq X \leq A$. Then $X \in \underline{PI}(A)$ because $\underline{E_A}$ has only trivial idempotents. So $X \leq_e A$. Therefore, A is uniform.

(ii) Clearly if A is CS, then A is $\underline{PI}$-extending. Conversely, assume $X \leq A$ and $e = e^2 \in \underline{E_A}$. Then $e(X) = e\sum_{i\in I} h_i A = \sum_{i\in I} h_i(e(A)) \subseteq X$. So $X \in \underline{PI}(A)$. Therefore, A is CS. To see the particular case for $A = R$, observe that $X = \sum_{x\in R} Rx$ and the map $h_x : R \to R$ defined by $h_x(r) = rx$ for all $r \in R$ is an R-endomorphism.

(iii) follows from Proposition 4.98 (ii). □

Using Proposition 4.108, Proposition 4.109 generalizes [ABT09, Prop. 1.8 (i)] and [BT07, Cor. 1.4 (i)].

We shall show that if A is $\underline{C}$-extending, where $\underline{C}$ is closed under direct sums and satisfies a weak hereditary condition, then $A = K \oplus T$, where T is maximal with respect to containing no nonzero $\underline{C}$-submodules of A, but every submodule of K is an essential extension of an element of $\underline{C}$. Applications are made by considering $\underline{C}$ sets consisting of projective submodules, or sets of submodules determined by a preradical.

Recall that in general a direct sum of CS-modules is not CS. However, from Theorem 4.26 and [BMR02, Thm. 1.3] the classes of C_{11}-modules and FI-extending modules are closed under direct sums. This motivates us to ask when certain classes of modules satisfying a generalized extending condition, such as the class of $\underline{P}\underline{I}$-extending modules, are closed under direct sums. Furthermore, we characterize the $\underline{P}\underline{I}$-extending Abelian groups.

Definition 4.110.

(i) Let $T \leq A$. We say T is totally non-$\underline{C}$ if T contains no nonzero submodules of A in $\underline{C}$.

(ii) $\underline{C}$ is called (weakly) hereditary if for each $0 \neq Y \in \underline{C}$ and $0 \neq B \leq Y$, then (there is $0 \neq C \in \underline{C}$ such that $C \leq B$) $B \in \underline{C}$. For example, assume that $\operatorname{soc} A \leq_{\mathrm{e}} A$ and $\emptyset \neq \underline{C} \subseteq L(A)$. If every simple submodule is contained in $\underline{C}$, then $\underline{C}$ is weakly hereditary.

Lemma 4.111. *Assume that A is a module and $\underline{C} \subseteq L(A)$ is such that $\underline{C}$ is closed under direct sums and weakly hereditary. Then there exists $T \leq A$ such that T is maximal among totally non-$\underline{C}$ submodules of A, and K is a complement of T which is maximal with respect to the property that every nonzero submodule of K is an essential extension of an element of $\underline{C}$.*

Proof. First we claim that there exists $T \leq A$ such that T is maximal among totally non-$\underline{C}$ submodules of A. To see this, observe that $\{0\}$ is a totally non-$\underline{C}$ submodule of A. Let $\{T_i\}_{i \in I}$ be a chain of totally non-$\underline{C}$ submodules of A, and let U be the union of the chain. Assume that $0 \neq Y \leq U$ such that $Y \in \underline{C}$. Then there exist $0 \neq y \in Y$ and $0 \neq B \in \underline{C}$ such that $B \leq Ry \leq U$. But then $y \in T_i$ for some i, a contradiction. Thus U is totally non-$\underline{C}$. By Zorn's lemma, T exists. Note that if $T = A$, we are finished by taking $K = 0$. So assume $T \neq A$. There exists $X \in \underline{C}$ such that X is maximal among direct sums of elements of $\underline{C}$ by a Zorn lemma argument (see the proof of Proposition 1 pp. 59–60 [Lamb86]). Observe that $X \cap T = 0$. Let K be a complement of T in A which contains X. We claim that $X \leq_{\mathrm{e}} K$. To see this, assume there exists $0 \neq H \leq K$ such that $X \cap H = 0$. By the maximality of T, there exists $0 \neq B \leq A$ such that $B \in \underline{C}$ and $B \leq H \oplus T$. So $B \cap X = 0$. By the maximality of X, we have a contradiction. Thus $X \leq_{\mathrm{e}} K$. Now let $0 \neq V \leq K$. There exists $0 \neq W \leq V \cap X$ with $W \in \underline{C}$. As above, there exists $X_1 \in \underline{C}$ such that X_1 is maximal among direct sums of elements of $\underline{C}$ in V. Then $X_1 \leq_{\mathrm{e}} V$. □

Lemma 4.112. *Let $\underline{C} \subseteq L(A)$, $X \in \underline{C}$, and $A = A_1 \oplus A_2$. If A is $\underline{C}$-extending and any of the following conditions hold, then $X \leq A_i$ implies that there exists $D_i \leq_{\mathrm{d}} A_i$ such that $X \leq_{\mathrm{e}} D_i$, for $i = 1, 2$.*

(i) *Either A_1 (or A_2) $\trianglelefteq A$.*

(ii) *There exists $K \in \underline{PI}(A) \cap \underline{C}$ such that K has a unique closure in A and $K \leq_{\mathrm{e}} A_1$ (or $K \leq_{\mathrm{e}} A_2$).*

Proof. Since $X \in \underline{C}$, there exists $e = e^2 \in \underline{E}_A$ such that $X \leq_{\mathrm{e}} Ae$. Let $c_i = c_i^2 \in \underline{E}_A$ such that $1_{\underline{E}_A} = c_1 + c_2$ and $A_i = Ac_i$ for $i = 1, 2$.

(i) Assume, without loss of generality, that $A_1 \trianglelefteq A$. By [BMR02, Lem. 1.9], $c_1 \in S_{\mathrm{r}}(\underline{E}_A)$ so $c_2 \in S_{\mathrm{l}}(\underline{E}_A)$.

Case 1. Assume that $X \leq A_1$. Then $X \leq Ae \cap Ac_1$. Since $c_1 e = c_1 e c_1$, $c_1 e = (c_1 e)^2$, so $Ae \cap Ac_1 = Ac_1 e$. Hence $X \leq_{\mathrm{e}} Ac_1 e \leq Ac_1$. Take $D_1 = Ac_1 e$.

Case 2. Assume that $X \leq A_2$. Then $X \leq Aec_2$. Since $ec_2 = c_2 ec_2$, $ec_2 = (ec_2)^2$. Let $0 \neq aec_2 \in Aec_2$. There exists $r \in R$ such that $0 \neq rae \in X$. Then $0 \neq raec_2 = rae \in X$. Thus $X \leq_{\mathrm{e}} Aec_2$. Take $D_2 = Aec_2$.

(ii) Assume that there exists $K \in \underline{PI}(A) \cap \underline{C}$ such that K has a unique closure in A and, without loss of generality, that $K \leq_{\mathrm{e}} A_1$.

Case 1. Assume that $X \leq A_1$. Since $K \in \underline{PI}(A)$, $K = Ke \oplus K(1-e)$. But $Ke = K \cap Ae$. Since $X = Xe \leq_{\mathrm{e}} Ae$, given $0 \neq ae$ there exists $r \in R$ such that $0 \neq rae = x \in A_1$. As $K \leq_{\mathrm{e}} A_1$, there exists $s \in R$ such that $0 \neq sx \in K$. We conclude that $Ke \leq_{\mathrm{e}} Ae$. Hence $K \leq_{\mathrm{e}} Ae \oplus K(1-e) \leq A_1$, since K has a unique closure in A. Thus $Ae \subseteq A_1$, so take $D_1 = Ae$.

Case 2. Assume that $X \leq A_2$. Then $X \leq_{\mathrm{e}} Ae \cap Ac_2$. So $Ke = Ae \cap K \leq_{\mathrm{e}} Ae \cap Ac_1 = 0$. Hence $K = K \cap A(1-e) \subseteq A(1-e)$. Since Ac_1 is the unique closure of K in A, $Ac_1 \subseteq A(1-e)$. Thus $c_1 e = 0$, so $c_2 e = e$. Hence, $ec_2 = (ec_2)^2$. Observe that $Ae \cap Ac_2 \leq Aec_2$. Since $X \leq_{\mathrm{e}} Ae$, $Ae \cap Ac_2 \leq_{\mathrm{e}} Ae$. Let $0 \neq aec_2 \in Aec_2$. There exists $r \in R$ such that $0 \neq rae \in Ae \cap Ac_2$. Hence $0 \neq raec_2 = rae \in Ae \cap Ac_2$, and so $Ae \cap Ac_2 \leq_{\mathrm{e}} Aec_2$. Take $D_2 = Aec_2$. □

Recall from [ABT09], that A is called *G-extending* if for each $X \leq A$ there exists a direct summand D of A such that $X \cap D$ is essential in both X and D. For instance, we will focus on this new class of modules as well as rings as an outer generalization of CS-modules in next chapter. Now we obtain several results on decompositions of $\underline{C}$-extending modules which are related to G-extending notion. Note that, clearly, the CS and G-extending concepts coincide on the class of nonsingular modules (see Exercise 4.23).

Theorem 4.113. *Assume A is $\underline{C}$-extending, where $\underline{C}$ is closed under direct sums and $\underline{C}$ is weakly hereditary. Then $A = K \oplus T$, where T is maximal among the totally non-$\underline{C}$ submodules of A, and K is maximal with respect to the property that every nonzero submodule of K is an essential extension of an element of $\underline{C}$. Moreover, if $K \trianglelefteq A$ or $T \trianglelefteq A$, then K is G-extending.*

Proof. Let K_1 and T_1 be as K and T in Lemma 4.111. Then there exists $X \in \underline{C}$ such that $X \leq_e K_1$. Since A is $\underline{C}$-extending, there is a $e = e^2 \in \underline{E}_A$ such that $X \leq_e Ae$. Then Ae is a complement of T_1. Now Ae is maximal with respect to the property that every nonzero submodule of Ae is an essential extension of an element of $\underline{C}$. Suppose $0 \neq Y \in \underline{C}$ such that $Y \leq A(1-e)$. Then $(X \oplus Y) \cap T_1 \neq 0$, a contradiction since $X \oplus Y \in \underline{C}$. Hence, $A(1-e)$ is totally non-$\underline{C}$. In fact, $A(1-e)$ is maximal among totally non-$\underline{C}$ submodules of $\underline{C}$. Now take $K = Ae$ and $T = A(1-e)$.

Finally, assume that $K \trianglelefteq A$ or $T \trianglelefteq A$ and $B \leq K$. There exists $C \in \underline{C}$ such that $C \leq_e B$. By Lemma 4.112, $C \leq_e D$, where D is a direct summand of K. By [ABT09, Prop. 1.5], K is G-extending. □

Corollary 4.114. *Assume A is $\underline{C}$-extending, where $\underline{C} = \{X \leq A : X$ is projective$\}$ and R is a left semihereditary ring. Then $A = K \oplus T$, where T is maximal among the submodules of A which contain no nonzero projective submodule, and K is maximal with respect to each of its submodules being an essential extension of a projective module. If $T = Z_2(A)$, then K is an extending module.*

Proof. Clearly, $\underline{C}$ is closed under direct sums. From [Lam99, Thm. 2.29], $\underline{C}$ is weakly hereditary. By Theorem 4.113, $A = K \oplus T$. If $T = Z_2(A)$, then $T \trianglelefteq A$. Again by Theorem 4.113, K is G-extending, and by [ABT09, Prop. 1.8 (ii)], K is extending. □

Observe that every finitely generated module A over a Dedekind domain is $\underline{C}$-extending, where $\underline{C} = \{X \leq A : X$ is projective$\}$. So $A = K \oplus T$. In this case, K is projective and $T = Z_2(A)$. Recall that finitely generated modules over Dedekind domains, in general, are not CS (Exercise 3.5).

For the next result recall that a preradical ρ in the category of left R-modules is a subfunctor of the identity functor such that $\rho(A) \trianglelefteq A$ (see Section 2.4). For more details on preradicals, see [Ste75] and [BKN82].

Corollary 4.115. *Let A be a $\underline{C}$-extending module and ρ be a preradical.*

(i) *If $\underline{C} = \{X \leq A : \rho(X) = 0\}$, then $\underline{C}$ is hereditary and closed under direct sums. Hence $A = K \oplus T$, as in Theorem* 4.113. *If ρ is left exact, then $\rho(A) \leq_e T$ and $K \in \underline{C}$.*

(ii) *If $\underline{C} = \{X \leq A : \rho(X) = X\}$ and ρ is left exact, then $\underline{C}$ is hereditary and closed under direct sums. Hence $A = K \oplus T$, as in Theorem* 4.113 *and $\rho(A) \leq_e K$.*

(iii) *If ρ is stable and left exact, then $A = K \oplus T$ with $\rho(A) = T$ and $\rho(A)$ is K-injective in* (i), *while $\rho(A) = K$ in* (ii). *Therefore, K is G-extending in case* (i) *and extending in case* (ii).

Proof. The proof follows from Theorem 4.113, properties of preradicals (see [Ste75, pp. 137–138]) and [BKN82, p. 25], and the following argument that $\rho(A)$ is K-injective in (iii) when $\underline{C} = \{X \leq A : \rho(X) = 0\}$. Let $N \leq A$ such that $N \cap \rho(A) =$

0. Then $A = B \oplus C$, where $N \leq_e B$, so $B \cap \rho(A) = 0$. Hence $\rho(A) \subseteq C$. By the modular law, $A = B \oplus \rho(A) \oplus (C \cap K)$. Since $N \subseteq B \oplus (C \cap K)$, [DHSW94, Lem. 7.5] yields that $\rho(A)$ is K-injective. □

By taking $\rho = Z_2$ and recalling that a nonsingular G-extending module is extending, one can see that Corollary 4.115 generalizes [DS98, Thms. 4.1 (ii) and 4.3 (ii)].

Observe that $\rho(-) = \text{soc}(-)$ is a left exact preradical. Thus A is weak CS if and only if A in $\underline{C}$-extending for $\underline{C} = \{X \leq A : \text{soc}\, X = X\}$. Also, note that any stable left exact preradical is a radical [BKN82, p. 25].

Lemma 4.116. *Let ρ be a stable left exact preradical. Then A is extending if and only if $A = \rho(A) \oplus T$ for some $T \leq A$ such that $\rho(A)$ and T are extending and $\rho(A)$ is T-injective.*

Proof. One can use proof of [DHSW94, 7.11 on p. 80], where we replace $Z_2(A)$ with $\rho(A)$. □

Theorem 4.117. *Let ρ be a stable left exact preradical, $\underline{C}_0 = \{X \leq A : \rho(X) = 0\}$, and $\underline{C}_\rho = \{X \leq A : \rho(X) = X\}$. Then A is both $\underline{C}_0$-extending and $\underline{C}_\rho$-extending if and only if A is extending.*

Proof. Assume A is both $\underline{C}_0$-extending and $\underline{C}_\rho$-extending. Since A is $\underline{C}_\rho$-extending, $A = \rho(A) \oplus T$ and $\rho(A)$ is extending by Corollary 4.115 (iii). Let $X \leq T$. Then $\rho(X) = 0$. Since A is $\underline{C}_\rho$-extending, there exists $c = c^2 \in \underline{E}_A$ such that $X \leq_e Ac$. Also, there exists $e = e^2 \in \underline{E}_A$ such that $T = Ae$ and $\rho(A) = A(1-e)$. Since $A(1-e) \trianglelefteq A$, T and $\rho(A)$ are extending by Lemma 4.112. Now, by Corollary 4.115 (iii) and Lemma 4.116, A is extending. The converse is clear. □

Theorem 4.118. *Let $A = \bigoplus_{j \in J} A_j$, $\underline{C}_j \subseteq L(A_j)$, and $\underline{C} \subseteq L(A)$ such that for each $X \in \underline{C}$ with $\pi_j(X) \subseteq X$ and $\pi_j(X) \in \underline{C}_j$, where π_j are the canonical projections. If each A_j is $\underline{C}_j$-extending, then A is $\underline{C}$-extending.*

Proof. Let $X \in \underline{C}$. Then $X = \bigoplus_{j \in J} \pi_j(X)$, $\pi_j(X) \subseteq X$ and $\pi_j(X) \in \underline{C}_j$ for each $j \in J$. So there exists $e_j = e_j^2 \in \underline{E}_A$ such that $\pi_j(X) \leq_e A_j e_j$. Therefore, since $X \leq_e \bigoplus_{j \in J} A_j e_j$ and $\bigoplus_{j \in J} A_j e_j$ is a direct summand of A, A is $\underline{C}$-extending. □

Corollary 4.119. *Let $A = \bigoplus_{j \in J} A_j$, $\emptyset \neq \underline{H}_j \subseteq \underline{E}_{A_j}$, and $\emptyset \neq \underline{H} \subseteq \underline{E}_A$ be such that $\pi_j \in \underline{H}$ for all $j \in J$, where π_j are the canonical projections, and for each $\alpha_j \in \underline{H}_j$ there exists $\overline{\alpha_j} \in \underline{H}$ such that $\overline{\alpha_j}|_{A_j} = \alpha_j$. If A_j is $\underline{H}_j\underline{I}$-extending for all $j \in J$, then A is $\underline{H}\,\underline{I}$-extending.*

Proof. Let $X \in \underline{H}\,\underline{I}(A)$. Then $\pi_j(X) \subseteq X$, since $\pi_j(X) \in \underline{H}$ for all $j \in J$. Now let $\alpha_j \in \underline{H}_j$. Then there exists $\overline{\alpha_j} \in \underline{H}$ such that $\alpha_j(\pi_j(X)) = \overline{\alpha_j}(\pi_j(X)) \subseteq X$. But $\alpha_j(\pi_j(X)) = \pi_j(\alpha_j(\pi_j(X)) \subseteq \pi_j(X)$, so $\pi_j(X) \in \underline{H}_j\, \underline{I}(A)$. Therefore, by Theorem 4.118, A is $\underline{H}\,\underline{I}$-extending. □

The next corollary shows the versatility of Corollary 4.119.

Corollary 4.120. *Let* $A = \bigoplus_{j\in J} A_j$.

(i) *If each* A_j *is a* $\underline{P}\,\underline{I}$*-extending (resp., FI-extending) module, then* A *is a* $\underline{P}\,\underline{I}$*-extending (resp., FI-extending) module.*

(ii) *Let* $\underline{H} = \underline{P}_A \cup \underline{A}_A$ *and* $\underline{H}_j = \underline{P}_{A_j} \cup \underline{A}_{A_j}$ *for each* $j \in J$. *If each* A_j *is* $\underline{H}_j\,\underline{I}$*-extending, then* A *is* $\underline{H}\,\underline{I}$*-extending.*

Proof. (i) Take $\underline{H} = \underline{P}_A$ (resp., $\underline{H} = \underline{E}_A$) and $\underline{H}_j = \underline{P}_{A_j}$ (resp., $\underline{H}_j = \underline{E}_{A_j}$) in Corollary 4.119.

(ii) Observe that if $e_j \in \underline{P}_{A_j}$, then $e_j\pi_j \in \underline{P}_A$. Also $\bigoplus_{j\in J} a_j\pi_j \in \underline{A}_A$ where each $a_j \in \underline{A}_{A_j}$. Hence this part follows from Corollary 4.119. □

Note that the FI-extending case of Corollary 4.120 (i) is [BMR02, Thm. 1.3].

Proposition 4.121. *Let* ${}_RA$ *be a module and* $f^2 = f \in \underline{E}_A$. *Then the following statements are equivalent.*

(i) Af *is a fully invariant submodule of* A *(i.e.,* $Af \trianglelefteq A$*);*

(ii) Af *is a* $\underline{P}$*-invariant submodule of* A *(i.e.,* $Af \in \underline{P}\,\underline{I}(A)$*);*

(iii) $f \in S_{\mathrm{r}}(\underline{E}_A)$.

Proof. (i) $\Rightarrow$ (ii) This implication is clear.

(ii) $\Rightarrow$ (iii) Let $h \in \underline{E}_A$. Then $(f + fh - fhf)^2 = f + fh - fhf$. Let $a \in A$. Since Af is $\underline{P}$-invariant, $[f + fh - fhf]f(a) = [f + fh - fhf](a) \in f(A)$. Then $[f + fh - fhf](a) = ([f + fh - fhf]f)(a) = f(a)$, hence $fh(a) = fhf(a)$. So $fh = fhf$. Therefore, $f \in S_{\mathrm{r}}(\underline{E}_A)$.

(iii) $\Rightarrow$ (i) This implication is established in [BMR02, Lem. 1.9]. □

Lemma 4.122. *Assume that* $A = A_1 \oplus A_2$ *with* $A_1 \in \underline{P}\,\underline{I}(A)$. *If* $X \in \underline{P}\,\underline{I}(A_2)$ *(resp.,* $X \trianglelefteq A_2$*), then* $A_1 \oplus X \in \underline{P}\,\underline{I}(A)$ *(resp.,* $A_1 \oplus X \trianglelefteq A$*).*

Proof. By Proposition 4.121, $A_1 = Af$ and $A_2 = A(1 - f)$ with $f \in S_{\mathrm{r}}(\underline{E}_A)$. Assume $X \in \underline{P}\,\underline{I}(A_2)$. Let $e = e^2 \in \underline{E}_A$ and $a + x \in A_1 \oplus X$. Then $e(a + x) = e(a) + ef(x) + e(1 - f)(x)$. Now $e(a) = fe(a) = fef(a) \in f(A)$ since $f \in S_{\mathrm{r}}(\underline{E}_A)$. Then $e(a) + ef(x) \in f(A)$. Also, $e(1 - f) = (1 - f)e(1 - f) = [e(1 - f)]^2$, since $(1 - f) \in S_{\mathrm{l}}(\underline{E}_A)$. Hence, $e(1 - f) = [e(1 - f)]^2 \in \underline{E}_{A_2}$. Thus $e(1 - f)(x) \in X$. Consequently, $A_1 \oplus X \in \underline{P}\,\underline{I}(A)$. By a similar argument, we see that $X \trianglelefteq A_2$ implies $A_1 \oplus X \trianglelefteq A$. □

Proposition 4.123. *Let* $K \in \underline{P}\,\underline{I}(A_2)$ *(resp.,* $K \trianglelefteq A$*) such that* K *is complement in* A. *Then* A *is* $\underline{P}\,\underline{I}$*-extending if and only if* $A = K \oplus N$, *where* K *and* N *are* $\underline{P}\,\underline{I}$*-extending (resp.,* FI*-extending).*

Proof. Assume A is $\underline{P}\,\underline{I}$-extending (the case where A is FI-extending is similar). Clearly, $A = K \oplus N$. By Proposition 4.121, there exists $f \in S_{\mathrm{r}}(\underline{E}_A)$ such that $K = Af$ and $N = A(1 - f)$. Suppose $X \in \underline{P}\,\underline{I}(K)$ and $e = e^2 \in \underline{E}_A$. Then $fe = fef = (fe)^2 \in \underline{E}_K$. Hence, $e(X) = fe(X) \subseteq X$, and so $X \in \underline{P}\,\underline{I}(A)$. By Lemma 4.122, K is $\underline{P}\,\underline{I}$-extending. Now suppose $Y \in \underline{P}\,\underline{I}(N)$. By Lemma 4.122,

$K \oplus Y \in \underline{P}\underline{I}(A)$. Hence there exists $D \leq_{\text{d}} A$ such that $K \oplus Y \leq_{\text{e}} D$. By the modular law, $D = D \cap A = D \cap (K \oplus N) = K \oplus (D \cap N)$. So $Y \leq_{\text{e}} D \cap N \leq_{\text{d}} N$. Therefore, N is $\underline{P}\underline{I}$-extending. The converse follows from Corollary 4.120 (i). □

Corollary 4.124. *A is $\underline{P}\underline{I}$-extending (resp., FI-extending) if and only if $A = A_1 \oplus A_2$, where each A_i is $\underline{P}\underline{I}$-extending (resp., FI-extending) and $A_1 = Z_2(A)$.*

Proof. This result follows from Proposition 4.123 since $Z_2(A) \trianglelefteq A$ and $Z_2(A)$ is the closure of $Z(A)$. □

We note that the FI-extending version of Proposition 4.123 is [BCFG01, Lem. 1.2] and the FI-extending ring version of Corollary 4.124 is [BMR02, Prop. 1.11].

Corollary 4.125. *Assume that A is polyform and $K \in \underline{P}\underline{I}(A)$. Then A is $\underline{P}\underline{I}$-extending if and only if $A = A_1 \oplus A_2$, where each A_i is $\underline{P}\underline{I}$-extending and $K \leq_{\text{e}} A_1$.*

Proof. This corollary is a consequence of Propositions 4.101 and 4.123. □

Note that if in Corollary 4.125 $K = \text{soc}\, A$, then A_1 has essential socle and A_2 has zero socle. Recall from [Fu73], that an Abelian group A is said to be *separable* if every finite subset $\{a_1, \ldots, a_n\}$ of A can be embedded in a direct summand S of A such that S is a direct sum of groups of rank 1.

Lemma 4.126. *Let A be a torsion Abelian group. Then A is $\underline{P}\underline{I}$-extending if and only if it is a direct sum of a divisible group and separable p-groups.*

Proof. Assume that A is $\underline{P}\underline{I}$-extending. Then A is FI-extending. From [BCFG01, Thm. 2.3], A is a direct sum of a divisible group and separable p-groups. Conversely, assume that A is a direct sum of a divisible group and separable p-groups. By [BCFG01, Thms. 2.3 and 3.2], each separable p-group is FI-extending. From [Me77], each projection invariant subgroup of a separable p-group is fully invariant. Hence each separable p-group is $\underline{P}\underline{I}$-extending. Corollary 4.120 (i) yields that A is $\underline{P}\underline{I}$-extending. □

We can now give a characterization of $\underline{P}\underline{I}$-extending Abelian groups.

Theorem 4.127. *Let A be an Abelian group. Then A is $\underline{P}\underline{I}$-extending if and only if $A = D \oplus T \oplus F$, where D is a divisible group, T is a direct sum of separable p-groups, and F is a torsion free group such that each of its projection invariant pure subgroups is a direct summand.*

Proof. Assume that A is $\underline{P}\underline{I}$-extending. By Corollary 4.124, $A = A_1 \oplus A_2$, where A_1 is the torsion subgroup of A and A_2 is torsion free and each A_i is $\underline{P}\underline{I}$-extending. By Lemma 4.126, Corollary 4.106, and the fact that a submodule of A is pure if and only if it is a complement, $A = D \oplus T \oplus F$ with the indicated properties. The converse follows from Lemma 4.126 and Corollaries 4.106 and 4.120. □

In the rest of this section, we investigate when various $\underline{C}$-extending conditions are inherited by direct summands or submodules. Let $\underline{C} \subseteq L(A)$ and $X \leq A$. Then we use $\underline{C}_X$ to denote $\underline{C} \cap L(X)$. The next theorem was stated in [Oshi83, Prop. 1.4].

Theorem 4.128. *Let $\underline{C} \subseteq L(A)$ be closed under closures. If A is $\underline{C}$-extending and $D \leq_{\mathrm{d}} A$, then D is $\underline{C}_D$-extending.*

Proof. Let $X \leq D$ with $X \in \underline{C}$, and let K be an essential closure of X with $K \leq D$. By Proposition 4.104, $K \leq_{\mathrm{d}} A$. Hence, D is $\underline{C}_D$-extending. □

Corollary 4.129. *A direct summand of an extending or uniform extending module A is extending or uniform extending, respectively.*

Proof. In Theorem 4.128, for A extending take $\underline{C} = L(A)$ and for A uniform extending take $\underline{C} = \{X \in L(A) : X \text{ is uniform}\}$. □

Recall from [Smi90] that a module is called a CESS-*module* if every complement submodule with essential socle is a direct summand. We say a module A is $\underline{Z_0}$-*extending* if it is $\underline{C}$-extending, where $\underline{C} = \{X \in L(A) : Z(X) = 0\}$. Our next corollary is an application of Theorem 4.128.

Corollary 4.130. *Let A be either a* CESS*-module or a $\underline{Z_0}$-extending module and $D \leq_{\mathrm{d}} A$. Then D is a* CESS*-module or a $\underline{Z_0}$-extending module, respectively.*

Proof. Let $\underline{C}_1 = \{X \in L(A) : \operatorname{soc} X \leq_{\mathrm{e}} X \text{ and } X \text{ is complement}\}$, and $\underline{C}_2 = \{X \in L(A) : Z(X) = 0\}$. It is readily seen that $\underline{C}_1$ and $\underline{C}_1$ are closed under essential extensions in A and that A is a CESS-module if and only if A is a $\underline{C}_1$-extending module. The result is now a consequence of Theorem 4.128. □

The CESS version of Corollary 4.130 is [Smi90, Cor. 1.3]. Observe that over a commutative principal ideal domain, every finitely generated module is $\underline{Z_0}$-extending but, in general, not extending (Example 4.2).

Corollary 4.131. *Let A be a polyform module and $\emptyset \neq \underline{H} \subseteq \underline{E}_A$ (e.g., $\underline{H} = \underline{P}$). If A is $\underline{H}\,\underline{I}$-extending and $D \leq_{\mathrm{d}} A$, then D is $\underline{H}\,\underline{I}(A) \cap L(D)$-extending.*

Proof. This is a consequence of Proposition 4.101 and Theorem 4.128. □

Observe that for $D \leq_{\mathrm{d}} A$, $\underline{P}\,\underline{I}(A) \cap L(D) \subseteq \underline{P}\,\underline{I}(D)$. In general, this inclusion is proper, as is seen in the following example of a $\underline{P}\,\underline{I}$-extending module A which has a direct summand D that is FI-extending but not $\underline{P}\,\underline{I}$-extending. Hence the class of $\underline{P}\,\underline{I}$-extending modules is not closed under direct summands.

Example 4.132. Let $n \geq 3$ be an odd integer. Let $\mathbb{R}$ be the real field and S the polynomial ring $\mathbb{R}[x_1, x_2, \ldots, x_n]$. Then the ring $R = S/Ss$, where $s = (\sum_{i=1}^{n} x_i^2) - 1$, is a commutative Noetherian domain. Consider the module $_RA = R^{(n)}$. Then $_RA$ is a $\underline{P}\,\underline{I}$-extending module which contains a direct summand that is FI-extending but not $\underline{P}\,\underline{I}$-extending.

To see this, observe that A is $\underline{P}\underline{I}$-extending by Corollary 4.120 (i). Let $\varphi : A \to R$ be the homomorphism defined by $\varphi(a_1 + Ss, a_2 + Ss, \ldots, a_n + Ss) = a_1x_1 + a_2x_2 + \cdots + a_nx_n + Ss$ for all $a_1, a_2, \ldots, a_n \in S$. Clearly, φ is an epimorphism and hence $A = \ker\varphi \oplus Y$, where $Y \cong R$. Take $D = \ker\varphi$. Then ${}_RD$ has uniform dimension $n-1 \geq 2$, so ${}_RD$ is not uniform. But ${}_RD$ is indecomposable (see Example 4.33). By Proposition 4.109 (i), ${}_RD$ is not $\underline{P}\underline{I}$-extending. Thus, by Corollary 4.131, $\underline{P}\underline{I}(A) \cap L(A) \neq \underline{P}\underline{I}(D)$. Since $Z({}_RA) = 0$, ${}_RD$ is FI-extending by [BPR02, Prop. 1.5 and Thm. 2.4]. Also observe that ${}_RD$ is an essential extension of a $\underline{P}\underline{I}$-extending module K, where K is the direct sum of $n-1$ uniform submodules of ${}_RD$. Hence the class of $\underline{P}\underline{I}$-extending modules is not closed under essential extensions.

Note that the following problem, posed in [BMR02, p. 1414], remains open: Show that any direct summand of an FI-extending module is FI-extending, or provide a counterexample.

Proposition 4.133. *Let A be a $\underline{C}$-extending module and $D \leq_{\mathrm{d}} A$. If A satisfies any of the following conditions, then A is $\underline{C}_D$-extending.*

(i) *A has* SIP.
(ii) *A satisfies C_2.*

Proof. Let $X \in \underline{C} \cap L(D)$. Then there exists $e = e^2 \in \underline{E}_A$ such that $X \leq_{\mathrm{e}} Ae$. If A has SIP, then $X \leq_{\mathrm{e}} Ae \cap D \leq_{\mathrm{d}} D$. Therefore, A is $\underline{C}_D$-extending. Now assume that A satisfies C_2. Let $\pi : A \to D$ be the projection homomorphism. Then $\pi|_{Ae}$ is a monomorphism which extends the identity map on X. Hence $X \leq_{\mathrm{e}} \pi(Ae) \leq D$. By C_2, $\pi(Ae) \leq_{\mathrm{d}} D$. Therefore, A is $\underline{C}_D$-extending. □

Proposition 4.134. *Let A be a $\underline{C}$-extending module where $\underline{C} \subseteq L(A)$ and $N \leq A$. If any of the following conditions hold, then N is $\underline{C}_N$-extending.*

(i) *For each $e = e^2 \in \underline{E}_A$, there exists $f = f^2 \in \underline{E}_N$ such that $N \cap Ae \leq_{\mathrm{e}} Nf$.*
(ii) *For each $e = e^2 \in \underline{E}_A$, there exists $f = f^2 \in \underline{E}_A$ such that $Ae = Af$ and $Nf \subseteq N$. In particular, if $N \in \underline{P}\underline{I}(A)$, then N satisfies the condition.*

Proof. For $X \in \underline{C}_N$, there exists $e = e^2 \in \underline{E}_A$ such that $X \leq_{\mathrm{e}} Ae$. If (i) holds, then $X \leq_{\mathrm{e}} N \cap Ae \leq_{\mathrm{e}} Nf$, for some $f = f^2 \in \underline{E}_N$. Therefore, N is $\underline{C}_N$-extending. If (ii) holds, then $X \leq_{\mathrm{e}} N \cap Ae = N \cap Af = Nf \subseteq N$, for some $f = f^2 \in \underline{E}_A$. Therefore, N is $\underline{C}_N$-extending. □

Exercises

4.21. Let A be an injective module. Prove the following:

(i) all fully invariant submodules of A are quasi-injective;
(ii) all projection invariant submodules of A are quasi-continuous.

4.22. Prove that every nonsingular module is polyform.

4.23. Show that for nonsingular modules, the extending and G-extending notions coincide.

4.24. Assume $A = \bigoplus_{i \in I} X_i$ and $S \in \underline{P}\underline{I}(A)$. Prove that $S = \bigoplus_{i \in I} \pi_i(S) = \bigoplus_{i \in I}(X_i \cap S)$, where π_i is the ith projection homomorphism.

4.25. Let A be a module. Show that $\underline{P}\underline{I}(A)$ is a complete sublattice of $(L(A), +, \cap)$.

4.26. Let A be a module. Let $\underline{C} \subseteq L(A)$ and $\underline{C}' = \underline{C} \cup \{D \leq A : D \leq_{\text{d}} A\}$. Prove that A is $\underline{C}$-extending if and only if A is $\underline{C}'$-extending.

4.7 Essential extensions with the $\underline{C}$-extending property

In this section we continue the investigation of $\underline{C}$-extending modules with respect to essential extensions.

In general, an essential extension of a $\underline{C}$-extending module is not $\underline{C}$-extending. For example, let ${}_RA = (\mathbb{Z}/\mathbb{Z}p) \oplus (\mathbb{Z}p/\mathbb{Z}p^3)$ and $K = (\mathbb{Z}/\mathbb{Z}p) \oplus (\mathbb{Z}/\mathbb{Z}p^3)$, where $R = \mathbb{Z}$ and p is a prime. Then A is extending, and A is essential in K; but K is not extending. Example 4.132 shows that in general an essential extension of a $\underline{P}\underline{I}$-extending module is not $\underline{P}\underline{I}$-extending. We show how to construct essential extensions of a module A which are $\underline{C}$-extending by using a set of representatives of an equivalence relation $\gamma_{\underline{C}}$ on $\{e = e^2 \in \underline{E}_{E(A)}\}$, where $E(A)$ denotes the injective hull of A. We also characterize when the rational hull of A is $\underline{C}$-extending in terms of such a set of representatives. Moreover, we show that several well-known types of $\underline{C}$-extending conditions (e.g., extending, uniform extending, $\underline{P}\underline{I}$-extending, FI-extending, etc.) transfer from A to its rational hull (see Theorem 3.6 and Corollary 4.97). Recall that the rational hull of A, $\widetilde{E}(A)$, is $\{s \in E(A) :$ for all $h \in \underline{E}_A, h(A) = 0 \Rightarrow h(s) = 0\}$. Note that $\widetilde{E}(R_R) = Q(R)_R$, where $Q(R)$ is the maximal right ring of quotients of R.

Definition 4.135. Define the equivalence relation γ on $\{e = e^2 \in \underline{E}_{E(A)}\}$ by $c\gamma e$ if $E(A)c = E(A)e$. Let $\underline{C} \subseteq L(A)$. Then $\gamma_{\underline{C}}$ denotes the restriction γ to $\{e = e^2 \in \underline{E}_{E(A)} : X \leq_{\text{e}} E(A)e$ for some $X \in \underline{C}\}$.

Lemma 4.136. *Let A be a module and $e = e^2 \in \underline{E}_{E(A)}$ such that $Ae \leq \widetilde{E}(A)$. Then $\widetilde{E}(A)e \leq \widetilde{E}(A)$.*

Proof. By [Lam99, Prop. 8.6 and Thm. 8.24], $e : A \to \widetilde{E}(A)$ can be extended to $\overline{e} : \widetilde{E}(A) \to \widetilde{E}(A)$. Then there exists $\overline{\overline{e}} : E(A) \to E(A)$ extending $\overline{e}$. Since $[e - \overline{\overline{e}}](A) = 0$, the definition of $\widetilde{E}(A)$ yields that $e|_{\widetilde{E}(A)} = \overline{e}|_{\widetilde{E}(A)} = \overline{\overline{e}}|_{\widetilde{E}(A)}$. Therefore $\widetilde{E}(A)e \leq \widetilde{E}(A)$. □

Recall that if γ is an equivalence relation on a set S, then $X \subseteq S$ is called a *set of representatives* of γ if each $x \in X$ is in exactly one equivalence class of γ and each equivalence class contains exactly one element of X.

Theorem 4.137. *Let A be a module and $\underline{C} \subseteq L(A)$ with $A \in \underline{C}$. Then*

(i) *A is $\underline{C}$-extending if and only if $A = AV$, where V is some set of representatives of $\gamma_{\underline{C}}$.*

(ii) *$\widetilde{E}(A)$ is $\underline{C}$-extending if and only if there exists a set of representatives for $\gamma_{\underline{C}}$, V, such that $Ae \leq \widetilde{E}(A)$, for each $e \in V$. In particular, if $A \leq N \leq \widetilde{E}(A)$ and N is $\underline{C}$-extending, then $\widetilde{E}(A)$ is $\underline{C}$-extending.*

Proof. (i) Assume that A is $\underline{C}$-extending. By Theorem 4.102, V can be chosen such that $1 \in V$ and $V \subseteq \{e = e^2 \in \underline{E}_{E(A)} : Ae \leq A\}$. Then $A = AV$. Conversely, let V be a set of representatives of $\gamma_{\underline{C}}$ such that $A = AV$. Let $X \in \underline{C}$. There exists $c = c^2 \in \underline{E}_{E(A)}$ such that $X \leq_{\rm e} E(A)c = E(X)$. Also, there exists $e \in V$ such that $c\gamma_{\underline{C}}e$. Then $X \leq_{\rm e} E(A)c = E(A)e$ and $Ae \leq AV = A$. By Theorem 4.102, A is $\underline{C}$-extending.

(ii) Assume that $\widetilde{E}(A)$ is $\underline{C}$-extending. By Theorem 4.102, there exists a set of representatives for $\gamma_{\underline{C}}$, V, such that $\widetilde{E}(A)V = \widetilde{E}(A)$. Hence $Ae \leq \widetilde{E}(A)$, for each $e \in V$. Conversely, assume that there exists a set of representatives for $\gamma_{\underline{C}}$, V, such that $Ae \leq \widetilde{E}(A)$ for each $e \in V$. Let $X \in \underline{C}$. Then $X \leq_{\rm e} E(X) = E(A)c$ for some $c = c^2 \in \underline{E}_{E(A)}$. There is an $e \in V$ such that $X \leq_{\rm e} E(A)c = E(A)e$ and $Ae \leq \widetilde{E}(A)$. By Lemma 4.136, $\widetilde{E}(A)e \leq \widetilde{E}(A)$. By Theorem 4.102, $\widetilde{E}(A)$ is $\underline{C}$-extending. □

Example 4.138. Let R be a semiprime ring and take R_R. Let $\underline{C} = \{X \trianglelefteq R\} = \{\text{ideals of } R\}$. Then $V = \{\text{central idempotents in } \underline{E}_{E(R)}\}$ and $\langle V \rangle R$ is the FI-extending hull of R_R. For more details, see [BPR13].

Theorem 4.139. *Let A be a module, $\widetilde{\underline{C}} \subseteq L(\widetilde{E}(A))$, and $\underline{C} = \{X \cap A : X \in \widetilde{\underline{C}}\}$. If A is $\underline{C}$-extending, then $\widetilde{E}(A)$ is $\widetilde{\underline{C}}$-extending.*

Proof. Let $X \in \widetilde{\underline{C}}$. Then $X \cap A \in \underline{C}$. By Theorem 4.102, there is $e = e^2 \in \underline{E}_{E(A)}$ such that $X \cap A \leq_{\rm e} E(A)e$ and $Ae \leq A$. By Lemma 4.136, $\widetilde{E}(A)e \leq \widetilde{E}(A)$. There exists $c = c^2 \in \underline{E}_{E(A)}$ such that $E(X) = E(A)c$. Assume that there exists $x \in X$ such that $[c - e](x) \neq 0$. Then there exists $r \in R$ such that $0 \neq r[c - e](x)$ and $rx \in A$ since A is dense in $\widetilde{E}(A)$ and $[c - e](x) \in \widetilde{E}(A)$. But $rx \in X \cap A \subseteq E(A)c \cap E(A)e$, contrary to $0 \neq [c - e](rx)$. Hence $e|_X = c|_X$. So $Xe = X$ and $X \leq_{\rm e} E(A)e$. By Theorem 4.102, $\widetilde{E}(A)$ is $\widetilde{\underline{C}}$-extending. □

From Theorem 4.139, one can deduce that $\widetilde{E}(A)$ has a stronger property than A since the set $\widetilde{\underline{C}}$ contains $\underline{C}$.

The next corollary shows the versatility of our results, in that they can be applied to many well-known generalizations of the extending property.

Corollary 4.140. *Let A be a module. If A satisfies any of the following conditions, then so does $\widetilde{E}(A)$ (in particular, if R_R is $\underline{C}$-extending then so is $Q(R)$):*

(i) *uniform extending;*
(ii) *extending;*
(iii) *weak* CS*;*
(iv) $\underline{P}\,\underline{I}$*-extending;*
(v) FI*-extending;*
(vi) CESS*;*
(vii) $\underline{Z}_0$*-extending;*
(viii) $\underline{H}\,\underline{I}$*-extending, where* $\underline{H} = \underline{P} \cup \underline{A}_A$.

Proof. The result follows from Theorem 4.139 letting $\widetilde{\underline{C}}$ and $\underline{C}$ be the indicated sets (see below) for each condition. Note that in each case $\underline{\widetilde{C}} = \{X \cap A : X \in \underline{C}\}$.

(i) Let $\widetilde{\underline{C}}$ be the set of uniform submodules of $\widetilde{E}(A)$. Then $\underline{C}$ is the set of uniform submodules of A.

(ii) Let $\widetilde{\underline{C}} = L(\widetilde{E}(A))$. Then $\underline{C} = L(A)$.

(iii) Since the socle of a module is the intersection of all essential submodules, $\operatorname{soc} A = \operatorname{soc} \widetilde{E}(A)$. Hence, $\widetilde{\underline{C}} = \{S \leq \widetilde{E}(A) : S \text{ is semisimple}\} = \{S \leq A : S \text{ is semisimple}\} = \underline{C}$.

(iv) Let $\widetilde{\underline{C}} = \{X \leq \widetilde{E}(A) : X \in \underline{P}\,\underline{I}(\widetilde{E}(A))\}$ and $\underline{C} = \{X \cap A : X \in \widetilde{\underline{C}}\}$. We need to show that $\underline{C} \subseteq \underline{P}\,\underline{I}(A)$. Let $e = e^2 \in \underline{E}_A$. By [Lam99, Prop. 8.6 and Thm. 8.24], there exists $\overline{e} \in \underline{E}_{\widetilde{E}(A)}$ such that $e|_A = \overline{e}|_A$. To see that $\overline{e} = \overline{e}^2$, suppose that $[\overline{e} - \overline{e}^2](y) \neq 0$ for some $y \in \widetilde{E}(A)$. Then there exists $r \in R$ such that $r[\overline{e} - \overline{e}^2](y) \neq 0$ and $ry \in A$. Then $[\overline{e} - \overline{e}^2](ry) = [e - e^2](ry) = 0$, a contradiction. We conclude that $e(X \cap A) = \overline{e}(X \cap A) \subseteq X \cap A$, for $X \in \widetilde{\underline{C}}$, and so $X \cap A \in \underline{P}\,\underline{I}(A)$.

(v) Let $\widetilde{\underline{C}} = \{X \trianglelefteq \widetilde{E}(A)\}$ and $\underline{C} = \{X \cap A : X \in \widetilde{\underline{C}}\}$. Similar to (iv), $X \cap A \trianglelefteq A$ for $X \in \widetilde{\underline{C}}$.

(vi) Let $\widetilde{\underline{C}} = \{X \leq \widetilde{E}(A) : \operatorname{soc} X \leq_{\mathrm{e}} X \text{ and } X \text{ is complement}\}$. By [Go76, p. 20 Ex. 17], $\underline{C} = \{Y \leq A : \operatorname{soc} Y \leq_{\mathrm{e}} Y \text{ and } Y \text{ is complement}\}$.

(vii) Let $\widetilde{\underline{C}} = \{X \leq \widetilde{E}(A) : Z(X) = 0\}$. Then $\underline{C} = \{Y \leq A : Z(Y) = 0\}$.

(viii) Let $\widetilde{\underline{C}} = \underline{H}\,\underline{I}(\widetilde{E}(A))$ and $\underline{C} = \{X \cap A : X \in \widetilde{\underline{C}}\}$. Use a proof similar to that in (iv) to show that if $X \in \widetilde{\underline{C}}$ is such that $X \in \underline{P}\,\underline{I}(\widetilde{E}(A))$, then $X \cap A \in \underline{P}\,\underline{I}(A)$. To show that $X \cap A \in \underline{A}_A\,\underline{I}(A)$, let $h \in \operatorname{Aut}({}_R A)$. From [Lam99, Prop. 8.6 and Thm. 8.24], there exists $\overline{h} \in \underline{E}_{\widetilde{E}(A)}$ which extends h. Since $1 \in \underline{E}_A$ extends uniquely to $1 \in \underline{E}_{\widetilde{E}(A)}$ by [Lam99, Thm. 8.24], then $\overline{h} \in \operatorname{Aut}(\widetilde{E}(A))$. So $h(X \cap A) = \overline{h}(X \cap A) \subseteq X \cap A$. Hence $X \cap A \in (\underline{P} \cup \underline{A}_A)\underline{I}(A)$. □

Note that Corollary 4.140 (ii) (using the proof of Theorem 4.38) corrects several errors in the proof of [ABT09, Thm. 5.3]. For $A_R = R_R$ and $\widetilde{E}(A) = Q(R)$, Corollary 4.140 (i), (ii), and (v) are in [BPR06, Cor. 1.8]. Also, Corollary 4.140 (viii) may be useful in the investigation of modules A for which $\underline{E}_A$ has a certain property (see Open Problems and Questions 6). The following example illustrates Corollary 4.140.

Example 4.141. Let $R = \mathbb{Z}$ and p a prime.

(i) Take $A = (\mathbb{Z}/\mathbb{Z}p) \oplus (\mathbb{Z}p/\mathbb{Z}p^3)$ and K as in the beginning of this section. Then $A = \widetilde{E}(A) \leq_e K$, where $A = \widetilde{E}(A)$ is extending, but K is not extending.

(ii) Let $A = \mathbb{Z} \oplus (\bigoplus_{i=1}^{\infty} H_i)$, where each $H_i = \mathbb{Z}/\mathbb{Z}p$. Then $\widetilde{E}(A) = \mathbb{Z}_{(p)} \oplus (\bigoplus_{i=1}^{\infty} H_i)$ (see [Lam99, Exa. 8.21]), where $\mathbb{Z}_{(p)}$ is the localization at p. By [ABT09, Cor. 3.17], A and $\widetilde{E}(A)$ are weak CS and $\widetilde{E}(A) \leq_e \mathbb{Z}_{(p)} \oplus (\bigoplus_{i=1}^{\infty} K_i)$, where $K_i = \mathbb{Z}/\mathbb{Z}p^i$, but $\mathbb{Z}_{(p)} \oplus (\bigoplus_{i=1}^{\infty} K_i)$ is not weak CS.

(iii) Corollary 4.140 naturally leads to the question: If $A \leq N \leq \widetilde{E}(A)$ and A is $\underline{P}\underline{I}$-extending, must N be $\underline{P}\underline{I}$-extending? Surprisingly, Example 4.132 (also see [BTY14, Exa. 5.5]) provides a negative answer. Using the notation as in Example 4.132, we observe that since A is nonsingular, D is a dense extension of K, where K is $\underline{P}\underline{I}$-extending, but D is not $\underline{P}\underline{I}$-extending. To this end, in the next section, we will show that if R_R is $\underline{P}\underline{I}$-extending, then $Q(R)_R$ and $Q(R)_{Q(R)}$ are $\underline{P}\underline{I}$-extending.

Exercises

4.27. Let R be a ring. Show that $\widetilde{E}(R_R) = Q(R)_R$.

4.28. Find more examples which illustrate Theorem 4.137.

4.8 Projection invariant extending rings

In this section, we are concerned with $\underline{P}\underline{I}$-extending rings. Recall that a ring R is said to be right $\underline{P}\underline{I}$-extending if every projection invariant right ideal of R is essential in a direct summand of R. We show how the $\underline{P}\underline{I}$-extending condition transfers from a ring R to its various ring extensions. More specifically, we characterize the right $\underline{P}\underline{I}$-extending generalized 2-by-2 triangular matrix rings, and we show that if R_R is $\underline{P}\underline{I}$-extending, then so is T_T, where T is an overring of R which is an essential extension of R, a n-by-n upper triangular matrix ring of R, a column finite or column and row finite matrix ring over R, or a certain type of trivial extension of R.

Let R be a ring and A a right R-module. If $X \subseteq A$, then $X \trianglelefteq_{\rm p} A$ and $Z^{\mathcal{K}}(A)$ denote X is a projection invariant submodule of A and the $\mathcal{K}$-singular submodule of A, respectively. Recall that $Z^{\mathcal{K}}(A) = \{\sum_{\varphi} \operatorname{im} \varphi : \varphi \in \operatorname{End}(A_R) \text{ and } \ker \varphi \leq_e A\}$ and $Z^{\mathcal{K}}(A) \leq Z(A)$ [RR04] and [RR07].

We note that in several of the ring extensions that we consider, if R is a prime ring, then so is the corresponding extension. So we begin by considering the $\underline{P}\underline{I}$-extending condition in prime rings.

Lemma 4.142. *Let $X \trianglelefteq_{\mathrm{p}} A$ and $X \leq_{\mathrm{e}} eA$, for some $e = e^2 \in \mathrm{End}(A_R)$. Then $eA + Z^{\mathcal{K}}(A)$ and $eA + Z(A)$ are both projection invariant in A. In particular, if $Z^{\mathcal{K}}(A) \leq eA$, then $eA \trianglelefteq A$ and $e \in \mathcal{S}_{\mathrm{l}}(\mathrm{End}(A_R))$.*

Proof. The first part of the proof is similar for both $Z^{\mathcal{K}}(A)$ and $Z(A)$, so we let Z denote either one. Let $c = c^2 \in \mathrm{End}(A_R)$. Assume that $c(eA+Z) \not\subseteq eA+Z$. Since $Z^{\mathcal{K}}(A)$ and $Z(A) \trianglelefteq A$ [RR07, Prop. 2.11], there exists $a \in A$ such that $cea \notin eA$. Observe $ce = ece+(1-e)ce$. If $(1-e)ce = 0$, then $ce = ece$. Hence $cea = ecea \in eA$, a contradiction. Thus $(1-e)ce \neq 0$. However, $(1-e)ce(X \oplus (1-e)A) = 0$ since $X \trianglelefteq_{\mathrm{p}} A$. But $X \oplus (1-e)A \leq_{\mathrm{e}} A$, so $(1-e)ceA \leq Z^{\mathcal{K}}(A)$. Also, $eceA \leq eA$. Hence, $ceA = eceA \oplus (1-e)ceA \subseteq eA + Z^{\mathcal{K}}(A)$, a contradiction. Therefore, $eA+Z^{\mathcal{K}}(A) \trianglelefteq_{\mathrm{p}} A$. Now assume $Z = Z(A)$. Then $L = \{r \in R : ear \in X\} \leq_{\mathrm{e}} R$. Also from above, $(1-e)cea \neq 0$. Hence, $(1-e)ceaL \leq (1-e)cX \leq (1-e)X \leq X = eX$. Then $(1-e)ceaL = 0$, so $(1-e)cea \in Z(A)$. Hence $cea = ecea + (1-e)cea \in eA + Z(A)$, a contradiction. Therefore $eA + Z(A) \trianglelefteq_{\mathrm{p}} A$. By [RR07, Prop. 2.11], $Z^{\mathcal{K}}(A) \leq Z(A)$. Then $Z(A) \leq eA$ implies that $Z^{\mathcal{K}}(A) \leq eA$. So assume $Z^{\mathcal{K}}(A) \leq eA$. The reasoning above shows that $eA \trianglelefteq_{\mathrm{p}} A$. By Proposition 4.121, $eA \trianglelefteq A$. From [BMR02, Lem. 1.9], $e \in \mathcal{S}_{\mathrm{l}}(\mathrm{End}(A_R))$. □

Lemma 4.142 generalizes [RR04, Prop. 2.28]. In next proposition we consider prime rings. First recall that a ring R is called semicentral reduced if $\mathcal{S}_{\mathrm{l}}(R) = \{0, 1\}$ (equivalently $\mathcal{S}_{\mathrm{r}}(R) = \{0, 1\}$) see [BPR13].

Proposition 4.143.

(i) *Let R be a prime right nonsingular ring. Then R_R is $\underline{P}\underline{I}$-extending if and only if $0 \neq X \trianglelefteq_{\mathrm{p}} R$ implies $X_R \leq_{\mathrm{e}} R_R$.*

(ii) *If R_R is $\underline{P}\underline{I}$-extending, right nonsingular, and semicentral reduced, then R is a prime ring.*

Proof. By [RR07, Prop. 2.7], $Z^{\mathcal{K}}(R_R) = 0$ if and only if $Z(R_R) = 0$.

(i) Assume R_R is $\underline{P}\underline{I}$-extending and $0 \neq X \trianglelefteq_{\mathrm{p}} R$. By Lemma 4.142, there exists $e = e^2 \in \mathcal{S}_{\mathrm{l}}(R)$ such that $X \leq_{\mathrm{e}} eR$. Since R is prime, $e = 1$. The converse is obvious.

(ii) Let $0 \neq X \trianglelefteq_{\mathrm{p}} R$. By Lemma 4.142, there exists $e \in \mathcal{S}_{\mathrm{l}}(R)$ such that $X \leq_{\mathrm{e}} eR$. Since R is semicentral reduced, $e = 1$ and $X \leq_{\mathrm{e}} R$. Hence $\ell(X) \leq Z(R_R) = 0$. Therefore, R is a prime ring. □

Proposition 4.144. *Let R be a semiprime right nonsingular (i.e., $Z(R_R) = 0$) right $\underline{P}\underline{I}$-extending ring. Let $S \leq R$. Then there exists $X \trianglelefteq_{\mathrm{p}} R$ such that $S = S_1 \oplus S_2$, where $X \leq_{\mathrm{e}} S_1$ and S_2 contains no nonzero projection invariant right ideals of R.*

Proof. Since the set of projection invariant right ideals is a complete sublattice of the lattice of right ideals of R, there exists $X_R \leq S_R$ which is maximal with respect to being projective invariant in R. So there exists $e = e^2 \in R$ such that $X \leq_{\mathrm{e}} eR$. By Proposition 4.121 and Lemma 4.142, $eR \trianglelefteq R$. Since R is semiprime, e

is a central idempotent. Then $S = eS \oplus (1-e)S$. Take $S_1 = eS$ and $S_2 = (1-e)S$. Then S_1 and S_2 have the desired properties. □

Let $\langle \mathbf{I}(R) \rangle$ denote the subring of R generated by its idempotent elements. Recall from [FU64] that a ring extension T of R is said to be *right intrinsic* over R if $X \cap R \neq 0$ for each nonzero right ideal X of T, denoted $R \leq_{\mathrm{r}}^{\mathrm{int}} T$.

Proposition 4.145. *If $\langle \mathbf{I}(R) \rangle \leq_{\mathrm{r}}^{\mathrm{int}} R$ and $\langle \mathbf{I}(R) \rangle$ is right* FI*-extending, then $\langle \mathbf{I}(R) \rangle$ and R are right $\underline{P}\underline{I}$-extending.*

Proof. Since $\langle \mathbf{I}(R) \rangle$ is generated by idempotents, it is right FI-extending if and only if it is right $\underline{P}\underline{I}$-extending. By [BL14, Thm. 3.19], R is right $\underline{P}\underline{I}$-extending. □

Some examples of rings for which $\langle \mathbf{I}(R) \rangle \leq_{\mathrm{r}}^{\mathrm{int}} R$ are rings such that every nonzero right ideal contains a nonzero idempotent, among them regular, π-regular rings with zero Jacobson radical, semiprime rings with essential socle [BPR13, Exa. 3.127].

Corollary 4.146. *Let R be a semiprime Baer ring such that $\langle \mathbf{I}(R) \rangle \leq_{\mathrm{r}}^{\mathrm{int}} R$. Then R is right $\underline{P}\underline{I}$-extending. In particular, every regular Baer ring is right and left $\underline{P}\underline{I}$-extending.*

Proof. By [Kap68, p. 16], $\langle \mathbf{I}(R) \rangle$ is a Baer ring since it contains all idempotents of R. We claim that $\langle \mathbf{I}(R) \rangle$ is semiprime. So assume $X \trianglelefteq \langle \mathbf{I}(R) \rangle$ such that $X^2 = 0$. Then $r_{\langle \mathbf{I}(R) \rangle}(X) = e\langle \mathbf{I}(R) \rangle$ for some $e = e^2 \in \langle \mathbf{I}(R) \rangle$. Since $e\langle \mathbf{I}(R) \rangle \trianglelefteq \langle \mathbf{I}(R) \rangle$, then $e \in S_l(\langle \mathbf{I}(R) \rangle)$. By [HT00], $e \in S_l(R)$. Now $X \leq r_{\langle \mathbf{I}(R) \rangle}(X)$, so $eX = X$. Then $(XR)^2 = XReXR = XeReXR = 0$. Since R is semiprime, $X = 0$. Hence $\langle \mathbf{I}(R) \rangle$ is semiprime. Then $\langle \mathbf{I}(R) \rangle$ is a semiprime Baer ring. By [BMR02, Thm. 4.7], $\langle \mathbf{I}(R) \rangle$ is right and left FI-extending. By Proposition 4.145, R and $\langle \mathbf{I}(R) \rangle$ are right $\underline{P}\underline{I}$-extending. □

Observe that if R is both left and right extending, then R is directly finite [Lam99, Thm. 6.48]. Thus if R is a regular Baer ring which is not directly finite, then R is right and left $\underline{P}\underline{I}$-extending (Corollary 4.146) but not extending on at least one side.

Example 4.147.

(i) Let $R = R_1 \oplus R_2$ where $R_1 = \mathrm{End}(V_F)$ and $R_2 = \mathrm{End}({}_K W)$, where V_F is an infinite-dimensional right vector space over a field F and W is an infinite-dimensional left vector space over a field K. From [Goo79, Thm. 9.12], R_1 is a prime regular right self-injective ring and R_2 is a prime regular left self-injective ring and both have a nonzero proper socle. Thus neither ring is simple, so neither ring is directly finite [Goo79, Cor. 9.30]. By [Goo79, Cors. 13.4 and 13.20], R_1 is not left extending and R_2 is not right extending. Thus R is neither right extending, nor left extending, but R is a regular Baer ring [Lam99, Cor. 7.53]. However, by Corollary 4.146, R is right and left $\underline{P}\,\underline{I}$-extending.

(ii) The ring $M_2(R)$ in [Goo79, Exa. 13.8] is regular, but not Baer. Hence it satisfies neither the left nor the right extending condition by [CK80, Thm. 2.1]. However since R is continuous, it is left and right $\underline{P}\,\underline{I}$-extending. From Example 4.77 and Proposition 4.108, $M_2(R)$ is right and left $\underline{P}\,\underline{I}$-extending.

Proposition 4.148. *Let $R = \prod_{i \in I} R_i$ be a direct product of rings. Then R is a right $\underline{P}\,\underline{I}$-extending if and only if each R_i is right $\underline{P}\,\underline{I}$-extending.*

Proof. It is routine (Exercise 4.30).

Proposition 4.149. *Let S be a right essential overring of R.*

(i) *If R_R is $\underline{P}\,\underline{I}$-extending, then S_R and S_S are $\underline{P}\,\underline{I}$-extending.*

(ii) *If S is a subring of $Q(R)$, then S_R is $\underline{P}\,\underline{I}$-extending if and only if S_S is $\underline{P}\,\underline{I}$-extending.*

Proof. (i) The proof that S_S is $\underline{P}\,\underline{I}$-extending is included in [BL14, Thm. 3.9]. Now, let $Y_R \trianglelefteq_{\mathrm{p}} S_R$ and $X_R = R \cap Y$. Then $X \trianglelefteq_{\mathrm{p}} R_R$. By hypothesis, there exists $e = e^2 \in R$ such that $X_R \leq_{\mathrm{e}} eR_R$.

Claim. $Y_R \leq_{\mathrm{e}} eS_R$.

Proof of the Claim. Let $y \in Y$. Then $y = ey + (1-e)y$. If $(1-e)y \neq 0$, then there exists $r \in R$ such that $0 \neq (1-e)yr \in R \cap Y = X \subset eR$, a contradiction. Hence $(1-e)y = 0$. It follows that $Y \leq eS$. Let $0 \neq es \in eS$. Then there exists $r_1 \in R$ such that $0 \neq esr_1 \in R \cap eS \leq eR$. Now $R \cap Y \leq R \cap eS \leq eR$. So, there exists $r_2 \in R$ such that $0 \neq es(r_1 r_2) \in R \cap Y \leq Y$. Therefore $Y_R \leq_{\mathrm{e}} eS_R$. □

(ii) By (i), if S_R is $\underline{P}\,\underline{I}$-extending, then so is S_S. Conversely, assume S_S is $\underline{P}\,\underline{I}$-extending. Let $Y_R \trianglelefteq_{\mathrm{p}} S_R$. Assume $c = c^2 \in S$. Observe that $c \in \mathrm{End}(S_R)$. Hence $cY \subseteq Y$. Then $YS \trianglelefteq_{\mathrm{p}} S_S$. By hypothesis, there exists $e = e^2 \in S$ such that $YS_S \leq_{\mathrm{e}} eS_S$. Let $0 \neq es \in eS$. There exists $t \in S$ such that $0 \neq est \in YS_S$. Since R_R is dense in S_R, there exists $r_0 \in R$ such that $0 \neq es(tr_0)$ and $tr_0 \in R$. So $estr_0 = y_1 s_1 + y_2 s_2 + \cdots + y_n s_n$ for some $y_i \in Y$ and $s_i \in S$. Then there exists $r_1 \in R$ such that $0 \neq estr_0 r_1$ and $s_1 r_1 \in R$. So $estr_0 r_1 = y_1(s_1 r_1) + y_2(s_2 r_1) + \cdots + y_n(s_n r_1)$. It follows that $tr_0 r_1 \in R$ and $y_1 s_1 r_1 \in Y$. Again there exists $r_2 \in R$ such that $0 \neq estr_0 r_1 r_2$ and $s_2 r_1 r_2 \in R$. So $0 \neq es(tr_0 r_1 r_2) = y_1(s_1 r_1 r_2) + y_2(s_2 r_1 r_2) + \cdots + y_n(s_n r_1 r_2)$ which gives that $tr_0 r_1 r_2 \in R$, $y_1 s_1 r_1 r_2$ and $y_2 s_2 r_1 r_2 \in Y$. Continuing in this manner, we obtain $r = tr_0 r_1 r_2 \cdots r_n \in R$ such that $0 \neq esr \in Y$. Thus $Y_R \leq_{\mathrm{e}} eS_R$. Therefore S_R is $\underline{P}\,\underline{I}$-extending. □

Theorem 4.150. *Let R be a ring, M an ideal of R, and $S = S(R, M)$.*

(i) *If S_S is $\underline{P}\,\underline{I}$-extending and $MX \subseteq X$ for all $X \trianglelefteq_{\mathrm{p}} R$, then R_R is $\underline{P}\,\underline{I}$-extending.*

(ii) *If S_S is* FI*-extending, then R_R is* FI*-extending.*

(iii) *If ${}_RM$ is faithful and R_R is $\underline{P}\,\underline{I}$-extending* (FI*-extending*)*, then S_S is $\underline{P}\,\underline{I}$-extending (FI-extending).*

Proof. First observe that for any (R,R)-bimodule M and $(e,m)=(e,m)^2 \in S$ one has $e=e^2 \in R$, $em+me=m$, and $eme=0$. So $(e,me)=(e,me)^2$, $(e,m)(e,me)=(e,me)$ and $(e,me)(e,m)=(e,m)$. Hence, $(e,m)S=(e,me)S$.

(i) Assume that S_S is $\underline{P}\underline{I}$-extending and $MX \subseteq X$ for all $X \trianglelefteq_{\rm p} R$. Let $X \trianglelefteq_{\rm p} R$. Since $(X, X\cap M) \trianglelefteq_{\rm p} S$, there exists $(e,m)=(e,m)^2 \in S$ such that $(X, X\cap M)_S \leq_{\rm e} (e,m)S_S$. Observe that for $x \in X$, $(x,0)=(e,m)(x,0)=(ex,mx)$ implies that $ex=x$ and $mx=0$. Let $0 \neq er \in eR$. To show that $X_R \leq_{\rm e} eR$, it is sufficient to prove that there exists $t \in R$ such that $0 \neq ert \in X$. Since $(X, X\cap M)_S \leq_{\rm e} (e,m)S_S=(e,me)S$, there exists $(s,k) \in S$ such that $0 \neq (e,me)(r,0)(s,k)=(ers, erk+mers) \in (X, X\cap M)$. If $ers=0=erk$, then $0 \neq (e,me)(r,0)(s,k)=(0,mers)$. But $0 \neq (e,me)(0,mers)=(0,(eme)rs)$, a contradiction. So $0 \neq ers \in X$, or $0 \neq erk \in X \cap M$. Therefore $X_R \leq_{\rm e} eR_R$, so R_R is $\underline{P}\underline{I}$-extending.

(ii) The proof is similar to that of (i) where we now take $X \trianglelefteq R$.

(iii) First we prove the $\underline{P}\underline{I}$-extending case. Recall from [Birk90, Lem. 4 (iv)], that $(0\times M)_S \leq_{\rm e} S_S$. Let $0 \neq Y \trianglelefteq_{\rm p} S$.

Claim 1. There exists $f=f^2 \in S$ such that $(Y \cap (0\times M)_S) \leq_{\rm e} fS_S$.

Proof of Claim 1. Consider the set $X=\{m \in M : (0,m) \in Y\}$. It is straightforward to show that $X \trianglelefteq_{\rm p} R$. Since R_R is $\underline{P}\underline{I}$-extending, there exists $e=e^2 \in R$ such that $X_R \leq_{\rm e} eR_R$. Observe that $Y\cap(0\times M) \subseteq (e,0)S$. For if $(0,m) \in Y\cap(0\times M)$, then $(e,0)(0,m)=(0,em)=(0,m)$ because $m=em$ since $m \in X \subseteq eR$. Now let $0 \neq (e,0)(r,m)=(er,em) \in (e,0)S$. We want to find (t,n) such that $0 \neq (er,em)(t,n) \in Y \cap (0\times M)$. First assume that $er=0$. Then $em \neq 0$, so there exists $v \in R$ such that $0 \neq emv \in X$. Hence $0 \neq (0,em)(v,0)=(0,emv) \in Y \cap (0\times M)$. Now if we assume $er \neq 0$, there exists $w \in R$ such that $0 \neq erw \in X$. Since $\ell_R(M)=0$, there exists a $k \in M$ such that $0 \neq erwk \in X$. Thus $0 \neq (er,em)(0,wk)=(0,erwk) \in Y\cap(0\times M)$. Therefore taking $f=(e,0)$, we obtain the claim.

Claim 2. $Y \subseteq (e,0)S_S$.

Proof of Claim 2. Let $(y_1,y_2) \in Y$. Note that $(y_1,y_2)=(e,0)(y_1,y_2)+(1-e,0)(y_1,y_2)$. Assume to the contrary that $(1-e,0)(y_1,y_2) \neq 0$. Note that $(1-e,0)(y_1,y_2) \in Y \trianglelefteq_{\rm p} S$. Since $(Y\cap(0\times M))_S \leq_{\rm e} Y_S$, there exists $(r,m) \in S$ such that $0 \neq (1-e,0)(y_1,y_2)(r,m) \in Y\cap(0\times M)$. This contradicts the fact that $(e,0)$ is a left identity for $Y\cap(0\times M)$. Hence $(y_1,y_2)=(e,0)(y_1,y_2) \in (e,0)S$, proving the claim.

Since $Y\cap(0\times M) \subseteq Y \subseteq (e,0)S$ and $(Y\cap(0\times M))_S \leq_{\rm e} (e,0)S_S$, we have shown that $Y_S \leq_{\rm e} (e,0)S_S$. Therefore, S_S is $\underline{P}\underline{I}$-extending.

The proof for the FI-extending case is similar to the $\underline{P}\underline{I}$-extending case except that we take $Y \trianglelefteq S$ and X becomes an ideal of R. □

Lemma 4.151. *Let S and R be rings, M an (S,R)-bimodule, and* $T=\begin{bmatrix} S & M \\ 0 & R \end{bmatrix}$, *the* 2×2 *generalized triangular matrix ring. Assume that* $V_T \leq T_T$ *such that*

$$V \subseteq \begin{bmatrix} S & M \\ 0 & 0 \end{bmatrix}, \quad and \quad \begin{bmatrix} e & x \\ 0 & b \end{bmatrix} = \begin{bmatrix} e & x \\ 0 & b \end{bmatrix}^2 \in T$$

such that $V_T \leq_{\text{e}} \begin{bmatrix} e & x \\ 0 & b \end{bmatrix} T_T$. *Then* $V_T \leq_{\text{e}} \begin{bmatrix} e & 0 \\ 0 & 0 \end{bmatrix} T_T$.

Proof. There exists $L_T \leq_{\text{e}} T_T$ such that $\begin{bmatrix} e & x \\ 0 & b \end{bmatrix} L \subseteq V$. Let $\overline{L} = \{r \in R :$ there exist $s \in S$ and $m \in M$ such that $\begin{bmatrix} s & m \\ 0 & r \end{bmatrix} \in L\}$. Either $\overline{L} = R$, or there is a $w \in R$ such that $w \notin \overline{L}$. If $w \notin \overline{L}$, then there exists $\begin{bmatrix} s_1 & m_1 \\ 0 & r_1 \end{bmatrix} \in T$ such that $0 \neq \begin{bmatrix} 0 & 0 \\ 0 & w \end{bmatrix} \begin{bmatrix} s_1 & m_1 \\ 0 & r_1 \end{bmatrix} = \begin{bmatrix} 0 & 0 \\ 0 & wr_1 \end{bmatrix} \in L$. Hence $0 \neq wr_1 \in \overline{L}$, so in all cases $\overline{L}_R \leq_{\text{e}} R_R$. Then $\begin{bmatrix} e & x \\ 0 & b \end{bmatrix} L \subseteq V$ implies that $b\overline{L} = 0$, so $b \in Z(R_R)$. Since $b = b^2 \in R$, $b = 0$. Observe that $e = e^2 \in S$ and $ex = x$. Therefore, $\begin{bmatrix} e & x \\ 0 & b \end{bmatrix} T = \begin{bmatrix} e & x \\ 0 & 0 \end{bmatrix} T = \begin{bmatrix} e & 0 \\ 0 & 0 \end{bmatrix} T.$ □

Theorem 4.152. *Let* S *and* R *be rings,* M *an* (S,R)*-bimodule and* $T = \begin{bmatrix} S & M \\ 0 & R \end{bmatrix}$, *the* 2×2 *generalized triangular matrix ring. Then* T_T *is* $\underline{P}\underline{I}$*-extending if and only if all of the following conditions hold:*

(i) $\ell_S(M) = eS$ *where* $e = e^2 \in S$ *and* eS_S *is* $\underline{P}\underline{I}$*-extending.*

(ii) *For* $N_R \leq M_R$ *with* $aN \leq N$ *for all* $a = a^2 \in S$, *there exists* $f = f^2 \in S$ *such that* $N_R \leq_{\text{e}} fM_R$.

(iii) R_R *is* $\underline{P}\underline{I}$*-extending*

Proof. ($\Rightarrow$) Assume that T_T is $\underline{P}\underline{I}$-extending. Then there exists

$$\begin{bmatrix} e & x \\ 0 & b \end{bmatrix} = \begin{bmatrix} e & x \\ 0 & b \end{bmatrix}^2 \in T \quad \text{such that} \quad \begin{bmatrix} \ell_S(M) & 0 \\ 0 & 0 \end{bmatrix} \leq_{\text{e}} \begin{bmatrix} e & x \\ 0 & b \end{bmatrix} T_T,$$

since $\begin{bmatrix} \ell_S(M) & 0 \\ 0 & 0 \end{bmatrix} \trianglelefteq_{\text{p}} T$ because $\begin{bmatrix} \ell_S(M) & 0 \\ 0 & 0 \end{bmatrix} \trianglelefteq T$. By Lemma 4.151,

$$\begin{bmatrix} \ell_S(M) & 0 \\ 0 & 0 \end{bmatrix} \leq_{\text{e}} \begin{bmatrix} e & 0 \\ 0 & 0 \end{bmatrix} T = \begin{bmatrix} eS & eM \\ 0 & 0 \end{bmatrix}.$$

Now since $\begin{bmatrix} 0 & eM \\ 0 & 0 \end{bmatrix} \cap \begin{bmatrix} \ell_S(M) & 0 \\ 0 & 0 \end{bmatrix} = 0$, $eM = 0$. So $e \in \ell_S(M)$. Thus $\ell_S(M) = eS$ and $e = e^2 \in S$. Since $\begin{bmatrix} e & 0 \\ 0 & 0 \end{bmatrix} T \trianglelefteq_{\text{p}} T$ and T_T is $\underline{P}\,\underline{I}$-extending, Proposition 4.123 yields that $\begin{bmatrix} eS & eM \\ 0 & 0 \end{bmatrix}$ is $\underline{P}\,\underline{I}$-extending. Let $X_S \leq eS_S$ such that X_S is projection invariant. By Lemma 4.151, there exists $c = c^2 \in S$ such that $\begin{bmatrix} X & 0 \\ 0 & 0 \end{bmatrix}_T \leq_{\text{e}} \begin{bmatrix} c & 0 \\ 0 & 0 \end{bmatrix} T_T$. It follows that $X_S \leq_{\text{e}} cS$. Since $eS \trianglelefteq S$, we have $e \in S_l(S)$, so $ceS = eceS \subseteq eS$. Further, since $ce = (ce)^2$ and $ceS = cS \cap eS$, then $X_S \leq_{\text{e}} ceS$. Thus eS_S is $\underline{P}\,\underline{I}$-extending. Therefore condition (i) is satisfied.

Assume that $N_R \leq M_R$ such that $aN \leq N$, for all $a = a^2 \in S$. Then $\begin{bmatrix} 0 & N \\ 0 & 0 \end{bmatrix}$ is projection invariant in T. Using Lemma 4.151, there exists $f = f^2 \in S$ such that $\begin{bmatrix} 0 & N \\ 0 & 0 \end{bmatrix} \leq_{\text{e}} \begin{bmatrix} f & 0 \\ 0 & 0 \end{bmatrix} T_T = \begin{bmatrix} fS & fM \\ 0 & 0 \end{bmatrix}$. Let $0 \neq k \in fM$. Then there exists $\begin{bmatrix} s & m \\ 0 & r \end{bmatrix} \in T$ such that $0 \neq \begin{bmatrix} 0 & k \\ 0 & 0 \end{bmatrix} \begin{bmatrix} s & m \\ 0 & r \end{bmatrix} = \begin{bmatrix} 0 & kr \\ 0 & 0 \end{bmatrix} \in \begin{bmatrix} 0 & N \\ 0 & 0 \end{bmatrix}$. So $N_R \leq_{\text{e}} fM_R$. Therefore condition (ii) is satisfied.

Observe that $\begin{bmatrix} 1 & 0 \\ 0 & 0 \end{bmatrix} T = \begin{bmatrix} S & M \\ 0 & 0 \end{bmatrix} \trianglelefteq T$. Hence $\begin{bmatrix} S & M \\ 0 & 0 \end{bmatrix} \trianglelefteq_{\text{p}} T$. Proposition 4.123 yields that $\begin{bmatrix} 0 & 0 \\ 0 & 1 \end{bmatrix} T = \begin{bmatrix} 0 & 0 \\ 0 & R \end{bmatrix}_T$ is $\underline{P}\,\underline{I}$-extending. A routine argument shows that R_R is $\underline{P}\,\underline{I}$-extending. Therefore condition (iii) is satisfied.

($\Leftarrow$) Conversely, assume conditions (i)–(iii) hold. A straightforward argument yields that $T = \begin{bmatrix} e & 0 \\ 0 & 0 \end{bmatrix} T \oplus \begin{bmatrix} 1-e & 0 \\ 0 & 1 \end{bmatrix} T$, $\begin{bmatrix} e & 0 \\ 0 & 0 \end{bmatrix} T_T \trianglelefteq T_T$ is $\underline{P}\,\underline{I}$-extending, M is a (S', R)-bimodule and ${}_{S'}M$ is faithful, where $S' = (1-e)S = (1-e)S(1-e)$ since $1-e \in S_r(S)$. Let $T' = \begin{bmatrix} 1-e & 0 \\ 0 & 1 \end{bmatrix} T = \begin{bmatrix} S' & M \\ 0 & R \end{bmatrix}$ and $a = a^2 \in S$. Then $(1-e)a = (1-e)a(1-e) = ((1-e)a)^2$ and $aM = [ea + (1-e)a]M = (1-e)aM$. So now condition (ii) holds when S is replaced with S'.

We claim that $T'_{T'}$ is $\underline{P}\,\underline{I}$-extending. Using condition (iii) and Proposition 4.123, we only need to show that $\begin{bmatrix} 1-e & 0 \\ 0 & 0 \end{bmatrix} T'_{T'}$ is $\underline{P}\,\underline{I}$-extending. Let $V \trianglelefteq_{\text{p}} \begin{bmatrix} 1-e & 0 \\ 0 & 0 \end{bmatrix} T'$. Then $V = \begin{bmatrix} X & N \\ 0 & 0 \end{bmatrix}$, where $X \trianglelefteq_{\text{p}} S'$ and $N_R \leq M_R$ such that $aN \subseteq N$ for all $a = a^2 \in S'$ and $XM \subseteq N$. Since condition (ii) holds for S', there is $f = f^2 \in S'$ such that $N_R \leq_{\text{e}} fM_R$. Then $XM \subseteq N \subseteq fM$, implies that $n = fn$ for all $n \in N$, in particular, $xm = fxm$ for any $x \in X$ and $m \in M$. Thus $(x - fx)M = 0$, so $(x - fx) = 0$ for any $x \in X$. Hence $V_{T'} \leq \begin{bmatrix} f & 0 \\ 0 & 0 \end{bmatrix} T'_{T'}$.

Now let $0 \neq \begin{bmatrix} fs_1 & fm \\ 0 & 0 \end{bmatrix} \in \begin{bmatrix} f & 0 \\ 0 & 0 \end{bmatrix} T'$. If $fm \neq 0$, then there exists $r \in R$ such that $0 \neq fmr \in N$. Thus $0 \neq \begin{bmatrix} fs_1 & fm \\ 0 & 0 \end{bmatrix} \begin{bmatrix} 0 & 0 \\ 0 & r \end{bmatrix} = \begin{bmatrix} 0 & fmr \\ 0 & 0 \end{bmatrix} \in V$. If $fm = 0$, then $fs_1 \neq 0$. Since ${}_{S'}M$ is faithful, there is a $k \in M$ such that $0 \neq fs_1k \in M$. Now there is $r_1 \in R$ such that $0 \neq fs_1kr_1 \in N$. So $0 \neq \begin{bmatrix} fs_1 & fm \\ 0 & 0 \end{bmatrix} \begin{bmatrix} 0 & kr_1 \\ 0 & 0 \end{bmatrix} = \begin{bmatrix} 0 & fs_1kr_1 \\ 0 & 0 \end{bmatrix} \in V$. Hence $V_{T'} \leq_{\mathrm{e}} \begin{bmatrix} f & 0 \\ 0 & 0 \end{bmatrix} T'_{T'}$.

Therefore, $\begin{bmatrix} 1-e & 0 \\ 0 & 0 \end{bmatrix} T'_{T'}$ is $\underline{P}\underline{I}$-extending, so the claim is true. A routine argument now shows that T'_T is $\underline{P}\underline{I}$-extending. Therefore, by Proposition 4.123, T_T is $\underline{P}\underline{I}$-extending. □

Corollary 4.153. *Let* $T = \begin{bmatrix} \mathrm{End}(M_R) & M \\ 0 & R \end{bmatrix}$, *where* M_R *is a right* R*-module. Then* T_T *is right* $\underline{P}\underline{I}$*-extending* (FI-*extending*) *if and only if* M_R *and* R_R *are* $\underline{P}\underline{I}$*-extending* (FI-*extending*).

Proof. The proof follows immediately from Theorem 4.152 and [BiL14, Thm. 1.16]. □

From [BPR02, Prop. 4.8] it follows that, if $Z(M_R) = 0$ and M_R is $\underline{P}\underline{I}$-extending in Corollary 4.153, then $\mathrm{End}(M_R)$ is a quasi-Baer ring.

Corollary 4.154. *Let* A *be a ring and* n *a positive integer. The following statements are equivalent:*

(i) A_A *is* $\underline{P}\underline{I}$*-extending.*
(ii) $T_n(A)$ *is right* $\underline{P}\underline{I}$*-extending for every* n.
(iii) $T_n(A)$ *is right* $\underline{P}\underline{I}$*-extending for some* n.

Proof. (i) ⇒ (ii) If $n = 1$, we are finished. So assume the result holds for all $k < n$, where k is a positive integer. We decompose $T_n(A)$ into blocks according to Theorem 4.152, by taking $S = A$, M is the 1-by-$(n-1)$ row over A, and $R = T_{n-1}(A)$. Since ${}_SM$ is faithful, condition (i) of Theorem 4.152 is satisfied. Also condition (iii) of Theorem 4.152 is satisfied. So we focus on condition (ii) of Theorem 4.152. Let $N_R \leq M_R$ such that $aN \leq N$ for all $a = a^2 \in S$. Then N is a 1-by-$(n-1)$ row of the form $[X_1X_2 \cdots X_{n-1}]$, where each $X_i \trianglelefteq_{\mathrm{p}} A$ and $X_1 \leq X_2 \leq \cdots \leq X_{n-1}$. So there exist $e_i = e_i^2 \in A$ such that $X_i \leq_{\mathrm{e}} e_iA$. Let $f = e_{n-1}$. Then $N \leq fM$. Let $0 \neq fm \in M$. Then $fm = [fa_1fa_2 \cdots fa_{n-1}]$. Since $fm \neq 0$, there exists $1 \leq h \leq n-1$ such that $fa_h \neq 0$. There exists $b \in A$ such that $0 \neq fa_hb \in X_{n-1}$. Let $t \in R$ be the matrix with b in the $(h, n-1)$-position and zero elsewhere. Then $0 \neq fmt = [0 \cdots 0fa_hb] \in N$. Hence $N_R \leq_{\mathrm{e}} fM_R$, so condition (ii) of Theorem 4.152 is satisfied. Therefore $T_n(A)$ is right $\underline{P}\underline{I}$-extending for every n.

(ii) $\Rightarrow$ (iii) Clear.

(iii)$\Rightarrow$ (i) The ring $T_n(A)$ can be block decomposed into the form of T in Theorem 4.152 so that $R = A$. Therefore A_A is $\underline{P}\underline{I}$-extending. □

Lemma 4.155. *Let F be a free right R-module and set $\Lambda = \text{End}(F_R)$. Let $U, V \leq \Lambda$ with $U \leq V$. Then $U_\Lambda \leq_{\text{e}} V_\Lambda$ if and only if $UF_R \leq_{\text{e}} VF_R$.*

Proof. See [CK80, Lem. 3.4]. □

The next lemma and part of (i) of the following theorem are adaptations of [BPR02, Lem. 4.5 and Thm. 4.6].

Lemma 4.156. *Let $\Lambda = \text{End}(M_R)$. Then for any $X \trianglelefteq_{\text{p}} \Lambda$, $XM \trianglelefteq_{\text{p}} M$. Conversely, for every $K \trianglelefteq_{\text{p}} M$ there exists $Y = \{y \in \Lambda : yM \leq K\} \trianglelefteq_{\text{p}} \Lambda$.*

Proof. Let $e = e^2 \in \Lambda$. Assume $X \trianglelefteq_{\text{p}} \Lambda$. Then $e(XM) = (eX)M \leq XM$. Conversely, let $K \trianglelefteq_{\text{p}} M$ and $h \in \Lambda$. Then $Yh \subseteq Y$ and $eY \subseteq Y$. Therefore $Y \trianglelefteq_{\text{p}} \Lambda$. □

Theorem 4.157. *Let R be a right $\underline{P}\underline{I}$-extending ring.*

(i) *$\text{End}(F_R)$ is right $\underline{P}\underline{I}$-extending, where F_R is a free right R-module.*
(ii) *$CFM_\Gamma(R)$ and $CRFM_\Gamma(R)$ are right $\underline{P}\underline{I}$-extending.*

Proof. (i) By Corollary 4.120, F_R is $\underline{P}\underline{I}$-extending. Let $X \trianglelefteq_{\text{p}} \Lambda = \text{End}(F_R)$. By Lemma 4.156, $XF \trianglelefteq_{\text{p}} F$. So there exists $e = e^2 \in \Lambda$ such that $XF \leq_{\text{e}} eF = e\Lambda F$. Using Lemma 4.155, it follows that $X_\Lambda \leq_{\text{e}} e\Lambda_\Lambda$. Therefore Λ is right $\underline{P}\underline{I}$-extending.

(ii) From part (i), $CFM_\Gamma(R)$ is right $\underline{P}\underline{I}$-extending, since $CFM_\Gamma(R) \cong \text{End}(F_R)$, where F_R is a free right R-module with a basis of cardinality Γ. To see that $CRFM_\Gamma(R)$ is right $\underline{P}\underline{I}$-extending, assume that Γ is well ordered. Let $S = CRFM_\Gamma(R)$ and $X \leq_{\text{p}} e_{11}S_S$, where e_{ii} denotes the matrix with 1_R in the (i,i)th position and zero elsewhere. Then there exists $V_R \trianglelefteq_{\text{p}} R_R$ such that $X_S \leq e_{11}\overline{V}_S$, where $\overline{V}_S = CRFM_\Gamma(V)$ and $e_{11}\overline{V}e_{jj} \leq X$, for all possible j. Since R_R is $\underline{P}\underline{I}$-extending, there exists $c = c^2 \in R$ such that $V_R \leq_{\text{e}} cR_R$. Let $K = ce_{11}S$. Then K_S is a direct summand of $e_{11}S_S$ and $X_S \leq K_S$. Let $0 \neq k = ce_{11}s \in K$. There exists an ordinal j such that $0 \neq ke_{jj}$. Let b be a nonzero entry in $ke_{jj} = ce_{11}se_{jj}$. Hence $b \in cR$. There exists $r \in R$ such that $0 \neq br \in V$. So $0 \neq ke_{jj}r \in X$. Hence $X_S \leq_{\text{e}} S_K$. Thus $e_{11}S_S$ is $\underline{P}\underline{I}$-extending. Now each row of S is isomorphic to $e_{11}S_S$ and S_S is the direct sum of its rows. By Corollary 4.120, S_S is $\underline{P}\underline{I}$-extending. □

Recall that $\langle \mathbf{I}(R) \rangle$ denotes the subring of R generated by its idempotent elements. Note that if $X_R \leq M_R$, then $X \trianglelefteq_{\text{p}} M$ if and only if $SX_R \leq X_R$, where $S = \langle \mathbf{I}(\text{End}(M_R)) \rangle$.

Proposition 4.158. *Let M be an R-module such that $End(M_R) = \langle \mathbf{I}(End(M_R)) \rangle$.*

(i) *Assume that $X \leq M$. Then $X \trianglelefteq_{\text{p}} M$ if and only if $X \trianglelefteq M$.*
(ii) *M is $\underline{P}\underline{I}$-extending if and only if M is* FI-*extending.*

Proof. The proof is straightforward. □

Corollary 4.159.

(i) *Assume* $R = \langle \mathbf{I}(R) \rangle$. *Then* R_R *is* $\underline{P}\underline{I}$*-extending if and only if* R_R *is* FI-*extending.*

(ii) *Assume* n *be a positive integer with* $n > 1$. *Then* $M_n(R)$ *is right* $\underline{P}\underline{I}$*-extending if and only if* $M_n(R)$ *is right* FI-*extending.*

Proof. The result is a consequence of Proposition 4.158 by observing that for $n > 1$, $M_n(R) = \langle \mathbf{I}(M_n(R)) \rangle$. □

The next example provides a class of rings that are right $\underline{P}\underline{I}$-extending but not extending.

Example 4.160. Let R be a right Ore domain. Then R_R is uniform hence extending.

(i) From Corollary 4.154, $T_n(R)$ is right $\underline{P}\underline{I}$-extending. However if $n > 1$ and R is not a division ring, then $T_n(R))$ is not right extending by [CK80, Thm. 2.1] and [Kap68, p. 16, Exc. 2].

(ii) From Theorem 4.157, $M_n(R)$ is right $\underline{P}\underline{I}$-extending. If R is commutative but not Prüfer, then $M_n(R)$ is not right extending by [DHSW94, pp. 108–109].

Exercises

4.29. Let A be a module. Prove that

(i) $Z^{\mathcal{K}}(A) \leq Z(A)$,

(ii) $Z^{\mathcal{K}}(A) \trianglelefteq A$.

4.30. Write out the proof of Proposition 4.148.

4.9 Weak version of C_{11}-modules

In this section, we explore a weak version of the C_{11} property. To this end, we introduce weak C_{11}-modules as a proper generalization of both C_{11}-modules and weak CS-modules. While most of the material can be found in [Er99], [Ter01], [Ter03] and [Ter04].

Definition 4.161. A module A is said to be a *weak* C_{11}*-module* if every semisimple submodule has a complement in A which is a direct summand of A. We denote this class by WC_{11}.

Note that the following implications hold for a module A:

$$\begin{array}{ccc} \text{CS} & \Rightarrow & \text{weak CS} \\ \Downarrow & & \Downarrow \\ C_{11} & \Rightarrow & WC_{11} \end{array}$$

No other implications can be added to this table, in general. In particular, Example 3.102 shows that WC_{11} does not imply C_{11} (see also Example 4.180). On the other hand, Example 4.3 shows that there exists a module with C_{11} which is not weak CS.

In a similar way to Proposition 4.25, we characterize weak C_{11}-modules.

Proposition 4.162. *The following statements are equivalent for a module A.*

(i) *A is WC_{11}.*

(ii) *For any semisimple submodule N of A, there exists a direct summand K of A such that $N \cap K = 0$ and $N \oplus K \leq_e A$.*

(iii) *For any complement L in A with essential socle, there exists a complement of L which is a direct summand of A.*

(iv) *For any complement L in A with essential socle, there exists a direct summand K of A such that $K \cap L = 0$ and $K \oplus L \leq_e A$.*

Proof. (i) $\Leftrightarrow$ (ii) and (iii) $\Leftrightarrow$ (iv) follow from Lemma 4.24.

(ii) $\Rightarrow$ (iv) Let L be a complement in A with essential socle. By assumption, there exists a direct summand K of A such that $K \cap \operatorname{soc} L = 0$ and $K \oplus \operatorname{soc} L \leq_e A$. So $K \cap L = 0$ and, since $\operatorname{soc} L \oplus K \leq L \oplus K$, $K \oplus L \leq_e A$.

(iii) $\Rightarrow$ (i) Let N be a semisimple submodule of A. Let L be the closure of N in A. Then L is a complement with essential socle. By hypothesis, there is a complement K of L which is a direct summand of A. Since $N \leq_e L$, K is also a complement of N. □

We may conjecture that a WC_{11}-module with essential socle is a C_{11}-module. Our next result provides this conjecture.

Proposition 4.163. *Let R be a ring and A a left R-module. If A is a WC_{11}-module with essential socle, then A is a C_{11}-module.*

Proof. Let X be a submodule of A. If $X = 0$, then A will do. Assume $X \neq 0$. Since $\operatorname{soc} X$ is a semisimple submodule of A, there exists a direct summand L of A such that $\operatorname{soc} X \cap L = 0$ and $\operatorname{soc} X \oplus L \leq_e A$. It is clear that $\operatorname{soc} X \leq_e X$. Hence $X \cap L = 0$. So we have $\operatorname{soc} X \oplus L \leq X \oplus L \leq A$. It follows that $X \oplus L \leq_e A$. By Proposition 4.25, A is a C_{11}-module. □

Modifying the proof of Theorem 4.26 we obtain the next result.

Theorem 4.164. *Any direct sum of WC_{11}-modules satisfies WC_{11}.*

Proof. Left as an exercise for the reader. □

Proposition 4.165. *Let A be a WC_{11}-module. Then $A = A_1 \oplus A_2$, where A_1 is a submodule of A with essential socle and A_2 a submodule of A with zero socle.*

Proof. Let S denote the socle of A. There exist submodules A_1 and A_2 of A such that $A = A_1 \oplus A_2$, $S \cap A_2 = 0$, and $S \oplus A_2 \leq_e A$. Now $S = \operatorname{soc} A = (\operatorname{soc} A_1) \oplus (\operatorname{soc} A_2)$. Clearly $\operatorname{soc} A_2 = 0$ so that $S \leq A_1$. Then $S \oplus A_2 \leq_e A$ implies $S \leq_e A_1$, whence the result follows. □

The following example shows that the converse of Proposition 4.165 is not true, in general.

Example 4.166. Let S be a ring and let V be a S-S-bimodule. Assume that S has zero socle and V is semisimple, but not simple. Let R be the trivial extension of S and the S-module V. Then $R = S \oplus V$ has the following addition and multiplication operations: $(s,a)+(t,b) = (s+t, a+b)$ and $(s,a)(t,b) = (st, sb+at)$. Let $A_1 = R_R$. Then $\operatorname{soc} A_1 = 0 \oplus V$ is essential in A_1. Set $I = 0 \oplus V$ and let $A_2 = R/I$. Then $\operatorname{soc} A_2 = 0$. Now consider the module $A = A_1 \oplus A_2$. Let N be a simple submodule of A. Then $N = (0 \oplus X) \oplus 0$ for some submodule X of V. Suppose there exists a direct summand L of A such that $N \cap L = 0$ and $N \oplus L$ is essential in A. Now

$$L = \{((s,a),(t,0)+I) : s,t \in R, a \in V\}.$$

Then $N \leq L$. But $N \cap L = N = 0$, a contradiction. Hence $L = 0$. However, N is not essential in V. It follows that A is not WC_{11}-module.

Recall that the WC_{11} property is closed under direct sums (Theorem 4.164). However, we do not know whether it is inherited by direct summands (Open Problems and Questions 9). Next we deal with when a direct summand of a WC_{11}-module is a WC_{11}-module. We first prove an easy result.

Lemma 4.167. *Let R be a ring and let A be an indecomposable left R-module such that* $\operatorname{soc} A \neq 0$. *Then A is a WC_{11}-module if and only if A is uniform.*

Proof. The sufficiency is clear. Conversely, suppose that A satisfies WC_{11}. Thus $\operatorname{soc} A \leq_e A$. Let $0 \neq X \leq A$. Then there exists an $L \leq_d A$ such that $\operatorname{soc} X \cap L = 0$ and $\operatorname{soc} X \oplus L \leq_e A$. If $L = A$, then $X = 0$, a contradiction. Hence $L = 0$. It follows that $X \leq_e A$. So A is uniform. □

Proposition 4.168. *Let R be a ring such that the left R-module R is WC_{11}-module and such that every direct summand of a WC_{11}-module is a WC_{11}-module. Then every indecomposable projective left R-module which has a nonzero socle is uniform.*

Proof. Let P be an indecomposable projective left R-module such that $\operatorname{soc} P \neq 0$. Then there exists a free R-module F such that $F = P \oplus N$ for some submodule N of F. By Theorem 4.164, F satisfies WC_{11} and, by hypothesis, so does P. Now, by Lemma 4.167, P is uniform. □

In view of Proposition 4.168, if R is a left WC_{11} R-module such that $\operatorname{soc} R$ is nonzero and P is any indecomposable projective left R-module of rank $n \geq 2$, then by Theorem 4.164 there exists a free left R-module A which satisfies WC_{11}.

Now, P is a direct summand of A and P is not a WC_{11}-module by Lemma 4.167. However, so far, we do not know whether such modules A exist or not.

Lemma 4.169. *Let a module $A = A_1 \oplus A_2$ be a direct sum of submodules A_1 and A_2. Then the module A_1 satisfies WC_{11} if and only if for every semisimple submodule N of A_1 there exists a direct summand K of A such that $A_2 \subseteq K$, $K \cap N = 0$, and $K \oplus N$ is an essential submodule of A.*

Proof. Suppose that A_1 satisfies WC_{11}. Let N be any semisimple submodule of A_1. There exists a direct summand L of A_1 such that $N \cap L = 0$ and $N \oplus L$ is an essential submodule of A_1. Clearly, $L \oplus A_2$ is a direct summand of A, $A_2 \subseteq L \oplus A_2$, $(L \oplus A_2) \cap N = 0$, and $(L \oplus A_2) \oplus N$ is essential in A. Conversely, suppose that A_1 has the stated property. Let H be a semisimple submodule of A_1. By hypothesis, there exists a direct summand K of A such that $A_2 \subseteq K$, $K \cap H = 0$, and $K \oplus H$ is an essential submodule of A. Now $K = K \cap (A_1 \oplus A_2) = (K \cap A_1) \oplus A_2$, so that $K \cap A_1$ is a direct summand of A, and hence also of A_1, $H \cap (K \cap A_1) = 0$ and $H \oplus (K \cap A_1) = A_1 \cap (H \oplus K)$, which is an essential submodule of A_1. It follows that A_1 is a WC_{11}-module. □

Theorem 4.170. *Let the WC_{11}-module $A = A_1 \oplus A_2$ be the direct sum of submodules A_1, and A_2 such that* $\operatorname{soc} A_2$ *is essential in A_2, and for every direct summand K of A with $K \cap A_2 = 0$, $K \oplus A_2$ is a direct summand of A. Then A_1 is a WC_{11}-module.*

Proof. Let N be any semisimple submodule of A_1. Then $N \oplus \operatorname{soc} A_2$ is a semisimple submodule of A. By hypothesis, there exists a direct summand K of A such that $(N \oplus \operatorname{soc} A_2) \cap K = 0$ and $N \oplus \operatorname{soc} A_2 \oplus K$ is an essential submodule of A. Since $\operatorname{soc} A_2$ is essential in A_2, we have that $N \cap A_2 = 0$ and $N \oplus A_2 \oplus K$ is essential in A. Moreover, $A_2 \oplus K$ is a direct summand of A. Now the result follows by Lemma 4.169. □

Corollary 4.171. *Let the WC_{11}-module $A = A_1 \oplus A_2$ be the direct sum of submodules A_1 and A_2 such that* $\operatorname{soc} A_2$ *is essential in A_2 and A/A_1 is A_1-injective. Then A_1 is a WC_{11}-module.*

Proof. By hypothesis, A_2 is A_1-injective. Let L be a direct summand of A such that $L \cap A_2 = 0$. By [DHSW94, Lem. 7.5], there exists a submodule H of A such that $H \cap A_2 = 0$, $A = H \oplus A_2$, and $L \subseteq H$. Now L is a direct summand of H and hence $L \oplus A_2$ is a direct summand of $A = H \oplus A_2$. By Theorem 4.170, A_1 is a WC_{11}-module. □

Corollary 4.172. *Let the module $A = A_1 \oplus A_2$ be the direct sum of a submodule A_1 and an injective submodule A_2 with essential socle. Then A satisfies WC_{11} if and only if A_1 satisfies WC_{11}.*

Proof. If A satisfies WC_{11}, then A_1 satisfies WC_{11} by Corollary 4.171. Conversely, if A_1 satisfies WC_{11}, then A satisfies WC_{11} by Theorem 4.164. □

Our next aim is to obtain decompositions of WC_{11}-modules. First recall that a module A is called WC_{11}^+-module if every direct summand of A is a WC_{11}-module (see §4.2). For example, if a module A has injective socle, then A satisfies WC_{11}^+. In particular, if R is a (commutative) Dedekind domain, then any finitely generated R-module is a WC_{11}^+-module. Moreover, we have the following.

Proposition 4.173. *Let R be a Dedekind domain and A an R-module with finite uniform dimension. Then A is a WC_{11}^+-module.*

Proof. Let $A = A_1 \oplus A_2$ be the direct sum of the submodules A_1 and A_2. Let us show that A_1 is a WC_{11}-module. First assume A_1 is torsion-free. Then $\operatorname{soc} A_1 = 0$ and, in this case A_1 is WC_{11}. Next we assume that A_1 is not torsion-free. By [Kap52, Thm. 9], it follows that $A_1 = N_1 \oplus N_2 \oplus N_3$ for some finitely generated module N_1, injective module N_2, and torsion-free module N_3. By Theorem 4.164, A_1 is WC_{11}. Hence A is WC_{11}^+. □

Lemma 4.174. *Let $A = U \oplus V$, where U and V are uniform modules. Then A has WC_{11}^+.*

Proof. Let $0 \neq K$ be a direct summand of A. If $K = A$ then K has WC_{11}. If $K \neq A$, then K is uniform, hence K has WC_{11}. Thus A has WC_{11}^+, as claimed. □

Theorem 4.175. *Let R be any ring and let A be a finitely generated WC_{11}^+ R-module. Let N be a semisimple submodule of A such that A/N has finite uniform dimension. Then N is finitely generated.*

Proof. Let $n < \infty$ be the uniform dimension of A/N. Suppose that N is not finitely generated. Then there exist non-finitely generated submodules N_1 and N_2 such that $N = N_1 \oplus N_2$. By hypothesis, there exist submodules A_1, A' of A such that $A = A_1 \oplus A'$, $N_1 \cap A' = 0$, and $N_1 \oplus A'$ is essential in A. Now let $\pi_1 : A \to A_1$ be the canonical projection. Since $A_1 \cap A' = 0$, we have $\pi_1(N_1) \oplus A' = N_1 \oplus A'$. Thus $\pi_1(N_1)$ is essential in A_1. Also $\operatorname{soc} A = \operatorname{soc}(N_1 \oplus A') = N_1 \oplus \operatorname{soc} A'$. Hence, $N = N_1 \oplus (N \cap \operatorname{soc} A')$. Now $N_2 \cong N \cap \operatorname{soc} A'$ so that $N \cap \operatorname{soc} A'$ is not finitely generated. Repeating this argument, there exist $\pi_i(N_i) \leq A_i \leq A$ $(2 \leq i \leq n+1)$ such that for each $2 \leq i \leq n+1$, N_i is not finitely generated, $A = A_1 \oplus A_2 \oplus \cdots \oplus A_{n+1}$. Let $L = \pi_1(N_1) \oplus \cdots \oplus \pi_{n+1}(N_{n+1})$. Then $A/L \cong (A_1/\pi_1(N_1)) \oplus (A_2/\pi_2(N_2)) \oplus \cdots \oplus (A_{n+1}/\pi_{n+1}(N_{n+1}))$. Since A/L has finite uniform dimension, there exists $1 \leq i \leq n+1$ such that $A_i = \pi_i(N_i)$. But A_i is finitely generated, and hence so is $\pi_i(N_i)$, a contradiction. Thus N is finitely generated. □

The next example shows that in Theorem 4.175 the WC_{11}^+ condition is not superfluous.

Example 4.176. Let K be a field and V an infinite-dimensional vector space over K. Let

$$R = \begin{bmatrix} K & & V \\ & \diagdown & \\ 0 & & K \end{bmatrix} = \left\{ \begin{bmatrix} k & v \\ 0 & k \end{bmatrix} : k \in K, v \in V \right\}.$$

Then R_R is not a WC_{11}^{+}-module and contains a semisimple submodule I such that R/I has finite uniform dimension, but I is not finitely generated.

Proof. It is straightforward to check that R_R is not a WC_{11}^{+}-module. Let $I = \operatorname{soc} R_R = \begin{bmatrix} 0 & V \\ 0 & 0 \end{bmatrix}$. Define $\varphi : R \to K$ by, $\varphi\left(\begin{bmatrix} k & v \\ 0 & k \end{bmatrix}\right) = k$ $(k \in K, v \in V)$. Then φ is an epimorphism with kernel I. Thus R/I has uniform dimension 1. Since V is infinite dimensional, I is not finitely generated. □

Now we return to general modules over arbitrary rings. We first prove:

Lemma 4.177. *Let A be a module such that A satisfies WC_{11}^{+} and $A/(\operatorname{soc} A)$ has finite uniform dimension. Suppose that $\operatorname{soc} A$ is contained in a finitely generated submodule of A. Then A has finite uniform dimension.*

Proof. Suppose A does not have finite uniform dimension. Then $\operatorname{soc} A$ is not finitely generated. There exist submodules S_1 and S_2 of $\operatorname{soc} A$ such that S_i is not finitely generated for $i = 1, 2$, and $socA = S_1 \oplus S_2$. By hypothesis, there exist submodules K, K' of A such that $A = K \oplus K'$, $S_1 \cap K = 0$, and $S_1 \oplus K$ is essential in A. Note that $S_1 \oplus S_2 = \operatorname{soc} A = \operatorname{soc}(S_1 \oplus K) = S_1 \oplus (\operatorname{soc} K)$. Thus $\operatorname{soc} K \cong S_2$ and hence $\operatorname{soc} K$ is not finitely generated. Also $\operatorname{soc} K \oplus \operatorname{soc} K' = \operatorname{soc} A = S_1 \oplus (\operatorname{soc} K)$ so that $\operatorname{soc} K' \cong S_1$, and hence $\operatorname{soc} K'$ is not finitely generated. By hypothesis, there exists a finitely generated submodule N of A such that $\operatorname{soc} A \leq N$. Suppose that $K = \operatorname{soc} K$. Then $\operatorname{soc} K$ is a direct summand of A and hence also a direct summand of N. It follows that $\operatorname{soc} K$ is finitely generated, which is a contradiction. Thus $K \neq \operatorname{soc} K$. Similarly, $K' \neq \operatorname{soc} K'$. Now $A/\operatorname{soc} A \cong [K/(\operatorname{soc} K)] \oplus [K'/(\operatorname{soc} K')]$. It follows that the modules $K/(\operatorname{soc} K)$ and $K'/(\operatorname{soc} K')$ each have smaller uniform dimension than $A/\operatorname{soc} A$. By induction on the uniform dimension of $A/\operatorname{soc} A$, we conclude that K and K' both have finite uniform dimension, and hence so does $A = K \oplus K'$, a contradiction. Thus A has finite uniform dimension. □

Next we prove a decomposition theorem for WC_{11}^{+}-modules.

Theorem 4.178. *Suppose the module A satisfies WC_{11}^{+} and $A/(\operatorname{soc} A)$ has finite uniform dimension. Then A contains a semisimple submodule A_1 and a submodule A_2 with finite uniform dimension such that $A = A_1 \oplus A_2$.*

Proof. If $A = \operatorname{soc} A$, then there is nothing to prove. Suppose that $A \neq \operatorname{soc} A$. Let $a \in A$, $a \notin \operatorname{soc} A$. By hypothesis, there exist submodules K and K' of A such that $A = K \oplus K'$, $\operatorname{soc}(Ra) \cap K = 0$ and $\operatorname{soc}(Ra) \oplus K$ is essential in A. Let $\pi : A \to K'$ be the canonical projection. Then $\operatorname{soc}(Ra) \oplus K = \pi(\operatorname{soc}(Ra)) \oplus K$. It follows that

$\pi(\operatorname{soc}(Ra))$ is an essential submodule of K' and hence $\operatorname{soc} K' = \pi(\operatorname{soc}(Ra)) \leq \pi(Ra) = R\pi(a)$, and so $\operatorname{soc} K' \leq R\pi(a)$. By Lemma 4.177, K' has finite uniform dimension. Note that $\pi(a) \in K'$ and $\pi(a) \notin \operatorname{soc} K'$. Thus, $K' \neq \operatorname{soc} K'$. Now $A/\operatorname{soc} A \cong [K/(\operatorname{soc} K)] \oplus [K'/(\operatorname{soc} K')]$ implies that the module $K/(\operatorname{soc} K)$ has smaller uniform dimension than $A/\operatorname{soc} A$. By induction on the uniform dimension of $A/\operatorname{soc} A$, there exist submodules K_1 and K_2 of K such that $K = K_1 \oplus K_2$, K_1 is semisimple, and K_2 has finite uniform dimension. Then A is the direct sum of the semisimple submodule K_1 and the submodule $K_2 \oplus K'$, which has finite uniform dimension. □

Corollary 4.179. *Let A be a module which satisfies WC_{11}^+ and* acc *(respectively,* dcc*) on essential submodules. Then $A = A_1 \oplus A_2$ for some semisimple submodule A_1 and Noetherian (respectively, Artinian) submodule A_2.*

Proof. We prove the result in the acc case; the dcc case is similar. Suppose A satisfies acc on essential submodules. By Proposition 2.13, $A/(\operatorname{soc} A)$ is Noetherian. Hence, by Theorem 4.178, $A = A_1 \oplus A_2$ for some semisimple submodule A_1 and submodule A_2 with finite uniform dimension. Now $\operatorname{soc} A = A_1 \oplus (\operatorname{soc} A_2)$ and hence $A/(\operatorname{soc} A) \cong A_2/(\operatorname{soc} A_2)$. Thus $A_2/(\operatorname{soc} A_2)$ is Noetherian. But $\operatorname{soc} A_2$ is Noetherian, because A_2 has finite uniform dimension. Thus A_2 is Noetherian. □

The following example shows that in Theorem 4.178 the module A need not be a direct sum of uniform modules.

Example 4.180. Let R be a principal ideal domain. If R is not a complete discrete valuation ring, then there exists an indecomposable torsion-free R-module A of rank 2 by [Kap69, Thm. 9]. For A, $\operatorname{soc} A = 0$, so that A satisfies WC_{11}^+ and ${}_RA$ has finite uniform dimension, namely 2. But A is not a direct sum of uniform modules.

The next few results deal with the endomorphism ring of WC_{11}-modules. For instance, we study WC_{11}-modules with a conditional direct summand property. We keep using the terminology of Exercise 3.45 and Theorem 4.60. We begin with the following observation.

Proposition 4.181. *If A is a WC_{11}-module satisfying C_3 and $A = A_1 \oplus A_2$ with $\operatorname{soc} A_2$ essential in A_2, then A_1 is WC_{11}.*

Proof. Immediate by Theorem 4.170. □

Theorem 4.182. *Let A be a module with essential socle. If A satisfies WC_{11} and C_2, then $S/\triangle$ is a regular ring and $\triangle = J(S)$.*

Proof. Let $\alpha \in S$ and let $K = \operatorname{soc}(\ker \alpha)$. By WC_{11}, there exists a direct summand L of A such that L is a complement of K in A. Since $\operatorname{soc} A$ is essential in A, $\ker \alpha \cap L = 0$ and hence $\alpha|_L$ is a monomorphism. By C_2, $\alpha(L)$ is a direct summand of A. The rest of the proof is the same as in Theorem 4.60. □

Note that the assumption that the module has essential socle in Theorem 4.182, is necessary as the following example shows.

Example 4.183. Let R be as in Example 3.102. Let A denote the R-module R. Then A satisfies WC_{11} and C_2, but $J(S) \neq \triangle$.

Proof. First note that R is a commutative local ring. Thus $S/J(S)$ is a (von Neumann) regular ring. Since $\operatorname{soc} R = 0$, A satisfies WC_{11}. By Example 3.102, A also satisfies C_2. It is straightforward to check that $\triangle \neq J(S)$. □

Corollary 4.184. *Let A be a nonsingular module with essential socle. If A satisfies WC_{11} and C_2, then S is a regular ring.*

Proof. Since A is nonsingular, $\triangle = 0$, by Lemma 4.61. Hence the result follows from Theorem 4.182. □

Finally, we consider WC_{11}-rings. A ring R is a right WC_{11}*-ring* if that the right R-module R is WC_{11}. Recall that a polynomial ring over a right CS-ring need not be a CS-ring, in general (Example 3.78). The following well-known lemma shows that a polynomial ring over any ring is a WC_{11}-ring.

Lemma 4.185. *Let R be any ring and let $S = R[x]$ be the polynomial ring. Then* $\operatorname{soc} S = 0$.

Proof. Suppose $\operatorname{soc} S \neq 0$. Let L be a minimal right ideal of S and $f(x)$ of minimal degree in L. Note that $(f(x))x \in L$ and hence $f(x)$ generates L. It follows that the leading coefficient of f is a zero divisor in R. Hence its annihilator annihilates all of f. It follows that there is no polynomial $g(x)$ such that $[(f(x))x]g(x) = f(x)$. Thus $\operatorname{soc} S = 0$. □

Theorem 4.186. *Let A be a left R-module where $R = ReR$ for some idempotent e in R and $S = eRe$. Then*

(i) *the left R-module A has WC_{11} if and only if the left S-module eA has WC_{11}.*
(ii) *R_R has WC_{11} if and only if the eRe-module Re has WC_{11}.*

Proof. Follows from Propositions 2.77 and 2.78. □

Corollary 4.187. *Let R be any ring. Then $M_m(R)$ is a right WC_{11}-ring if and only if the free right R-module $R^{(m)}$ has WC_{11}.*

Proof. By Theorem 4.186. □

Corollary 4.188. *If R is a right WC_{11}-ring, then so does the full matrix ring $M_m(R)$.*

Proof. Observe that $R^{(m)}$ is a right WC_{11}-module by Theorem 4.164. Now Corollary 4.187 gives the result. □

Exercises

4.31. Write out the proof of Theorem 4.164.

4.32. Prove that the Specker group $\prod_{i=1}^{\infty} \mathbb{Z}$ is a WC_{11}-module.

4.33. Find modules $A = A_1 \oplus A_2$ such that $\operatorname{soc} A_1 = 0$, $\operatorname{soc} A_2 = A_2$, and A does not satisfy the C_{11} property.

Remarks. Weak CS-modules were introduced by Smith [Smi90]. Example 4.3 is due to Zhou [Zh99], which shows that a direct sum of weak CS-modules need not be weak CS, in general. Amongst other results, Theorems 4.12 and 4.21 were obtained by Smith [Smi90].

C_{11}-modules were first asked as an open problem in [MM90, p. 106]. Smith and Tercan [ST93] investigated the class of modules with the property that every submodules has a complement, which is a direct summand under the name C_{11}-modules. It is shown that the class of modules with C_{11} is closed under direct sums but not closed under direct summands. To this end, Theorems 4.26, 4.29, and 4.31 are due to Smith and Tercan [ST93]. As shown Example 4.33, C_{11} property does not inherited by direct summands. Note that the aforementioned example appeared in [ST04] and originated in [Swan62]. A module is FI-extending if every fully invariant submodule is essential in a direct summand. FI-extending modules were introduced in [BMR02] and [BCFG01]. Every C_{11}-module enjoys with FI-extending property. Propositions 4.67, 4.68, and 4.78 are due to Birkenmeier and Tercan [BT07]. Several decompositions results are obtained on C_{11}-modules with conditional direct summand properties. In particular, Theorems 4.60 and 4.64 are due to Tercan [Ter94] and [Ter01]

It is also obtained the transference of the C_{11} condition from a given ring or module to various ring or module extensions. In particular, Theorems 4.83 and 4.87 show that if R is right C_{11}-ring, then the ring of column finite matrices of size Γ over R, the ring of m-by-m upper triangular matrices over R, and any right essential overring T of R are all right C_{11}-rings. Theorem 4.84 is proven in [BT07], which provides necessary and sufficient conditions to make the generalized triangular matrix ring is a right C_{11}-ring. Propositions 4.91, 4.93 and Corollary 4.97 were proved by Birkenmeier and Tercan [BT07]. The former results yield that for a module A all essential extensions of A satisfying C_{11} are essential extensions of C_{11}-modules constructed from A and certain subsets of endomorphisms of the injective hull of A and if A is a C_{11}-module, then so is its rational hull.

Birkenmeier, Tercan, and Yücel [BTY14] introduced a framework which encompasses of the generalizations of the CS property. Let $\emptyset \neq \underline{C} \subseteq L(A)$, where $L(A)$ is the set of submodules of A. Then A is called $\underline{C}$-extending if for each $X \in \underline{C}$ there exists a direct summand D of A such that X is essential in D. This concept was introduced by Oshiro in [Oshi83]. However, in the results of his article, he assumes that $\underline{C}$ is closed under isomorphisms and essential extensions. In [DS98], Dogrouz and Smith introduced type 1 χ-extending and type 2 χ-extending modules, where they assumed that χ is a class of modules containing the zero submodule and closed under isomorphisms. In contrast to [DS98] and [Oshi83], we do not implicitly assume that $\underline{C}$ is closed under isomorphisms or essential extensions. For example, we take $\underline{C} = \{$fully invariant submodules$\}$. Lemma 4.111 was obtained by Birkenmeier and Lennon [BiL14]. Theorems 4.102, 4.113, 4.117, 4.118, 4.127, and 4.128 are due to Birkenmeier, Tercan, and Yücel [BTY14]. $\underline{C}$-extending modules with respect to essential extensions are investigated and obtained Theorems 4.137 and 4.129 in [BTY16].

The projection (fully) invariant and extending concepts are naturally related in the quasi-continuous (quasi-injective) notions. By [MM90, Thm. 2.8] ([Lam99, Thm. 6.74]), A is quasi-continuous (quasi-injective) if and only if $A \trianglelefteq_{\text{p}} E(A)$ ($A \trianglelefteq E(A)$). Moreover if A is an injective module, then each $X \trianglelefteq_{\text{p}} A$ ($X \trianglelefteq A$) is quasi-continuous (quasi-injective). A ring R is right projection invariant extending ($\underline{PI}$-extending) if R_R is $\underline{PI}$-extending. In [BTY16], the authors used "π-extending" to denote $\underline{PI}$-extending. It is obtained the transfer of the $\underline{PI}$-extending condition between a ring R and its various ring extensions. More specifically, it is characterized the right $\underline{PI}$-extending generalized triangular matrix rings, and it is shown that if R_R is $\underline{PI}$-extending, then so is T_T, where T is an overring of R, which is an essential extension of R, a n-by-n upper triangular matrix ring of R, a column finite or column and row finite matrix ring over R, or a certain type of trivial extension of R. Proposition 4.149, Theorems 4.150 4.152 and 4.157 were proved by Birkenmeier, Tercan, and Yücel [BTY16].

Weak C_{11}-modules were introduced by Er [Er99] and investigated in [Ter01], [Ter03], and [Ter04]. While most of the material on weak C_{11}-modules can be found in the aforementioned papers.

Chapter 5

Outer Generalizations of Extending Modules

Chapter 5 treats generalized forms of extending modules which are not included in the previous chapter. Specifically, we consider generalized extending forms based on technical machinery conditions like existence of a homomorphism into a direct summand or an equivalence relation on the lattice of submodules etc. (so-called outer generalization). We then apply our results to matrix rings. In this way we build a basis for further research in the area. Moreover, we obtain implications between inner and outer generalizations of extending modules. We support our results with several interesting examples.

5.1 C_{12}-modules

We show that for any ring R the class of R-modules with the C_{12} property contains the class of modules which satisfy C_{11}, and we establish some properties of modules with C_{12}. A module A satisfies C_{12} if, for each submodule N of A, there exist a direct summand K of A and a monomorphism $\alpha : N \to K$ such that $\alpha(N)$ is an essential submodule of K.

Lemma 5.1. *A module A satisfies C_{12} if and only if for each complement N in A there exist a direct summand K of A and a monomorphism $\alpha : N \to K$ such that $\alpha(N)$ is an essential submodule of K.*

Proof. The necessity is clear. Conversely, suppose A satisfies the stated property for complements. Let L be any submodule of A. There exists a complement N in A such that L is an essential submodule of N. By hypothesis, there exist a direct summand K of A and a monomorphism $\alpha : N \to K$ such that $\alpha(N)$ is an essential submodule of K. But $\alpha(L)$ is essential in $\alpha(N)$, and this implies that $\alpha(L)$ is essential in K. It follows that A satisfies C_{12}. □

Lemma 5.1 makes it clear that modules with C_1 (or extending) satisfy C_{12}. We show next that modules with C_{11} satisfy C_{12}.

Proposition 5.2. *If the module A satisfies C_{11}, then A satisfies C_{12}.*

Proof. Let N be a submodule of A. Then there exist submodules K and K' of A such that $A = K \oplus K'$, $N \cap K' = 0$, and $N \oplus K'$ is essential in A (Proposition 4.25). Let $\pi : A \to K$ be the canonical projection mapping and α the restriction of π to the submodule N. Then $\alpha : N \to K$ is a monomorphism. Let $0 \neq k \in K$. Then there exists $r \in R$ such that $0 \neq rk = x + k'$ for some $x \in N$, $k' \in K'$. Now $rk = \pi(rk) = \pi(x + k') = \pi(x) = \alpha(x)$. Thus $Rk \cap \alpha(N) \neq 0$ for all $0 \neq k \in K$. Thus $\alpha(N)$ is essential in K. □

Proposition 5.3. *Every module A is isomorphic to a direct summand of a module which satisfies C_{12}.*

Proof. Let $A' = E(E(A) \oplus E(A) \oplus E(A) \oplus \cdots)$. Note that A' is injective. Let $A'' = A \oplus A'$. So A is isomorphic to the direct summand $A \oplus 0$ of A''. Now we show that A'' satisfies C_{12}. Note that $E(A'') = E(A) \oplus A' = E(E(A) \oplus E(A) \oplus E(A) \oplus \cdots)$, which is isomorphic to A' and hence there exists a monomorphism $\beta : A'' \to A'$. Let N be a submodule of A''. Then $\beta(N)$ is a submodule of A'. But A' is injective, so there exists a direct summand K of A' (and hence of A'') such that $\beta(N)$ is an essential submodule of K. Thus A'' satisfies C_{12}. □

We next give an example of a $\mathbb{Z}$-module which satisfies C_{12}, but not C_{11}. First we prove.

Lemma 5.4. *The Specker group does not satisfy C_{12}.*

Proof. Let A be the Specker group $\prod_{i=1}^{\infty} \mathbb{Z}$ and let N be the subgroup $\bigoplus_{i=1}^{\infty} \mathbb{Z}$ of A. Suppose that there exist a direct summand K of A and a monomorphism $\alpha : N \to K$ such that $\alpha(N)$ is essential in K. Note that N is isomorphic to $\alpha(N)$. By Nunke's Theorem [Nu62, Thm. 5], K is isomorphic to A. This implies K has uncountable rank. But $\alpha(N)$ has countable rank. Thus $\alpha(N)$ cannot be essential in K. □

Corollary 5.5. *There exists a $\mathbb{Z}$-module A satisfying C_{12} and which has a direct summand K that does not satisfy C_{12}.*

Proof. By Proposition 5.3 and Lemma 5.4. □

Proposition 5.6. *There exists a $\mathbb{Z}$-module A which satisfies C_{12} but not C_{11}.*

Proof. By Proposition 5.3, if A is the Specker group $\prod_{i=1}^{\infty} \mathbb{Z}$, then there exists an injective $\mathbb{Z}$-module A' such that $A'' = A \oplus A'$ satisfies C_{12}. By Propositions 4.35 and 5.2 and Lemma 5.4, A'' does not satisfy C_{11}. □

Related to Proposition 5.6, we have the following easy lemma.

Lemma 5.7. *Let R be a left Noetherian ring and A a nonzero indecomposable left R-module. Then the following conditions are equivalent.*

(i) *A is uniform.*
(ii) *A has C_{11}.*
(iii) *A has C_{12}.*
(iv) *A is $\underline{P}\,\underline{I}$-extending.*

Proof. (i) $\Rightarrow$ (ii) is clear.

(ii) $\Rightarrow$ (iii) By Proposition 5.2.

(iii) $\Rightarrow$ (iv) Since A is indecomposable, every submodule of A is projection invariant in A. Let X be a nonzero submodule of A. Then there exists a nonzero element $x \in X$. Now $R/\ell(x) \cong Rx$. Thus there exists a uniform submodule U of A such that $U \leq Rx \leq A$. By hypothesis, there exists a monomorphism $\varphi : U \to A$ such that $\varphi(U)$ is essential in A. Since $U \cong \varphi(U)$, we have that U is essential in A, and hence X is essential in A.

(iv) $\Rightarrow$ (i) Let N be a nonzero submodule of A. Then N is a projection invariant submodule of A. By (iv), there exists a direct summand K of A such that N is essential in K. Hence $K = A$. So A is uniform. □

The next result shows that the C_{12} property is closed under direct sums.

Theorem 5.8. *Any direct sum of modules with property C_{12} satisfies C_{12}.*

Proof. Let A_λ $(\lambda \in \Lambda)$ be a nonempty collection of modules, each satisfying C_{12}. Let $A = \bigoplus_{\lambda\in\Lambda} A_\lambda$ and Λ' be a nonempty subset of Λ containing λ such that $A' = \bigoplus_{\lambda\in\Lambda'} A_\lambda$. Assume that A' satisfies C_{12} property. Then for any submodule N' of A' there exist a direct summand K' of A' and a monomorphism $\alpha_1 : N' \to K'$ such that $\alpha_1(N')$ is essential in K'. Suppose $\Lambda \neq \Lambda'$. Then there exists $\mu \in \Lambda$, $\mu \notin \Lambda'$. Let $\Lambda'' = \Lambda \cup \{\mu\}$ and $A'' = \bigoplus_{\lambda\in\Lambda''} A_\lambda = A' \oplus A_\mu$. Since A_μ is a C_{12}-module, for any submodule N_μ of A_μ there exist a direct summand K_μ of A_μ and a monomorphism $\alpha_2 : N_\mu \to K_\mu$ such that $\alpha_2(N_\mu)$ is essential in K_μ. It is clear that $K' \cap K_\mu = 0$. Let $K'' = K' \oplus K_\mu$. Note that K'' is a direct summand of A''. Consider the submodule $N' \oplus N_\mu$ of A''. Define $\beta : N' \oplus N_\mu \to K' \oplus K_\mu$ by $\beta(n) = \beta(n_1 + n_2) = \alpha_1(n_1) + \alpha_2(n_2)$, where $n \in N' \oplus N_\mu$, $n_1 \in N'$, and $n_2 \in N_\mu$. It is easy to check that β is a monomorphism. Furthermore, $\beta(N' \oplus N_\mu) = \alpha_1(N') + \alpha_2(N_\mu)$ is an essential submodule of $K' \oplus K_\mu$ by Proposition 2.1. Now, let N be any submodule of A. Using a transfinite induction argument, there exist a direct summand K of A and a monomorphism $\gamma : N \to K$ such that $\gamma(N)$ is an essential in K. Thus A satisfies C_{12}. □

Corollary 5.9. *Any direct sum of modules which satisfy C_{11} (extending and uniform, respectively) satisfies C_{12}.*

Proof. This is an immediate consequence of Theorem 5.8. □

In contrast to direct sums, the C_{12} property is not inherited by direct summands. Observe that over the ring of integers this is shown by Corollary 5.5. However, the following example exhibits a C_{12}-module with a direct summand that does not satisfy C_{12} (see also Example 4.33).

Example 5.10. Let $\mathbb{R}$ be the real field and n any odd integer with $n \geq 3$. Let S be the ring $\mathbb{R}[x_1, x_2, \ldots, x_n]$ of polynomials in the indeterminates $x_1, x_2, \ldots, x_n$. Let R be the ring S/Ss, where $s = x_1^2 + \cdots + x_n^2 - 1$. Let P be the R-module with generators $s_1, s_2, \ldots, s_n$ and relation $\Sigma_{i=1}^n x_i s_i = 0$. Then the R-module $P \oplus R$ satisfies C_{12}, but P does not satisfy C_{12}.

Proof. Let $A = P \oplus R$. Then it is clear that A is a free R-module. Note that P is an indecomposable R-module by [Swan62, Thm. 3]. Now by Theorem 5.8, A is a C_{12}-module. Since P has uniform dimension $n-1$, it follows that P is not uniform. By Lemma 5.7, P is not a C_{12}-module. □

Observe that in the previous example, the module P has uniform dimension $n-1$. Hence P contains $n-1$ uniform submodules U_i $(1 \le i \le n-1)$ such that $\bigoplus_{i=1}^{n-1} U_i$ is essential in P. By Corollary 5.9, $\bigoplus_{i=1}^{n-1} U_i$ has C_{12}. Hence in general the C_{12} property is not closed under essential extensions.

In view of Example 5.10, we have the following.

Corollary 5.11. *Let R be a left Noetherian ring such that the left R-module R is a C_{12}-module and every direct summand of a C_{12}-module is a C_{12}-module. Then every indecomposable projective left R-module is uniform.*

Proof. Let P be an indecomposable projective R-module. Then there exists a free R-module F such that $F = P \oplus P'$ for some submodule P'. By Theorem 5.8, F satisfies C_{12}. By hypothesis, P is a C_{12}-module. Therefore P is uniform, thanks to Lemma 5.7. □

Then next result provides a condition which ensures that a direct summand of a module satisfies the C_{12} condition.

Theorem 5.12. *Let $A = A_1 \oplus A_2$. Then A_1 satisfies C_{12} if and only if for every submodule N of A_1, there exists a direct summand K of A and a monomorphism φ on N such that $A_2 \subseteq K$, $\varphi(N) \cap K = 0$ and $\varphi(N) \oplus K$ is an essential submodule of A.*

Proof. Suppose A_1 satisfies C_{12}. Let N be any submodule of A_1. There exist a direct summand L of A_1 and a monomorphism $\varphi : N \to L$ such that $\varphi(N)$ is an essential submodule of L. So $A_1 = L \oplus L'$ for some submodule L' of A. Clearly, $L' \oplus A_2$ is a direct summand of A, $(L' \oplus A_2) \cap \varphi(N) = 0$ and, $(L' \oplus A_2) \oplus \varphi(N)$ is an essential submodule of A. Conversely, suppose A_1 has the stated property. Let H be a submodule of A_1. By hypothesis, there exist a direct summand K of A and a monomorphism on H such that $A_2 \subseteq K$, $\varphi(H) \cap K = 0$, and $\varphi(H) \oplus K$ is an essential submodule of A. Now $K = K \cap (A_1 \oplus A_2) = (K \cap A_1) \oplus A_2$. Hence, $K \cap A_1$ is a direct summand of A, and so also of A_1. Let $A_1 = (K \cap A_1) \oplus X$ for some $X \le A_1$ and $\pi : A \to X$ be the canonical projection with kernel K. Define $f : H \to X$ by $f(h) = \pi(\varphi(h))$, where $h \in H$. It is easy to check that f is a monomorphism. Let $0 \neq x \in X$. Then there exists $r \in R$ such that $0 \neq rx \in \varphi(H) \oplus K$. It follows that $rx = \pi(rx) = f(h) + \pi(k)$ for some $h \in H$, $k \in K$. So $0 \neq rx = f(h) \in f(H)$. We see that $f(H)$ is an essential submodule of X. Thus, A_1 satisfies C_{12}. □

Recall that a module A is continuous if and only if A satisfies Q_1 and C_1 (Theorem 3.100). We shall show that A is continuous if it satisfies Q_1 and C_{12}.

Lemma 5.13. *Suppose that A satisfies Q_1. Then, for every submodule K of A which is isomorphic to a complement in A and any direct summand L of A such that $K \cap L = 0$, $K \oplus L \in \mathrm{Lift}_A(A)$.*

Proof. Let $\varphi : K \oplus L \to A$ be a homomorphism. $A = L \oplus L'$ for some submodule L' of A. There exists a homomorphism $\alpha : A \to A$ such that $\varphi(y) = \alpha(y)$ $(y \in L)$. Let $\chi = \varphi - \alpha$; then $\chi : K \oplus L \to A$ is a homomorphism and $\chi(L) = 0$. Let $\pi : A \to L'$ denote the canonical projection. Then $\pi(K) \cong K$. Define $\beta : \pi(K) \to A$ by $\beta(\pi(k)) = \chi(k)$ $(k \in K)$. By hypothesis, there exists a homomorphism $\gamma : A \to A$ such that $\gamma(\pi(k)) = \chi(k)$ $(k \in K)$. Now define a homomorphism $\theta : A \to A$ by $\theta(x + x') = \gamma(x')$ $(x \in L, x' \in L')$. Let $a \in K \oplus L$, so that $a = y + z$ for some $y \in K, z \in L$. Then, $a = \pi(y) + y - \pi(y) + z$, and so $\theta(a) = \gamma(\pi(y)) = \chi(y) = \chi(a)$. Thus χ lifts to A, and φ lifts to A. □

The next result generalizes Lemma 3.97.

Lemma 5.14. *Let K be a submodule of the module A such that K is isomorphic to a complement in A. Then K is a direct summand of A if and only if there exists a complement L of K in A such that $K \oplus L \in \mathrm{Lift}_A(A)$.*

Proof. Suppose that K is a direct summand of A. Then $A = K \oplus K'$ for some submodule K' of A. Thus $L = K'$ will do. Conversely suppose that K is isomorphic to a complement submodule K' of A. Hence there exists an isomorphism $\alpha : K \to K'$. Define $\beta : K \oplus L \to K'$ by $\beta(x + y) = \alpha(x)$ $(x \in K, y \in L)$. By hypothesis, there exists a homomorphism $\theta : A \to A$ such that $\theta|_{K \oplus L} = \beta$. Note that $K' = \beta(K) = \theta(K)$, which is a submodule of $\theta(A)$. Let $0 \neq a \in \theta(A)$. Therefore, $a = \theta(a')$ for some $a' \in A$ and $a' \notin \ker\theta$. But $L \subseteq \ker\theta$, so we have $a' \notin L$. This implies that $K \cap (L + Ra') \neq 0$. Let $0 \neq x \in K \cap (L + Ra')$. There exist $y \in L$, $r \in R$ such that $x = y + ra'$. Thus $\alpha(x) = 0 + ra \neq 0$. Hence, $K' \cap Ra \neq 0$ for all $0 \neq a \in \theta(A)$, i.e., K' is an essential submodule of $\theta(A)$. Since K' is a complement in A, we have $K' = \theta(A)$. Let $a \in A$. Then $\theta(a) = \theta(x)$ for some $x \in K$. Therefore $A = K + (\ker\theta)$. Since $K \cap (\ker\theta) = 0$, $A = K \oplus (\ker\theta)$. □

Theorem 5.15. *A module A is continuous if and only if A satisfies Q_1 and C_{12}.*

Proof. The necessity is clear by Theorem 3.100. Conversely, suppose that A satisfies Q_1 and C_{12}. Let K be a complement in A. There exist a direct summand L of A and a monomorphism $\alpha : K \to L$ such that $\alpha(N)$ is an essential submodule of L. Further, there exists a submodule L' of A such that $A = L \oplus L'$. Consider the submodule $\alpha(K) \oplus L'$. Let $\varphi : \alpha(K) \oplus L' \to A$ be a homomorphism. By Lemma 5.13, φ lifts to A. Now apply Lemma 5.14 to conclude that $\alpha(K)$ is a direct summand of A. By Proposition 2.85, K is a direct summand of A. It follows that A satisfies C_1. Also by Proposition 2.85, A satisfies C_2. Thus A is continuous. □

We now consider conditions on a module A which imply that A is a direct sum of uniform modules. We begin with the following theorem for modules over the ring of integers which are direct sums of uniform modules.

Theorem 5.16. *Let A be a $\mathbb{Z}$-module which is a direct sum of uniform modules. Then A satisfies C_{12}^{+}.*

Proof. Let N be a direct summand of A. Then N is also a direct sum of uniform modules, by Theorem 4.45. By Corollary 5.9, N satisfies C_{12}. □

Proposition 5.17. *Let A be a non-zero module with finite uniform dimension. Then the following statements are equivalent.*

(i) *Every direct summand of A is a (finite) direct sum of uniform modules.*
(ii) *A satisfies C_{11}^{+}.*
(iii) *A satisfies C_{12}^{+}.*

Proof. (i) ⇒ (ii) By Corollary 4.28.

(ii) ⇒ (iii) By Proposition 5.2.

(iii) ⇒ (i) Every direct summand of A satisfies C_{12}^{+} and has finite uniform dimension. Thus, it suffices to prove that A is a direct sum of uniform submodules. Let n denote the uniform dimension of A. If $n = 1$, then A is uniform. Suppose $n > 1$. Let U be a uniform submodule of A. Since A satisfies C_{12}, there exist submodules K and K' of A such that $A = K \oplus K'$ and a monomorphism $\alpha : U \to K$ such that $\alpha(U)$ is an essential submodule of K. Then $\alpha(U)$, and hence K, are uniform. Since K' has uniform dimension $n-1$ and satisfies C_{12}^{+}, induction on n shows that K' is a direct sum of uniform submodules. Thus A is a direct sum of uniform modules, as required. □

Lemma 5.18. *Let A be a module which satisfies (A). Suppose that A satisfies C_{12} and C_2. Then every local summand of A is a direct summand.*

Proof. Let $X = \sum_{\lambda\in\Lambda} X_\lambda$ be a local summand of A. Because A satisfies C_{12}, there exist a direct summand K of A and a monomorphism $\alpha : X \to K$ such that $\alpha(X)$ is essential in K. Now $A = K \oplus K'$ for some submodule K' of A. Consider $\alpha(X) \oplus K' = (\sum_{\lambda\in\Lambda} \alpha(X_\lambda)) \oplus K'$. For any finite subset Λ' of Λ, $Y = \bigoplus_{\lambda\in\Lambda'} \alpha(X_\lambda)$ is a direct summand of A, because A satisfies C_2. Thus $\alpha(X) \oplus K'$ is a local summand of A. By Lemma 4.50, $\alpha(X) \oplus K'$ is a complement in A. But $\alpha(X) \oplus K'$ is an essential submodule of A. Thus $A = \alpha(X) \oplus K'$. By hypothesis, X is a direct summand of A. □

Theorem 5.19. *Let A be a module which satisfies (A), C_{12} and C_2. Then A is a direct sum of indecomposable modules.*

Proof. By Lemma 5.18 and [MM90, Thm. 2.17]. □

Corollary 5.20. *Let A be a module which satisfies (A). Suppose that A satisfies C_{12} and C_2. Then the finite exchange property implies the full exchange property.*

Proof. By Theorem 5.19 and [ZH84, Cor. 6]. □

Corollary 5.21. *Let R be a left Noetherian ring and A a left R-module which satisfies C_{12}^+ and C_2. Then A is a direct sum of uniform modules. Moreover, A satisfies C_{11}^+.*

Proof. By Theorem 5.19, there exists an index set Λ such that $A = \bigoplus_{\lambda\in\Lambda} A_\lambda$, where A_λ is an indecomposable submodule of A for each $\lambda \in \Lambda$. By hypothesis, A_λ satisfies C_{12}. Lemma 5.7 yields that A_λ is uniform. For the last statement, apply Corollary 4.28. □

Next we focus on decompositions of a C_{12}^+-module as a direct sum of a module of finite uniform dimension and a semisimple module. For analogous results in the extending, C_{11}, and WC_{11} cases we refer to [CY91], Theorem 4.64, and Theorem 4.178. Since we will use it in our results, we need to recall the following definition. For a module A, we say that $\operatorname{soc} A$ is *cancellable* whenever $socA = B\oplus C \cong B\oplus D$ for some submodules, then $C \cong D$, i.e., the socle of A is cancellable as an R-module (see [Lam99] and [TT09]). Let R be a ring as in Example 4.176 with $\dim_K V = 2$. Then $R/(\operatorname{soc} R)$ has finite uniform dimension, namely 1, and $\operatorname{soc} R$ is cancellable. However R does not have C_{12}^+. One can ask whether a C_{12}^+-module A such that $A/(\operatorname{soc} A)$ has finite uniform dimension has to have cancellable socle. The next easy example shows that in general this is not the case.

Example 5.22. Let p be any prime integer. Let A be the $\mathbb{Z}$-module $\bigoplus_i \mathbb{Z}/\mathbb{Z}p$. Then A has C_{12}^+ by Theorem 5.16. It is clear that $A/(\operatorname{soc} A)$ has finite uniform dimension, namely zero. However, it can be seen that $\operatorname{soc} A$ is not cancellable.

Lemma 5.23. *Let A be a module such that A satisfies C_{12}^+ and $A/(\operatorname{soc} A)$ has finite uniform dimension. Suppose that $\operatorname{soc} A$ is cancellable and contained in a finitely generated submodule of A. Then A has finite uniform dimension.*

Proof. Suppose A does not have finite uniform dimension. Then $\operatorname{soc} A$ is not finitely generated. There exist submodules S_1, S_2 of $\operatorname{soc} A$ such that S_i is not finitely generated for $i = 1, 2$, and $\operatorname{soc} A = S_1 \oplus S_2$. There exist a direct summand K' of A and a monomorphism $\alpha : S_1 \to K'$ such that $\alpha(S_1)$ is essential in K'. Hence, there exists a submodule K of A such that $A = K \oplus K'$ and $\alpha(S_1) \oplus K \leq_e K' \oplus K = A$. Note that $\operatorname{soc} A = S_1 \oplus S_2 = \operatorname{soc}(\alpha(S_1) \oplus K) = \operatorname{soc}(\alpha(S_1)) \oplus \operatorname{soc} K$. Since $\operatorname{soc}(\alpha(S_1)) \oplus \operatorname{soc} K \cong \alpha(S_1) \oplus \operatorname{soc} K$ and $\operatorname{soc} A$ is cancelable, $S_2 \cong \operatorname{soc} K$, and hence $\operatorname{soc} K$ is not finitely generated. On the other hand, $\operatorname{soc} A = \operatorname{soc} K \oplus \operatorname{soc} K' \cong \operatorname{soc} K \oplus \alpha(S_1)$, so that $S_1 \cong \alpha(S_1) \cong \operatorname{soc} K'$, and hence $\operatorname{soc} K'$ is not finitely generated. By hypothesis, there exists a finitely generated submodule N of A such that $\operatorname{soc} A \leq N$. Suppose that $K' = \operatorname{soc} K'$. Then $\operatorname{soc} K'$ is a direct summand of A and hence also of N, and it follows that $\operatorname{soc} K$ is finitely generated, a contradiction. Thus $K' \neq \operatorname{soc} K'$. Similarly, $K \neq \operatorname{soc} K$. But $A/(\operatorname{soc} A) = K/(\operatorname{soc} K) \oplus K'/(\operatorname{soc} K')$. It follows that the modules $K/(\operatorname{soc} K)$ and $K'/(\operatorname{soc} K')$ each have smaller uniform dimension than $A/(\operatorname{soc} A)$. By induction on the uniform dimension of $A/\operatorname{soc}(A)$, we conclude that K and K' both have finite uniform dimension, and hence so does $A = K \oplus K'$, a contradiction. Thus A has finite uniform dimension. □

Theorem 5.24. *Let A be a module which satisfies C_{12}^{+} and for which $A/(\operatorname{soc} A)$ has finite uniform dimension. Suppose that $\operatorname{soc} A$ is cancellable. Then A contains a semisimple submodule A_1 and a submodule A_2 with finite uniform dimension such that $A = A_1 \oplus A_2$.*

Proof. If $A = \operatorname{soc} A$, then there is nothing to prove. Suppose that $A \neq \operatorname{soc} A$. Let $a \in A$, $a \notin \operatorname{soc} A$. There exist a direct summand K' of A and a monomorphism $\alpha : Ra \to K'$ such that $R\alpha(a)$ is essential in K'. Hence there exists a submodule K of A such that $A = K \oplus K'$. Let $\pi' : A \to K'$ denote the canonical projection. Clearly, $R\alpha(a) \oplus K = R\pi'(\alpha(a)) \oplus K$. It follows that $R\pi'(\alpha(a))$ is an essential submodule of K' and hence $\operatorname{soc} K' \leq R\pi'(\alpha(a))$. By Lemma 5.23, K' has finite uniform dimension. Note that $R\pi'(\alpha(a)) \cong R\alpha(a) \cong Ra$ and hence $\pi'(\alpha(a)) \in K'$, $\pi'(\alpha(a)) \notin \operatorname{soc} K'$. Thus $K' \neq \operatorname{soc} K'$. Since $A/(\operatorname{soc} A) \cong K/(\operatorname{soc} K) \oplus K'/(socK')$, it follows that the module $K/\operatorname{soc} K$ has smaller uniform dimension than $A/(\operatorname{soc} A)$. By induction on the uniform dimension of $A/(\operatorname{soc} A)$, there exist submodules K_1, K_2 of K such that $K = K_1 \oplus K_2$, K_1 is semisimple, and K_2 has finite uniform dimension. Thus A is the direct sum of the semisimple submodule K_1 and the submodule $K_2 \oplus K'$, which has finite uniform dimension. □

Corollary 5.25. *Let A be a module which satisfies C_{12}^{+} and* acc *(respectively,* dcc*) on essential submodules. Assume that* $\operatorname{soc} A$ *is cancellable. Then $A = A_1 \oplus A_2$ for some semisimple submodule A_1 and Noetherian (respectively, Artinian) submodule A_2.*

Proof. We prove the result in the acc case; the dcc case is similar. Suppose A satisfies acc on essential submodules. By Proposition 2.13, $A/(\operatorname{soc} A)$ is Noetherian. Hence, by Theorem 5.24, $A = A_1 \oplus A_2$ for some semisimple submodule A_1 and submodule A_2 with finite uniform dimension. Note that $\operatorname{soc} A = A_1 \oplus (\operatorname{soc} A_2)$, and hence $A/(\operatorname{soc} A) \cong A_2/(\operatorname{soc} A_2)$. Thus $A_2/(\operatorname{soc} A_2)$ is Noetherian. But $\operatorname{soc} A_2$ is Noetherian, because A_2 has finite uniform dimension. Therefore, A_2 is Noetherian. □

Note that there are modules which satisfy condition WC_{11} but not condition C_{12} (see Example 4.180). Now we show that the properties WC_{11} and C_{12} are independent.

Proposition 5.26. *Let R be a left Noetherian ring with left essential socle which does not satisfy C_{12}. Then there exists a left R-module which satisfies C_{12}, but does not satisfy WC_{11}.*

Proof. By the proof of Proposition 5.3, if A is the ring R, then $A'' = R \oplus A'$, where $A' = E(E(R) \oplus E(R) \oplus \cdots)$, satisfies C_{12} but does not satisfy C_{11}. Since $\operatorname{soc} A''$ is essential in A'', A'' does not satisfy WC_{11}, by Proposition 4.163. □

Let us give an example of a ring R which is covered by Proposition 5.26, and shows that we cannot remove the C_{12}^+ assumption in Theorem 5.24 and Corollary 5.25.

Example 5.27. Let R be the finite ring $\begin{bmatrix} \mathbb{Z}_4 & 2\mathbb{Z}_4 \\ 0 & \mathbb{Z}_4 \end{bmatrix}$. Then, $\operatorname{soc} R_R = \begin{bmatrix} 2\mathbb{Z}_4 & 2\mathbb{Z}_4 \\ 0 & 2\mathbb{Z}_4 \end{bmatrix}$ is cancellable and essential in R_R. It is clear that $R/(\operatorname{soc} R)$ has finite uniform dimension. Now, assume that R_R has C_{12}^+. Let $D = \begin{bmatrix} \mathbb{Z}_4 & 2\mathbb{Z}_4 \\ 0 & 0 \end{bmatrix}$, which is a direct summand of R. Then there exists a monomorphism $\varphi : N \to D$, where $N = \begin{bmatrix} 2\mathbb{Z}_4 & 0 \\ 0 & 0 \end{bmatrix}$ such that $\varphi(N)$ is essential in D. Hence $N \cong \varphi(N)$ is uniform which yields a contradiction. Hence there is no such monomorphism φ. We see that R_R does not satisfy C_{12}^+.

We know that a WC_{11}-module decomposes into a direct sum of a module with essential socle and a module with zero socle (Proposition 4.165). We have the following analogous result for C_{12}-modules with cancellable socle.

Proposition 5.28. *Let A be a C_{12}-module. If* $\operatorname{soc} A$ *is cancellable, then $A = A_1 \oplus A_2$ for some submodules A_1, A_2 such that* $\operatorname{soc} A_1$ *is essential in A_1 and A_2 has zero socle.*

Proof. Let $N = \operatorname{soc} A$. Then there exist a direct summand A_1 of A and a monomorphism $\alpha : N \to A_1$ such that $\alpha(N)$ is essential in A_1. Hence $A = A_1 \oplus A_2$ for some submodule A_2 of A. So $\alpha(N) \oplus \operatorname{soc} A_2$ is essential in $\operatorname{soc} A$. Therefore, $\operatorname{soc} N \oplus \operatorname{soc} A_2 \cong \operatorname{soc} A_1 \oplus \operatorname{soc} A_2 \oplus \operatorname{soc} A_2$. By assumption, $\operatorname{soc} A_2 = 0$. Hence A has the required decomposition. □

Finally, we construct a counterexample to the converse of Proposition 5.28. Specifically, we exhibit a module over a commutative ring R, that is a direct sum of a module with essential socle and a nonzero module with zero socle that neither satisfies condition C_{12}, nor has cancellable socle. This counterexample also clarifies the pattern of Example 4.166.

Let S be a commutative domain which is not a field and whose Jacobson radical $J(S) = 0$. Let V be a faithful semisimple S-module. Note that, since $J(S) = 0$, such a module exists and it has infinite uniform dimension, because it can contain an infinite direct sum of pairwise non-isomorphic simple S-modules. Let $R = \{\left[\begin{smallmatrix} s & v \\ 0 & s \end{smallmatrix}\right] : s \in S, v \in V\}$ denote the trivial extension of S by V. Let I be the ideal of R, $I = \{\left[\begin{smallmatrix} 0 & v \\ 0 & 0 \end{smallmatrix}\right] : v \in V\}$. Since V is semisimple and S has zero socle, we have $\operatorname{soc} R = I$ and $\operatorname{soc}(R/I) = 0$. Since V is faithful, I is an essential ideal of R. Thus R is a commutative ring with essential socle I.

Let $A = R \oplus (R/I)$. Note that $\operatorname{soc} A = I \oplus 0$ is not cancellable. We want to determine the direct sum decompositions of A, and to this aim we compute the idempotents of $\operatorname{End}_R(A)$.

Lemma 5.29. *With the notations above:*

(i) $\mathrm{End}_R(A) \cong \begin{bmatrix} R & I \\ R/I & R/I \end{bmatrix}$.

(ii) *The nontrivial idempotents (i.e., idempotents* $\neq 0,1$) *of* $\mathrm{End}_R(A)$ *are either of the form* $\begin{bmatrix} 1 & y \\ 0 & 1+I \end{bmatrix}$, *or* $\begin{bmatrix} 1 & y \\ t+I & 0+I \end{bmatrix}$, *or* $\begin{bmatrix} 0 & 0 \\ t+I & 1+I \end{bmatrix}$, *where* $y \in I$ *and* $t = \begin{bmatrix} s & 0 \\ 0 & s \end{bmatrix} \in R$.

Proof. (i) By Exercise 5.2, we identify $\mathrm{End}_R(A)$ with

$$\begin{bmatrix} \mathrm{End}_R(R) & \mathrm{Hom}_R(R/I, R) \\ \mathrm{Hom}_R(R, R/I) & \mathrm{End}_R(R/I) \end{bmatrix}.$$

Using the morphism that assigns to each endomorphism "the image of one" we see that $\mathrm{End}_R(R) \cong R$, $\mathrm{End}_R(R/I) \cong R/I$, and $\mathrm{Hom}_R(R, R/I) \cong R/I$. If $f : R/I \to R$, then $f(R/I)I = 0$. As S is a domain, this implies that $f(R/I)I \subseteq I$. Hence, as I is an R/I-module, $\mathrm{Hom}_R(R/I, R) = \mathrm{Hom}_{R/I}(R/I, I) \cong I$. This finishes the proof of (i).

(ii) Assume that $\begin{bmatrix} s & v \\ 0 & s \end{bmatrix}$ is an idempotent of R. Then

$$\begin{bmatrix} s & v \\ 0 & s \end{bmatrix}^2 = \begin{bmatrix} s^2 & sv+vs \\ 0 & s^2 \end{bmatrix} = \begin{bmatrix} s & v \\ 0 & s \end{bmatrix}.$$

Hence, s is an idempotent of S and, since S is a domain, this implies that $s \in \{0,1\}$. It follows that $v = 0$. Therefore, R has only trivial idempotents. As $R/I \cong S$, it also follows that the only idempotents of R/I are the trivial ones.

Using (i), now we are ready to compute the idempotents of $\mathrm{End}_R(A)$. Let $\begin{bmatrix} r & y \\ t_1+I & t_2+I \end{bmatrix}$ be a nontrivial idempotent of $\mathrm{End}_R(A)$. Then, observing that $I \cdot R/I = R/I \cdot I = 0$, we have

$$\begin{bmatrix} r & y \\ t_1+I & t_2+I \end{bmatrix}^2 = \begin{bmatrix} r^2 & ry \\ t_1r+t_2t_1+I & t_2^2+I \end{bmatrix} = \begin{bmatrix} r & y \\ t_1+I & t_2+I \end{bmatrix}.$$

Then r, being an idempotent of R, must be either 0 or 1. The same holds for t_2+I. Since $t_1r+t_2t_1+I = t_1+I$, we have the following possibilities.

Case 1. If $t_1+I = 0$, then we obtain the idempotents $\begin{bmatrix} 1 & y \\ 0 & 1+I \end{bmatrix}$ and $\begin{bmatrix} 1 & y \\ 0 & 0+I \end{bmatrix}$, for any $y \in I$ and, finally $\begin{bmatrix} 0 & 0 \\ 0 & 1+I \end{bmatrix}$.

Case 2. It $t_1+I \neq 0$, then either $r_1 = 1$ and $t_2+I = 0+I$, or $r_1 = 0$ and $t_2+I = 1+I$. Note that in the latter case $y = 0$. Therefore, the idempotents we obtain in this situation are either of the form $\begin{bmatrix} 1 & y \\ t+I & 0+I \end{bmatrix}$, or of the form $\begin{bmatrix} 0 & 0 \\ t+I & 1+I \end{bmatrix}$ for $y \in I$ and $t + I \in R/I \backslash \{0\}$. □

As a consequence we have the following result.

Lemma 5.30. *Let L be a direct summand of A. Then either* $\operatorname{soc} L = \operatorname{soc} A$ *or* $\operatorname{soc} L = 0$.

Proof. We use the fact that the direct summands of A are precisely the images of the idempotent endomorphism of A.

For any endomorphism $f : A \to A$ of the from $\begin{bmatrix} 1 & \star \\ \star & \star \end{bmatrix}$, the image of $I \oplus 0$ under f is $I \oplus 0$, hence $f(A)$ contains the socle of A. If the endomorphism f is of the form $\begin{bmatrix} 0 & 0 \\ \star & \star \end{bmatrix}$, then $f(A) \subseteq 0 \oplus (R/I)$ and, hence, $\operatorname{soc}(f(A)) = 0$. In Lemma 5.29, we prove that all the idempotents of $\operatorname{End}_R(A)$ are in one of these two classes of endomorphisms, thus our claim follows from the previous observations. □

Corollary 5.31. *With the notations above, ${}_RA$ does not satisfy C_{12}.*

Proof. Assume that ${}_RA$ satisfies C_{12}. Let N be a simple submodule of A. Then there exists a monomorphism $\alpha : N \to L$, where L is a direct summand of A, such that $\alpha(N)$ is essential in L. Since $N \cong \alpha(N)$, $\operatorname{soc} L = \operatorname{soc}(\alpha(N)) = \alpha(N)$. By Lemma 5.30, either $\alpha(N) = 0$, or $\alpha(N) = \operatorname{soc} A$. In any case, we have a contradiction. It follows that A does not satisfy C_{12}. □

We conclude from Corollary 5.31 that in general the converse of Proposition 5.28 is not true.

In a similar way to weak CS-modules (or WC_{11}-modules) weak C_{12}-modules were introduced in [Er99] and investigated in [KaT16]. Recall that a module A is a *weak* C_{12} (or WC_{12})-module if each semisimple submodule of A can be essentially embedded in a direct summand of A. Most of the proof techniques in the setting of C_{12}-modules work well for WC_{12}-modules as well. Accordingly, the results for WC_{12}-modules are given without detailed proofs.

Theorem 5.32. *Any direct sum of with WC_{12}-modules satisfies WC_{12}.*

Proof. Similar to that of Theorem 5.8. □

Corollary 5.33. *Any direct sum of modules which satisfy WC_{11} (respectively, one of the extending, uniform, or C_{11} conditions) satisfies WC_{12}.*

Proof. Immediate by Theorem 5.32. □

Applying Theorem 5.32, we obtain the following easy fact on modules over Dedekind domains.

Corollary 5.34. *Let R a Dedekind domain and A be an R-module with finite uniform dimension. Then A is a WC_{12}-module.*

Proof. This follows from Corollary 4.13. □

Recall that direct summands of a C_{12}-module need not to be C_{12}-modules (Corollary 5.5 or Example 5.10). We do not know whether direct summands of a WC_{12}-module need to be WC_{12}-modules or not. We next deal with some special cases of this last question.

Corollary 5.35. *Let A be a left R-module and $A = U \oplus V$, where U and V are uniform submodules. Then every direct summand of A is WC_{12}-module.*

Proof. Let $0 \neq K$ be a direct summand of A. If $K = A$, then K has property WC_{12} thanks to Corollary 5.33. If $K \neq A$, then K is uniform. Hence K has WC_{12}. □

Theorem 5.36. *Let A be a $\mathbb{Z}$-module which is a direct sum of uniform modules. Then every direct summand of A is a WC_{12}-module.*

Proof. Let N be a direct summand of A. Then A is a direct sum of uniform modules by Theorem 4.45. Now, Corollary 5.33 gives that N satisfies WC_{12}. □

The following result provides a condition which ensures that a direct summand of a module is a WC_{12}-module.

Theorem 5.37. *Let $A = A_1 \oplus A_2$. Then A_1 satisfies WC_{12} if and only if for every semisimple submodule N of A_1, there exist a direct summand K of A and a monomorphism φ on N such that $A_2 \subseteq K$, $\varphi(N) \cap K = 0$, and $\varphi(N) \oplus K$ is an essential submodule of A.*

Proof. Modified proof of Theorem 5.12 gives the result. □

A question posed in [Er99, p. 496] is whether condition WC_{12} implies condition C_{12} or not? We answer this question negatively by providing several counterexamples. Observe that by the latter examples, we have then, for any module, the following strict implications;

$$\begin{array}{ccccc} \text{CS} & \Rightarrow & C_{11} & \Rightarrow & C_{12} \\ \Downarrow & & \Downarrow & & \Downarrow \\ Weak\,\text{CS} & \Rightarrow & WC_{11} & \Rightarrow & WC_{12} \end{array}$$

Example 5.38. The Specker group satisfies WC_{12}, but not C_{12}.

Proof. Let A be the Specker group $\prod_{i=1}^{\infty} \mathbb{Z}$. First note that, by Lemma 5.4, ${}_{\mathbb{Z}}A$ does not satisfy C_{12}. By [Go76, Prop. 1.22], ${}_{\mathbb{Z}}A$ is nonsingular. Hence, ${}_{\mathbb{Z}}A$ has zero socle by [Go76, Cor. 1.26]. So ${}_{\mathbb{Z}}A$ is a WC_{12}-module. □

Example 5.39. Let R and A be as in Example 4.180. Then A is indecomposable of rank 2 and $\operatorname{soc} A = 0$. Clearly ${}_RA$ is a WC_{12}-module. However, by Lemma 5.7, ${}_RA$ does not have property C_{12}.

Even though Examples 5.38 and 5.39 would suffice as counterexamples for the aforementioned question, we provide next more algebraic topology type examples which are interesting in their own right.

Theorem 5.40. *With the notations in Example* 4.33*,*

(i) *If* n *is* 1 *or* $2, 4, 8$ *then* ${}_RK$ *has* C_{12}.
(ii) *If* n *is any odd integer with* $n \geq 3$*, then* ${}_RK$ *has* WC_{12}*, but not* C_{12}.

Proof. (i) If $n = 1$, then ${}_RK \cong 0$. Further, if $n = 2, 4, 8$, then ${}_RK$ corresponds to a free module of finite rank (see [Swan62, Thm. 3]). It follows that ${}_RK$ is isomorphic to a (finite) direct sum of uniform modules. Hence ${}_RK$ has C_{12} by Theorem 5.8 (or Corollary 5.9).

(ii) Since ${}_RA$ has zero socle, the same is true for ${}_RK$. It follows that ${}_RK$ is a WC_{12}-module. Observe that ${}_RK$ has uniform dimension $n - 1$. Now, Lemma 5.7 yields that ${}_RK$ does not satisfy property C_{12}. □

Exercises

5.1. Write out the proof of Corollary 5.25 for the Artinian case.

5.2. Let $A = A_1 \oplus A_2$ be an R-module. Prove that

$$\operatorname{End}_R(A) \cong \begin{bmatrix} \operatorname{End}_R(A_1) & \operatorname{Hom}_R(A_2, A_1) \\ \operatorname{Hom}_R(A_1, A_2) & \operatorname{End}_R(A_2) \end{bmatrix}.$$

5.3. Determine $\operatorname{End}_R(A)$ if ${}_RA = (\mathbb{Z}/\mathbb{Z}p) \oplus \mathbb{Z}$ and ${}_RA = (\mathbb{Z}/\mathbb{Z}p) \oplus \mathbb{Q}$, where p is any prime integer.

5.4. Let $R = \mathbb{Z}_{(p)}$ be the localization of the integers $\mathbb{Z}$ at a prime p. Let ${}_RA = \mathbb{Z}_{(p)} \oplus \mathbb{Z}(p^\infty)$. Show that ${}_RA$ has the full exchange property.

5.2 Self-c-injective modules

In this section, we are concerned with modules A such that every homomorphism from a complement submodule of A to A can be lifted to A. Although such modules share some of the properties of extending modules, we show that they form a substantially bigger class of modules.

Recall that a module A is quasi-injective if, for any submodule N of A, any homomorphism $\alpha : N \to A$ can be lifted to a homomorphism $\beta : A \to A$. Continuous and quasi-continuous modules are other classes of modules that can be characterized by the lifting of homomorphisms from certain submodules to the module itself, as was shown in Theorems 3.99 and 3.100. In fact, in §3.4, we studied the following property for a module A:

P_n For every submodule K of A such that K can be written as a finite direct sum $K_1 \oplus K_2 \oplus \cdots \oplus K_n$ of complements $K_1, K_2, \ldots, K_n$ of A, every homomorphism $\alpha : K \to A$ can be lifted to a homomorphism $\beta : A \to A$, i.e., $K \in \mathrm{Lift}_A(A)$.

Here we investigate *self-c-injective modules*, i.e., modules A that satisfy P_1 in [SCS00]. Recall that property P_1 is a generalization of the extending property and serves a basis for further investigations in this direction. This section presents results with detailed proofs which are essentially contained in [SCS00].

Let A_1 and A_2 be modules. The module A_2 is A_1-*c-injective* if $K \leq_c A_1$ implies $K \in \mathrm{Lift}_{A_2}(A_1)$. Clearly, if A_2 is A_1-injective, then A_2 is A_1-c-injective. A module A is called *self-c-injective* (or P_1) when it is A-c-injective. Observe that, by Corollary 2.50, every extending module is self-c-injective. But not every self-c-injective module is extending (Example 2.51).

Compare the following result with [DHSW94, Lem. 7.5] and [SCS98, Lems. 3 and 4].

Lemma 5.41. *Let A_1 and A_2 be modules and let $A = A_1 \oplus A_2$. Then A_2 is A_1-c-injective if and only if for every (complement) submodule N of A such that $N \cap A_2 = 0$ and $\pi_1(N) \leq_c A_1$, where $\pi_1 : A \to A_1$ denotes, the canonical projection, there exists a submodule N' of A such that $N \leq N'$ and $A = N' \oplus A_2$.*

Proof. Assume that A_2 is A_1-c-injective and let N be a submodule of A such that $N \cap A_2 = 0$ and $\pi_1(N) \leq_c A_1$. As $N \cap A_2 = 0$, $\pi_1|_N : N \to \pi_1(N)$ is an isomorphism. Consider the homomorphism $\alpha : \pi_1(N) \to A_2$, $\alpha(x) = \pi_2(\pi_1|_N)^{-1}(x)$, where $\pi_2 : A \to A_2$ is the canonical projection. The map α can be extended to a homomorphism $\beta : A_1 \to A_2$, since A_2 is A_1-c-injective and $\pi_1(N) \leq_c A_1$. Define $N' = \{x + \beta(x) : x \in A_1\}$. Clearly, N' is a submodule of A and $A = N' \oplus A_2$. For every $x \in N$, $\beta\pi_1(x) = \alpha\pi_1(x) = \pi_2(x)$ and hence $x = \pi_1(x) + \beta\pi_1(x) \in N'$. Thus, $N \leq N'$.

Conversely, assume that for every submodule N of A such that $N \cap A_2 = 0$ and $\pi_1(N) \leq_c A_1$, there exists a submodule N' of A such that $N \leq N'$ and $A = N' \oplus A_2$. Let $K \leq_c A_1$ and let $\alpha : K \to A_2$ be a homomorphism. Define $N = \{x - \alpha(x) : x \in K\}$. Clearly, N is a submodule of A such that $N \cap A_2 = 0$. It is easy to see that $\pi_1(N) = K$, and so $\pi_1(N) \leq_c A_1$. Then, by hypothesis, there exists a submodule N' of A such that $N \leq N'$ and $A = N' \oplus A_2$. Let $\pi : A \to A_2$ denote the projection with kernel N' and let $\beta : A_1 \to A_2$ be the restriction of π to A_1. For any $x \in K$, $\beta(x) = \pi(x) = \pi((x - \alpha(x)) + \alpha(x)) = \alpha(x)$, and therefore, $K \in \mathrm{Lift}_{A_2}(A_1)$. Thus, A_2 is A_1-c-injective.

Finally, observe that, if N is a submodule of A such that $N \cap A_2 = 0$ and $\pi_1(N) \leq_c A_1$, then $N \leq_c A$. In fact, if $N \leq_e K$, then $K \cap A_2 = 0$ and π_1 gives an isomorphism between K and $\pi_1(K)$. Therefore, from $N \leq_e K$ we can conclude that $\pi_1(N) \leq_e \pi_1(K)$. On the other hand, $\pi_1(N) \leq_c A_1$, and so $\pi_1(N) = \pi_1(K)$. Thus, $N = K \leq_c A$. □

Lemma 5.42. *Let A_1 and A_2 be modules. If A_2 is A_1-c-injective, then for every complement submodule N of A_1, A_2 is N-c-injective and A_1/N-c-injective.*

Proof. Let $N \leq_c A_1$. As every complement submodule of N is also a complement submodule of A_1, it is obvious that A_2 is N-c-injective. Let us prove that A_2 is (A_1/N)-c-injective. Let $K/N \leq_c A_1/N$ and consider a homomorphism $\alpha : K/N \to A_2$. By Proposition 2.9, $K \leq_c A_1$. Let $\pi : A_1 \to A_1/N$ and $\pi' : K \to K/N$ be the canonical epimorphisms. As A_2 is A_1-c-injective, there exists a homomorphism $\beta : A_1 \to A_2$ that extends $\alpha\pi'$. Since $N \leq \ker\beta$, the existence of a homomorphism $\gamma : A_1/N \to A_2$ such that $\alpha\pi = \beta$ is guaranteed. For every $a \in K$, $\gamma(a+N) = \gamma\pi(a) = \beta(a) = \alpha\pi'(a) = \alpha(a+N)$. Therefore, $K \in \mathrm{Lift}_{A_2}(A_1/N)$ and A_2 is A_1/N-c-injective. □

Lemma 5.43. *Let A and $\{A_i : i \in I\}$ be modules. Then $\prod_{i\in I} A_i$ is A-c-injective if and only if A_i is A-c-injective, for every $i \in I$.*

Proof. The proof is the same as for injectivity. □

The modules A_1 and A_2 are *relatively* c-*injective* if A_i is A_j-c-injective, for every $i, j \in \{1, 2\}$, $i \neq j$.

Corollary 5.44. *Let A_1 and A_2 be modules. If $A_1 \oplus A_2$ is self-c-injective, then A_1 and A_2 are both self-c-injective and are relatively c-injective. In particular, a direct summand of a self-c-injective module is self-c-injective.*

Proof. This follows from Lemmas 5.42 and 5.43. □

The converse of Corollary 5.44 is not true in general. Consider, e.g., the $\mathbb{Z}$-modules $A_1 = (\mathbb{Z}/\mathbb{Z}p) \oplus 0$ for a prime p and $A_2 = 0 \oplus \mathbb{Z}$. Both A_1 and A_2 are uniform, so they are self-c-injective and relatively c-injective. By Example 2.52, $A_1 \oplus A_2$ is not self-c-injective.

In order to obtain sufficient conditions for a direct sum of two self-c-injective modules to be self-c-injective, we need the following module property and lemma.

Let A_1 and A_2 be modules. The module A_2 is *essentially A_1-injective* if every homomorphism $\alpha : K \to A_2$, where K is a submodule of A_1 and $\ker\alpha \leq_e K$, can be extended to a homomorphism $\beta : A_1 \to A_2$ (see [DHSW94]).

Lemma 5.45. *Let A_1 and A_2 be modules such that A_2 is essentially A_1-injective. If a module is A_1-c-injective and A_2-injective, then it is $(A_1 \oplus A_2)$-c-injective.*

Proof. Let $A = A_1 \oplus A_2$ and suppose that N is an A_1-c-injective and A_2-injective module. Let $K \leq_c A$ and consider a homomorphism $\alpha : K \to N$. Take $H \leq_c K$ such that $K \cap A_1 \leq_e H$. Then $H \cap A_1 = K \cap A_1 \leq_e H$ and, because A_2 is essentially A_1-injective by [SCS98, Lem. 4], there exists a submodule H' of A such that $A = H' \oplus A_2$ and $H \leq H'$. Clearly, $H \leq_c H'$ and, since A_1 and

H' are isomorphic, N is H'-c-injective. Thus, there exists a homomorphism $\beta : H' \to N$ that extends $\alpha|_H$. Obviously, β can be extended to the homomorphism $\beta\pi : A \to N$, where $\pi : A \to H'$ is the projection of A onto H' with kernel A_2. Consider the homomorphism $\alpha - \beta\pi : K \to N$. As $K \cap A_1 \leq H \leq \ker(\alpha - \beta\pi)$, $\alpha - \beta\pi$ can be lifted to a homomorphism $\gamma : K/(K \cap A_1) \to N$, defined by $\gamma(x + K \cap A_1) = \alpha(x) - \beta\pi(x)$. The homomorphism $\phi : K/(K \cap A_1) \to A_2$, $\phi(x + K \cap A_1) = \pi_2(x)$, where $\pi_2 : A \to A_2$ is the projection, is clearly injective. Since N is A_2-injective, there exists $\delta : A_2 \to N$ such that $\delta\phi = \gamma$. Consider, finally, the homomorphism $\theta = \beta\pi + \delta\pi_2 : A \to N$. For all $x \in K$, $\theta(x) = \beta\pi(x) + \delta\pi_2(x) = \beta\pi(x) + \delta\phi(x + K \cap A_1) = \beta\pi(x) + \gamma(x + K \cap A_1) = \beta\pi(x) + \alpha(x) - \beta\pi(x) = \alpha(x)$. Therefore, θ extends α and N is A-c-injective. □

Now we have the following.

Theorem 5.46. *Let $A_1, \ldots, A_n$ ($n \in \mathbb{N}$) be relatively injective modules. Then $A_1 \oplus \cdots \oplus A_n$ is self-c-injective if and only if A_i is self-c-injective, for every $i \in \{1, \ldots, n\}$.*

Proof. This follows from Lemmas 5.43 and 5.45, by using induction. □

Let A_1 and A_2 be modules. The module A_2 is A_1-cu-*injective* if every homomorphism $\alpha : K \to A_2$, where K is a complement uniform submodule of A_1, can be extended to a homomorphism $\beta : A_1 \to A_2$. Clearly, if A_2 is A_1-c-injective, then A_2 is A_1-cu-injective. However, the converse is false in general as we shall see in Corollary 5.49. A module A is called *self*-cu-*injective* when it is A-cu-injective.

Proposition 5.47. *Let A_1 be an extending module and let A_2 be a uniform module such that A_2 is essentially A_1-injective. Then $A_1 \oplus A_2$ is self-*c*-injective if and only if it is self-*cu*-injective.*

Proof. The necessity is obvious. Let us prove the converse. Suppose that $A = A_1 \oplus A_2$ is self-cu-injective. Let K be a complement submodule of A and let $\alpha : K \to A$ be a homomorphism. Take $H_1 \leq_c K$ such that $K \cap A_1 \leq_e H_1$. Then $H_1 \cap A_1 = K \cap A_1 \leq_e H_1$ and, since A_2 is essentially A_1-injective, by [SCS98, Lem. 4], there exists a submodule H of A such that $A = H \oplus A_2$ and $H_1 \leq H$. Clearly, $H_1 \leq_c H$ and, since A_1 and H are isomorphic, H is extending. Thus, $H = H_1 \oplus H_2$, for some submodule H_2 of H, so that $A = H_1 \oplus H_2 \oplus A_2$ and $K = H_1 \oplus L$, where $L = (H_2 \oplus A_2) \cap K$. Since $L \cap A_1 = (H_2 \oplus A_2) \cap K \cap A_1 \leq (H_2 \oplus A_2) \cap H_1 = 0$, L embeds in A_2 and hence is zero or uniform. By hypothesis, $\alpha|_L$ lifts to A and, in particular, there exists a homomorphism $\beta : H_2 \oplus A_2 \to A$ that extends $\alpha|_L$. Therefore, it is clear that α can be extended to A. We conclude that A is self-c-injective. □

Next we focus on self-c-injective modules over commutative domains.

Theorem 5.48. *Let R be a commutative domain which is not a field. Then every self-*c*-injective free R-module is finitely generated.*

Proof. Suppose that F is a self-c-injective free module that is not finitely generated. By Corollary 5.44, we can assume, without loss of generality, that $F = R \oplus R \oplus \cdots$. Consider the elements $e_n = (0, \dots, 0, 1, 0, 0, \dots) \in F$, where 1 is the nth component of e_n, for each positive integer n. Let Q denote the field of fractions of R and let $0 \neq c \in R$. We aim to prove that c is a unit in R. Define a homomorphism $\varphi : F \to Q$ by $\varphi(r_1, r_2, \dots) = r_1 + c^{-1}r_2 + c^{-1}r_3 + \cdots$, for all $(r_1, r_2, \dots) \in F$, and let $K = \ker \varphi$. Consider the elements $f_n = (0, \dots, 0, 1, -c, 0, 0, \dots) = e_n - ce_{n+1} \in K$, for each positive integer n, and let us prove that K is a free submodule of F with basis $\{f_1, f_2, \dots\}$. Clearly, $L = \sum_{n \in \mathbb{N}} Rf_n \leq K$.

In order to prove that $L = K$, let $(r_1, r_2, \dots, r_n, 0, 0, \dots) \in K$. Then $r_1 + c^{-1}r_2 + \cdots + c^{-n+2}r_{n-1} + c^{-n+1}r_n = 0$ and hence $r_n + cr_{n-1} + \cdots + c^{n-2}r_2 + c^{n-1}r_1 = 0$. Thus $(r_1, r_2, \dots, r_{n-1}, r_n, 0, 0, \dots) = (r_1, \dots, r_{n-1}, -cr_{n-1}, \dots, -c^{n-1}r_1, 0, 0, \dots) = r_1x_1 + \cdots + r_{n-1}x_{n-1}$, where $x_i = e_i - c^{n-i}e_n$, for $1 \leq i \leq n-1$. Clearly, $x_{n-1} = f_{n-1} \in L$. If, for some $2 \leq i \leq n-1$, $x_i \in L$, then $x_{i-1} = f_{i-1} + cx_i \in L$. By induction, $x_i \in L$, for all $1 \leq i \leq n-1$, and hence $(r_1, r_2, \dots, r_n, 0, 0, \dots) \in L$. It follows that $K \leq L$ and hence $K = L$.

Let us prove that the set $\{f_1, f_2, \dots\}$ is linearly independent. Suppose that, for a positive integer m and for some $s_i \in R$, $1 \leq i \leq m$, we have $s_1f_1 + \cdots + s_mf_m = 0$, i.e., $s_1(1, -c, 0, 0, \dots) + s_2(0, 1, -c, 0, 0, \dots) + \cdots + s_m(0, \dots, 0, 1, -c, 0, 0, \dots) = 0$, so that $s_1 = 0$, $s_2 - cs_1 = 0$, $s_3 - cs_2 = 0, \dots$, $s_m - cs_{m-1} = 0$, $-cs_m = 0$. Thus, $s_1 = s_2 = \cdots = s_m = 0$. It follows that K is a free module with basis $\{f_n : n \in \mathbb{N}\}$. Hence K is isomorphic to F. Moreover, F/K, being isomorphic to a submodule of ${}_RQ$, is a torsion-free module and hence K is a complement submodule of F. Then, by Corollary 2.50, K is a direct summand of F. Suppose that $F = K \oplus K'$. Now, K' is isomorphic to F/K, which in turn isomorphic to $\varphi(F)$, so that K' is a uniform submodule of F. Let $0 \neq u \in K'$. Then $u \in Re_1 \oplus \cdots \oplus Re_q$, for some positive integer q and hence $K' \leq Re_1 \oplus \cdots \oplus Re_q$. So $e_{q+1} = z + (v_1, \dots, v_q, 0, 0, \dots)$, for some $z \in K$, $v_i \in R$, $1 \leq i \leq q$. Hence $c^{-q} = \varphi(e_{q+1}) = \varphi(v_1, \dots, v_q, 0, 0, \dots) = v_1 + c^{-1}v_2 + \cdots + c^{-q+1}v_q$, so that $c^{-1} = v_q + cv_{q-1} + \cdots + c^{q-1}v_1 \in R$. Thus c is a unit in R. It follows that R is a field, a contradiction. □

Corollary 5.49. *Let R be a Prüfer domain which is not a field. Then any non-finitely generated free R-module is self-*cu*-injective, but not self-*c*-injective.*

Proof. Let A be a free R-module with infinite basis $\{a_i : i \in I\}$. Let U be a complement uniform submodule of A and let $0 \neq u \in U$. There exists a finite subset F of I such that $u \in \bigoplus_{i \in F} Ra_i$. Since U/Ru is a torsion module, it follows that $U \subseteq \bigoplus_{i \in F} Ra_i$. By [DHSW94, Cor. 12.10], U is a direct summand of $\bigoplus_{i \in F} Ra_i$ and hence also of A. Thus A is self-cu-injective. By Theorem 5.48, A is not self-c-injective. □

Lemma 5.50. *Let U be an essential submodule of a module X. For every $x \in X$, the subset $\{(r, rx) : r \in R,\ rx \in U\}$ is a complement submodule of the R-module $A = R \oplus U$.*

Proof. Let $x \in X$ and let $V = \{(r, rx) : r \in R,\ rx \in U\}$. Clearly, V is a submodule of A and $V \cap (0 \oplus U) = 0$. Let W be any submodule of A such that $V \leq W$, $V \neq W$. Then there exist $s \in R$ and $u \in U$ such that $(s, u) \in W$ and $u \neq sx$. Hence $u - sx \neq 0$ and since $U \leq_e X$, $R(u - sx) \cap U \neq 0$. Therefore, there exists $r \in R$ such that $0 \neq r(u - sx) \in U$. Now $rsx = ru - y \in U$ and $(0, y) = r(s, u) - (rs, rsx) \in W$, so that $W \cap (0 \oplus U) \neq 0$. Thus V, being a complement of $0 \oplus U$ in A, is a complement in A. □

Theorem 5.51. *Let I be a non-zero ideal of a commutative domain R such that $I \leq Rc$, for some non-unit $c \in R$. Then the R-module $A = R \oplus (R/I)$ is not self-*c*-injective.*

Proof. Let $U = R/I$ and $u = 1 + I$, so that $U = Ru$. Let Q denote the field of fractions of R. Note that $I \leq Rc \leq R \leq Rc^{-1} \leq Q$. Thus U is a submodule of the R-module $(Rc^{-1})/I$. Let $r \in R$ be such that $rc^{-1} + I$ is a non-zero element of $(Rc^{-1})/I$. If $r \in Rc$, then $0 \neq rc^{-1} + I \in U \cap R(rc^{-1} + I)$. If $r \notin Rc$, then $0 \neq r + I = c(rc^{-1} + I) \in U \cap R(rc^{-1} + I)$. It follows that U is an essential submodule of $(Rc^{-1})/I$. Let $V = \{(r, rc^{-1} + I) : r \in R,\ rc^{-1} + I \in U\}$. Then $V = \{(r, rc^{-1} + I) : r \in Rc\} = \{(sc, s + I) : s \in R\}$ and V is a complement submodule of the R-module $A = R \oplus U$, by Lemma 5.50, Define a mapping $\alpha : V \to A$ by $\alpha(sc, s + I) = (s, I)$, for all $s \in R$. Clearly, α is a homomorphism. Suppose that α can be lifted to a homomorphism $\beta : A \to A$. Note that $a\beta(0, 1 + I) = \beta(0, I) = (0, I)$, for every non-zero $a \in I$, and hence $\beta(0, 1 + I) \in 0 \oplus U$. Now $(1, I) = \alpha(c, 1 + I) = \beta(c, 1 + I) = \beta(c, I) + \beta(0, 1 + I) = c\beta(1, I) + \beta(0, 1 + I) = c(d, e + I) + (0, f + I)$, for some $d, e, f, \in R$. It follows that $(1, I) = (cd, ce + f + I)$, and hence $cd = 1$, a contradiction. Therefore A is not a self-c-injective module. □

Corollary 5.52. *Let R be a commutative domain and let c be a non-zero non-unit element of R. Then the R-module $R \oplus (R/Rc)$ is not self-*c*-injective.*

Proof. Take $I = Rc$ in the Theorem 5.51. □

Finally, we look at self-c-injective modules over Dedekind domains. In order to establish that, over a Dedekind domain, the direct sum of an indecomposable torsion-free injective module and a cyclic torsion module is self-c-injective, we need the following lemmas.

Given a positive integer n, the modules $A_1, A_2, \ldots, A_n$ are said to be *compatible* if, for all $1 \leq i \leq n$ and elements $a_j \in A_j$, $1 \leq j \leq n$, we have $\ell(a_i) + \ell(\{a_j : 1 \leq j \leq n,\ i \neq j\}) = R$ [SCS00].

Lemma 5.53. *Let R be a left hereditary ring and let A be an R-module such that $A = A_0 \oplus A_1 \oplus \cdots \oplus A_n$ for some positive integer n, nonsingular injective submodule A_0 and singular uniform submodules $A_i = Ra_i$, $1 \leq i \leq n$, with $E(A_1), \ldots, E(A_n)$ compatible. Let K be a non-zero complement submodule of A such that $K \cap (A_1 \oplus \cdots \oplus A_n) = 0$. Then $x_0 + x_1 + \cdots + x_n \in K$, for some $0 \neq x_0 \in A_0$ and $x_i \in \{0, a_i\}$, $1 \leq i \leq n$. Moreover, $K \subseteq A_0 \oplus (\bigoplus_{i=1}^{n} Rx_i)$.*

Proof. There exists $0 \neq a = a' + a'' \in K$, where $a' \in A_0$ and $a'' \in N = A_1 \oplus \cdots \oplus A_n$. Since $K \cap N = 0$, it follows that $a' \neq 0$. There exists an essential left ideal E of R such that $Ea'' = 0$. Then $Ea = Ea' \neq 0$. Thus $K \cap A_0 \neq 0$. There exists a submodule A_0' of A_0 such that $A_0 = (K \cap A_0)E \oplus A_0'$. Note that $K \cap A_0 \cap A_0' = 0$. Since $K \cap N = 0$, it follows that K embeds in A_0, and hence K is nonsingular. Suppose that $K \cap (A_0' \oplus N) \neq 0$ and let $0 \neq x \in K \cap (A_0' \oplus N)$. Then $Fx \subseteq K \cap A_0' = K \cap A_0 \cap A_0' = 0$, for some essential left ideal F of R. Thus $x = 0$, that is, K is a complement of $A_0' \oplus N$ in $A = (K \cap A_0)E \oplus A_0' \oplus N$. By Lemma 3.8, there exists a homomorphism $\varphi : (K \cap A_0)E \to A_0' \oplus E(A_1) \oplus \cdots \oplus E(A_n)$ such that $K = \{y + \varphi(y) : y \in (K \cap A_0)E, \varphi(y) \in A_0' \oplus A_1 \oplus \cdots \oplus A_n\}$. For each $1 \leq i \leq n$, let $\pi_i : A_0' \oplus E(A_1) \oplus \cdots \oplus E(A_n) \to E(A_i)$ be the canonical projection. Let $1 \leq i \leq n$ and consider the homomorphism $\pi_i\varphi : (K \cap A_0)E \to E(A_i)$. Suppose that $\pi_i\varphi \neq 0$. Because R is left hereditary, $\pi_i\varphi((K \cap A_0)E)$ is a non-zero injective submodule of the indecomposable module $E(A_i)$, and hence $\pi_i\varphi((K \cap A_0)E) = E(A_i)$. In particular, there exists $e_0 \in (K \cap A_0)E$ such that $\pi_i\varphi(e_0) = a_i$. Now $\varphi(e_0) = e' + e_1 + \cdots + e_n$, for some $e' \in A_0'$, $e_j \in E(A_j)$, $1 \leq j \leq n$, and $e_i = a_i$. There exist $s \in \ell(a_i)$, $t \in \ell(e_1, \ldots, e_{i-1}, e_{i+1}, \ldots, e_n)$ such that $1 = s + t$. Then $\varphi(te_0) = te' + ta_i = te' + (1 - s)a_i = te' + a_i$. It follows that $te_0 + te' + a_i \in K$. Let $z_i = te_0 + te'$. Then $z_i \in A_0$ and $z_i + a_i \in K$. If $\pi_i\varphi = 0$, choose any $z_i \in K \cap A_0$. In any case, $z_i + x_i \in K$, where $x_i \in \{0, a_i\}$.
We have shown that, for each $1 \leq i \leq n$, there exists $z_i \in A_0$ such that $z_i + x_i \in K$, where $x_i = 0$, if $\pi_i\varphi = 0$, and $x_i = a_i$, if $\pi_i\varphi \neq 0$. Then $z + x_1 + \cdots + x_n \in K$, where $z = z_1 + z_2 + \cdots + z_n \in A_0$. Since $K \cap N = 0$, it follows that $z \neq 0$. Finally, note that $K = \{y + \varphi(y) : y \in (K \cap A_0)E, \varphi(y) \in A_0' \oplus A_1 \oplus \cdots \oplus A_n\} \subseteq A_0 \oplus (\bigoplus_{i=1}^n Rx_i)$, because $x_i = 0$ if and only if $\pi_i\varphi = 0$. □

Lemma 5.54. *Let R be a principal ideal domain and let the R-module $A = A_1 \oplus A_2$ be the direct sum of a torsion-free injective submodule A_1 and a cyclic torsion submodule A_2. Then A is self-*cu*-injective.*

Proof. There exists an element $a \in A_2$ such that $A_2 = Ra$. Let $I = \ell(a)$. Then I is a non-zero ideal of R. If $I = R$, then $A_2 = 0$ and there is nothing to prove. Suppose that $I \neq R$. Note that $I = P_1^{k_1}P_2^{k_2} \cdots P_n^{k_n}$, for some positive integers $n, k_i, 1 \leq i \leq n$, and distinct maximal ideals P_i, $1 \leq i \leq n$, of R. It follows that R/I is isomorphic to $(R/P_1^{k_1}) \oplus (R/P_2^{k_2}) \oplus \cdots \oplus (R/P_n^{k_n})$ and the R-module R/I is extending, by [Ha82, Thm. 7]. Since A_2 is isomorphic to R/I, we have $A_2 = L_1 \oplus \cdots \oplus L_n$, where $a = l_1 + l_2 + \cdots + l_n$, $L_i = Rl_i$ and $P_i^{k_i} = \ell(l_i)$, for $1 \leq i \leq n$. For each $1 \leq i \leq n$ and each $w \in E(L_i)$, there exists a positive integer v such that $P_i^v w = 0$, by [SV72, Prop. 4.23]. It follows that $E(L_1), \ldots, E(L_n)$ are compatible. Let U be a maximal uniform submodule of A. If $U \cap A_1 = 0$, then $U \subseteq A_2$, because A_2 is the torsion submodule of A. Since A_2 is isomorphic to R/I, it follows that U is a complement submodule of the extending module A_2, and hence U is a direct summand of A_2. In this case, it is clear that any homomorphism $\varphi : U \to A$ can be lifted to A. Now suppose that $U \subseteq A_1$. Then U is a direct summand of A_1 and any homomorphism $\varphi : U \to A$ can be lifted to A. Otherwise, $U \cap A_1 \neq 0$ and $U \nsubseteq A_1$. Clearly $U \cap A = 0$ and hence, by rearranging the modules $L_1, \dots, L_n$ if necessary, Lemma 5.53 gives that $R(l_0+l_1+\cdots+l_k) \subseteq U \subseteq A_1 \oplus L_1 \oplus \cdots \oplus L_k$, for some $0 \neq L_0 \in A_1$, $1 \leq k \leq n$. Let $X = L_1 \oplus \cdots \oplus L_k$ and $x = l_1 + \cdots + l_k$, so that $X = Rx$. Let $T = \{c \in R : cX = X\}$. Then T is a multiplicatively closed subset of the domain R and we let S denote the subring $\{r/t : r \in R,\ t \in T\}$ of Q, the field of fractions of R. Given $b \in A_1$ and $c \in X$, we define $(r/t)(b+c) = rb/t + c'r$, where $c' \in X$ satisfies $c = c't$, for all $r \in R$, $t \in T$. This makes $A_1 \oplus X$ into an S-module. Note that, for each $d \in T$, $x = drx$ for some $r \in R$, and hence $(1-dr)x = 0$, i.e., $(1-dr)X = 0$. It follows that X is T-torsion-free. Also, A_1 is a vector space over Q. Now, we claim that $U = S(l_0+x)$. Let $V = \{a \in A_1 \oplus X : ta \in U, \text{ for some } t \in T\}$. Clearly, V is a submodule of $A_1 \oplus X$ and $U \leq V$. Since $A_1 \oplus X$ is T-torsion-free, it follows that U is essential in V. Thus $U = V$.

Let $r \in R, t \in T$. Then $t(r/t)(l_0+x) = r(l_0+x) \in U$ and hence $(r/t)(l_0+x) \in U$. Thus $S(l_0+x) \subseteq U$. Let $u \in U$. Then $R(l_0+x)+Ru \subseteq U$, so that $R(l_0+x)+Ru$ is a finitely generated uniform module over a principal ideal domain and hence is cyclic. Suppose that $R(l_0+x)+Ru = R(p+dx)$, for some $p \in A_1$, $d \in R$. There exists $c \in R$ such that $l_0 + x = c(p+dx)$ and hence $x = cdx$. It follows that $1 - cd \in \ell(x)$ and hence $X = cdX \subseteq cX \subseteq X$, i.e., $X = cX$ and $c \in T$. Therefore, $p + dx = (1/c)(l_0+x) \in S(l_0+x)$. It follows that $u \in S(l_0+x)$. Thus $U = S(l_0+x)$.

Let $y = l_{k+1} + \cdots + l_n$, so that $a = x + y$. Because $P_1, \dots, P_n$ are distinct maximal ideals, $R = (P_1^{k_1} \cap \cdots \cap P_k^{k_k}) + (P_{k+1}^{k_{k+1}} \cap \cdots \cap P_n^{k_n}) = \ell(x) + \ell(y)$. Then, there exists $c \in R$ such that $cx = x$ and $cy = 0$. Clearly, $c \in T$. Let $\varphi : U \to A_2$ be an R-homomorphism. Suppose that $\varphi[(1/c)(l_0+x)] = fa$, for some $f \in R$. Then $\varphi(l_0+x) = c\varphi[(1/c)(l_0+x)] = cfa = cf(x+y) = fx$. A similar argument shows that $\varphi(U) \leq X$. If $r \in R$ and $t \in T$, then $(r/t)(l_0+x) = (r/t)l_0 + rx'$, where $x' \in X$ and $tx' = x$. So, $t(\varphi[(r/t)(l_0+x)]) = \varphi[r(l_0+x)] = rfx = rftx'$. Since X is T-torsion-free, it follows that $\varphi[(r/t)(l_0+x)] = rfx'$. Note that $A = A_1 \oplus Rx \oplus Ry$ and let $\pi : A \to Rx$ be the canonical projection with kernel $A_1 \oplus Ry$. Define $\theta : Rx \to A_2$ by $\theta(z) = fz$, for all $z \in Rx$. Clearly, θ is an R-homomorphism. Then $\theta\pi : A \to A_2$ is a homomorphism and $\theta\pi(A) \leq Rx$. Also, for $r \in R$ and $t \in T$, $t(\theta\pi[(r/t)(l_0+x)]) = \theta\pi[r(l_0+x)] = \theta(rx) = rfx = rftx'$. Again, because X is T-torsion-free, it follows that $\theta\pi[(r/t)(l_0+x)] = rfx' = \varphi[(r/t)(l_0+x)]$. Thus φ can be lifted to A. We have shown that A_2 is A-c-injective. Since A_1 is an injective module, it now follows that A is self-cu-injective, by Lemma 5.43. □

Theorem 5.55. *Assume R is a Dedekind domain and the R-module $A = A_1 \oplus A_2$ is the direct sum of a torsion-free indecomposable injective submodule A_1 and a cyclic torsion submodule A_2. Then A is self-c-injective.*

Proof. Let $A_2 = Ra \neq 0$. Let $T = \{c \in R : A_2 = cA_2\}$. Then T is a multiplicatively closed subset of R. Let $S = \{r/t : r \in R,\ t \in T\}$, which is a subring of the field of fractions of R. Let us show that A is T-torsion-free. Let $x \in A$ be such that

$cx = 0$, for some $c \in T$. Then $x = (x_1, x_2)$, for some $x_1 \in A_1$, $x_2 \in A_2$. Hence $cx_1 = 0$, $cx_2 = 0$. Clearly, $x_1 = 0$. There exist $r, s \in R$ such that $x_2 = ra$ and $a = csa$, so that $x_2 = ra = rcsa = scx_2 = 0$. Since A_1 is injective, it is clear that $A = cA$, for all $c \in T$. Thus A is an S-module, and $A = A_1 \oplus A_2$ is a direct sum of S-modules A_1 and A_2, where A_1 is a torsion-free injective S-module and A_2 is a cyclic torsion S-submodule. Moreover, any complement R-submodule of A is a complement S-submodule of A. Thus, without loss of generality, we can pass to the ring S and suppose that every element in T is a unit in R. Let P be a maximal ideal of R. If $A_2 = PA_2$, then $a_2 = pa_2$, for some $p \in P$, and hence $A_2 = pA_2$, so that $p \in T$ and p is a unit in R, a contradiction. Thus $A_2 \neq PA_2$.
Let $I = \ell(a)$, which is a non-zero ideal of R. Then $R/I \neq (R/I)P = (P + I)/I$, so that $R \neq P + I$ and hence $I \leq P$. Thus every maximal ideal of R contains I. But R/I is a semilocal ring, so that R is a semilocal Dedekind domain, i.e., R is a PID, by [ZS58, Chap. V, Thm. 16]. Since A_2 is extending (see the proof of Lemma 5.54), combining Lemma 5.54 and Proposition 5.47, we have the result. $\square$

We arrive at the end of this section. It would be handy to provide some examples. For instance, recall that, for a prime $p \in \mathbb{Z}$ and for $m, n \in \mathbb{N}$, the $\mathbb{Z}$-module $(\mathbb{Z}/p^m\mathbb{Z}) \oplus (\mathbb{Z}/p^n\mathbb{Z})$ is extending if and only if $|m - n| \leq 1$, by [KM88, Cor. 23]. Now, we have the following.

Theorem 5.56. *Let $p \in \mathbb{Z}$ be any prime and let $m, n \in \mathbb{N}$. Then the $\mathbb{Z}$-module $(\mathbb{Z}/p^m\mathbb{Z}) \oplus (\mathbb{Z}/p^n\mathbb{Z})$ is self-*c*-injective.*

Proof. Suppose that $m \leq n$ and let $A = (\mathbb{Z}/p^m\mathbb{Z}) \oplus (\mathbb{Z}/p^n\mathbb{Z}) = A_1 \oplus A_2$, where $A_1 = \mathbb{Z}(1 + p^m\mathbb{Z}, p^n\mathbb{Z})$ and $A_2 = \mathbb{Z}(p^m\mathbb{Z}, 1 + p^n\mathbb{Z})$. Let K be a complement submodule of A and let $\alpha : K \to \mathbb{Z}/p^m\mathbb{Z}$ be a homomorphism. Let us prove that α can be lifted to A.

If the uniform dimension of K is 0 or 2, the result is obvious. Suppose, then, that K is uniform. Hence K is cyclic and $K = \mathbb{Z}(a + p^m\mathbb{Z}, b + p^n\mathbb{Z})$, with $a, b \in \mathbb{Z}$. Because K is complement, either p does not divide a, or p does not divide b. If p does not divide b, it is not hard to prove that $A = A_1 \oplus K$ and, clearly, α lifts to A. If p does not divide a, a and p^m are coprime, so that $A_1 = \mathbb{Z}(a + p^m\mathbb{Z}, p^n\mathbb{Z})$. Let $c \in \mathbb{Z}$ be such that $\alpha(a + p^m\mathbb{Z}, b + p^n\mathbb{Z}) = c + p^m\mathbb{Z}$. Define $\beta : A \to A$ by $\beta(ra + p^m\mathbb{Z}, s + p^n\mathbb{Z}) = rc + p^m\mathbb{Z}$, for every $r, s \in \mathbb{Z}$. Clearly, β is a well-defined homomorphism that lifts α. As A_2 is A-injective, we can conclude that A is self-c-injective. $\square$

Contrast Theorem 5.56 with the following example.

Example 5.57. Let $p \in \mathbb{Z}$ be any prime. The $\mathbb{Z}$-module $(\mathbb{Z}/p\mathbb{Z}) \oplus (\mathbb{Z}/p^2\mathbb{Z}) \oplus (\mathbb{Z}/p^3\mathbb{Z})$ is not self-cu-injective.

Proof. Let $A = (\mathbb{Z}/p\mathbb{Z}) \oplus (\mathbb{Z}/p^2\mathbb{Z}) \oplus (\mathbb{Z}/p^3\mathbb{Z})$. Let U denote the submodule $\mathbb{Z}(1 + p\mathbb{Z}, p + p^2\mathbb{Z}, p + p^3\mathbb{Z})$. Then U is a cyclic p-torsion module, and so is uniform. Suppose that U is essential in a submodule V of A. Thus, V is also uniform and

$V = \mathbb{Z}(a+p\mathbb{Z}, b+p^2\mathbb{Z}, c+p^3\mathbb{Z})$, for some $a, b, c \in \mathbb{Z}$. There exists $d \in \mathbb{Z}$ such that $(1+p\mathbb{Z}, p+p^2\mathbb{Z}, p+p^3\mathbb{Z}) = d(a+p\mathbb{Z}, b+p^2\mathbb{Z}, c+p^3\mathbb{Z})$. Hence $1 \equiv da \pmod{p}$, so that $d \notin p\mathbb{Z}$. Therefore, $U = V$ and we can conclude that U is a complement submodule of A. Define a homomorphism $\alpha : U \to A$ by $\alpha[r(1+p\mathbb{Z}, p+p^2\mathbb{Z}, p+p^3\mathbb{Z})] = (p\mathbb{Z}, r+p^2\mathbb{Z}, p^3\mathbb{Z})$, for all $r \in \mathbb{Z}$. Suppose that α can be lifted to a homomorphism $\beta : A \to A$, i.e., $U \in \mathrm{Lift}_A(A)$. Then $\beta(p\mathbb{Z}, p^2\mathbb{Z}, 1+p^3\mathbb{Z}) = (u+p\mathbb{Z}, v+p^2\mathbb{Z}, w+p^3\mathbb{Z})$, for some $u, v, w \in \mathbb{Z}$. Thus $\beta(p\mathbb{Z}, p^2\mathbb{Z}, 1+p^3\mathbb{Z}) = (p\mathbb{Z}, p^2\mathbb{Z}, wp^2+p^3\mathbb{Z})$, and $\alpha(p\mathbb{Z}, p^2\mathbb{Z}, p^2+p^3\mathbb{Z}) = \alpha[p(1+p\mathbb{Z}, p+p^2\mathbb{Z}, p+p^3\mathbb{Z})] = (p\mathbb{Z}, p+p^2\mathbb{Z}, p^3\mathbb{Z})$. Therefore, $\alpha(p\mathbb{Z}, p^2\mathbb{Z}, p^2+p^3\mathbb{Z}) \neq \beta(p\mathbb{Z}, p^2\mathbb{Z}, p^2+p^3\mathbb{Z})$ and $U \notin \mathrm{Lift}_A(A)$. It follows that A is not self-cu-injective. □

Exercises

5.5. Write out the proof of Lemma 5.43.

5.6. Show that being free is not superfluous in Theorem 5.48.

5.3 Modules whose z-closed submodules are direct summands

In this section, we consider CLS-modules as an outer type proper generalization of extending modules. A module A is called a CLS-*module* if every z-closed submodule is a direct summand of A. Since z-closed submodules are building bricks of CLS-modules, we provide some basic facts on these kind of submodules.

Following [Ter95], a submodule N of A is a z-*closed submodule* of A if A/N is nonsingular, i.e., $Z(A/N) = 0$ (see Exercises 2.16 and 2.17). Recall that z-closed submodules are the same as the closed submodules in [Go76].

The next lemma is taken from [Sand68, Lem. 2.3], and its proof is given for completeness.

Lemma 5.58. *Let A be a module.*

(i) *Every* z-*closed submodule is a complement.*
(ii) *If ${}_RA$ is nonsingular, then every complement is* z-*closed.*

Proof. (i) Suppose K is a z-closed submodule of A. Let N be a submodule of A such that K is essential in N. Then $N/K \leq Z(A/K)$, so that $N/K = Z(A/K)$. It follows that $K = N$. Thus K is a complement in A.

(ii) Suppose K is a complement submodule of A which is not a z-closed submodule. Then A/K is not nonsingular. There exists $a \in A$, $a \notin K$, such that $Ea \leq K$ for some essential left ideal E of R. Let $r \in R$, $k \in K$, and consider $ra+k$. Let $F = \{s \in R : sr \in E\}$. Then F is essential in ${}_RR$ and $F(ra+k) \leq K$. If $ra+k \neq 0$, then $F(ra+k) \neq 0$, and hence $K \cap R(ra+k) \neq 0$. Thus, K is essential in $Ra+K$, so that K is not a complement in A. □

The next example illustrates that in general a complement submodule need not be z-closed, (see Exercise 2.17).

Example 5.59. Let R, V, and I_v be as in Exercise 2.17. Then I_v is a complement submodule which is not z-closed submodule of R_R for all $v \in V$.

Proof. Let $E = \begin{bmatrix} 0 & V \\ 0 & 0 \end{bmatrix}$. Then E is essential in R_R. Note that $I_v = \begin{bmatrix} 0 & Fv \\ 0 & 0 \end{bmatrix}$ $(v \in V)$. Suppose that $G \leq R$ is such that I_v is essential in G. Thus, I_v is essential in $G \cap E$, and hence $I_v = G \cap E$. Let $\begin{bmatrix} k & w \\ 0 & k \end{bmatrix} \in G$ for some $w \in V$, $0 \neq k \in F$. Let $x \in V$ such that $x \notin Fv$. Thus, $\begin{bmatrix} k & w \\ 0 & k \end{bmatrix} \begin{bmatrix} 0 & (1/k)x \\ 0 & 0 \end{bmatrix} = \begin{bmatrix} 0 & x \\ 0 & 0 \end{bmatrix} \in G \cap E$. Therefore $x \in Fv$, a contradiction. Thus, $k = 0$. Hence $G \leq E$, so $I_v = G$. It follows that I_v is a complement in R_R for all $v \in V$. But $E^2 = 0$ so $E^2 \leq I_v$. However, E is not contained in I_v. Thus I_v is not a z-closed ideal of R. □

Corollary 5.60.

(i) *Every* CS-*module is a* CLS-*module.*

(ii) *Every nonsingular* CLS-*module is a* CS-*module.*

Proof. By Lemma 5.58. □

Clearly, over a commutative integral domain, any torsion module is a CLS-module. Note that CLS-modules actually differ from CS-modules. For example, let p be any prime integer, and let A be the $\mathbb{Z}$-module $(\mathbb{Z}/\mathbb{Z}p) \oplus (\mathbb{Z}/\mathbb{Z}p^3)$. Since $_{\mathbb{Z}}A$ is singular, A is a CLS-module. Now let $K = \mathbb{Z}(1 + \mathbb{Z}p, p + \mathbb{Z}p^3)$. Then K is a complement in A of order p^2 which is not a direct summand of A. Thus $_{\mathbb{Z}}A$ is not a CS-module (Exercise 3.5).

The following result shows that with respect to direct summands, CLS-modules behave like CS-modules.

Lemma 5.61. *Any direct summand of a* CLS-*module is a* CLS-*module.*

Proof. Let A be a CLS-module. Suppose $A = K \oplus K'$ for some submodules K and K' of A. Let L be a z-closed submodule of K. Since $A/(L \oplus K') = (K \oplus K')/(L \oplus K') \cong K/L$, then $L \oplus K'$ is a z-closed submodule of A and so $L \oplus K'$ is a direct summand of A. This yields that L is a direct summand of A. Hence L is a direct summand of K, therefore K is a CLS-module. □

Note that a direct sum of CLS-modules need not be a CLS-module in general, as the following example illustrates. Let A be the $\mathbb{Z}$-module $\mathbb{Z} \oplus \mathbb{Z}_{(2)}$, where $\mathbb{Z}_{(2)} = \{a/b : a, b \in \mathbb{Z}, b \text{ is odd}\}$. Obviously, $_{\mathbb{Z}}A$ is torsion-free and $\mathbb{Z}$, $\mathbb{Z}_{(2)}$ are CLS-modules. But A is not a CLS-module (see [MM90, p. 19]).

Proposition 5.62. *A left R-module A is a* CLS-*module if and only if there exists a submodule A' of A such that $A = Z_2(A) \oplus A'$ and A' is a* CS-*module. In this case, A' is $Z_2(A)$-injective.*

Proof. Suppose that A is a CLS-module. Thus $Z_2(A)$ is a direct summand of A, so that $A = Z_2(A) \oplus A'$ for some submodule A' of A. Note that A' is nonsingular and, by Lemma 5.61, a CLS-module and hence a CS-module by Corollary 5.60. Conversely, suppose $A = Z_2(A) \oplus A'$ for some CS-module A'. Let K be a z-closed submodule of A. Then $Z(A) \leq K$ and hence $Z_2(A) \leq K$. Thus $K = Z_2(A) \oplus (K \cap A')$. Now $A/K \cong A'/(K \cap A')$, so that $K \cap A'$ is a z-closed submodule of A'. Hence, by Corollary 5.60, $A' = (K \cap A') \oplus K'$ for some submodule K'. Then $A = K \oplus K'$. It follows that A is a CLS-module.

Now, let us show that A' is $Z_2(A)$-injective. For suppose N is a submodule of $Z_2(A)$ and $\varphi : N \to A'$ a homomorphism. If $x \in N \cap Z(A)$, then $Ex = 0$ for some essential left ideal E of R. Thus $E\varphi(x) = 0$, and hence $\varphi(x) = 0$, because A' is nonsingular. So $\varphi(N \cap Z(A)) = 0$. If $y \in N$, then $Fy \subseteq N \cap Z(A)$ for some essential left ideal F of R. Thus $F\varphi(y) = 0$ and hence $\varphi(y) = 0$. It follows that $\varphi(y) = 0$ for all $y \in N$. Thus $\varphi = 0$, and φ can be lifted to $Z_2(A)$. □

The next result shows a condition under which a direct sum of two CLS-modules is also a CLS-module.

Theorem 5.63. *Suppose that a left R-module A is a direct sum of $A_1 \oplus A_2$ of* CLS-*modules A_1 and A_2 such that A_1 is A_2-injective. Then A is a* CLS-*module.*

Proof. Let N be a z-closed submodule of A. Then A/N is nonsingular. Now $A_1/(N \cap A_1) \cong (A_1 + N)/N$ implies that $N \cap A_1$ is a z-closed submodule of A_1. Thus $N \cap A_1$ is a direct summand of A_1 and hence of A. It follows that $N \cap A_1$ is a direct summand of N, so $N = (N \cap A_1) \oplus K$ for some submodule K of N. Let $\pi_i : A \to A_i$, $i = 1, 2$, denote the canonical projections. Consider the diagram

$$\begin{array}{ccccc} 0 & \longrightarrow & K & \xrightarrow{\ \alpha\ } & A_2 \\ & & {\scriptstyle\beta}\big\downarrow & & \\ & & A_1 & & \end{array} \qquad\qquad \text{exact}$$

where $\alpha = \pi_2|_K$ and $\beta = \pi_1|_K$. Note that α is a monomorphism and A_1 is A_2-injective. Thus, there exists a homomorphism $\varphi : A_2 \to A_1$ such that $\varphi\alpha = \beta$. Let $L = \{x + \varphi(x) : x \in A_2\}$. Then it can be easily checked that L is a submodule of A and $L \cong A_2$. Moreover, $A = A_1 \oplus L$. If $k \in K$, then $k = a_1 + a_2$ for some $a_i \in A_i$, $i = 1, 2$. Then $a_1 = \beta(k) = \varphi\alpha(k) = \varphi(a_2)$, and this implies that $k = \varphi(a_2) + a_2 \in L$. Thus $K \subseteq L$. Since $A/N = (A_1/(N \cap A_1)) \oplus L/K$, we see that L/K is nonsingular, i.e., K is a z-closed submodule of L. But $L \cong A_2$, so that K is a direct summand of L. Thus, N is a direct summand of A. It follows that A is a CLS-module. □

Now we have the following fact on finite direct sums of CLS-modules.

Theorem 5.64. *Suppose R is a ring and A is a left R-module such that $A = A_1 \oplus A_2 \oplus \cdots \oplus A_n$ is a finite direct sum of relatively injective modules A_i, $1 \leq i \leq n$. Then A is a* CLS-*module if and only if A_i is a* CLS-*module for each $1 \leq i \leq n$.*

Proof. The necessity is clear by Lemma 5.61. The converse follows by induction on n and using Theorem 5.63. □

Corollary 5.65. *Suppose that the nonsingular left R-module A is the direct sum $A_1 \oplus A_2$ of* CS*-modules A_1 and A_2 such that A_1 is A_2-injective. Then A is a* CS*-module.*

Proof. By Corollary 5.60 and Theorem 5.63. □

The next result has also been proved in [HS93, Thm. 4].

Corollary 5.66. *Suppose that the left R-module A is the direct sum $A_1 \oplus A_2$ of* CS*-modules A_1 and A_2 such that A_1 is A_2-injective and A_2 is nonsingular. Then A is a* CS*-module.*

Proof. It is clear that $Z_2(A) = Z_2(A_1)$ is a direct summand of A_1. Thus $A_1 = Z_2(A) \oplus A_1'$ for some nonsingular submodule A_1' of A_1. Now $A = Z_2(A) \oplus A_1' \oplus A_2$. Note that A_1' is A_2-injective, A_1' is a CS-module, and $A_1' \oplus A_2$ is nonsingular. By Corollary 5.65, $A_1' \oplus A_2$ is a CS-module. But by Exercise 3.12, $Z_2(A)$ is A_1'-injective, and hence $Z_2(A)$ is $(A_1' \oplus A_2)$-injective. Again, by Exercise 3.12, A is a CS-module. □

Observe that the module in Example 2.51 is a CLS-module by Theorem 5.63, but is not CS. Furthermore, note that the aforementioned example shows that neither the sufficiency of Corollary 13 in [Ter95], nor Corollary 5.66 without the nonsingularity assumption hold.

Using the idea in Section 2.3, we now characterize the CLS condition in terms of lifting homomorphisms from z-closed submodules to the module. Let A be a module and let $\underline{Z}$ be the collection of all z-closed submodules of A. Then we have:

Proposition 5.67. *The following statements are equivalent for a module A.*

(i) *A is a* CLS*-module.*
(ii) $\underline{Z} \subseteq \mathrm{Lift}_X(A)$ *for all left R-modules X.*
(iii) $\underline{Z} \subseteq \mathrm{Lift}_X(A)$ *for all $X \in \underline{Z}$.*

Proof. Apply Theorem 2.48 to $\underline{A} = \underline{Z}$. □

There are some examples which show that there is no implication between CLS-modules and weak CS-modules, or modules with C_{11}. First note that the ring R in Example 5.59 is a CLS-module which is neither weak CS-module, nor a module with C_{11}. Further, recall that the $\mathbb{Z}$-module $A = \mathbb{Z} \oplus \mathbb{Z}_{(2)}$ is not a CLS-module which is weak CS. We conclude with the following example.

Example 5.68. Let $A = \bigoplus_{i=1}^{\infty} \mathbb{Z}$. Then ${}_{\mathbb{Z}}A$ satisfies C_{11}, but is not a CLS-module.

Proof. By Corollary 4.28, ${}_{\mathbb{Z}}A$ satisfies C_{11}. Suppose that $\varphi : A \to \mathbb{Q}$ is an epimorphism. Let $K = \ker \varphi$. Thus, $A/K \cong \mathbb{Q}$ which is nonsingular. Hence K is a z-closed submodule of A. If K were a direct summand of A, then we would have $A = K \oplus L$ for some submodule L of A. Thus, $L \cong \mathbb{Q}$, which is a contradiction. It follows that ${}_{\mathbb{Z}}A$ is not a CLS-module. □

In [PZ03], the behaviour of CS-modules with respect to excellent extensions of rings was investigated. The results therein were generalized in [CS14]. Here we use some of the results in [CS14] to derive corresponding facts for CLS-modules. Let R be a subring of a ring S such that they have the same identity element. S is called a *left excellent extension* of R [Pas77] if the following two conditions are satisfied:

(1) ${}_RS$ and S_R are free modules with a basis $\{1 = a_1, a_2, \ldots, a_n\}$ such that $a_iR = Ra_i$, for every $i \in \{1, 2, \ldots, n\}$.
(2) For every submodule ${}_SN$ of a module ${}_SA$, if ${}_RN$ is a direct summand of ${}_RA$, then ${}_SN$ is a direct summand of ${}_SA$.

In what follows, S will be a left excellent extension of R, with S_R and ${}_RS$ having a basis $\{1 = a_1, a_2, \ldots, a_n\}$. Let us recall some basic properties which will be used in sequel.

Lemma 5.69 ([PZ03, Props. 1.1 and 2.2, Lem. 2.1]).

(i) *Let ${}_SN$ be a submodule of a module ${}_SA$. Then ${}_RN \leq_{\mathrm{e}} {}_RA$ if and only if ${}_SN \leq_{\mathrm{e}} {}_SA$.*
(ii) *Let A be a left S-module. Then $Z({}_SA) = Z({}_RA)$.*
(iii) *Let A be a left R-module. Then $Z({}_S(S \otimes_R A)) = S \otimes_R Z({}_RA)$.*

Lemma 5.70 ([CS14, Lems. 5.2 and 5.3]).

(i) *Let ${}_SN$ be a submodule of a module ${}_SA$. Then ${}_SN$ is z-closed in ${}_SA$ if and only if ${}_RN$ is z-closed in ${}_RA$.*
(ii) *Let ${}_RN$ be a submodule of a module ${}_RA$. Then ${}_RN$ is z-closed in ${}_RA$ if and only if ${}_S(S \otimes_R N)$ is z-closed in ${}_S(S \otimes_R A)$.*

Proof. (i) is an immediate consequence of Lemma 5.69 (ii).

(ii) Suppose that ${}_RN$ is z-closed in ${}_RA$, that is $Z(A/N) = 0$. Now, consider the element $\sum_{i=1}^{n} a_i \otimes m_i + S \otimes_R N \in Z({}_S(S \otimes_R A)/{}_S(S \otimes_R N))$ for some $m_i \in A$. There exists ${}_SI \leq_{\mathrm{e}} {}_SS$ such that ${}_SI(\sum_{i=1}^{n} a_i \otimes m_i + S \otimes_R N) \subseteq S \otimes_R N$, and so ${}_RI \leq_{\mathrm{e}} {}_RS$, by Lemma 5.69 (i). Thus, ${}_RH = {}_RI \cap {}_RR \leq_{\mathrm{e}} {}_RR$, and so we have ${}_RH(\sum_{i=1}^{n} a_i \otimes m_i + S \otimes_R N) \subseteq S \otimes_R N$. It follows that ${}_RH(\sum_{i=1}^{n} a_i \otimes m_i) \subseteq S \otimes_R N$. Since S_R is a free module with basis $\{1 = a_1, a_2, \ldots, a_n\}$, every element in $S \otimes_R N$ is uniquely written as $\sum_{i=1}^{n} a_i \otimes m_i$ for elements $m_i \in A$. Notice that for every $i \in \{1, 2, \ldots, n\}$, there exists an automorphism σ_i of R such that $a_i\sigma_i(r) = ra_i$, for every $r \in R$. Thus, for every $h \in H$, we have $h(\sum_{i=1}^{n}(a_i \otimes m_i)) = \sum_{i=1}^{n}(ha_i) \otimes m_i = \sum_{i=1}^{n}(a_i\sigma_i(h)) \otimes m_i = \sum_{i=1}^{n} a_i \otimes (\sigma_i(h)m_i) \in S \otimes_R N$.

Since S_R is a free module with basis $\{1 = a_1, a_2, \ldots, a_n\}$, every element of $S \otimes_R N$ is uniquely written as $\sum_{i=1}^n a_i \otimes n_i$ for some elements $n_i \in N$. It follows that $\sigma_i(H)m_i \subseteq N$ for every $i \in \{1, 2, \ldots, n\}$. Since σ_i is an automorphism of R and ${}_RH \leq_e {}_RR$, we have $\sigma_i(H) \leq_e {}_RR$ for every $i \in \{1, 2, \ldots, n\}$. Then $m_i + N \in Z({}_RA/{}_RN) = 0$ for every $i \in \{1, 2, \ldots, n\}$. Thus $m_i \in N$ for every $i \in \{1, 2, \ldots, n\}$. Hence, $\sum_{i=1}^n a_i \otimes m_i \in S \otimes_R N$, and so $Z({}_S(S \otimes_R A)/{}_S(S \otimes_R N)) = 0$. Hence ${}_S(S \otimes_R N)$ is z-closed in ${}_S(S \otimes_R A)$.

Conversely, assume that ${}_S(S \otimes_R N)$ is z-closed in ${}_S(S \otimes_R A)$. Then $Z({}_R(S \otimes_R A)/{}_R(S \otimes_R N)) = 0$, by Lemma 5.69(ii). Let $a + N \in Z({}_RA/{}_RN)$. There exists ${}_RI \leq_e {}_RR$ such that ${}_RIa \subseteq {}_RN$. Therefore ${}_RI(1 \otimes a + S \otimes_R N) \subseteq S \otimes_R N$. Since $Z({}_R(S \otimes_R A)/{}_R(S \otimes_R N)) = 0$, we have that $1 \otimes a \in S \otimes_R N$. From this we have $a \in N$, which shows that $Z({}_RA/{}_RN) = 0$. □

Theorem 5.71.

(i) *Let A be a left S-module. If ${}_RA$ is a* CLS*-module, then so is ${}_SA$.*

(ii) *Let A be a left R-module. If ${}_S(S \otimes_R A)$ is a* CLS*-module, then so is ${}_RA$.*

Proof. (i) Assume that ${}_RA$ is a CLS-module. Let ${}_SN$ be a z-closed submodule of ${}_SA$. By Lemma 5.70, ${}_RN$ is a z-closed submodule of ${}_RA$. Hence ${}_RN$ is a direct summand of ${}_RA$. Since S is a left excellent extension of R, ${}_SN$ is a direct summand of ${}_SA$. It follows that ${}_SA$ is a CLS-module.

(ii) Now, assume that ${}_S(S \otimes_R A)$ is a CLS-module. Let ${}_RN$ be a z-closed submodule of ${}_RA$. By Lemma 5.70, ${}_S(S \otimes_R N)$ is a z-closed submodule of ${}_S(S \otimes_R A)$. By hypothesis, ${}_S(S \otimes_R N)$ is a direct summand of ${}_S(S \otimes_R A)$. It follows that ${}_R(S \otimes_R N)$ is a direct summand of ${}_R(S \otimes_R A)$. Note that ${}_R(1 \otimes N)$ is a direct summand of ${}_R(S \otimes_R N)$ and ${}_R(1 \otimes N) \subseteq {}_R(1 \otimes A)$. Thus, ${}_R(1 \otimes N)$ is a direct summand of ${}_R(1 \otimes A)$. It follows that ${}_RN$ is a direct summand of ${}_RA$. Hence ${}_RA$ is a CLS-module. □

Now Lemma 5.69 and Theorem 5.71 yield the following consequence.

Corollary 5.72. *Every left R-module is* CLS *if and only if every left S-module is* CLS.

Exercises

5.7. Let A be the $\mathbb{Z}$-module $(\mathbb{Z}/\mathbb{Z}p) \oplus \mathbb{Q}$, where p is a prime integer. Show that all z-closed submodules of A are $(\mathbb{Z}/\mathbb{Z}p) \oplus 0$ and A.

5.8. Let R be a Dedekind domain and A be an R-module. Prove that A is a CLS-module if and only if $A = A_1 \oplus A_2$ for some torsion CS-module A_1 and torsion-free module A_2.

5.4 Goldie extending modules

In Section 4.6, we defined a module A to be G-extending if for each $X \leq A$ there exists a direct summand D of A such that $X \cap D$ is essential in both X and D. As we promised in there, we continue to look at this new class of modules. To this end, we obtain several characterizations of the G-extending condition and we locate the G-extending condition with respect to several generalizations of the extending condition. Moreover we investigate various conditions for a direct sum of G-extending modules to be G-extending, or for a direct summand of a G-extending module to be G-extending. Our results enable us to obtain a characterization of the G-extending Abelian groups.

Definition 5.73. Let A be a module. We consider the following relations on the set of submodules of A:

(i) $X\alpha Y$ if and only if there exists $N \leq A$ such that $X \leq_{\mathrm{e}} N$ and $Y \leq_{\mathrm{e}} N$.
(ii) $X\beta Y$ if and only if $X \cap Y \leq_{\mathrm{e}} X$ and $X \cap Y \leq_{\mathrm{e}} Y$. (Equivalently, $X\beta Y$ if and only if $X \cap N = 0$ implies $Y \cap N = 0$ and $Y \cap K = 0$ implies $X \cap K = 0$ for all $N, K \leq A$.)

Observe that α is reflexive and symmetric; but it may not be transitive. However, it can be seen that α is transitive if and only if A is a UC-module. β is an equivalence relation and is defined in [Gol60] for right ideals of a ring. Note that if $X, Y \leq A$ are such that $X\alpha Y$, then $X\beta Y$. Thus, A is a UC-module if and only if $\alpha = \beta$.

Proposition 5.74. *A is extending if and only if for each $X \leq A$ there exists a direct summand D of A such that $X\alpha D$.*

Proof. The proof is routine. □

Motivated by Proposition 5.74, and Goldie's and Smith's use of the β equivalence relation in [Gol60] and [Smi93], respectively, we give the following definition.

Definition 5.75. We say A is G-extending (i.e., Goldie extending) if for each $X \leq A$, there exists a direct summand D of A such that $X\beta D$. Equivalently, A is G-extending if for each complement submodule C of A, there is a direct summand D of A such that $C\beta D$. Moreover, we say A is G^{+}-extending provided that every direct summand of A is G-extending.

Thus every extending module is G^{+}-extending. As we shall see in the sequel, the finitely generated Abelian groups provide a large class of G-extending modules which are not, in general, extending.

Definition 5.76. On $\mathbf{I}(\underline{E}_{E(A)})$, we define the equivalence relation β_{IE} by $c\beta_{IE}e$ if $E(A)c\beta E(A)e$.

The next result gives several conditions equivalent to being G-extending.

Proposition 5.77. *Let A be a module. The following conditions are equivalent.*

(i) *A is* G*-extending.*

(ii) *For each $Y \leq A$, there exist $X \leq A$ and a direct summand D of A such that $X \leq_e Y$ and $X \leq_e D$.*

(iii) *For each $Y \leq A$, there exist a complement L of Y and a complement K of L such that $Y\beta K$ and every homomorphism $f : K \oplus L \to A$ extends to a homomorphism $\overline{f} : A \to A$.*

(iv) *For each $c \in \mathbf{I}(\underline{E}_{E(A)})$ there exists $e \in \mathbf{I}(\underline{E}_{E(A)})$ such that $c\beta_{IE}e$ and $Ae \leq A$.*

Proof. (i) ⇒ (ii) Let $Y \leq A$. Hence there exists a direct summand D of A such that $Y\beta D$. Now take $X = Y \cap D$.

(ii) ⇒ (iii) From (ii), there exist D, D' such that $Y \cap D \leq_e Y$, $Y \cap D \leq_e D$, and $A = D \oplus D'$. Take $D = K$ and $D' = L$.

(iii) ⇒ (i) By (iii) and Lemma 3.97, K is a direct summand of A. Hence A is G-extending.

(i) ⇒ (iv) Let $c \in \mathbf{I}(\underline{E}_{E(A)})$ and $Y = A \cap E(A)c$. There exists $f \in \mathbf{I}(\underline{E}_A)$ such that $Y\beta Af$. Let $e : E(A) \to E(Af)$ be the projection. Then $Af\beta E(A)e$. Since $Y\beta E(A)c$, $c\beta_{IE}e$ and $Ae = Af \leq A$.

(iv) ⇒ (i) Let $Y \leq A$. There exists $c \in \mathbf{I}(\underline{E}_{E(A)})$ such that $Y \leq_e E(A)c$. Then $c\beta_{IE}e$ for some $e \in \mathbf{I}(\underline{E}_{E(A)})$ such that $Ae \leq A$. So $e|_A \in \mathbf{I}(\underline{E}_A)$. Therefore, Ae is a direct summand of A and $Y\beta Ae$. □

Now we explain how the G-extending condition is related with several generalizations of the extending condition. First recall that a module is said to be FI-*extending* if every fully invariant submodule is essential in a direct summand [BPR02] (see also Section 4.4).

Proposition 5.78. *Let A be a module and consider the following conditions.*

(i) *A is extending,*

(ii) *A is* G*-extending,*

(iii) *A has C_{11},*

(iv) *A is* FI*-extending,*

(v) *A is weak* CS.

Then (i) ⇒ (ii) ⇒ (iii) ⇒ (iv) *and* (ii) ⇒ (v). *In general, the reverse implications do not hold.*

Proof. (i) ⇒ (ii) This implication is clear.

(ii) ⇒ (iii) Let $Y \leq A$. There exist $D, D' \leq A$ such that $A = D \oplus D'$ and $Y\beta D$. Then D' is a complement of Y and a direct summand of A. Thus A has C_{11}.

(iii) ⇒ (iv) This implication is shown in Proposition 4.67.

(ii) ⇒ (v) Let $Y \leq A$ be semisimple. There exists $D \leq A$ such that $Y\beta D$ where D is a direct summand of A. Since Y is semisimple, $Y \cap D = Y$. Thus A is weak CS. An example of a right G-extending ring which is neither left nor right extending is provided in Example 4.77. By Proposition 4.67, (iv) $\not\Rightarrow$ (iii). To see that (iii) $\not\Rightarrow$ (ii) and (v) $\not\Rightarrow$ (ii), let $R = T_2(D)$, where D is a right Ore domain that is not a division ring. By Theorem 4.26, R is a right C_{11}-module. Since $\operatorname{soc}(R_R) = 0$, R is weak CS. Let $u = \begin{bmatrix} 0 & 1 \\ 0 & x \end{bmatrix} \in R$, where $0 \neq x \in D$ and $x \neq 1$. Then uR is a uniform submodule of R. Suppose there exists a nonzero submodule X of uR. By direct calculation, it can be verified that X is not essential in a direct summand of R. By Proposition 5.77, R_R is not a right G-extending module. Observe that Example 4.3 shows that (iii) $\not\Rightarrow$ (v). Also a domain which is not right Ore is weak CS, but not C_{11}. □

Lemma 5.79. *Let A be a module. Then A is (complement bounded) strongly bounded if and only if every nonzero (complement) submodule of A is an essential extension of a fully invariant submodule of A.*

Proof. Assume A is complement bounded (the strongly bounded case is similar). Let C be a nonzero complement submodule of A. Let X be the sum of the fully invariant submodules of A in C. So $0 \neq X \trianglelefteq A$. Let K be the complement of X in C. Then K is complement in A. Hence $K = 0$ (otherwise K contains a nonzero fully invariant submodule of A which is contained in X, a contradiction). Therefore, $X \leq_e C$. The converse is clear. □

Proposition 5.80. *Let A be a module.*

(i) *Let A be indecomposable. Then A is* G*-extending if and only if A is uniform.*
(ii) *Let A be a* UC*-module (e.g., A is nonsingular). Then A is* G*-extending if and only if A is extending.*
(iii) *Let $\operatorname{End}(_RA)$ be Abelian such that if $X \leq A$, then $X = \sum_{i\in I} h_i(A)$, where each $h_i \in \operatorname{End}(_RA)$. Then A is* G*-extending if and only if A is extending.*
(iv) *Let A be complement bounded. Then A is* G*-extending if and only if A is* FI*-extending.*
(v) *Assume $\operatorname{soc} A \leq_e A$. Then A is* G*-extending if and only if A is weak* CS.

Proof. (i) The proof of this part is straightforward.

(ii) and (iii) These statements are consequences of Proposition 5.78 and Proposition 4.68.

(iv) Assume A is complement bounded and FI-extending. Let $0 \neq Y \leq A$, and let K be an essential closure of Y in A. By Lemma 5.79, there exists $B \trianglelefteq A$ such that $B \leq_e K$ and $B \leq_e D$, where D is a direct summand of A. Hence $Y\beta D$. Thus A is G-extending. The converse follows from Proposition 5.78.

(v) Assume $\operatorname{soc} A \leq_e A$ and A is weak CS. Let $Y \leq A$. Then there exists a direct summand D of A such that $\operatorname{soc} Y \leq_e D$. But $\operatorname{soc} Y \leq_e Y$. Therefore, A is G-extending. The converse follows from Proposition 5.78. □

Proposition 5.81. *If A is* G*-extending, $X \trianglelefteq A$, and $Z(A/X) = 0$, then A/X is extending.*

Proof. Let $Y/X \leq A/X$. There exists $d \in \mathbf{I}(\underline{E}_A)$ such that $Y\beta Ad$. Since $X \trianglelefteq A$, $X = (X \cap Ad) \oplus (X \cap A(1-d))$. Then $X \leq Ad$ because $X \cap Ad = X \cap (Y \cap Ad) \leq_{\mathrm{e}} X \cap Y = X$. Hence, Ad/X is a direct summand of A/X. Let $0 \neq ad + X \in Ad/X$. There exists $L \leq_{\mathrm{e}} R$ such that $Lad \leq Y \cap Ad$. Since $Z(A/X) = 0$, $Lad \nleq X$. So $0 \neq L(ad+X) \leq (Y/X) \cap (Ad/X) \leq_{\mathrm{e}} Ad/X$. Similarly, $(Y/X) \cap (Ad/X) \leq_{\mathrm{e}} Y/X$. Hence $(Y/X)\beta(Ad/X)$. Therefore, A/X is G-extending. By Proposition 5.80 (ii), A/X is extending. □

There are nonsingular modules $A = A_1 \oplus A_2$ in which A_1 and A_2 are extending, but A is not extending (e.g., let $R = \mathbb{Z}[x]$ and take $A = \mathbb{Z}[x] \oplus \mathbb{Z}[x]$). By Proposition 5.80 (ii), these modules A also show that the class of G-extending modules is not closed under direct sums. Moreover, the authors have, so far, not been able to determine if the class of G-extending modules is closed under direct summands (see Open Problems and Questions 2). Now, we investigate various conditions for a direct sum of G-extending modules to be G-extending, or for a direct summand of a G-extending module to be G-extending. An application of our results yields a characterization of the G-extending Abelian groups.

Theorem 5.82. *Let $A = A_1 \oplus A_2$ be a direct sum of modules. Then*

(i) *If A_1 is A_2-ejective (or A_2 is A_1-ejective) and A_1 and A_2 are* G*-extending, then A is* G*-extending.*

(ii) *If A_1 is A_2-injective and A is* G*-extending, then A_2 is* G*-extending.*

Proof. (i) Let $Y \leq A$. If $Y \cap A_1 = 0$, then, by Theorem 2.59, there exists $A_3 \leq A$ such that $A = A_1 \oplus A_3$ and $Y \cap A_3 \leq_{\mathrm{e}} Y$. Since $A_3 \cong A_2$, A_3 is G-extending. By Proposition 5.77, there exist $X \leq_{\mathrm{e}} Y \cap A_3$ and D, a direct summand of A_3, such that $X \leq_{\mathrm{e}} D$. Since A_3 is a direct summand of A, D is a direct summand of A. Now assume $Y \cap A_1 \neq 0$. Then there exists $K \leq Y$ such that $(Y \cap A_1) \oplus K \leq_{\mathrm{e}} Y$. But $K \cap A_1 = 0$. Again by Theorem 2.59, there exists $A_4 \leq A$ such that $A = A_1 \oplus A_4$ and $K \cap A_4 \leq_{\mathrm{e}} K$. Since $A_4 \cong A_2$, A_4 is G-extending. There exist $X_1 \leq_{\mathrm{e}} Y \cap A_1$, $X_2 \leq_{\mathrm{e}} K \cap A_4$, a direct summand D_1 of A_1, such that $X_1 \leq_{\mathrm{e}} D_1$, and D_2 a direct summand of A_4 such that $X_2 \leq_{\mathrm{e}} D_2$. Hence $X_1 \oplus X_2 \leq_{\mathrm{e}} Y$ and $X_1 \oplus X_2 \leq_{\mathrm{e}} D_1 \oplus D_2$ and $D_1 \oplus D_2$ is a direct summand of $A_1 \oplus A_4 = A$. Therefore, A is G-extending.

(ii) Let $Y \leq A_2$. By Proposition 5.77, there exist $X \leq_{\mathrm{e}} Y$ and $D \leq A$ such that $X \leq_{\mathrm{e}} D$ and D is a direct summand of A. Then $D \cap A_1 = 0$. By [DHSW94, Lem. 1.5], there exists $A' \leq A$ such that $A = A_1 \oplus A'$ and $D \leq A'$. There exists an isomorphism $h : A' \to A_2$ so that $h(x) = x$ for all $x \in X$, i.e., $h = \pi_2|_{A'}$. Let $D_2 = h(D)$. Then $X \leq_{\mathrm{e}} D_2$ and D_2 is a direct summand of A_2. Hence A_2 is G-extending. □

Corollary 5.83. *Let $A = \bigoplus_{i=1}^{n} A_i$ be a finite direct sum.*

(i) *If A_i is A_j-ejective for all $j > i$ and each A_i is* G*-extending, then A is* G*-extending.*

(ii) *If all the A_i are relatively injective and A is* G*-extending, then each A_i is* G*-extending.*

Proof. (i) Proceeding by induction, assume $A = \bigoplus_{i=1}^{n+1} A_i$, where each A_i is G-extending, A_i is A_j-ejective for all $1 \leq i < j \leq n+1$, and $\bigoplus_{i=1}^{k} A_i$ is G-extending for $1 \leq k \leq n$. Let $K \leq A_{n+1}$ and let $f : K \to \bigoplus_{i=1}^{n} A_i$ be a homomorphism. Let $\pi_i : A \to A_i$ be the ith projection. Define $f_i = \pi_i \circ f$. Then there exist $X_i \leq_{\text{e}} K$ and $\overline{f_i} : A_{n+1} \to A_i$, for $1 \leq i \leq n+1$, where $\overline{f_i}(x) = f_i(x)$ for all $x \in X_i$. Let $X = \bigcap_{i=1}^{n} X_i$, then $X \leq_{\text{e}} K$. Define $\overline{f} : A_{n+1} \to \bigoplus_{i=1}^{n} A_i$ by $\overline{f} = (\overline{f_1}, \ldots, \overline{f_n})$. Then $\overline{f}(x) = (\overline{f_1}(x), \ldots, \overline{f_n}(x)) = (f_1(x), \ldots, f_n(x)) = f(x)$, for all $x \in X$. Hence $\bigoplus_{i=1}^{n} A_i$ is A_{n+1}-ejective. From Theorem 5.82 (i), A is G-extending.

(ii) The proof follows by using an induction argument similar to that in part (i) and Theorem 5.82 (ii). □

Corollary 5.84. *Let $A = A_1 \oplus A_2$. Then*

(i) *If A_1 is injective, then A is* G*-extending if and only if A_2 is* G*-extending.*
(ii) *If A_1 is* G*-extending and A_2 is semisimple, then A is* G*-extending.*

Proof. This result is an immediate consequence of Theorem 5.82 since in parts (i) and (ii) A_1 is A_2-injective. □

Example 5.85. Let A_1 and A_2 be Abelian groups (i.e., $\mathbb{Z}$-modules) with A_1 divisible and $A_2 = \mathbb{Z}/\mathbb{Z}p^n$, where p is a prime and n is a positive integer. Corollary 5.84 (i) yields that $A = A_1 \oplus A_2$ is G-extending, but A is not extending, when A_1 is torsion-free, by [KaM88] and [KM88].

Example 5.86. Let A_1 be a G-extending module with a finite composition series, $0 = X_0 \leq X_1 \leq \cdots \leq X_m = A_1$ (e.g., a finite uniform module such as $\mathbb{Z}/\mathbb{Z}p^m$, for a prime p and a positive integer $m > 1$). Let $A_2 = (X_m/X_{m-1}) \oplus \cdots \oplus (X_1/X_0)$. Then $A = A_1 \oplus A_2$ is G-extending by Corollary 5.84 (ii). In general, A is not extending [DHSW94, Cor. 1.4]. Also observe that if $A_1 = \mathbb{Z}/\mathbb{Z}p^k$, where p is a prime and $k \geq 3$, then A is a G-extending $\mathbb{Z}$-module that is not extending by [KaM88] and [KM88].

Proposition 5.87. *Let A be a* G*-extending module and $N \leq A$.*

(i) *If for each $e \in \mathbf{I}(\underline{E}_A)$, there exists $f \in \mathbf{I}(\underline{E}_N)$ such that $N \cap Ae \leq_{\text{e}} Nf$, then N is* G*-extending.*
(ii) *If for each $e \in \mathbf{I}(\underline{E}_A)$, there exists $f \in \mathbf{I}(\underline{E}_A)$ such that $Ae\beta Af$ and $Nf \subseteq N$, then N is* G*-extending. In particular, all projection invariant submodules of A are* G*-extending.*

Proof. (i) Let $Y \leq N$. By Proposition 5.77, there exist $X \leq_{\text{e}} Y$ and $e \in \mathbf{I}(\underline{E}_A)$ such that $X \leq_{\text{e}} Ae$. Then $X \leq_{\text{e}} Ae \cap N \leq_{\text{e}} Nf$ for some $f \in \mathbf{I}(\underline{E}_N)$. Hence N is G-extending.

(ii) Let $Y \leq N$. Then there exists $e \in \mathbf{I}(\underline{E}_A)$ such that $Y\beta Ae$. Hence $Y\beta Af$. Since $Nf \subseteq N$, N is G-extending. □

Theorem 5.88. *Let K be a projection invariant submodule of A.*

(i) *If A is* G*-extending, there exist $A_1, A_2 \leq A$ such that $A = A_1 \oplus A_2$ and $K \leq_e A_2$.*

(ii) *If A is* G*-extending and K has a unique essential closure, then there exist $A_1, A_2 \leq A$ such that $A = A_1 \oplus A_2$, $K \leq_e A_2$, and A_1 and A_2 are* G*-extending.*

(iii) *If $A = A_1 \oplus A_2$, where A_1 and A_2 are* G*-extending and $\rho(A) \leq_e A_2$, for some left exact preradical ρ, then A is* G*-extending.*

Proof. (i) There exists $e \in \mathbf{I}(\underline{E}_A)$ such that $K\beta Ae$. But $K = Ke \oplus K(1-e)$, $Ke = K \cap Ae$, and $K(1-e) = K \cap A(1-e)$ because K is projection invariant. Since $K\beta Ae$, then $Ke \leq_e Ae$ and $Ke \leq_e K$. Hence $K \cap A(1-e) = 0$. So $K = Ke \leq_e Ae$. Now put $A_1 = A(1-e)$ and $A_2 = Ae$.

(ii) By part (i), $A = A_1 \oplus A_2$, where $A_1 = A(1-e)$, $A_2 = Ae$, $K \leq_e A_2$, and $e \in \mathbf{I}(\underline{E}_A)$.

Claim 1. A_2 is G-extending.

Proof of Claim 1. Let $Y \leq A_2$. There exists $d \in \mathbf{I}(\underline{E}_A)$ such that $Y\beta Ad$. So $K = Kd \oplus K(1-d)$, where $Kd = K \cap Ad$ and $K(1-d) = K \cap A(1-d)$. Since $Y \cap K \leq_e Y$ and $Y\beta Ad$, $Y \cap K \cap Ad \leq_e Y \cap Ad \leq_e Ad$. Thus $Kd = K \cap Ad \leq_e Ad$, and so $K \leq_e Ad \oplus K(1-d)$. Since Ae is the unique closure of K, $Ad \subseteq Ae = A_2$. Therefore, A_2 is G-extending.

Claim 2. A_1 is G-extending.

Proof of Claim 2. Let $X \leq A_1 = A(1-e)$. There exists $d \in \mathbf{I}(\underline{E}_A)$ such that $X\beta Ad$. So $X \cap Ad = X \cap Ad \cap A(1-e) \leq_e Ad \cap A(1-e) \leq_e Ad$. Since $Ad \cap A(1-e) \leq_e Ad$, $Ad \cap Ae = 0$. Thus $Kd = Ad \cap K \leq Ad \cap Ae = 0$. Hence $K = K \cap A(1-d) \subseteq A(1-d)$. Let C be a closure of K in $A(1-d)$. Since $A(1-d)$ is complement in A, C is a closure of K in A. By uniqueness, $C = Ae \subseteq A(1-d)$. So $ed = 0$ and $(1-e)d = d$. Hence $[d(1-e)]^2 = d(1-e)$. Observe that $Ad \cap A(1-e) \leq Ad(1-e)$, and recall that $Ad \cap A(1-e) \leq_e Ad$. Let $0 \neq ad(1-e) \in Ad(1-e)$. There exists $r \in R$ such that $0 \neq rad \in Ad \cap A(1-e)$. Hence, $0 \neq rad = rad(1-e) \in Ad \cap A(1-e)$. So $Ad \cap A(1-e) \leq_e Ad(1-e)$. Hence $Ad\beta Ad(1-e)$. Then $X\beta Ad(1-e)$. Therefore, A_1 is G-extending. □

Corollary 5.89. *If A is* G*-extending and $A = \bigoplus_{i\in I} A_i$, where each A_i is projection invariant, then each A_i is* G*-extending.*

Corollary 5.90. *Let ρ be a left exact preradical such that $\rho(A)$ has a unique essential closure. Then A is* G*-extending if and only if $A = A_1 \oplus A_2$, where A_1 and A_2 are* G*-extending and A_2 is the essential closure of $\rho(A)$.*

The next result is an application of Corollary 5.90.

Corollary 5.91. *Let ρ be the radical for a stable hereditary torsion theory (e.g., $\rho = Z_2$). Then a module A is* G*-extending if and only if $A = A_1 \oplus A_2$, where A_1 and A_2 are* G*-extending and $A_2 = \rho(A)$.*

Proof. This result is a consequence of Corollary 5.90 and [Ste75, Prop. 7.2]. □

Observe that A in Example 5.85 can be decomposed as $A = A_1 \oplus A_2$, but $A_1 = B \oplus C$, where B is torsion-free and C is torsion. Then $A = B \oplus C \oplus A_2$ and $Z_2(A) = C \oplus A_2$. Hence Example 5.85 provides a nonextending module A that illustrates Corollary 5.91. Moreover, comparing Corollary 5.91 with the analogous extending result [DHSW94, p. 60], one can see that the condition that $Z_2(A)$ be A_1-injective is not required in Corollary 5.91 when $\rho = Z_2$.

Corollary 5.92. *Let A be a* G*-extending module. Then*

(i) *If D is a direct summand of A such that $Z(D) = 0$, then D is extending.*
(ii) *A is* G$^+$*-extending if and only if $Z_2(A)$ is* G$^+$*-extending.*

Proof. (i) There exists $C \leq A$ such that $A = C \oplus D$. Then $Z_2(A) = Z_2(C) = C \cap Z_2(A)$. By Corollary 5.91, $C = Z_2(A) \oplus B$ for some $B \leq C$, and $A = Z_2(A) \oplus B \oplus D$, where $B \oplus D$ is extending by Proposition 5.80 (ii). Therefore, D is extending.

(ii) Since A is G-extending, Corollary 5.91 yields that $Z_2(A)$ is a G-extending direct summand of A. Since every direct summand of $Z_2(A)$ is a direct summand of A, A being G$^+$-extending implies that $Z_2(A)$ is G$^+$-extending. Conversely, assume $Z_2(A)$ is G$^+$-extending. Let D be a direct summand of A. There exists $C \leq A$ such that $A = C \oplus D$. Then $Z_2(A) = Z_2(D) \oplus Z_2(C)$. Hence $D = Z_2(D) \oplus D_1$ for some $D_1 \leq D$. By (i), D_1 is extending. Since $Z_2(A)$ is G$^+$-extending, $Z_2(D)$ is G-extending. By Corollary 5.91, D is G-extending. Therefore, A is G$^+$-extending. □

Proposition 5.93. *Let $A = \bigoplus_{i \in I} A_i$ be a direct sum where each A_i is* G*-extending. If $\operatorname{soc} A \leq_{\mathrm{e}} A$ and each simple submodule of A is contained in some A_i, then A is* G*-extending.*

Proof. Let S be a semisimple submodule of A. Then $S = \bigoplus_{i \in I} X_i$, where $X_i = S \cap A_i$. By Proposition 5.80 (v), $A_i = B_i \oplus C_i$, where $X_i \leq_{\mathrm{e}} B_i$. Hence $S \leq_{\mathrm{e}} \bigoplus_{i \in I} B_i$. Since $A = (\bigoplus_{i \in I} B_i) \oplus (\bigoplus_{i \in I} C_i)$, Proposition 5.80 (v) yields that A is G-extending. □

Let R be a Dedekind domain and A an R-module. For any maximal ideal P of R, let

$$A(P) = \{a \in A : aP^n = 0, \text{ for some positive integer } n\}.$$

Corollary 5.94. *Let R be a Dedekind domain and A a torsion module. Then A is* G*-extending if and only if $A(P)$ is* G*-extending for each maximal ideal P of R.*

Proof. It is well known that $\operatorname{soc} A \leq_{\mathrm{e}} A$, $A = \bigoplus A(P)$ (P varies through all maximal ideals of R), and each simple submodule of A is contained in some $A(P)$. Therefore, this corollary is a consequence of Corollary 5.89 and Proposition 5.93. □

Lemma 5.95. *If G is a pure subgroup of an Abelian group $A = B \oplus C$, such that $G\beta C$, then $A = B \oplus G$.*

Proof. See [Fu70, p. 120, Exc. 12]. □

Our next result characterizes the G-extending Abelian groups. By Proposition 5.80 (ii), the G-extending concept coincides with the extending concept in the class of torsion-free Abelian groups. However, for the class of torsion Abelian groups the two concepts are vastly different. For example, a direct sum of finite uniform p-groups is extending if and only if the exponents of their orders differ by at most 1. By contrast, such a direct sum is G-extending if and only if their orders are bounded. In particular, our characterization shows that every finitely generated Abelian group is G-extending.

Theorem 5.96. *The following conditions are equivalent for an Abelian group A:*

(i) *A is* G*-extending.*
(ii) *Every pure subgroup of A is a direct summand.*
(iii) *$A = D \oplus T \oplus F$, where D is a divisible group, T is a reduced torsion group each of whose p-component, T_p, is a bounded p-group, and F is a direct sum of a finite number of copies of a fixed subgroup X of the rational numbers.*

Proof. (i) $\Rightarrow$ (ii) This is a direct consequence of Lemma 5.95.

(ii) $\Rightarrow$ (iii) Since the torsion subgroup of an Abelian group is pure, $A = D \oplus T \oplus F$, where D is a divisible group, T is a reduced torsion group, and F is a torsion-free group. In a torsion-free group every subgroup is essential in a pure subgroup [Fu70, p. 116]. Hence, F is extending. By [KM88], F has the indicated decomposition. Assume that T_p is not bounded for some prime p. By [Fu70, Cor. 35.4, p. 149], T_p has at least two nonzero basic subgroups. Hence there exists a basic subgroup B of T_p such that $0 \neq B \lneqq T_p$. Since B is pure in T_p, and T_p is pure in A, B is a direct summand of T_p. But T_p/B is divisible, a contradiction. We conclude that T_p is bounded.

(iii) $\Rightarrow$ (i) By [Fu70, p. 119, Exc. 5], T_p is weak CS. Proposition 5.80 (v) shows that T_p is G-extending. The remainder of the proof follows from Corollaries 5.84, 5.91 and 5.94. □

Corollary 5.97.

(i) *The class of* G*-extending Abelian groups is closed under direct summands.*
(ii) *The class of* G*-extending torsion Abelian groups is closed under finite direct sums.*

Proof. (i) follows from Theorem 5.96 and the fact that a direct summand of an Abelian group is pure.

(ii) is an immediate consequence of Theorem 5.96. □

Corollary 5.98. *Let A be an Abelian group. Then A is weak* CS *if and only if $A = T \oplus F$, where F is torsion-free and T is a torsion Abelian group with each p-component, T_p, a direct sum of a bounded Abelian group and a divisible Abelian group.*

Proof. ($\Rightarrow$) Since A is weak CS, $A = T \oplus F$, where $\operatorname{soc} A \leq_{\mathrm{e}} T$. Hence T is a torsion Abelian group and F is torsion-free. Now Proposition 5.80 (v) and Theorem 5.96 yield this implication.

($\Leftarrow$) This implication follows from Proposition 5.80 (v), Theorem 5.96, and [Smi90, Thm. 1.12]. □

The following lemma exhibits a connection between the extending, UC, and SIP conditions.

Lemma 5.99. *Let A be a* G*-extending module. If A has UC, then A has* SIP.

Proof. By Proposition 5.80 (ii), A is extending. Let K, L be direct summands of A. Let C, D denote the closures of $K \cap L$ in K and L, respectively. Then $C = D$. It follows that $C = D = K \cap L$. Hence, $K \cap L$ is a direct summand of A, and so A has SIP. □

Proposition 5.100. *Let A be a* G*-extending module. If A has* SIP *or satisfies the C_3 condition, then A is* G$^+$*-extending.*

Proof. Using Proposition 5.87 (i), where N is taken to be a direct summand of A, and applying the SIP yields that A is G$^+$-extending. Now assume that A satisfies condition C_3. Let D be a direct summand of A. Then $A = D \oplus D'$ for some submodule D' of A. Let $\pi : A \to D$ be the canonical projection. For any submodule K of D, there exists a direct summand L of A such that $K \cap L \leq_{\mathrm{e}} K$ and $K \cap L \leq_{\mathrm{e}} L$. Since A satisfies C_3, $D' \oplus L$ is a direct summand of A. By Lemma 2.71, $D' \oplus L = D' \oplus \pi(L)$. It follows that $\pi(L)$ is a direct summand of D. For any $0 \neq x \in \pi(L)$, $x = \pi(l)$ for some $0 \neq l \in L$. There exists an $r \in R$ such that $0 \neq rl \in K \cap L$. Then $rl = k = l_1$, where $k \in K$ and $l_1 \in L$. Now $0 \neq rx = r\pi(l) = k = \pi(l_1) \in K \cap \pi(L)$. Thus $K \cap \pi(L) \leq_{\mathrm{e}} \pi(L)$. Clearly, $\pi(L) = D \cap (D' \oplus \pi(L)) = D \cap (D' \oplus L)$. Hence $K \cap \pi(L) = K \cap (D' \oplus L) \leq_{\mathrm{e}} K$. Thus D is G-extending, and consequently A is G$^+$-extending. □

Corollary 5.101. *Let A be a* G*-extending module. If A satisfies condition C_2, then A is* G$^+$*-extending.*

Proof. By Lemma 2.72, A satisfies condition C_3, and so the result follows by Proposition 5.100. □

Now we give examples illustrating Proposition 5.100 and Corollary 5.101.

Example 5.102.

(i) Let $R = \begin{bmatrix} \mathbb{Z}_2 & \mathbb{Z}_2 \\ 0 & \mathbb{Z} \end{bmatrix}$ be the 2-by-2 generalized triangular matrix ring. Note that the only direct summands (as right ideals) of R are 0, R, $\begin{bmatrix} 0 & 0 \\ 0 & \mathbb{Z} \end{bmatrix}$, $\begin{bmatrix} 0 & 1 \\ 0 & 1 \end{bmatrix} R$, and $\begin{bmatrix} \mathbb{Z}_2 & \mathbb{Z}_2 \\ 0 & 0 \end{bmatrix}$. It can be checked that R_R is G-extending and satisfies C_3. However, R_R is not extending (Exercise 3.34) and satisfies neither C_2, nor SIP.

(ii) Let A be the $\mathbb{Z}$-module, $A = \mathbb{Q} \oplus (\mathbb{Z}/\mathbb{Z}p)$, where p is a prime. Since A is a special case of Example 5.85, A is G-extending, but not extending. By [Wil86, Prop. 10] and Example 4.46, A has SIP and satisfies condition C_2.

Since the extending property is closed under direct summands, if A is extending and has an indecomposable decomposition, then A is a direct sum of uniform modules. From Proposition 5.80 (i), the same is true if A is a G^+-extending module (see Proposition 5.100, Corollary 5.101 or [DY12, Thm. 2.1]).

Proposition 5.103. *If A is a weak* CS*-module and has* SIP *or satisfies condition C_3, then $A = A_1 \oplus A_2$, where* $\operatorname{soc} A \leq_{\mathrm{e}} A_2$ *and A_2 is* G^+*-extending.*

Proof. Since A is a weak CS-module, $A = A_1 \oplus A_2$, where $\operatorname{soc} A \leq_{\mathrm{e}} A_2$. If A has SIP or satisfies C_3, then A_2 has SIP or satisfies C_3 (Lemma 2.74). Now Proposition 5.80 (v) and Proposition 5.100 yield the result. □

Corollary 5.104. *If A is a weak* CS*-module and satisfies condition C_2, then $A = A_1 \oplus A_2$, where* $\operatorname{soc} A \leq_{\mathrm{e}} A_2$ *and A_2 is* G^+*-extending.*

Proof. This follows from Lemma 2.72 and Proposition 5.103. □

Exercises

5.9. Let α and β be as in Definition 5.73. Prove the following statements.

(i) α is reflexive, symmetric, but not transitive.

(ii) β is an equivalence relation.

(iii) A module A is UC if and only if $\alpha = \beta$.

5.10. Write out the proof of Proposition 5.74.

5.11. Let β_{IE} be as in Definition 5.76. Prove that β_{IE} is an equivalence relation.

5.5 Goldie extending modules via direct sums of uniform modules and essential extensions

Uniform modules are often considered to be fundamental objects in the structure theory of various classes of modules, e.g., the class of modules with finite uniform dimension, or the class of extending modules. Hence, for a given class, solving the following problems is important.

(1) Determine conditions under which a module from a given class is a direct sum of uniform modules.
(2) Determine when a direct sum of uniform modules is in a given class.

We begin this section with results providing solutions to the first problem and we continue with solutions to the second problem, where the given class is that of the G-extending modules. We finish with results on G-extending essential extensions of a module.

Proposition 5.105. *Let A be a module with uniform dimension 2. Then*

(i) *A satisfies C_{11} if and only if A is a direct sum of uniform submodules.*
(ii) *If A is G-extending, then A is a direct sum of uniform submodules.*

Proof. (i) Assume A satisfies C_{11}, and let $0 \neq U$ be a uniform submodule of A. Then $A = B \oplus K$, where K is a complement of U. Then K must be uniform since it is a complement of U, and B must be uniform since it is a complement of K. The converse is clear.

(ii) is a consequence of Proposition 5.78 and (i). □

Observe that in Proposition 5.105, since A has uniform dimension 2, the C_{11} and G-extending conditions imply the C_{11}^{+} and G^{+}-extending conditions, respectively. Our next result is analogous of [DHSW94, 8.2 (i)], which implies that if A is an extending module over a left Noetherian ring, then A is a direct sum of uniform submodules.

Theorem 5.106. *Let R be a ring and let A be an R-module such that R satisfies (A). Then*

(i) *If A is G-extending, then for each $a \in A$ such that $\ell(a)$ is maximal in $\{\ell(x) : 0 \neq x \in A\}$, there exists a primitive idempotent $e \in \text{End}(_RA)$ with $Ra\beta Ae$.*
(ii) *If A is G^{+}-extending, then Ra is uniform for each $a \in A$ such that $\ell(a)$ is maximal in $\{\ell(x) : 0 \neq x \in A\}$. Hence, any direct summand of A contains a uniform direct summand.*
(iii) *If A is G-extending and has SIP or satisfies C_3, then A is a direct sum of uniform submodules.*

Proof. (i) Let $0 \neq a \in A$ such that $\ell(a)$ is maximal in $\{\ell(x) : 0 \neq x \in A\}$. By Proposition 5.77, there exists $e \in \mathbf{I}(\underline{E}_A)$ such that $Ra \cap Ae \leq_e Ae$ and $Ra \cap Ae \leq_e Ra$. Assume that e is not primitive. Then $Ae = K_1 \oplus K_2$, for some $0 \neq K_1, K_2 \leq$

Ae. Now $a = a_1 + a_2 + a'$, where $a_i \in K_i$ and $a' \in A(1-e)$. If $a_1 = 0$, then $a \in K_2 \oplus A(1-e)$ and $Ra \cap K_1 = 0$. Then $K_1 = 0$, since $Ra \cap Ae \leq_e Ae$. But this is a contradiction, so $a_1 \neq 0$. Clearly, $\ell(a) \subseteq \ell(a_1)$. Hence $\ell(a) = \ell(a_1)$, by the choice of a. Similarly, $\ell(a_2) \neq 0$ and $\ell(a) = \ell(a_2)$. Since $0 \neq a_1 \in K_1$, there exists $r_1 \in R$ such that $0 \neq r_1 a_1 \in Ra$. Hence there exists $r_2 \in R$ such that $0 \neq r_1 a_1 = r_2 a = r_2 a_1 + r_2 a_2 + r_2 a'$. Thus $r_2 a_2 + r_2 a' = r_1 a_1 - r_2 a_1 \in K_1 \cap K_2 \oplus A(1-e) = 0$, and so $r_2 a_2 = -r_2 a' \in K_2 \cap A(1-e) = 0$. Consequently, $r_2 \in \ell(a_2) \backslash \ell(a)$, a contradiction. Therefore, e is primitive.

(ii) follows from (i) and Proposition 5.80 (i).

(iii) Assume A is G-extending. First we consider the case when A has SIP. By Proposition 5.100, A is G$^+$-extending. By Zorn's lemma, A contains a maximal local direct summand $N = \bigoplus_{i \in I} N_i$, where each N_i is uniform. By Proposition 5.77, there exists $D \leq A$ such that $A = D \oplus D'$ for some $D' \leq A$ such that $N \beta D$. Let $\overline{N_i} = N_i \cap D$. By the SIP, $\overline{N} = \bigoplus_{i \in I} \overline{N_i}$ is a local summand of D. Since $\overline{N_i} \leq_e N_i$, $\overline{N} \leq_e N \cap D \leq_e D$. By [DHSW94, Lem. 8.1 (i)], $\overline{N} = D$, so $D = N$. Thus $A = N \oplus D'$. If $D' \neq 0$, then by the above argument, $D' = U \oplus U'$ for some U, $U' \leq D'$ with U uniform. Then $N \oplus U$ is a local direct summand of A, contradicting the choice of N. Therefore, $D' = 0$, so $A = N = \bigoplus_{i \in I} N_i$, a direct sum of uniform modules. The case when A satisfies C_3 follows from Theorem 4.52 and Proposition 5.78. □

The following examples show how Proposition 5.105 and Theorem 5.106 (iii) are related to analogous results for the extending and C_{11} conditions [DHSW94, 8.2 (i)], Theorem 4.52, respectively.

Example 5.107.

(i) By Example 5.102 (ii), $A = \mathbb{Q} \oplus (\mathbb{Z}/\mathbb{Z}p)$ illustrates Theorem 5.106 (iii), but, since A is not extending, it is not covered by [DHSW94, 8.2 (i)].

(ii) If we take $A = R = T_2(F)$ (i.e., the ring of 2-by-2 upper triangular matrices over a field F), then A is extending and has SIP by Lemma 5.99, but it does not satisfy C_3 (Exercise 2.44). Thus it is covered by Theorem 5.106 (iii), but it is not included in Theorem 4.52.

The following corollary is an interesting special case of Theorem 5.106 (iii).

Corollary 5.108. *For any ring R, any locally Noetherian* G*-extending module which has* SIP *or satisfies C_3 is a direct sum of uniform submodules.*

Proof. Let A be a locally Noetherian G-extending module which has SIP or satisfies C_3. Let $a \in A$. Then $R/\ell(a) \cong Ra$ is a Noetherian R-module. It follows that R satisfies acc on left ideals of the form $\ell(x)$, where $x \in A$. By Theorem 5.106 (iii), A is a direct sum of uniform submodules. □

Lemma 5.109. *Let A be a module. Then*

(i) *If A is strongly bounded, then each semisimple submodule of A is fully invariant.*

(ii) *Let* $\operatorname{soc} A = \bigoplus_{i \in I} X_i$, *where each* X_i *is simple and* $X_i \not\cong X_j$ *for* $i \neq j$.

(a) *If* S *is a simple submodule of* A, *then* $S = X_i$ *for some* $i \in I$ *and* $S \trianglelefteq A$.

(b) *If* $\operatorname{soc} A \leq_{\mathrm{e}} A$, *then* A *is strongly bounded.*

Proof. (i) is an immediate consequence of the pertinent definitions.

(ii) (a) Let $f \in \operatorname{End}({}_R A)$ such that $0 \neq f(S)$. Let $0 \neq s \in S$. Then $S = Rs$ and $f(s) = \sum_{j \in J} x_j$, where $0 \neq x_j \in X_j$, for all $j \in J$, and J is finite subset of I. Assume $|J| > 1$. Then there exist $j_1, j_2 \in J$ such that $j_1 \neq j_2$, $0 \neq x_{j_1} \in X_{j_1}$, and $0 \neq x_{j_2} \in X_{j_2}$. Let $\pi_i : \operatorname{soc} A \to X_i$ be the projection map. Then $f(S) \cong \pi_{j_1}(f(S)) = X_{j_1}$ and $f(S) \cong \pi_{j_2}(f(S)) = X_{j_2}$, a contradiction. Hence $f(S) = X_n$ for some $n \in I$. Taking f to be the identity endomorphism yields that $S = X_m$ for some $m \in I$. But $S \cong f(S)$ for any $f \in \operatorname{End}({}_R A)$ such that $f(S) \neq 0$. Hence $m = n$ and $S = f(S)$. Thus $S \trianglelefteq A$.

(b) is an immediate consequence of part (ii) (a). □

The next result and example show that R-Mod has a minimal cogenerator which is G-extending and strongly bounded but, in general, it is not extending.

Theorem 5.110.

(i) *Let* $A = \bigoplus_{i \in I} A_i$, *where each* A_i *is a subdirectly irreducible module and* $H(A_i) \not\cong H(A_j)$ *for* $i \neq j$, *where* $H(\cdot)$ *denotes the heart of* $\cdot$. *Then* A *is* G*-extending.*

(ii) *Let* $\{X_i : i \in I\}$ *be a set of representatives of the isomorphism classes of all simple* R*-modules, and* $C = \bigoplus_{i \in I} E(X_i)$. *Then* C *is a minimal cogenerator for* R*-Mod and* C *is* G*-extending. Moreover, minimal cogenerators and minimal injective cogenerators are strongly bounded.*

Proof. (i) Since each A_i is uniform (hence G-extending), this part follows from Lemma 5.109 and Proposition 5.93.

(ii) It is well known that C is a minimal cogenerator for R-Mod [Lam99, pp. 508–509]. Observe that each $E(X_i)$ is subdirectly irreducible. By part (i), C is G-extending. By Lemma 5.109, if $Y \leq E(C)$ is such that $\operatorname{soc}(E(C)) \leq Y$, Y is strongly bounded. By [Lam99, p. 510], every minimal injective cogenerator is isomorphic to $E(C)$, hence strongly bounded. Also, every minimal cogenerator K is isomorphic to some Y, where $\operatorname{soc} C \leq Y \leq C$. Hence K is strongly bounded. □

Example 5.111. There exist rings R such that C is not extending, where C is as in Theorem 5.110. Osofsky [Osof91] indicates that if C is quasi-injective, then C is the unique (up to isomorphism) minimal cogenerator of R-Mod. She then produces several rings for which C is not the unique (up to isomorphism) cogenerator. Hence, for such rings C is not quasi-injective. By [DHSW94, Cor. 8.10], C is not uniform-extending (hence not extending) for these rings.

The modules $T_2(D)$ in the proof of Proposition 5.78 and $A = \mathbb{Z}[x] \oplus \mathbb{Z}[x]$, where $R = \mathbb{Z}[x]$ show that the converse of Proposition 5.105 (ii) does not hold, in general, for a direct sum of two uniform modules. Now we consider this case.

Definition 5.112. Let B and A be modules. On $\mathrm{Hom}(B, A)$ define the equivalence relation γ by $f\gamma g$ if there exists $N \leq_{\mathrm{e}} B$ such that $f(x) = g(x)$ for all $x \in N$.

Theorem 5.113. *Let $A = A_1 \oplus A_2$, where A_1 and A_2 are uniform modules. If $V \subseteq \mathrm{Hom}(E(A_2), E(A_1))$ is such that V contains a set of representatives of the equivalence classes of γ and $A_2V \subseteq A_1$, then A_1 is A_2-ejective, so A is G-extending.*

Proof. Let $0 \neq K \leq A_2$ and $0 \neq f : K \to A_1$. First assume that $\mathrm{ker} f \neq 0$. Let $\overline{f} : A_2 \to A_1$ be the zero homomorphism. Then $f|_{\mathrm{ker} f} = \overline{f}|_{\mathrm{ker} f}$ and $\mathrm{ker} f \leq_{\mathrm{e}} K$. Now assume $\mathrm{ker} f = 0$. There exists $\overline{f} : E(A_2) \to E(A_1)$ such that $f|_K = \overline{f}|_K$. Observe that $\overline{f}$ is an isomorphism. There exists $h \in V$ such that $h\gamma\overline{f}$. Note that h is also an isomorphism. There exists $X \leq_{\mathrm{e}} E(A_2)$ such that $h|_X = \overline{f}|_X$. Hence $f|_{X\cap K} = \overline{f}|_{X\cap K}$ and $X \cap K \leq_{\mathrm{e}} K$. But $h(A_2) \subseteq A_1$. Therefore A_1 is A_2-ejective. By Theorem 5.82, A is G-extending. □

Observe that for γ defined on $\mathrm{Hom}(E(A_2), E(A_1))$, the γ equivalence class of the zero homomorphism is $\{g \in \mathrm{Hom}(E(A_2), E(A_1)) : \mathrm{ker}\, g \neq 0\}$. Thus for a choice of a set of representatives of the γ equivalence classes, one can take the zero homomorphism for the zero γ equivalence class, and then all other representatives will be isomorphisms.

Corollary 5.114. *Let U be a uniform module and let $V \subseteq \underline{E}_{E(U)}$ be such that V contains a set of representatives of the equivalence classes of γ. Then UV is U-ejective, and $A = U \oplus UV$ is G-extending.*

Proof. This result is an immediate consequence of Theorem 5.113. □

Corollary 5.115. *Let $A = \bigoplus_{i=1}^{m} A_i$, where each A_i is uniform. If any of the following conditions hold, then A is G-extending.*

(i) *$E(A_i) \not\cong E(A_j)$ for $i \neq j$;*

(ii) *For each $0 \neq K_j \leq A_j$ and monomorphism $f_i : K_j \to A_i$, where $i < j$ for $1 \leq i < m$ and $1 < j \leq m$, there exist $0 \neq X_j \leq K_j$ and $\overline{f}_j : A_j \to A_i$ such that $f_j|_{X_j} = \overline{f}_j|_{X_j}$.*

Proof. (i) Let $i < j$. By the proof of Theorem 5.113 and the comment after it, we can choose the set consisting of the zero homomorphism to be a set of representatives of the equivalence classes of γ on $\mathrm{Hom}(E(A_j), E(A_i))$. Hence A_i is A_j-ejective. By Corollary 5.83, A is G-extending.

(ii) Let $i < j$. Observe that each $\overline{f}_j$ can be extended to an isomorphism from $E(A_j)$ onto $E(A_i)$. Now Theorem 5.113, the comment after its proof, and Corollary 5.83 yield the result. □

From Corollary 5.115 (i) it can be seen that to determine if $A = A_1 \oplus A_2$ is G-extending, where A_1 and A_2 are uniform, one only needs to consider when the direct sum is external and $E(A_1) = E(A_2)$. So if we further assume that A_1 and A_2 are subdirectly irreducible, then by Corollary 5.115 (ii), we only need to show that for each $0 \neq f \in \text{End}(H)$ there exists $\overline{f} : A_2 \to A_1$ such that $f|_H = \overline{f}|_H$, where H is the common heart of A_1 and A_2, to conclude that A is G-extending. However not every such f can be extended to $\overline{f}$. Example 4.3 provides a module $A = A_1 \oplus A_2$, where A_1 and A_2 are subdirectly irreducible, uniserial and $E(A_1) \cong E(A_2)$. However, A is not G-extending. In this example, we note that R is commutative. In contrast, we have the following result.

Corollary 5.116. *Let R be a commutative ring, S a simple module, and $A_1, A_2 \leq E(S)$. If $IA_2 \leq A_1$ for some $I \subseteq R$ which contains at least one element from each coset of $R/\ell(S)$, then A_1 is A_2-ejective. So $A = A_1 \oplus A_2$ is* G*-extending. In particular, if $A_2 \leq A_1$, then A_1 is A_2-ejective and $A = A_1 \oplus A_2$ is* G*-extending.*

Proof. Let $0 \neq f \in \text{End}(S)$ and $0 \neq s \in S$. Then $Rs = S$ and $f(s) = xs$ for some $x \in R$. Then $f(ts) = x(ts)$ for all $t \in R$. There exists $r \in I$ such that $x - r \in \ell(S)$. Thus $f(s) = xs = rs$. So $f(ts) = r(ts)$ for all $t \in R$. Define $\overline{f} \in \underline{E}_{E(S)}$ by $\overline{f}(k) = rk$ for all $k \in E(S)$. Then $\overline{f}|_S = f|_S$, and $\overline{f}(A_2) = rA_2 \subseteq A_1$. Observe that if $0 \neq g \in \underline{E}_{E(S)}$ then $g(S) \subseteq S$. Thus each g is γ-related to an $\overline{f}$ as described above. Now the result is a consequence of Theorem 5.113. □

Corollary 5.117. *Let R be a commutative ring and S a subdirectly irreducible uniserial module. Let $A = \bigoplus_{i=1}^{n} A_i$ where each $A_i \leq S$. Then A is* G*-extending.*

Proof. This result is a consequence of Corollaries 5.116 and 5.83. □

In Corollary 5.117, G-extending cannot be replaced by extending. To see this, let $S = \mathbb{Z}/\mathbb{Z}p^n$ for $n \geq 3$, p a prime and $A = \bigoplus_{i=1}^{n} A_i$, where each $A_i = \mathbb{Z}/\mathbb{Z}p^i$.

Corollary 5.118. *Let S be a simple module and $A_1, A_2 \leq E(S)$. If there exists a homomorphism $h : A_2 \to S$ such that $h(S) \neq 0$, then A_1 is A_2-ejective, so $A = A_1 \oplus A_2$ is* G*-extending.*

Proof. Let $0 \neq f \in \text{End}(S)$. Define $g \in \text{End}(S)$ by $g = h|_S$. Then $f \circ g^{-1} \circ h|_S = f$. Let $\overline{f} \in \underline{E}_{E(S)}$ be an extension of $f \circ g^{-1} \circ h$. Hence $\overline{f}|_S = f$. By the comment after the proof of Corollary 5.115, A_1 is A_2-ejective, so $A = A_1 \oplus A_2$ is G-extending. □

Example 5.119. Let K be a field and $R = K[x, y]$ the commutative local Frobenious K-algebra [Lam99, p. 419] defined by the relations $xy = x^2 - y^2 = 0$. Then R_R is a uniform injective module with simple submodule Kx^2. Let $A_2 = xR = \{k_1x + k_2x^2 : k_i \in K\}$, and let $h : xR \to Kx^2$ be the R-homomorphism, defined by $h(k_1x + k_2x^2) = k_2x^2$. Then $h(Kx^2) \neq 0$. Thus, by Corollary 5.118, $A = A_1 \oplus xR$ is G-extending for any $A_1 \leq R_R$. Observe that each $f \in \text{End}(Kx^2)$ is a scalar multiplication map by an element of K. This multiplication map is also in $\text{End}(R_R)$. Thus in Corollary 5.116, $K \subseteq I$ so $xRI = xR \nleq yR$. Hence Corollary 5.116 cannot be used to show that $A = xR \oplus yR$ is G-extending.

It is clear that every module has at least one G-extending essential extension, namely its injective hull. However, since the G-extending property is a generalization of injectivity, one may expect to find G-extending essential extensions intermediate between a module and its injective hull. Now we focus on G-extending essential extensions of a module. It is well known that the extending property is not closed under essential extensions. However, we show that if a module is G-extending, then so is its rational hull (see Exercise 3.6 and Theorem 3.6).

Theorem 5.120. *Let A be a module and $A \leq N \leq E(A)$. Then N is G-extending if and only if $N = NV$, where V is some set of representatives for β_{IE} on $\mathbf{I}(\underline{E}_{E(A)})$.*

Proof. Assume N is G-extending. Observe that $\underline{E}_{E(N)} = \underline{E}_{E(A)}$. By Proposition 5.77, a set of representatives for β_{IE} on $\mathbf{I}(\underline{E}_{E(A)})$, V, can be chosen so that $V \subseteq \{e \in \mathbf{I}(\underline{E}_{E(A)}) : Ne \leq N\}$. Observe that $\{1_{\underline{E}}\}$ is the β_{IE} equivalence class for $1_{\underline{E}}$, the unity for $\underline{E}_{E(A)}$. Therefore, $NV = N$. Conversely, assume that $N = NV$, where V is some set of representatives for β_{IE} on $\mathbf{I}(\underline{E}_{E(A)})$. Let $c \in \mathbf{I}(\underline{E}_{E(N)})$. Observe that $\underline{E}_{E(N)} = \underline{E}_{E(A)}$. Hence there exists $e \in V$ such that $c\beta_{IE}e$ and $Ne \leq NV = N$. By Proposition 5.77, N is G-extending. □

Corollary 5.121.

(i) *$A\langle V\rangle$ is G-extending, where V is any set of representatives for β_{IE} on $\mathbf{I}(\underline{E}_{E(A)})$.*

(ii) *$\widetilde{E}(A)$ is G-extending if and only if there exists a set of representatives for β_{IE} on $\mathbf{I}(\underline{E}_{E(A)})$, V, such that $Ae \leq \widetilde{E}({}_RA)$ for each $e \in V$. In particular, if $A \leq N \leq \widetilde{E}({}_RA)$ and N is G-extending, then $\widetilde{E}(A)$ is G-extending.*

(iii) *Assume $\operatorname{soc} A \leq_e A$ and $c, e \in \mathbf{I}(\underline{E}_{E(A)})$. Then $c\beta_{IE}e$ if and only if $\operatorname{soc}(Ac) = \operatorname{soc}(Ae)$. Hence a set V of representatives for β_{IE} on $\mathbf{I}(\underline{E}_{E(A)})$ can be formed by choosing exactly one $e \in \mathbf{I}(\underline{E}_{E(A)})$ for each semisimple S on A such that $E(A)e$ is an injective hull of S and including $0_{\underline{E}}$ and $1_{\underline{E}} \in V$.*

Proof. (i) Let $N = A\langle V\rangle$. Then $NV = (A\langle V\rangle)V \subseteq A\langle V\rangle = N$. By Theorem 5.120, $A\langle V\rangle$ is G-extending.

(ii) If $\widetilde{E}(A)$ is G-extending then, by Theorem 5.120, there exists a set of representatives for β_{IE} on $\mathbf{I}(\underline{E}_{\widetilde{E}(A)}) = \mathbf{I}(\underline{E}_{E(A)})$, V, such that $\widetilde{E}(A)V = \widetilde{E}(A)$. Hence $Ae \leq \widetilde{E}(A)$ for each $e \in V$. Conversely, assume there exist a set of representatives for β_{IE} on $\mathbf{I}(\underline{E}_{E(A)})$, V, such that $Ae \leq \widetilde{E}(A)$ for each $e \in V$. Let $c \in \mathbf{I}(\underline{E}_{\widetilde{E}(A)}) = \mathbf{I}(\underline{E}_{E(A)})$. Then $c\beta_{IE}e$ for some $e \in V$. By [Lam99, Prop. 8.6 and Thm. 8.24], there exists $\overline{e} : \widetilde{E}(A) \to \widetilde{E}(A)$ such that $e|_A = \overline{e}|_A$. Assume there exists $y \in \widetilde{E}(A)$ such that $[\overline{e} - e](y) \neq 0$. Then there exists $r \in R$ such that $r[\overline{e} - e](y) \neq 0$ and $ry \in A$. But $[\overline{e} - e](ry) = \overline{e}(ry) - e(ry) = 0$, a contradiction. Hence $\widetilde{E}(A)e = \widetilde{E}(A)\overline{e} \leq \widetilde{E}(A)$. By Proposition 5.77, $\widetilde{E}(A)$ is G-extending. So if $A \leq N \leq \widetilde{E}(A)$ and N is G-extending, then, by Theorem 5.120,

$Ae \leq Ne \leq N \leq \widetilde{E}(A)$ for all $e \in V$, where V is some set of representatives for β_{IE} on $\mathbf{I}(\underline{E}_{E(A)})$. Therefore, $\widetilde{E}(A)$ is G-extending.

(iii) Observe that $\operatorname{soc}(Ae) = \operatorname{soc}(E(A)e) = E(A)e \cap \operatorname{soc} A = \bigcap\{L \leq_{\mathrm{e}} E(A)e\}$. Hence $\operatorname{soc}(Ae) \leq_{\mathrm{e}} E(A)e$ for all $e \in \mathbf{I}(\underline{E}_{E(A)})$. If $c\beta_{IE}e$, then both $\operatorname{soc}(Ac)$ and $\operatorname{soc}(Ae)$ are essential in $E(A)c \cap E(A)e$. Since the socle of a module is semisimple, $\operatorname{soc}(Ac) = \operatorname{soc}(Ae)$. Conversely, if $\operatorname{soc}(Ac) = \operatorname{soc}(Ae) = S$, then $S \leq_{\mathrm{e}} E(A)c$. So $E(A)c \cap E(A)e \leq_{\mathrm{e}} E(A)c$ and $E(A)c \cap E(A)e \leq_{\mathrm{e}} E(A)e$. Thus $c\beta_{IE}e$. □

Exercises

5.12. Find examples that illustrate Theorem 5.110 (i).

5.13. Let γ be as in Definition 5.112. Show that γ is an equivalence relation.

5.6 Goldie extending rings

In this section, we study the behavior of the G-extending property under various ring extensions. We show that if R_R is G-extending and S is a right essential overring, then S_R and S_S are G-extending. For split-null extensions, we obtain that if $M \trianglelefteq R$ and M is left faithful, then R_R is G-extending if and only if S_S is G-extending, where $S = S(R, M)$ (see Theorem 3.95 and Proposition 4.86).

A ring R is called a *right* G*-extending ring* if R_R is a G-extending module, i.e., for every right ideal I of R there exists an idempotent e of R such that $I \cap eR$ is essential in both I and eR. The class of right G-extending rings properly contains the class of right extending rings. For example, if R is right self-injective, then the 2-by-2 upper triangular matrix ring $T_2(R)$ is right G-extending; but it is not, in general, right extending (e.g., take $R = \mathbb{Z}/\mathbb{Z}4$).

Theorem 5.122. *Let S be a right essential overring of R.*

(i) *If R_R is* G*-extending, then S_R and S_S are* G*-extending,*
(ii) *If S is a subring of $Q(R)$, then S_R is* G*-extending if and only if S_S is* G*-extending.*

Proof. (i) Let $Y_R \leq S_R$ and $X = Y \cap R$. By Proposition 5.77, there exist $K_R \leq R_R$ and $e \in \mathbf{I}(R)$ such that $K_R \leq_{\mathrm{e}} X_R$ and $K_R \leq_{\mathrm{e}} eR_R$. Observe that $K_R \leq_{\mathrm{e}} Y_R$. We claim that $K_R \leq_{\mathrm{e}} eS_R$. Let $0 \neq es \in eS$. There exists $r_1 \in R$ such that $0 \neq esr_1 \in R$. Hence $0 \neq esr_1 \in eR$, so there exists $r_2 \in R$ such that $0 \neq esr_1r_2 \in K$. Thus $K_R \leq_{\mathrm{e}} eS_R$. By Proposition 5.77, S_R is G-extending. A proof similar to the above shows that $KS_S \leq_{\mathrm{e}} Y_S$ and $KS_S \leq_{\mathrm{e}} eS_S$. Thus, S_S is G-extending.

(ii) By (i), if S_R is G-extending, then so is S_S. Conversely, assume S_S is G-extending. Let $Y_R \leq S_R$. There exists $e \in \mathbf{I}(S)$ such that $YS_S\beta eS_S$. By [BPR06, Lem. 1.4 (i)], $(YS \cap eS)_R \leq_{\mathrm{e}} YS_R$ and $(YS \cap eS)_R \leq_{\mathrm{e}} eS_R$. Since R_R is rational in S_R, $Y_R \leq_{\mathrm{e}} YS_R$. To see this, let $0 \neq y \in YS$. Then $y = \sum_{i=1}^{n} y_i s_i$ for some $y_i \in Y$

and $s_i \in S$, and $y_i s_i \neq 0$ for each i. Then there exists $r_1 \in R$ such that $0 \neq yr_1$ and $s_1 r_1 \in R$. Again there exists $r_2 \in R$ such that $0 \neq yr_1 r_2$ and $s_2 r_1 r_2 \in R$. Continuing this process, we obtain $r \in R$ such that $0 \neq yr \in Y$. So $Y_R \leq_{\mathrm{e}} YS_R$, and consequently $(Y \cap eS)_R \leq_{\mathrm{e}} Y_R$ and $(Y \cap eS)_R \leq_{\mathrm{e}} eS_R$. Therefore, S_R is G-extending. □

Corollary 5.123. *Let $T = T_m(R)$ and $M = M_m(R)$. If T_T is* G*-extending, then M_T and M_M are* G*-extending.*

Proof. This result is a consequence of Theorem 5.122 and the fact that M_T is a rational extension of T_T (Exercise 5.14). □

According to [CHN07], there exists a commutative local uniform (hence extending) ring R such that $E(R)$ has a ring multiplication that extends its R-module scalar multiplication, but $E(R)$ is not right self-injective. The next result shows that for such a ring R, $E(R)$ must be at least right G-extending (see Corollary 4.89).

Corollary 5.124. *Let R_R be G-extending. If $E(R)$ has a ring multiplication which extends its right R-module scalar multiplication, then $E(R)_{E(R)}$ is* G*-extending.*

In the following theorem, $S(R, M)$ denotes the split-null (or the trivial) extension of R by M, where M is a (R, R)-bimodule. We refer to Theorem 3.95 and Proposition 4.86 for corresponding results on extending and C_{11} cases, respectively.

Theorem 5.125. *Let R be a ring, M be an ideal of R, and denote $S = S(R, M)$.*

(i) *If S_S is* G*-extending, then R_R is* G*-extending,*
(ii) *If ${}_RM$ is faithful and R_R is* G*-extending, then S_S is* G*-extending,*
(iii) *If ${}_RM$ is faithful, R_R is extending, and R is Abelian, then S_S is extending and Abelian.*

Proof. First observe that for any (R, R)-bimodule M and $(e, m) = (e, m)^2 \in S$, we have $e = e^2 \in R$, $em + me = m$ and $eme = 0$. So $(e, me) = (e, me)^2$, $(e, m)(e, me) = (e, me)$, and $(e, me)(e, m) = (e, m)$. Hence $(e, m)S = (e, me)S$.

(i) Assume that S_S is G-extending. Let $X_R \leq R_R$. Then there exists $(e, m) = (e, m)^2 \in S$ such that $[(X, X \cap M) \cap (e, m)S]_S \leq_{\mathrm{e}} (e, m)S_S$ and $[(X, X \cap M) \cap (e, m)S]_S \leq_{\mathrm{e}} (X, X \cap M)_S$. To show that R_R is G-extending, let $0 \neq er \in eR$ and $0 \neq y \in X$. We need to find $t, v \in R$ such that $0 \neq ert \in X \cap eR$ and $0 \neq yv \in X \cap eR$. Since $[(X, X \cap M) \cap (e, m)S]_S \leq_{\mathrm{e}} (e, m)S_S$, there exists $(t, k) \in S$ such that $0 \neq (e, me)(r, 0)(t, k) = (ert, erk + mert) \in (X, X \cap M) \cap (e, m)S$. If $0 \neq ert$, then $0 \neq ert \in X \cap eR$. Otherwise, $ert = mert = 0$ and $0 \neq erk \in X \cap eR$. Therefore, $(X \cap eR)_R \leq_{\mathrm{e}} eR$. Since $[(X, X \cap M) \cap (e, m)S]_S \leq_{\mathrm{e}} (X, X \cap M)_S$, there exists $(a, h) \in S$ such that $0 \neq (y, 0)(a, h) = (ya, yh) \in (X, X \cap M) \cap (e, m)S$. Then $0 \neq (ya, yh) = (e, me)(ya, yh) = (eya, eyh + meya)$. If $0 \neq eya$, then $0 \neq eya \in X \cap eR$. Otherwise, $eya = meya = 0$ and $0 \neq eyh \in X \cap eR$. Therefore $(X \cap eR)_R \leq_{\mathrm{e}} X_R$. Consequently, R_R is G-extending.

(ii) Assume ${}_RM$ is faithful and R_R is G-extending. Recall from [Birk90, Lem. 4 (iv)] that $(\{0\}\times M)_S \leq_e S_S$. Let $0 \neq Y_S \leq S_S$. Observe that $[Y\cap(\{0\}\times M)]_S \leq_e Y_S$. Consider the set $X = \{m \in M : (0,m) \in Y\}$. Then $X_R \leq R_R$. Since R_R is G-extending, there exists $e^2 = e \in R$ such that $(X\cap eR)_R \leq_e X_R$ and $(X\cap eR)_R \leq_e eR_R$. Let $V = \{(0,x) \in S : x \in X \cap eR\}$. Note that $V_S \leq Y_S$. We will show that $V_S \leq_e Y_S$ and $V_S \leq_e (e,0)S$. Let $0 \neq (0,y) \in Y\cap(\{0\}\times M)$. Then $y \in X$, so there exists $a \in R$ such that $0 \neq ya \in X\cap eR$. Then $0 \neq (0,y)(a,0) = (0,ya) \in V$. Thus $V_S \leq_e [Y\cap(\{0\}\times M)]_S$, so $V_S \leq_e Y_S$. Now let $0 \neq (e,0)(r,k) = (er,ek) \in (e,0)S$. If $0 \neq er$, there exists $b \in R$ such that $0 \neq erb \in X \cap eR$. Since ${}_RM$ is faithful, there exists $m_1 \in M$ such that $0 \neq erbm_1 \in X\cap eR$. Then $0 \neq (er,ek)(0,bm_1) = (0,erbm_1) \in V$. So assume $0 = er$. Then $0 \neq ek$, so there exists $d \in R$ such that $0 \neq ekd \in X \cap eR$. It follows that $0 \neq (0,ek)(d,0) = (0,ekd) \in V$, and so $V_S \leq_e (e,0)S_S$. By Proposition 5.77, S_S is G-extending.

(iii) If $f = f^2 \in S$, then $f = (e,m)$ where $e = e^2 \in R$ and $m = em + me$. Since R is Abelian and $eme = 0$, we have $m = 0$, and so $f = (e,0)$. Hence, f is central in S, and therefore S is Abelian. By (ii) and Proposition 5.80 (iii), S_S is extending. □

Observe that in Theorem 5.125 the condition that M is left faithful is not superfluous. For example, if $R = \mathbb{Z}/\mathbb{Z}4$ and $M = \mathbb{Z}2/\mathbb{Z}4$, then $S(R,M)$ is not right G-extending, even though R is self-injective. The next corollary shows that the ideal M cannot be replaced with an arbitrary bimodule in Theorem 5.125 either. Moreover, the second part of the following corollary shows that [ABT09, Cor. 2.5 (ii)] is not valid.

Corollary 5.126. *Let V be a vector space over a field F with* $\dim_F(V) = 2$ *and let*

$$S = S(F,V) = \left\{ \begin{bmatrix} a & v \\ 0 & a \end{bmatrix} : a \in F, v \in V \right\},$$

the trivial extension of F by V. Then S_S is neither G*-extending nor S_S-ejective.*

Proof. Since S_S is indecomposable with uniform dimension 2, it is not G-extending by Proposition 5.80 (i). Now let $N = \begin{bmatrix} 0 & V \\ 0 & 0 \end{bmatrix}$. Then $N_S \leq_e S_S$ by routine computation. Let $\{u_1, u_2\}$ be a fixed basis of V, and write $F = Fu_1 \oplus Fu_2$. Let $\pi : V \to Fu_1$ be the canonical projection. Define $h \in \mathrm{Hom}(N_S, S_S)$ by $h\left(\begin{bmatrix} 0 & v \\ 0 & 0 \end{bmatrix}\right) = \begin{bmatrix} 0 & \pi(v) \\ 0 & 0 \end{bmatrix}$ for $v \in V$. Assume on the contrary that S_S is S_S-ejective. Then there exists $\varphi \in \mathrm{Hom}(S_S, S_S)$ and $A_S \leq_e N_S$ such that $\varphi|_A = h|_A$. Say $A = \begin{bmatrix} 0 & W \\ 0 & 0 \end{bmatrix}$. Then W is an essential subspace of V, so $W = V$. Thus $A = N$ and therefore φ is an extension of h. Put $\varphi\left(\begin{bmatrix} 1 & 0 \\ 0 & 1 \end{bmatrix}\right) = \begin{bmatrix} a_0 & v_0 \\ 0 & a_0 \end{bmatrix}$, where $a_0 \in F$ and $v_0 \in V$. Then $\varphi\left(\begin{bmatrix} a & 0 \\ 0 & a \end{bmatrix}\right) = \begin{bmatrix} a_0a & av_0 \\ 0 & a_0a \end{bmatrix}$ for all $a \in F$.

On the other hand, since φ is an extension of h, $\varphi\left(\begin{bmatrix}0 & v\\0 & 0\end{bmatrix}\right) = h\left(\begin{bmatrix}0 & v\\0 & 0\end{bmatrix}\right) = \begin{bmatrix}0 & \pi(v)\\0 & 0\end{bmatrix}$ for all $v \in V$. Thus $\varphi\left(\begin{bmatrix}a & 0\\0 & a\end{bmatrix}\right) = \begin{bmatrix}a_0a & av_0 + \pi(v)\\0 & a_0a\end{bmatrix}$ for all $a \in F$ and $v \in V$. Since $\varphi \in \mathrm{Hom}(S_S, S_S)$, $\varphi\left(\begin{bmatrix}a & 0\\0 & a\end{bmatrix}\begin{bmatrix}0 & w\\0 & 0\end{bmatrix}\right) = \left(\varphi\left(\begin{bmatrix}a & 0\\0 & a\end{bmatrix}\right)\right)\begin{bmatrix}0 & w\\0 & 0\end{bmatrix} = \begin{bmatrix}a_0a & av_0 + \pi(v)\\0 & a_0a\end{bmatrix}\begin{bmatrix}0 & w\\0 & 0\end{bmatrix} = \begin{bmatrix}0 & a_0aw\\0 & 0\end{bmatrix}$ for all $a \in F$ and $v, w \in V$. Also note that $\varphi\left(\begin{bmatrix}a & v\\0 & a\end{bmatrix}\begin{bmatrix}0 & w\\0 & 0\end{bmatrix}\right) = \varphi\left(\begin{bmatrix}0 & aw\\0 & 0\end{bmatrix}\right) = \begin{bmatrix}0 & a\pi(w)\\0 & 0\end{bmatrix}$ for all $a \in F$ and $v, w \in V$. Therefore $a\pi(w) = a_0aw$ for all $a \in F$ and $w \in V$. In particular, $a\pi(u_2) = a_0au_2$, so $0 = a_0au_2$ for all $a \in F$. Thus $a_0 = 0$. It follows that $\varphi\left(\begin{bmatrix}1 & 0\\0 & 1\end{bmatrix}\right) = \begin{bmatrix}0 & v_0\\0 & 0\end{bmatrix}$, and hence $\varphi\left(\begin{bmatrix}a & v\\0 & a\end{bmatrix}\right) = \begin{bmatrix}0 & av_0 + \pi(v)\\0 & 0\end{bmatrix}$ for all $a \in F$ and $v \in V$. We note that $\varphi\left(\begin{bmatrix}a & v\\0 & a\end{bmatrix}\begin{bmatrix}0 & w\\0 & 0\end{bmatrix}\right) = \varphi\left(\begin{bmatrix}0 & aw\\0 & 0\end{bmatrix}\right) = \begin{bmatrix}0 & a\pi(w)\\0 & 0\end{bmatrix}$ and $\left(\varphi\left(\begin{bmatrix}a & v\\0 & a\end{bmatrix}\right)\right)\begin{bmatrix}0 & w\\0 & 0\end{bmatrix} = \begin{bmatrix}0 & av_0 + \pi(v)\\0 & 0\end{bmatrix}\begin{bmatrix}0 & w\\0 & 0\end{bmatrix} = \begin{bmatrix}0 & 0\\0 & 0\end{bmatrix}$ for all $a \in F$ and $v, w \in V$. Thus $a\pi(w) = 0$ for all $a \in F$ and $w \in V$, a contradiction.

Therefore there is no $\varphi \in \mathrm{Hom}(S_S, S_S)$ such that $\varphi|_N = h|_N$. We conclude that S_S is not S_S-ejective. □

Note that Corollary 5.126 holds true when $\dim_K(V) > 1$.

Proposition 5.127. *Let $R = ReR$ and $S = eRe$, for some $e \in \mathbf{I}(R)$ and let A be a left R-module. Then*

(i) *${}_RA$ is* G*-extending if and only if ${}_S(eA)$ is* G*-extending.*
(ii) *R_R is* G*-extending if and only if $(Re)_S$ is* G*-extending.*

Proof. The proof is a routine application of Proposition 2.77. □

Theorem 5.128. *$M_m(R)$ is* G*-extending if and only if $(\bigoplus_{i=1}^n R_i)_R$ is* G*-extending, where each $R_i = R$.*

Proof. Observe that $M_m(R) = M_m(R)eM_m(R)$, where e is the matrix unit with 1 in the $(1, 1)$th position and zero elsewhere. The result now follows from Proposition 5.127. □

Recall that there are uniform rings over which there are finitely generated free modules that are not G-extending (Exercise 3.7). So Theorem 5.128 yields that being G-extending is not a Morita invariant property.

Corollary 5.129. *If $T_m(R)$ is* G*-extending, then $\bigoplus_{i=1}^n R_i$ is* G*-extending, where each $R_i = R$.*

Proof. This result is a consequence of [BPR06, Cor. 1.8 (ii)], the statement and proof of Corollary 5.123, and Theorem 5.128. □

Let C be a commutative ring with unity. If T is a C-algebra not necessarily with unity, we use T^1 to denote the Dorroh extension of T by C. Thus $T^1 = T \times C$, where addition is componentwise and multiplication is defined by $(t_1, c_1)(t_2, c_2) = (t_1t_2 + c_1t_2 + c_2t_1,\ c_1c_2)$.

Theorem 5.130. *Let C be a commutative domain with unity and T a C-algebra with left unity u that is not a unity, $\ell_C(T) \neq 0$. Suppose that for each algebra right ideal Y of T there exist a algebra right ideal X of T and $e \in \mathbf{I}(T)$ such that $X_T \leq_e Y_T$ and $X_T \leq_e eT_T$. Then $T^1_{T^1}$ is* G-*extending. Moreover, if every $0 \neq c \in \mathbf{I}(T)$ is a left unity, and $\ell_T(T) \neq 0$, then $T^1_{T^1}$ is not extending.*

Proof. Let $V \leq T^1$, $Y = V \cap (T, 0)$, and $K = V \cap (0, xC)$, where $0 \neq x \in \ell_C(T)$. Observe that $(Y \oplus K)_{T^1} \leq_e V_{T^1}$. To see this, let $0 \neq (w, m) \in V$. If $m = 0$, then $(w, m) \in Y$. Otherwise, $0 \neq (w, m)(0, x) = (0, mx) \in K$. There exists $V_1 \leq T$ such that $Y = (V_1, 0)$. Hence there exist $X \leq T$ and $e \in \mathbf{I}(T)$ such that $X_T \leq_e (V_1)_T$ and $X_T \leq_e eT_T$. So $(X, 0)_{T^1} \leq_e Y_{T^1}$ and $(X, 0)_{T^1} \leq_e (e, 0)T^1_{T^1}$. Hence $((X, 0) \oplus K)_{T^1} \leq_e V_{T^1}$. If $K = 0$, we are finished by Proposition 5.77. So assume $K \neq 0$. Then $K_{T^1} \leq_e (-u, 1)T^1_{T^1}$. Since $(-u, 1)^2 = (-u, 1)$, we have $(eu - u, 1)^2 = (eu - u, 1)$ and $(e, 0)T^1 \oplus (-u, 1)T^1 = (eu - u, 1)T^1$. Then $((X, 0) \oplus K)_{T^1} \leq_e (eu - u)T^1_{T^1}$. By Proposition 5.77, T^1 is G-extending.

Now assume that every $0 \neq c \in \mathbf{I}(T)$ is a left unity. Let $0 \neq a \in \ell_T(T)$. Observe that $(a, x)T^1 \cap (T, 0) = 0$, where $0 \neq x \in \ell_T(T)$. Hence $(a, x)T^1_{T^1}$ is not essential in $T^1_{T^1}$. Let $f \in \mathbf{I}(T^1)$ such that $f \neq 1$. Then $f = (b, 0)$, where $b = b^2 \in T$, or $f = (-e, 1)$ where $e = e^2 \in T$. But $(b, 0)(a, x) \neq (a, x)$ and $(-e, 1)(a, x) \neq (a, x)$. Thus $(a, x)T^1_{T^1}$ is not essential in a direct summand of $T^1_{T^1}$. Therefore, $T^1_{T^1}$ is not extending. □

Corollary 5.131.

(i) *Let M be a finite Abelian group, $R = \mathrm{End}(M_{\mathbb{Z}})$, and $T = \begin{bmatrix} R & M \\ 0 & 0 \end{bmatrix}$ an ideal of the generalized matrix ring $\begin{bmatrix} R & M \\ 0 & \mathbb{Z} \end{bmatrix}$. Then T is a $\mathbb{Z}$-algebra and $T^1_{T^1}$ is* G-*extending, but $T^1_{T^1}$ is not, in general, extending.*

(ii) *Let $T = (\mathbb{Z}/\mathbb{Z}p)[A]$ be the semigroup ring, where p is a prime and $A = \{a, b\}$ with a binary operation defined by $xy = y$ for all $x, y \in A$. Then T is a $\mathbb{Z}$-algebra and T^1 is strongly right bounded and $T^1_{T^1}$ is* G-*extending, but $T^1_{T^1}$ is not extending.*

Proof. (i) Observe that $\begin{bmatrix} 1 & 0 \\ 0 & 0 \end{bmatrix}$ is a left unity for T. Let $0 \neq Y \leq T$, $\overline{Y} = Y \cap \begin{bmatrix} 0 & M \\ 0 & 0 \end{bmatrix}$, and $0 \neq \begin{bmatrix} a & y \\ 0 & 0 \end{bmatrix} \in Y$. If $a = 0$, then $\overline{Y} \neq 0$. If $a \neq 0$, there exists

$m \in M$ such that $a(m) \neq 0$. Then $0 \neq \begin{bmatrix} a & y \\ 0 & 0 \end{bmatrix} \begin{bmatrix} 0 & m \\ 0 & 0 \end{bmatrix} = \begin{bmatrix} 0 & a(m) \\ 0 & 0 \end{bmatrix} \in \overline{Y}$. Hence $\overline{Y} \leq_e Y_T$. There exists $V_{\mathbb{Z}} \leq M_{\mathbb{Z}}$ such that $\overline{Y} = \begin{bmatrix} 0 & V \\ 0 & 0 \end{bmatrix}$. So by Theorem 5.96, there exist $X_{\mathbb{Z}} \leq M_{\mathbb{Z}}$ and $e \in \mathbf{I}(R)$ such that $X_{\mathbb{Z}} \leq_e V_{\mathbb{Z}}$ and $X_{\mathbb{Z}} \leq_e eM_{\mathbb{Z}}$. Then $\begin{bmatrix} 0 & X \\ 0 & 0 \end{bmatrix}_T \leq_e \overline{Y}_T$ and $\begin{bmatrix} 0 & X \\ 0 & 0 \end{bmatrix}_T \leq_e \begin{bmatrix} e & 0 \\ 0 & 0 \end{bmatrix} T_T$. Therefore, by Theorem 5.130, $T^1_{T^1}$ is G-extending.

If we take $M = \mathbb{Z}/\mathbb{Z}p^n$, for some prime p, then T^1 is G-extending. Since $R \cong \mathbb{Z}/\mathbb{Z}p^n$, every nonzero idempotent of T is a left unity of T. Also, $\begin{bmatrix} 0 & M \\ 0 & 0 \end{bmatrix} T = 0$. Therefore, by Theorem 5.130, $T^1_{T^1}$ is not extending.

(ii) Let $t, x \in T$. Then there exist $t_1, t_2, x_1, x_2 \in \mathbb{Z}/\mathbb{Z}p$ such that $t = t_1 a + t_2 b$ and $x = x_1 a + x_2 b$. Note that $tx = (t_1 a + t_2 b)(x_1 a + x_2 b) = (t_1 + t_2)x$.

Claim 1. $N(T) = \{t_1 a + t_2 b : t_1 + t_2 = \overline{0}\} = T(a + \overline{(p-1)}b)$ is a minimal right ideal of T, where $N(-)$ denotes the set of nilpotent elements of $-$.

Proof of Claim 1. Clearly $\{t_1 a + t_2 b : t_1 + t_2 = \overline{0}\}$ is a nilpotent ideal of T of index two. Let $x \in N(T)$ such that $x^k = 0$. Then $(x_1 + x_2)^{k-1}(x_1 a + x_2 b) = 0$, so $(x_1 + x_2)^{k-1} x_1 = \overline{0} = (x_1 + x_2)^{k-1} x_2$. Hence $(x_1 + x_2)^k = \overline{0}$. But $\mathbb{Z}/\mathbb{Z}p$ is a field, so $x_1 + x_2 = \overline{0}$. Hence $N(T) = \{t_1 a + t_2 b : t_1 + t_2 = \overline{0}\}$. Clearly $T(a + \overline{(p-1)}b) \subseteq N(T)$. Let $t_1 a + t_2 b \in N(T)$. Then $t_1 + t_2 = \overline{0}$. So $t_1 a + t_2 b = t_1 a + (-t_1 b) = t_1(a + \overline{(p-1)}b) \in T(a + \overline{(p-1)}b)$. Therefore, $N(T) = T(a + \overline{(p-1)}b)$. Observe that $T(a + \overline{(p-1)}b)$ is the additive subgroup of T generated by $a + \overline{(p-1)}b$, and $|T(a + \overline{(p-1)}b)| = p$. Hence $N(T)$ is a minimal right ideal of T.

Since $|T| = p^2$, $N(T)$ is the unique nonzero proper right ideal of T. Thus $N(T)_T \leq_e aT_T = bT_T = T_T$, but $N(T)T = 0$. By Theorem 5.130, $T^1_{T^1}$ is G-extending, but is not extending.

Claim 2. If $Y \leq T^1$ such that $Y \nleq (T, 0)$, then $(0, pv\mathbb{Z}) \leq Y$, for some $0 \neq v \in \mathbb{Z}$.

Proof Claim 2. There exists $(y, v) \in Y$ such that $v \neq 0$. Then $0 \neq (y, v)(0, p) = (0, pv)$. Hence $(0, pv\mathbb{Z}) \leq (y, v)T^1 \leq Y$. Since $(0, pv\mathbb{Z}) \trianglelefteq T^1$ and $N(T) \trianglelefteq T$, T^1 is strongly right bounded. □

Observe that Corollary 5.131 (ii) remains true if $\mathbb{Z}/\mathbb{Z}p$ is replaced by $\mathbb{Z}/\mathbb{Z}n$, where $n = p_1 \cdots p_m$ and the p_i are distinct primes.

Exercises

5.14. Let R be a ring. Let $T = T_m(R)$ and $M = M_m(R)$. Show that M_T is a rational extension of T_T.

5.15. Find examples that illustrate Theorem 5.122.

5.7 Goldie extending modules over commutative domains

We characterize G-extending modules over Dedekind domains and use the G-extending property to characterize the modules over a principal ideal domain whose pure submodules are direct summands. Moreover, we show that if R is a principal ideal domain, then the class of G-extending modules is closed under direct summands and that if R is a Dedekind domain, then the class of G-extending torsion modules is closed under finite direct sums.

Let R be a ring and A an R-module. For any ideal I of R, $[I]A$ denotes $\{a \in A : Ia = 0\}$. There are several notions of purity currently in the literature. We state these notions for a ring R which is not necessarily commutative, for A a left R-module and K a submodule of A:

(1) [Co59] K is a *pure* submodule of A if the sequence $0 \to K \otimes_R E \to A \otimes_R E$ is exact for every left R-module E.
(2) [AF74] K is a *pure* submodule of A if $KI = K \cap AI$ for every left ideal I of R.
(3) [Ri72] K is a *pure* submodule of A if $Kr = K \cap Ar$ for each $r \in R$.
(4) [Ma74] K is a P^n*-pure* in A if $KP^m = K \cap AP^m$ for each natural number $m \leq n$ and each prime ideal P of R; K is P^w*-pure* in A if K is P^n-pure for every natural number n.

It can be seen that (1) $\Rightarrow$ (2) $\Rightarrow$ P^w-pure and (1) $\Rightarrow$ (2) $\Rightarrow$ (3). Also, if R is a PID, then (1) $\Leftrightarrow$ (2) $\Leftrightarrow$ (3) $\Leftrightarrow$ P^w-pure. In what follows we use (3) as our definition of purity.

A module A is called *bounded* if $\gamma A = 0$ for some nonzero $\gamma \in R$. According to [Ma74, p. 537], if P is a prime ideal of R, then a submodule B of a module A is called a P*-basic* submodule if it satisfies the following conditions.

(i) B is a direct sum of uniform right ideals and modules of the form R/P^m $(m = 1, 2, \ldots)$,
(ii) B is P^w-pure in A,
(iii) A/B is P-divisible (i.e., $PA = A$ for a prime ideal P of R).

Lemma 5.132. *Let A be an R-module and $\{X_1, \ldots, X_n\}$ a set of pairwise comaximal ideals. Then $[X_1 \cdots X_n]A = \sum_{i=1}^n [X_i]A$.*

Proof. Clearly $\sum_{i=1}^n [X_i]A \subseteq [X_1 \cdots X_n]A$. We write out the case $n = 3$, then the result follows from a routine induction argument. Let $X_1 = X$, $X_2 = Y$, and $X_3 = Z$. By comaximality, $1 = x+y = x_1+z = y_1+z_1$, where $x, x_1 \in X$, $y, y_1 \in Y$, and $z, z_1 \in Z$. Let $a \in [XYZ]A$. Then $a = 1a = (y_1 + z_1)(x_1 + z)(x + y)a = yx_1xa + y_1x_1ya + y_1zya + z_1x_1xa + z_1zxa + z_1zya \in [Z]A + [X]A + [Y]A$. Hence, $[X_1X_2X_3]A = \sum_{i=1}^3 [X_i]A$. By induction, $[X_1 \cdots X_n]A = \sum_{i=1}^n [X_i]A$. □

Proposition 5.133. *Let $A = D' \oplus D$ be a module over a principal ideal ring* (PIR) *R such that every ideal contains a finite product of maximal ideals. If K is a pure submodule of A such that $K \cap D \leq_{\rm e} D$ and $K \cap D \leq_{\rm e} K$, then $A = D' \oplus K$.*

Proof. If $K = 0$ or $K = D$, we are finished. So assume $K \neq 0$ and $K \neq D$. Since $K \cap D \leq_{\rm e} K$ and $K \cap D \leq_{\rm e} D$, we have $K \cap D' = 0$. To show that the sum $D' + K$ is A, it is enough to show that D is contained in $D' + K$. Let $0 \neq d \in D$. Since $K \cap D \leq_{\rm e} D$, we have $Rd \cap (K \cap D) \neq 0$, so there exists $r \in R$ such that $0 \neq k = rd \in K$. By purity, $rd = rk'$ for some $k' \in K$. Then $r(d - k') = 0$. So $d - k' \in [I]A$, where $I = Rr$.

Now $[I]A = [I]D' \oplus [I]D$, and so $d - k' = d_1 + d_2$ with $d_1 \in [I]D'$ and $d_2 \in [I]D$. Hence, it is enough to show that $[I]D$ is contained in $D' + K$ for each ideal I of R. By hypothesis, $P_1^{t_1} \cdots P_s^{t_s} \subseteq I$, where each P_j is a maximal ideal and $P_i \neq P_j$ for $i \neq j$. Therefore, by Lemma 5.132, it is enough to show that for each maximal ideal $P = P_j$ and positive integer n, the submodule $[P^n]D$ is inside $D' + K$, since the P^{t_j} are pairwise comaximal. We apply induction on n. Note that since $K \cap D \leq_{\rm e} D$ and $K \cap D \leq_{\rm e} K$ we have $[P]D = [P]K = [P](D \cap K)$. To see this observe that $[P](D \cap K) \subseteq [P]D$ and $[P](D \cap K) \subseteq [P]K$. Now we need to show the reverse inclusions. If $[P]D = 0$, then $[P]D = [P](D \cap K)$. So assume $0 \neq a \in [P]D$. Then $Pa = 0$ and $a \in D$. Since $K \cap D \leq_{\rm e} D$, there exists $s \in R$ such that $0 \neq sa \in K$. Set $Rs = J$. Then $Ja \subseteq K$. Now since P is a maximal ideal, either $J + P = P$, or $J + P = R$. In the first case $J \subseteq P$ and $Ja = 0$, which is a contradiction. Therefore, $J + P = R$ and we have $Ra = (J + R)a = Ja \subseteq K$, and so $a \in K$. Thus $a \in [P](K \cap D)$. Similarly we can show the other inclusion. Therefore, $[P]D \subseteq K \subseteq D' + K$ and we are done for $n = 1$. Suppose $[P^n]D \subseteq K$ for some $n > 1$. Let b be an element of $[P^{n+1}]D$. Then $P^n b \subseteq [P]D \subseteq K$. We know that $P^n = Rg$ for some $g \in R$, since R is a PIR. Thus $gp \in [P]D \subseteq K$. By purity, $gb = gk_1$ for some $k_1 \in K$. Then $Rg(b - k_1) = P^n(b - k_1) = 0$ (i.e., $b - k_1 \in [P^n]A = [P^n]D' \oplus [P^n]D$, which is contained in $D' + K$). Then $b \in D' + K$. We conclude that $A = D' + K$. □

Corollary 5.134. *Let R be a* PIR *such that every ideal contains a finite product of maximal ideals and let A be an R-module. If A is* G*-extending, then every pure submodule of A is a direct summand.*

Proof. This is an immediate consequence of Proposition 5.133. □

Any PID is an example of a commutative PIR ring such that every ideal contains a finite product of maximal ideals. We provide the following example to exhibit such a ring which is not a domain.

Example 5.135. Let $R = D/P^n$, where D is a Dedekind domain, $n > 1$, and P is a maximal ideal. Then R is a commutative PIR such that every ideal contains a finite product of maximal ideals, but R is not a domain.

Lemma 5.136. *Let R be a domain and A a torsionfree module. Let $S \leq A$ and $N = \bigcap\{K \leq A : S \leq K$ and K is pure in $A\}$. Then N is pure in A and $S \leq_e N$. In fact, N is the essential closure of S in A.*

Proof. If $S = 0$, we are done. So assume $S \neq 0$. First we show that N is pure in A. Take $n \in N$ and $0 \neq r \in R$ such that $ra = n$ for some $a \in A$. Let $K \leq A$ such that $S \leq K$ and K is pure in A. Then $rK = K \cap rA$. Therefore $n \in K \cap rA$, so $n = rk$, where $k \in K$. Then $ra = rk$ so $r(a - k) = 0$. Since A is torsionfree, $a - k = 0$ or $a = k \in K$. Thus $rN = N \cap rA$.

To prove that $S \leq_e N$, let $H = \{x \in N : rx \in S$ for some $0 \neq r \in R\}$. Clearly, $S \leq H \leq A$ and $S \leq_e H$. We claim that H is pure in A. Indeed, let $h \in H$ and $0 \neq r \in R$ such that $h = ra$ for some $a \in A$. There exists $0 \neq t \in R$ such that $tra = th \in S \leq N$, because $h \in H$. Therefore there exists $n \in N$ such that $tra = trn = th \in S$, because N is pure. Then $m = n$ because A is torsionfree. Thus $h = rn$ and $trn \in S$ implies that $n \in H$. So H is pure in A. Now since $S \leq N \leq H$ and $S \leq_e H$, we have $S \leq_e N$. □

Recall that we provided a characterization for the G-extending Abelian groups (Theorem 5.96). Since the proof of that characterization relied heavily on group-theoretic techniques, it was a natural problem to generalize this characterization to a principal ideal domain (PID) or Dedekind domain. By using module theoretic techniques, we are able to show that our characterization generalizes exactly to a PID and that for a Dedekind domain we have that (i) $\Leftrightarrow$ (iii) and (ii) $\Rightarrow$ (iii) in the aforementioned theorem.

Theorem 5.137. *Let R be a domain and A an R-module. Consider the following conditions.*

(i) *A is G-extending.*
(ii) *Every pure submodule of A is a direct summand.*
(iii) *$A = D \oplus T \oplus F$, where D is a divisible module, T is a reduced torsion submodule of A where each P-component, T_P, is bounded for a maximal ideal P of R, and F is a reduced torsionfree extending module, hence $F \cong \bigoplus_{j=1}^{h} I_j N$, where N is a proper R-submodule of the field of fractions of R and the I_j are fractional ideals of R. Then*
   (1) *If R is a Dedekind domain, then (i) $\Leftrightarrow$(iii) and (ii) $\Rightarrow$ (iii).*
   (2) *If R is a PID, then (i) $\Leftrightarrow$ (ii) $\Leftrightarrow$ (iii).*

Proof. (1) First we assume that R is a Dedekind domain.

(i) $\Rightarrow$ (iii) Suppose that A is G-extending. By Corollary 5.84 (i) and [Kap52, Thm. 8], $A = D \oplus E$, where D is divisible and E is a reduced G-extending module. By Corollaries 5.91 and 5.92, $E = T \oplus F$, where T is the torsion submodule of E, T is G-extending, and F is extending. The decomposition $F \cong \bigoplus_{j=1}^{h} I_j N$, follows from [KaM88, Thm. 14]. By Corollary 5.94 and [Kap52, p. 332], $T = \bigoplus T_P$ (P varies through all maximal ideals of R) and each T_P is a G-extending R-module. However, since T_P is a P-primary module (see, [Kap52, p. 332]), it embeds into its module of fractions, V, localized at P as an R-module via the canonical homomorphism $f : T_P \to V$ defined by $f(a) = \frac{a}{1}$. Hence T_P can be identified with an R-submodule of V. Moreover, it can be seen that V is a G-extending R_P-module, where R_P is R localized at P. So R_P is a discrete valuation domain (hence R_P is a PID). Observe that T_P is a bounded R-module if and only if V is bounded R_P-module.

Now assume that T_P is not bounded and identify T_P with V and R with R_P in the following claim. Thus we are thinking of T_P as a module over the PID R_P (see [Kap52, p. 332]). This allows us to use this claim in proving that (ii) $\Rightarrow$ (iii) without changing notation.

Claim 1. There exists a P-basic submodule B of T_P such that $0 \neq B < T_P$ and T_P/B is divisible.

Proof of Claim 1. By [Ma74, p. 541], T_P has at least two P-basic submodules. We have $T_P Q = T_P$ for every maximal ideal $Q \neq P$ because if we take any $t \in T_P$ we have $P^n t = 0$ for some n. Then $(Q + P^n)t = Rt$ and $Qt = Rt$ and so $T_P \subseteq QT_P$. However, $PT_P \neq T_P$ because otherwise T_P would be divisible, which is a contradiction. Thus T_P is not P-divisible, so 0 is not a P-basic submodule of T_P. Hence, there exists a P-basic submodule B of T_P such that $0 \neq B < T_P$. By the definition of a P-basic submodule, B is P^w-pure in T_P. Now if we can show that B is Q^w-pure in T_P for any maximal ideal Q of R, then P is pure in T_P. Take any maximal ideal $Q \neq P$ in R and an element $b \in B$. We have $P^n b = 0$ for some n and $Q + P^n = R$. Thus $Rb = (Q + P^n b) = Qb$ and so $Rb = Qb$ and $b = qb$ for some $q \in Q$. Therefore $B \subseteq QB$ and $QB = B$. Thus B is Q^w-pure in T_P and so B is pure in T_P. Since B is a P-basic submodule of T_P, T_P/B is P-divisible. The claim is proved.

By Corollary 5.134 and the proof of Claim 1, B is a direct summand of T_P. But, using Claim 1, this is contrary to the fact that T_P is reduced. Thus V is a bounded R_P-module, and so T_P is a bounded R-module. Therefore T_P is bounded for all maximal ideals P of R.

(ii) $\Rightarrow$ (iii) By [Kap52, Thm. 8], $A = D \oplus E$, where D is a divisible module and E is reduced. Then $E = T \oplus F$, where T is the torsion submodule of E and F is torsionfree, because the torsion submodule of E is pure [Kap52, p. 328]. Thus $A = D \oplus T \oplus F$. By assumption and Lemma 5.136, F is extending, and as above, $T = \bigoplus T_P$. Assume that T_P is not bounded for some maximal ideal of R. Then again by Claim 1, T_P/B is divisible and B is pure in T_P. Since T_P is pure in A, we see that B is a direct summand of T_P, contrary to the fact that T is reduced. Therefore T_P is bounded for all maximal ideals P of R.

(iii) $\Rightarrow$ (i) By Corollary 5.84, we need only show that $T \oplus F$ is G-extending. Taking $\rho(A)$ in Corollary 5.91 to be T, we see that we only need to show that T is G-extending. Corollary 5.94 reduces the proof to showing that T_P is G-extending for each maximal ideal P of R. From [SV72, Thm. 6.11], $T_P \cong \bigoplus_{i \in \Lambda} R/P^{n_i}$. By [SV72, Lems. 6.6 and 6.8], R/P^n is uniserial and quasi-injective.

Claim 2. R/P^n is R/P^k-injective for $k \leq n$.

Proof of Claim 2. If $k = n$ we are finished, since R/P^k is quasi-injective. So assume that $k < n$. Let $0 \neq J \leq R/P^k$ and $f : J \to R/P^n$ be a nonzero homomorphism. Again by [SV72, Lem. 6.6], there exists $p \in R$ such that $P = Rp$ and the ideals of R/P^b are of the form $R(p^i + P^b)$ for $0 \leq i \leq n$, where $p^0 = 1$. So there exists $0 \leq j \leq k$ such that $J = R(p^j + P)$. By straightforward calculation the homomorphism $f : J \to R/P^n$ has one of the forms f_a, $1 \leq a \leq k - j$, defined by $f_1(p^j + P^k) = p^{n-k+j} + P^n$, $f_2(p^j + P^k) = p^{n-k+j+1} + P^n$, $\ldots$, $f_{k-j}(p^j + P^k) = p^{n-1} + P^n$. The corresponding extension homomorphisms $\overline{f_\alpha}$: $R/P^k \to R/P^n$ are defined by $\overline{f_1}(1+P^k) = p^{n-k}+P^n$, $\overline{f_2}(1+P^k) = p^{n-k+1}+P^n$, $\ldots$, $\overline{f}_{k-j}(1 + P^k) = p^{n-j-1} + P^n$. Therefore, the claim is proved.

Let k_1 be the least exponent among the $\{R/P^{n_i}\}_{i\in\Lambda}$. Define $A_1 = \bigoplus\{R/P^{n_i} : n_i = k_1\}$. Similarly, define $A_2 = \bigoplus\{R/P^{n_i} : n_i = k_2\}$, where k_2 is the next smallest exponent among the $\{R/P^{n_i}\}_{i\in\Lambda}$. Since T_P has bounded order, this procedure terminates after a finite number of steps, say k, to yield $T_P \cong \bigoplus_{\alpha=1}^{k} A_\alpha$. Since each R/P^{n_i} is quasi-injective, it follows that each A_α is quasi-continuous by [MR83, Thm. 12]. Thus each A_α is G-extending. By Claim 2, R/P^{k_β} is R/P^{k_α}-injective for $\alpha \leq \beta$. By Lemma 2.27, $\prod\{R/P^{n_i} : n_i = k_\beta\}$ is R/P^{k_α}-injective. Observe that $A_\beta = \bigoplus\{R/P^{n_i} : n_i = k_\beta\}$. Let ${}_RX \leq {}_R(R/P^{k_\alpha})$. Then ${}_RX$ is finitely generated. Let $f : X \to A_\beta$ be an R-homomorphism. Then $f(X)$ is contained in a sum of only finitely many of the direct summands R/P^{n_i} of A_β. Let S be this finite direct sum. There is an R-homomorphism $\overline{f} : \prod\{R/P^{n_i} : n_i = k_\beta\} \to A_\beta$ such that $\overline{f}|_X = f$. Let $h : \prod\{R/P^{n_i} : n_i = k_\beta\} \to S$ be the projection R-homomorphism. Then $f = h\overline{f}$. So A_β is R/P^{k_α}-injective. By Proposition 2.20, A_β is A_α-injective. Now the result follows from Corollary 5.83.

(2) Now assume that R is PID.

(i) $\Rightarrow$ (ii) This implication follows from Corollary 5.134.

(ii) $\Rightarrow$ (iii) and (iii) $\Rightarrow$ (i) Since any PID is a Dedekind domain, these implications follow from part (1). □

Corollary 5.138.

(i) *If R is a* PID, *then the class of* G-*extending modules is closed under direct summands.*

(ii) *If R is a Dedekind domain, then class of* G-*extending torsion modules is closed under finite direct sums.*

Proof. (i) follows from Theorem 5.137(2) and the fact that a direct summand of a domain is pure.

(ii) The fact that the class of G-extending torsion modules over a Dedekind domain is closed under finite direct sums is an immediate consequence of Theorem 5.137. □

The following corollary also appears in Section 5.8, with a different proof.

Corollary 5.139. *Every finitely generated module over a Dedekind domain is* G-*extending.*

Proof. This result follows from Theorem 5.137 and [SV72, Thm. 6.16]. □

Corollary 5.140. *Let A be a* G-*extending module over a Dedekind domain. Then A is a direct sum of uniform submodules.*

Proof. By Theorem 5.137, $A = D \oplus T \oplus F$. Further, by Exercise 1.120, D is a direct sum of indecomposable submodules. Since each of these indecomposable submodules is injective, it is uniform. By Theorems 5.137 and [SV72, Thms. 6.11 and 6.14], each P-component, T_P, is a direct sum of uniform submodules. Since T is a direct sum of its P-components, T is a direct sum of uniform submodules. By Theorem 5.137, F is a direct sum of uniform submodules. Therefore, A is a direct sum of uniform submodules. □

Observe that if a domain R is not right Ore, then R_R is weak CS, but not G-extending. Next we give a characterization of weak CS-modules over Dedekind domains.

Proposition 5.141. *Let R be a Dedekind domain and A an R-module. Then A is weak* CS *if and only if $A = T \oplus F$, where F is torsionfree and T is the torsion submodule of A such that for each maximal ideal P of R, T_P is a direct sum of a bounded submodule and a divisible submodule.*

Proof. ($\Rightarrow$) Since A is weak CS, $A = T \oplus F$, where $\operatorname{soc} A \leq_e T$. By [Kap52, Thm. 9], F is torsion-free. Now Proposition 5.80 and Theorem 5.137 yield the implication.

($\Leftarrow$) This implication follows from Proposition 5.80, Theorem 5.137, and [Smi90, Thm. 1.12]. □

Exercises

5.16. Find examples of modules that contain pure submodules, which are not direct summands.

5.17. Prove the implications between the different kind of pure submodules which were defined at the beginning of this section.

5.8 Goldie extending matrix rings

In this section, we obtain a characterization of the right G-extending generalized triangular matrix rings. This result and its corollaries improve and generalize the existing results on right extending generalized triangular matrix rings. It is well known that the ring of n-by-n triangular matrices over a right self-injective ring is not, in general, right extending. One application of our characterization shows that such rings are right G-extending. Connections to operator theory and a characterization of the class of right extending right SI-rings are also obtained. We provide several examples to illustrate and delimit the theory.

Recall that a 2-by-2 generalized triangular matrix ring (also called a formal triangular matrix ring) T has the form

$$T = \begin{bmatrix} R & A \\ 0 & S \end{bmatrix},$$

where R and S are associative rings with unity and A is an (R, S)-bimodule. They are useful in many areas of algebra (e.g., finite-dimensional algebras and Morita contexts) and functional analysis (e.g., operator algebras, see [BL04, pp. 118–119]). Observe that when $R = \mathrm{End}(A_S)$, then T provides a convenient entity to encode information about A. Since A is an (R, S)-bimodule, there is always a natural ring homomorphism from R to $\mathrm{End}(A_S)$. This ring homomorphism is injective if and only if ${}_RA$ is faithful.

Throughout this section, let R and S be rings with unity, A an (R, S)-bimodule, and $T = \begin{bmatrix} R & A \\ 0 & S \end{bmatrix}$ be the generalized triangular matrix ring. We use $\mathrm{Cen}(R)$ to denote the center of R.

Our first proposition shows that for commutative or right nonsingular rings the concepts of extending and G-extending are identical. The second result is useful in working with generalized triangular matrices.

Proposition 5.142. *Let R be Abelian (i.e., every idempotent element is central) or $Z(R_R) = 0$. Then R_R is extending if and only if R_R is* G*-extending.*

Proof. This result is an immediate consequence of Proposition 5.80 (ii) and (iii). □

Thus to see a distinction between the extending and G-extending properties one must consider noncommutative rings that are not nonsingular.

Proposition 5.143.

(i) *Let $c \in \mathcal{S}_l(\mathrm{End}(A_S))$ and $e = e^2 \in \mathrm{End}(A_S)$ such that $cA_S \beta eA_S$. Then $cA = eA$, $c\mathrm{End}(A_S) = e\mathrm{End}(A_S)$, and $e \in \mathcal{S}_l(\mathrm{End}(A_S))$.*

(ii) $\begin{bmatrix} c & k \\ 0 & e \end{bmatrix} \in \mathcal{S}_l(T)$ *if and only if $c \in \mathcal{S}_l(R)$, $e \in \mathcal{S}_l(S)$, $k = ck$, $ke = 0$, and $cae = ae$ for all $a \in A$.*

(iii) *Let* $\begin{bmatrix} c & k \\ 0 & e \end{bmatrix} = \begin{bmatrix} c & k \\ 0 & e \end{bmatrix}^2$. *Then $k = ck$ if and only if* $\begin{bmatrix} c & 0 \\ 0 & e \end{bmatrix} T = \begin{bmatrix} c & k \\ 0 & e \end{bmatrix} T$.

Proof. (i) Since $c \in \mathcal{S}_l(\mathrm{End}(A_S))$, $ec = cec = ecec = (ec)^2$. Clearly, $cA \cap eA = ecA$. Hence $cA = ecA = eA$. Let $ca \in cA$. There exists $n \in A$, such that $ca = en$. Then $eca = e^2n = en = ca$. Thus $ec = c$, so $c\mathrm{End}(A_S) \subseteq e\mathrm{End}(A_S)$. Similarly,

$e\mathrm{End}(A_S) \subseteq c\mathrm{End}(A_S)$. So $e\mathrm{End}(A_S) = c\mathrm{End}(A_S)$. Let $r \in \mathrm{End}(A_S)$. Then $re = (rc)e = (crc)e = ecrce$, so $re = ere$. Therefore, $e \in S_l(\mathrm{End}(A_S))$.

(ii) Assume $\begin{bmatrix} c & k \\ 0 & e \end{bmatrix} \in S_l(T)$. Let $\begin{bmatrix} r & a \\ 0 & s \end{bmatrix} \in T$. Then

$$\begin{bmatrix} rc & rk+ae \\ 0 & se \end{bmatrix} = \begin{bmatrix} r & a \\ 0 & s \end{bmatrix}\begin{bmatrix} c & k \\ 0 & e \end{bmatrix} = \begin{bmatrix} c & k \\ 0 & e \end{bmatrix}\begin{bmatrix} r & a \\ 0 & s \end{bmatrix}\begin{bmatrix} c & k \\ 0 & e \end{bmatrix} = \begin{bmatrix} crc & crk+cae+kse \\ 0 & ese \end{bmatrix}.$$

Thus $c \in S_l(T)$ and $e \in S_l(S)$. For $r = 1$, $a = 0$, and $s = 0$, then $k = ck$. For $r = 0$ and $s = 0$, then $cae = ae$. For $r = 0$, $m = 0$, and $s = 1$, then $ke = 0$. Conversely, assume $c \in S_l(R)$, $e \in S_l(S)$, $k = ck$, $ke = 0$, and $cae = ae$ for all $a \in A$. Let $\begin{bmatrix} r & a \\ 0 & s \end{bmatrix} \in T$. Then

$$\begin{aligned}\begin{bmatrix} c & k \\ 0 & e \end{bmatrix}\begin{bmatrix} r & a \\ 0 & s \end{bmatrix}\begin{bmatrix} c & k \\ 0 & e \end{bmatrix} &= \begin{bmatrix} crc & crk+cae+kse \\ 0 & ese \end{bmatrix} = \begin{bmatrix} rc & (crc)k+ae+(ke)se \\ 0 & se \end{bmatrix} \\ &= \begin{bmatrix} rc & r(ck)+ae \\ 0 & se \end{bmatrix} = \begin{bmatrix} rc & rk+ae \\ 0 & se \end{bmatrix} = \begin{bmatrix} r & a \\ 0 & s \end{bmatrix}\begin{bmatrix} c & k \\ 0 & e \end{bmatrix}.\end{aligned}$$

Therefore, $\begin{bmatrix} c & k \\ 0 & e \end{bmatrix} \in S_l(T)$.

(iii) Assume $k = ck$. Let $\begin{bmatrix} r & a \\ 0 & s \end{bmatrix} \in T$. Then

$$\begin{aligned}\begin{bmatrix} c & k \\ 0 & e \end{bmatrix}\begin{bmatrix} r & a \\ 0 & s \end{bmatrix} &= \begin{bmatrix} cr & ca+ks \\ 0 & es \end{bmatrix} = \begin{bmatrix} cr & ca+(ck)s \\ 0 & es \end{bmatrix} \\ &= \begin{bmatrix} c & 0 \\ 0 & e \end{bmatrix}\begin{bmatrix} r & a+ks \\ 0 & s \end{bmatrix} \in \begin{bmatrix} c & 0 \\ 0 & e \end{bmatrix} T.\end{aligned}$$

Thus $\begin{bmatrix} c & k \\ 0 & e \end{bmatrix} T \subseteq \begin{bmatrix} c & 0 \\ 0 & e \end{bmatrix} T$. Now let $\begin{bmatrix} r_1 & a_1 \\ 0 & s_1 \end{bmatrix} \in T$. Then

$$\begin{aligned}\begin{bmatrix} c & 0 \\ 0 & e \end{bmatrix}\begin{bmatrix} r_1 & a_1 \\ 0 & s_1 \end{bmatrix} &= \begin{bmatrix} cr_1 & ca_1 \\ 0 & es_1 \end{bmatrix} = \begin{bmatrix} cr_1 & ca_1-(ck)s_1+ks_1 \\ 0 & es_1 \end{bmatrix} \\ &= \begin{bmatrix} c & k \\ 0 & e \end{bmatrix}\begin{bmatrix} r_1 & a_1-ks \\ 0 & s_1 \end{bmatrix} \in \begin{bmatrix} c & k \\ 0 & e \end{bmatrix} T.\end{aligned}$$

So $\begin{bmatrix} c & 0 \\ 0 & e \end{bmatrix} T \subseteq \begin{bmatrix} c & k \\ 0 & e \end{bmatrix} T$. Therefore, $\begin{bmatrix} c & 0 \\ 0 & e \end{bmatrix} T = \begin{bmatrix} c & k \\ 0 & e \end{bmatrix} T$. Conversely, assume $\begin{bmatrix} c & 0 \\ 0 & e \end{bmatrix} T = \begin{bmatrix} c & k \\ 0 & e \end{bmatrix} T$. Then there is $\begin{bmatrix} r & a \\ 0 & s \end{bmatrix} \in T$ such that

$$\begin{bmatrix} c & k \\ 0 & e \end{bmatrix} = \begin{bmatrix} c & 0 \\ 0 & e \end{bmatrix}\begin{bmatrix} r & a \\ 0 & s \end{bmatrix} = \begin{bmatrix} cr & ca \\ 0 & es \end{bmatrix}.$$

Hence $k = ca$, so $ck = c^2a = ca = k$. □

Proposition 5.144. *Let H be a ring with unity such that $H = cH \oplus (1-c)H$ (right ideal decomposition), where $c \in S_l(H)$ and $1 - c \neq 0$.*

(i) *$H' = (1-c)H$ is a ring with unity $1-c$ such that H' is ring isomorphic and H-isomorphic to H/cH. If $Z(H_H) \subseteq cH$, then H'_H and $H'_{H'}$ are nonsingular.*
(ii) *H'_H is* G*-extending if and only if $H'_{H'}$ is* G*-extending.*
(iii) *Assume $Z(H_H) \subseteq cH$. Then $H'_{H'}$ is* G*-extending if and only if $H'_{H'}$ is extending.*

Proof. This result follows from [Birk83, Lem. 2], Proposition 5.81, and routine arguments. □

Proposition 5.145. *The following conditions are equivalent.*

(i) *If $Y_S \subseteq A_S$, then there exists $f = f^2 \in R$ such that $Y_S \beta f A_S$ ($Y_S \leq_{\rm e} fA_S$).*
(ii) *A_S is* G*-extending (extending), and if D_S is a direct summand of A_S then there is $f = f^2 \in R$ such that $D_S \beta f A_S$ ($D_S = fA_S$).*

Proof. The proof is straightforward (see Exercise 5.18). □

Note that if R is a subring of $\mathrm{End}(A_S)$, then the direct summand condition of Proposition 5.145 (ii) is satisfied whenever R contains all the idempotents of $\mathrm{End}(A_S)$.

Since the ejective concept will play an important role in our main results in this section we recall some basic properties in Proposition 2.61. In conjunction with Proposition 2.61, we provide the following example.

Example 5.146.

(i) Let $R = \mathbb{Z}$ and $N = (\mathbb{Z}/\mathbb{Z}p^m) \oplus (\mathbb{Z}/\mathbb{Z}p^n)$, where $m > n+1$. By Theorem 5.96, ${}_RN$ is G-extending; and by Proposition 2.61 (vi), ${}_RN$ is ${}_RR$-ejective. However, ${}_RN$ is not extending, so ${}_RN$ is not ${}_RR$-injective.
(ii) Let $R = \mathbb{Z}$. Then ${}_RR$ is extending (hence G-extending); but, by Proposition 2.61 (ii), ${}_RR$ is not ${}_RR$-ejective, since ${}_RR$ is not injective.

It is well known that a finitely generated torsion $\mathbb{Z}$-module is not, in general extending (e.g., let $N = (\mathbb{Z}/\mathbb{Z}p) \oplus (\mathbb{Z}/\mathbb{Z}p^3)$, for some prime p). However, as our next result shows, such modules are indeed G-extending. Recall that this result was also obtained as Corollary 5.139, with a different proof.

Theorem 5.147. *Let S be a domain. Then every finitely generated module is* G*-extending.*

Proof. Let N be a finitely generated module. By [SV72, Thm. 6.20], $N = T \oplus F$, where T is the torsion submodule of N and F is torsionfree. By Corollary 5.94, T is G-extending. Since every fractional ideal is invertible [Gil72, p. 442, Thm. 37.1] and every invertible ideal is projective [Lam99, p. 31, Thm. 2.17], F is projective. By [DHSW94, p. 112, 12.17], ${}_SS$ is finitely $\sum$-extending. Hence ${}_SF$ is extending. By Corollary 5.91, ${}_SN$ is G-extending. □

Since every G-extending module is a weak CS-module (Proposition 5.78), Theorem 5.147 sharpens Theorem 4.12.

Recall that $\ell_R(A) = 0$ (i.e., A is left faithful) if and only if the left multiplication map from R into $\text{End}(A_S)$ is an embedding.

Proposition 5.148. *Assume that ${}_RA$ is faithful and that if D_S is a direct summand of A_S, then there is an $f = f^2 \in R$ such that $D_S \beta f A_S$. If A_S and S_S are G-extending, then there exist $c = c^2 \in R$ and $e \in S_l(S)$ such that*

(i) $Z_2(A_S) = cA$;

(ii) $Z_2(S_S) = eS$; and

(iii) $Z_2(T_T) = \begin{bmatrix} c & 0 \\ 0 & e \end{bmatrix} T = \begin{bmatrix} cR & Z_2(A_S) \\ 0 & Z_2(S_S) \end{bmatrix}$.

Proof. (i) and (ii) By Corollary 5.91, $Z_2(A_S)$ and $Z_2(S_S)$ are direct summands of A_S and S_S, respectively. Since $Z_2(A_S) \trianglelefteq A_S$ and $Z_2(S_S) \trianglelefteq S$, [BMR02, Lem. 1.9] shows that $Z_2(A_S) = c'A$ and $Z_2(S_S) = eS$, where $c' \in S_l(\text{End}(A_S))$ and $e \in S_l(S)$, respectively. From the hypothesis and Proposition 5.143 (i), $Z_2(A_S) = cA$, where $c \in R$ with $c \in S_l(\text{End}(A_S))$.

(iii) Assume $\begin{bmatrix} r & a \\ 0 & s \end{bmatrix} \in Z(T_T)$. Then there is $L_T \leq_e T_T$ such that $\begin{bmatrix} r & a \\ 0 & s \end{bmatrix} L = 0$. By [Go76, Prop. 4.2], there exist $N_S \leq_e A_S$ and $E_S \leq_e S_S$ such that $\begin{bmatrix} 0 & N \\ 0 & E \end{bmatrix} \subseteq L$. So $rN + aE = 0$ and $sE = 0$. Hence $rN = 0$ and $aE = 0$. Thus $a \in Z(A_S)$ and $s \in Z(S_S)$. If $rA = 0$, then $r = 0$, so $\begin{bmatrix} r & a \\ 0 & s \end{bmatrix} \in \begin{bmatrix} c & 0 \\ 0 & e \end{bmatrix} T$. Assume $rA \neq 0$. Let $k \in A$ such that $rk \neq 0$. Then $k \notin N$. Since $N_S \leq_e A_S$, there exists $X_S \leq_e S_S$ such that $kX \subseteq N$. So $r(kX) \subseteq rN = 0$. Hence $rk \in Z(A_S) \subseteq cA$. Thus for any $y \in A$, $ry \in Z(A_S) \subseteq cA$. Since ${}_RA$ is faithful, $cr = r$. Then $\begin{bmatrix} c & 0 \\ 0 & e \end{bmatrix} \begin{bmatrix} r & a \\ 0 & s \end{bmatrix} = \begin{bmatrix} cr & ca \\ 0 & es \end{bmatrix} = \begin{bmatrix} r & a \\ 0 & s \end{bmatrix} \in \begin{bmatrix} c & 0 \\ 0 & e \end{bmatrix} T$. Hence $Z(T_T) \subseteq \begin{bmatrix} c & 0 \\ 0 & e \end{bmatrix} T$. Therefore, $Z_2(T_T) \subseteq \begin{bmatrix} c & 0 \\ 0 & e \end{bmatrix} T$. Now through a series of claims, we show that $\begin{bmatrix} 0 & Z(A_S) \\ 0 & Z(S_S) \end{bmatrix} \subseteq Z(T_T)$ and $\begin{bmatrix} 0 & Z(A_S) \\ 0 & Z(S_S) \end{bmatrix}_T \leq_e \begin{bmatrix} c & 0 \\ 0 & e \end{bmatrix} T$.

Claim 1. $\begin{bmatrix} 0 & Z(A_S) \\ 0 & Z(S_S) \end{bmatrix} \subseteq Z(T_T)$.

Proof of Claim 1. Let $0 \neq \begin{bmatrix} 0 & k \\ 0 & v \end{bmatrix} \in \begin{bmatrix} 0 & Z(A_S) \\ 0 & Z(S_S) \end{bmatrix}$. There exist $X_S \leq_e S_S$ and $Y_S \leq_e S_S$ such that $kX = 0$ and $vY = 0$. Observe that $(X \cap Y)_S \leq_e S_S$. So $\begin{bmatrix} 0 & k \\ 0 & v \end{bmatrix} \begin{bmatrix} R & A \\ 0 & X \cap Y \end{bmatrix} = 0$. Since $\begin{bmatrix} R & A \\ 0 & X \cap Y \end{bmatrix} \leq_e T_T$, $\begin{bmatrix} 0 & k \\ 0 & v \end{bmatrix} \in Z(T_T)$. Therefore $\begin{bmatrix} 0 & Z(A_S) \\ 0 & Z(S_S) \end{bmatrix} \subseteq Z(T_T)$.

Claim 2. $\begin{bmatrix} 0 & Z(A_S) \\ 0 & Z(S_S) \end{bmatrix} \leq_{\rm e} \begin{bmatrix} c & 0 \\ 0 & e \end{bmatrix} T_T$.

Proof of Claim 2. Let $0 \neq \begin{bmatrix} c & 0 \\ 0 & e \end{bmatrix} \begin{bmatrix} r & a \\ 0 & s \end{bmatrix} = \begin{bmatrix} cr & ca \\ 0 & es \end{bmatrix} \in \begin{bmatrix} c & 0 \\ 0 & e \end{bmatrix} T$. If $cr = 0$, we are done. So assume $cr \neq 0$. Since ${}_RA$ is faithful, there exists $k \in A$ such that $0 \neq crk = c(rk) \in cA = Z_2(A_S)$. So there is a $v \in S$ such that $0 \neq crkv \in Z(A_S)$. Hence $\begin{bmatrix} cr & ca \\ 0 & es \end{bmatrix} \begin{bmatrix} 0 & kv \\ 0 & 0 \end{bmatrix} = \begin{bmatrix} 0 & crkv \\ 0 & 0 \end{bmatrix} \in \begin{bmatrix} 0 & Z(A_S) \\ 0 & Z(S_S) \end{bmatrix}$. Therefore, $\begin{bmatrix} 0 & Z(A_S) \\ 0 & Z(S_S) \end{bmatrix}_T \leq_{\rm e} \begin{bmatrix} c & 0 \\ 0 & e \end{bmatrix} T_T$. From the above, we have that $\begin{bmatrix} 0 & Z(A_S) \\ 0 & Z(S_S) \end{bmatrix} \subseteq Z(T_T) \subseteq Z_2(T_T) \subseteq \begin{bmatrix} c & 0 \\ 0 & e \end{bmatrix} T$ and $\begin{bmatrix} 0 & Z(A_S) \\ 0 & Z(S_S) \end{bmatrix} \leq_{\rm e} \begin{bmatrix} c & 0 \\ 0 & e \end{bmatrix} T_T$. So $Z_2(T_T) \leq_{\rm e} \begin{bmatrix} c & 0 \\ 0 & e \end{bmatrix} T_T$, but $Z_2(T_T) \leq_{\rm c} T_T$. Therefore, $Z_2(T_T) = \begin{bmatrix} c & 0 \\ 0 & e \end{bmatrix} T$. □

Lemma 5.149. *Assume that there exists $x = x^2 \in R$ such that $xR = N$, where $N = \ell_R(A)$. If $f = f^2 \in R$, then $(1-x)f = [(1-x)f]^2$ and $fA = (1-x)fA$.*

Proof. Since $N \trianglelefteq R$, $x \in \mathcal{S}_\ell(R)$. By Proposition 5.144, $(1-x)f = (1-x)f(1-x)$. Hence $(1-x)f = [(1-x)f]^2$. Let $a \in A$. Then $fa = xfa + (1-x)fa = (1-x)fa$. We conclude that $fA = (1-x)fA$. □

Lemma 5.150. *Let $N = \ell_R(A)$. Assume $N = xR$, $R' = (1-x)R(1-x)$, and $B = \begin{bmatrix} R' & A \\ 0 & S \end{bmatrix}$, where $x = x^2 \in R$.*

(i) $T = \begin{bmatrix} x & 0 \\ 0 & 0 \end{bmatrix} T \oplus \begin{bmatrix} 1-x & 0 \\ 0 & 1 \end{bmatrix} T$, $\begin{bmatrix} x & 0 \\ 0 & 0 \end{bmatrix} T = \begin{bmatrix} N & 0 \\ 0 & 0 \end{bmatrix} \trianglelefteq T$, *$A$ is a (R', S)-bimodule, and* $\begin{bmatrix} 1-x & 0 \\ 0 & 1 \end{bmatrix} T = B$.

(ii) *B_T is* $\begin{bmatrix} x & 0 \\ 0 & 0 \end{bmatrix} T_T$*-injective.*

(iii) $\begin{bmatrix} x & 0 \\ 0 & 0 \end{bmatrix} T_T$ *is B_T-ejective.*

(iv) *If $x = x^2 \in$ Cen(R) (e.g., R is semiprime), then* $\begin{bmatrix} x & 0 \\ 0 & 0 \end{bmatrix} T_T$ *is B_T-injective.*

Proof. (i) Since $x = x^2 \in R$, $T = \begin{bmatrix} x & 0 \\ 0 & 0 \end{bmatrix} T \oplus \begin{bmatrix} 1-x & 0 \\ 0 & 1 \end{bmatrix} T$. Further, since $N = \ell_R(A)$ and $N = xR$, we have $\begin{bmatrix} x & 0 \\ 0 & 0 \end{bmatrix} T = \begin{bmatrix} xR & xA \\ 0 & 0 \end{bmatrix} = \begin{bmatrix} N & 0 \\ 0 & 0 \end{bmatrix}$. Since $N \trianglelefteq R$, $\begin{bmatrix} N & 0 \\ 0 & 0 \end{bmatrix} \trianglelefteq T$. Hence $x \in \mathcal{S}_\ell(R)$ and $\begin{bmatrix} x & 0 \\ 0 & 0 \end{bmatrix} \in \mathcal{S}_\ell(T)$. Again, since $N = \ell_R(A)$,

$(1-x)a = a$ for all $a \in A$. By Proposition 5.144 (i), $(1-x)R = R'$. Thus $\begin{bmatrix} 1-x & 0 \\ 0 & 1 \end{bmatrix} T = B$.

(ii) Let $\begin{bmatrix} K & 0 \\ 0 & 0 \end{bmatrix}_T \leq \begin{bmatrix} x & 0 \\ 0 & 0 \end{bmatrix} T$ and $h : \begin{bmatrix} K & 0 \\ 0 & 0 \end{bmatrix} \to B$ a T-homomorphism. Let $k \in K$. Then $h\left(\begin{bmatrix} k & 0 \\ 0 & 0 \end{bmatrix}\right) = \begin{bmatrix} (1-x)r & a \\ 0 & s \end{bmatrix} \in B$. Assume $(1-x)r \neq 0$. Since ${}_RA$ is faithful, there is an $a_1 \in A$ such that $(1-x)ra_1 \neq 0$. Hence $h\left(\begin{bmatrix} k & 0 \\ 0 & 0 \end{bmatrix}\begin{bmatrix} 0 & a_1 \\ 0 & 0 \end{bmatrix}\right) = h\left(\begin{bmatrix} 0 & 0 \\ 0 & 0 \end{bmatrix}\right) = \begin{bmatrix} (1-x)r & a \\ 0 & s \end{bmatrix}\begin{bmatrix} 0 & a_1 \\ 0 & 0 \end{bmatrix} = \begin{bmatrix} 0 & (1-x)ra_1 \\ 0 & 0 \end{bmatrix}$, a contradiction. So $(1-x)r = 0$. Moreover, $h\left(\begin{bmatrix} k & 0 \\ 0 & 0 \end{bmatrix}\begin{bmatrix} 0 & 0 \\ 0 & 1 \end{bmatrix}\right) = \begin{bmatrix} 0 & 0 \\ 0 & 0 \end{bmatrix} = \begin{bmatrix} 0 & a \\ 0 & s \end{bmatrix}\begin{bmatrix} 0 & 0 \\ 0 & 1 \end{bmatrix} = \begin{bmatrix} 0 & a \\ 0 & s \end{bmatrix}$. Hence $h = 0$. We see that B_T is $\begin{bmatrix} x & 0 \\ 0 & 0 \end{bmatrix} T_T$-injective.

(iii) Let $X_T \leq B_T$ and $h : X \to \begin{bmatrix} x & 0 \\ 0 & 0 \end{bmatrix} T$ a T-homomorphism. Since $\begin{bmatrix} 0 & A \\ 0 & S \end{bmatrix}_T \leq_{\text{e}} B_T$, $Y_T \leq_{\text{e}} X_T$, where $Y = X \cap \begin{bmatrix} 0 & A \\ 0 & S \end{bmatrix}$. Let $y \in Y$. Then there exist $a \in A$ and $s \in S$ such that $y = \begin{bmatrix} 0 & a \\ 0 & s \end{bmatrix}$. Then $h(y) = h\left(\begin{bmatrix} 0 & a \\ 0 & s \end{bmatrix}\begin{bmatrix} 0 & 0 \\ 0 & 1 \end{bmatrix}\right) = h(y)\begin{bmatrix} 0 & 0 \\ 0 & 1 \end{bmatrix} = 0$. Thus $h(Y) = 0$. Let $\overline{h} : B \to \begin{bmatrix} x & 0 \\ 0 & 0 \end{bmatrix} T$ be the zero homomorphism. We see that $\begin{bmatrix} x & 0 \\ 0 & 0 \end{bmatrix} T_T$ is B_T-ejective.

(iv) We use the notation of part (iii). Let $\begin{bmatrix} r & a_1 \\ 0 & s_1 \end{bmatrix} \in X$. Then $h\left(\begin{bmatrix} r & a_1 \\ 0 & s_1 \end{bmatrix}\right) = \begin{bmatrix} k & 0 \\ 0 & 0 \end{bmatrix}$ for some $k \in N$. Then $h\left(\begin{bmatrix} r & a_1 \\ 0 & s_1 \end{bmatrix}\begin{bmatrix} 1-x & 0 \\ 0 & 1 \end{bmatrix}\right) = \begin{bmatrix} k & 0 \\ 0 & 0 \end{bmatrix}\begin{bmatrix} 1-x & 0 \\ 0 & 1 \end{bmatrix} = \begin{bmatrix} k(1-x) & 0 \\ 0 & 0 \end{bmatrix} = \begin{bmatrix} 0 & 0 \\ 0 & 0 \end{bmatrix}$. Therefore, $\begin{bmatrix} x & 0 \\ 0 & 0 \end{bmatrix} T_T$ is B_T-injective. Note that if R is semiprime, then $S_{\text{l}}(R) \subseteq \text{Cen}(R)$. □

The next result indicates conditions imposed on R, A, and S under which T_T is G-extending, and is the G-extending counterpart to Theorem 3.88 (see, also [Te95, Thm. 2.4]).

Lemma 5.151. *Assume T_T is* G*-extending.*

(i) *If $Y_S \leq A_S$, then there is a $c = c^2 \in R$ such that $Y_S \beta c A_S$.*
(ii) *A_S and S_S are* G*-extending.*
(iii) *A_S is S_S-ejective.*
(iv) *Let $N = \ell_R(A)$. Then there exists $x = x^2 \in R$ such that $N = xR$, xR_R is*

G-extending, and $\begin{bmatrix} x & 0 \\ 0 & 0 \end{bmatrix} \in S_l(T)$. *Moreover,* $\begin{bmatrix} x & 0 \\ 0 & 0 \end{bmatrix} T_T$, B_T *and* B_B *are G-extending, where* B *denotes the generalized triangular matrix ring* $\begin{bmatrix} R' & A \\ 0 & S \end{bmatrix}$ *with* $R' = (1-x)R(1-x)$.

Proof. (i) Since T_T is G-extending, there exists $\begin{bmatrix} c & k \\ 0 & e \end{bmatrix} = \begin{bmatrix} c & k \\ 0 & e \end{bmatrix}^2 \in T$ such that

$$\left(\begin{bmatrix} 0 & Y \\ 0 & 0 \end{bmatrix} \cap \begin{bmatrix} c & k \\ 0 & e \end{bmatrix} T\right)_T \leq_e \begin{bmatrix} 0 & Y \\ 0 & 0 \end{bmatrix}_T \quad \text{and} \quad \left(\begin{bmatrix} 0 & Y \\ 0 & 0 \end{bmatrix} \cap \begin{bmatrix} c & k \\ 0 & e \end{bmatrix} T\right)_T \leq_e \begin{bmatrix} c & k \\ 0 & e \end{bmatrix} T_T.$$

There exists $X_S \leq Y_S$ such that $\begin{bmatrix} 0 & X \\ 0 & 0 \end{bmatrix} = \begin{bmatrix} 0 & Y \\ 0 & 0 \end{bmatrix} \cap \begin{bmatrix} c & k \\ 0 & e \end{bmatrix} T$. Observe that $X \subseteq cA$. Let $0 \neq ca \in cA$. Then $0 \neq \begin{bmatrix} c & k \\ 0 & e \end{bmatrix}\begin{bmatrix} 0 & a \\ 0 & 0 \end{bmatrix} = \begin{bmatrix} 0 & ca \\ 0 & 0 \end{bmatrix}$. Hence, there exists $0 \neq \begin{bmatrix} 0 & 0 \\ 0 & s \end{bmatrix} \in T$ such that $0 \neq \begin{bmatrix} 0 & ca \\ 0 & 0 \end{bmatrix}\begin{bmatrix} 0 & 0 \\ 0 & s \end{bmatrix} = \begin{bmatrix} 0 & cas \\ 0 & 0 \end{bmatrix} \in \begin{bmatrix} 0 & X \\ 0 & 0 \end{bmatrix}$. Hence, $0 \neq cas \in X$. Thus $(Y \cap cA)_S \leq_e cA_S$. Similarly, $(Y \cap cA)_S \leq_e Y_S$. Therefore, $Y_S \beta cA_S$.

(ii) Let $0 \neq X_S \leq A_S$. Since T_T is G-extending, there exists an idempotent $\begin{bmatrix} c & k \\ 0 & e \end{bmatrix} \in T$ such that

$$\begin{bmatrix} 0 & X \\ 0 & 0 \end{bmatrix}_T \cap \begin{bmatrix} c & k \\ 0 & e \end{bmatrix} T_T \leq_e \begin{bmatrix} 0 & X \\ 0 & 0 \end{bmatrix}, \quad \text{and} \quad \begin{bmatrix} 0 & X \\ 0 & 0 \end{bmatrix}_T \cap \begin{bmatrix} c & k \\ 0 & e \end{bmatrix} T_T \leq_e \begin{bmatrix} c & k \\ 0 & e \end{bmatrix} T_T.$$

Let $0 \neq x \in X$. Then there exists $s \in S$ such that $0 \neq \begin{bmatrix} 0 & x \\ 0 & 0 \end{bmatrix}\begin{bmatrix} 0 & 0 \\ 0 & s \end{bmatrix} = \begin{bmatrix} 0 & xs \\ 0 & 0 \end{bmatrix} \in \begin{bmatrix} c & k \\ 0 & e \end{bmatrix} T$. Hence $\begin{bmatrix} 0 & xs \\ 0 & 0 \end{bmatrix} = \begin{bmatrix} c & k \\ 0 & e \end{bmatrix}\begin{bmatrix} 0 & xs \\ 0 & 0 \end{bmatrix} = \begin{bmatrix} 0 & cxs \\ 0 & 0 \end{bmatrix}$. So $0 \neq xs = cxs \in X \cap cA$. Thus $X_S \cap cA_S \leq_e X_S$. Now let $0 \neq ca \in cA$. Since $\begin{bmatrix} 0 & ca \\ 0 & 0 \end{bmatrix} = \begin{bmatrix} c & k \\ 0 & e \end{bmatrix}\begin{bmatrix} 0 & a \\ 0 & 0 \end{bmatrix} \in \begin{bmatrix} c & k \\ 0 & e \end{bmatrix} T$, there is a $v \in S$ such that $0 \neq \begin{bmatrix} 0 & ca \\ 0 & 0 \end{bmatrix}\begin{bmatrix} 0 & 0 \\ 0 & v \end{bmatrix} = \begin{bmatrix} 0 & cav \\ 0 & 0 \end{bmatrix} \in \begin{bmatrix} 0 & X \\ 0 & 0 \end{bmatrix} \cap \begin{bmatrix} c & k \\ 0 & e \end{bmatrix} T$. It follows that $X_S \cap cA_S \leq_e cA_S$. Therefore, A_S is G-extending. The proof that S_S is G-extending is similar.

(iii) Let $X_S \leq S_S$ and let $\varphi : X_S \to A_S$ be a homomorphism. Let

$$F = \left\{ \begin{bmatrix} 0 & \varphi(x) \\ 0 & x \end{bmatrix} : x \in X \right\}.$$

Then $F_T \leq T_T$ and there exists $\begin{bmatrix} d & k \\ 0 & b \end{bmatrix} = \begin{bmatrix} d & k \\ 0 & b \end{bmatrix}^2 \in T$ such that $F_T \beta \begin{bmatrix} d & k \\ 0 & b \end{bmatrix} T_T$.

Assume $d \neq 0$. Then $0 \neq \begin{bmatrix} d & k \\ 0 & b \end{bmatrix}\begin{bmatrix} 1 & 0 \\ 0 & 0 \end{bmatrix} = \begin{bmatrix} d & 0 \\ 0 & 0 \end{bmatrix} \in \begin{bmatrix} d & k \\ 0 & b \end{bmatrix} T$. So there is a $\begin{bmatrix} r & a \\ 0 & s \end{bmatrix} \in T$ such that $0 \neq \begin{bmatrix} d & 0 \\ 0 & 0 \end{bmatrix}\begin{bmatrix} r & a \\ 0 & s \end{bmatrix} = \begin{bmatrix} dr & da \\ 0 & 0 \end{bmatrix} \in F$. Thus $\begin{bmatrix} dr & da \\ 0 & 0 \end{bmatrix} = \begin{bmatrix} 0 & \varphi(x) \\ 0 & x \end{bmatrix}$ for some $x \in S$. Hence $\begin{bmatrix} dr & da \\ 0 & 0 \end{bmatrix} = 0$, a contradiction. Therefore, $\begin{bmatrix} d & k \\ 0 & b \end{bmatrix} = \begin{bmatrix} 0 & k \\ 0 & b \end{bmatrix}$. Let $Y = \{y \in X : \begin{bmatrix} 0 & \varphi(y) \\ 0 & y \end{bmatrix} \in F \cap \begin{bmatrix} 0 & k \\ 0 & b \end{bmatrix} T_T\}$. Clearly, $Y_S \leq X_S$. Let $0 \neq x \in X$. Then $0 \neq \begin{bmatrix} 0 & \varphi(x) \\ 0 & x \end{bmatrix} \in F$. Since $F_T \beta \begin{bmatrix} 0 & k \\ 0 & b \end{bmatrix} T_T$, there is a $v \in S$ such that $0 \neq \begin{bmatrix} 0 & \varphi(x) \\ 0 & x \end{bmatrix}\begin{bmatrix} 0 & 0 \\ 0 & v \end{bmatrix} = \begin{bmatrix} 0 & \varphi(xv) \\ 0 & xv \end{bmatrix} \in F \cap \begin{bmatrix} 0 & k \\ 0 & b \end{bmatrix} T$. So $0 \neq xv \in Y$. Therefore, $Y_S \leq_{\rm e} X_S$. Now let $\begin{bmatrix} 0 & \varphi(w) \\ 0 & w \end{bmatrix} \in F \cap \begin{bmatrix} 0 & k \\ 0 & b \end{bmatrix} T$. Then $\begin{bmatrix} 0 & \varphi(w) \\ 0 & w \end{bmatrix} = \begin{bmatrix} 0 & k \\ 0 & b \end{bmatrix}\begin{bmatrix} 0 & \varphi(w) \\ 0 & w \end{bmatrix} = \begin{bmatrix} 0 & kw \\ 0 & bw \end{bmatrix}$. Hence $\varphi(w) = kw$. Define $\overline{\varphi} : S_S \to A_S$ by $\overline{\varphi}(s) = ks$, for all $s \in S$. Then $\overline{\varphi}(w) = \varphi(w)$ for all $w \in Y$. Therefore, A_S is S-ejective.

(iv) Since $\begin{bmatrix} N & 0 \\ 0 & 0 \end{bmatrix} \trianglelefteq T$, there exists $\begin{bmatrix} x & k \\ 0 & e \end{bmatrix} = \begin{bmatrix} x & k \\ 0 & e \end{bmatrix}^2 \in T$ such that $\begin{bmatrix} N & 0 \\ 0 & 0 \end{bmatrix}_T \beta \begin{bmatrix} x & k \\ 0 & e \end{bmatrix} T_T$. Let $\begin{bmatrix} X & 0 \\ 0 & 0 \end{bmatrix} = \begin{bmatrix} N & 0 \\ 0 & 0 \end{bmatrix} \cap \begin{bmatrix} x & k \\ 0 & e \end{bmatrix} T$.

Since $\begin{bmatrix} X & 0 \\ 0 & 0 \end{bmatrix} \leq_{\rm e} \begin{bmatrix} x & k \\ 0 & e \end{bmatrix} T_T$, we have $e = 0$. Assume $0 \neq k \in A$. Then $0 \neq \begin{bmatrix} x & k \\ 0 & 0 \end{bmatrix}\begin{bmatrix} 0 & 0 \\ 0 & 1 \end{bmatrix} = \begin{bmatrix} 0 & k \\ 0 & 0 \end{bmatrix} \in \begin{bmatrix} x & k \\ 0 & e \end{bmatrix} T$. So there is $\begin{bmatrix} r_1 & a_1 \\ 0 & s_1 \end{bmatrix} \in T$ such that $0 \neq \begin{bmatrix} 0 & k \\ 0 & 0 \end{bmatrix}\begin{bmatrix} r_1 & a_1 \\ 0 & s_1 \end{bmatrix} = \begin{bmatrix} 0 & ks_1 \\ 0 & 0 \end{bmatrix} \in \begin{bmatrix} X & 0 \\ 0 & 0 \end{bmatrix}$, a contradiction. Hence $k = 0$, and $\begin{bmatrix} x & k \\ 0 & e \end{bmatrix} = \begin{bmatrix} x & 0 \\ 0 & 0 \end{bmatrix}$.

Now $\begin{bmatrix} x & 0 \\ 0 & 0 \end{bmatrix} T = \begin{bmatrix} xR & xA \\ 0 & 0 \end{bmatrix}$. Assume $xA \neq 0$. Then $0 \neq \begin{bmatrix} x & 0 \\ 0 & 0 \end{bmatrix}\begin{bmatrix} 0 & a_2 \\ 0 & 0 \end{bmatrix} = \begin{bmatrix} 0 & xa_2 \\ 0 & 0 \end{bmatrix} \in \begin{bmatrix} x & 0 \\ 0 & 0 \end{bmatrix} T$, for some $a_2 \in A$. So there exists $\begin{bmatrix} r_3 & a_3 \\ 0 & s_3 \end{bmatrix} \in T$ such that $0 \neq \begin{bmatrix} 0 & xa_2 \\ 0 & 0 \end{bmatrix}\begin{bmatrix} r_3 & a_3 \\ 0 & s_3 \end{bmatrix} \in \begin{bmatrix} X & 0 \\ 0 & 0 \end{bmatrix}$, a contradiction. Thus $xR \subseteq N$. So $\begin{bmatrix} X & 0 \\ 0 & 0 \end{bmatrix}_T \leq \begin{bmatrix} x & 0 \\ 0 & 0 \end{bmatrix} T_T \leq \begin{bmatrix} N & 0 \\ 0 & 0 \end{bmatrix}_T$. Since $\begin{bmatrix} X & 0 \\ 0 & 0 \end{bmatrix}_T \leq_{\rm e} \begin{bmatrix} N & 0 \\ 0 & 0 \end{bmatrix}$, $\begin{bmatrix} N & 0 \\ 0 & 0 \end{bmatrix} = \begin{bmatrix} x & 0 \\ 0 & 0 \end{bmatrix} T = \begin{bmatrix} xR & 0 \\ 0 & 0 \end{bmatrix}$. Since $\begin{bmatrix} N & 0 \\ 0 & 0 \end{bmatrix}_T \trianglelefteq T$, then $\begin{bmatrix} x & 0 \\ 0 & 0 \end{bmatrix} \in \mathcal{S}_l(T)$. By Proposition 5.143, $x \in \mathcal{S}_l(R)$. The remaining conclusions of this part follow from Theorem 5.88 and Proposition 5.144. □

Lemma 5.152. *If ${}_RA$ is faithful and all of the following conditions are satisfied, then T_T is G-extending.*

(i) *If D_S is a direct summand of A_S, then there is an $f = f^2 \in R$ such that $D_S \beta f A_S$;*

(ii) *A_S and S_S are G-extending;*

(iii) *A_S is S_S-ejective.*

Proof. By Proposition 5.148, $T = Z_2(T_T) \oplus T'$, where $Z_2(T_T) = \begin{bmatrix} c & 0 \\ 0 & e \end{bmatrix} T$, $Z_2(A_S) = cA$, $Z_2(S_S) = eS$, and $T' = \begin{bmatrix} 1-c & 0 \\ 0 & 1-e \end{bmatrix} T$. Observe that $Z(T'_T) = 0$. First, we show that $Z_2(T_T)_T$ is G-extending.

Claim 1. $\begin{bmatrix} c & 0 \\ 0 & 0 \end{bmatrix} T_T = \begin{bmatrix} cR & cA \\ 0 & 0 \end{bmatrix}_T$ is G-extending.

Proof of Claim 1. If $c = 0$, we are finished. So assume $c \neq 0$. Let $X_T \leq \begin{bmatrix} c & 0 \\ 0 & 0 \end{bmatrix} T_T$. Observe that $\begin{bmatrix} 0 & cA \\ 0 & 0 \end{bmatrix} \leq_{\text{e}} \begin{bmatrix} c & 0 \\ 0 & 0 \end{bmatrix} T_T$. Let $Y = X \cap \begin{bmatrix} 0 & cA \\ 0 & 0 \end{bmatrix} = \begin{bmatrix} 0 & Y' \\ 0 & 0 \end{bmatrix}$. Note that $Y_T \leq_{\text{e}} X_T$, so $Y_T \beta X_T$. Then $Y'_S \leq A_S$. Since A_S is G-extending, there is $b = b^2 \in R$ such that $Y'_S \beta b A_S$. Observe that $Y_T \beta \begin{bmatrix} b & 0 \\ 0 & 0 \end{bmatrix} T_T$. To see this, let $0 \neq \begin{bmatrix} br & ba \\ 0 & 0 \end{bmatrix} \in \begin{bmatrix} b & 0 \\ 0 & 0 \end{bmatrix} T$. Assume that $br \neq 0$. Since ${}_RA$ is faithful, there exists $n \in A$ such that $0 \neq brn \in bA$ and $s \in S$ such that $0 \neq brns \in Y'$. Hence $0 \neq \begin{bmatrix} br & ba \\ 0 & 0 \end{bmatrix} \begin{bmatrix} 0 & ns \\ 0 & 0 \end{bmatrix} = \begin{bmatrix} 0 & brns \\ 0 & 0 \end{bmatrix} \in Y \cap \begin{bmatrix} b & 0 \\ 0 & 0 \end{bmatrix} T$. If $br = 0$, there is a $v \in S$ such that $0 \neq bav \in Y'$. So $0 \neq \begin{bmatrix} br & ba \\ 0 & 0 \end{bmatrix} \begin{bmatrix} 0 & 0 \\ 0 & v \end{bmatrix} = \begin{bmatrix} 0 & bav \\ 0 & 0 \end{bmatrix} \in Y \cap \begin{bmatrix} b & 0 \\ 0 & 0 \end{bmatrix} T$. Thus $(Y \cap \begin{bmatrix} b & 0 \\ 0 & 0 \end{bmatrix} T)_T \leq_{\text{e}} \begin{bmatrix} b & 0 \\ 0 & 0 \end{bmatrix} T_T$. Let $0 \neq \begin{bmatrix} 0 & y \\ 0 & 0 \end{bmatrix} \in Y$. There is a $w \in S$ such that $0 \neq yw \in bA$. Hence, $0 \neq \begin{bmatrix} 0 & y \\ 0 & 0 \end{bmatrix} \begin{bmatrix} 0 & 0 \\ 0 & w \end{bmatrix} = \begin{bmatrix} 0 & yw \\ 0 & 0 \end{bmatrix} \in Y \cap \begin{bmatrix} b & 0 \\ 0 & 0 \end{bmatrix} T$. Thus $(Y \cap \begin{bmatrix} b & 0 \\ 0 & 0 \end{bmatrix} T)_T \leq_{\text{e}} Y_T$. So $Y_T \beta \begin{bmatrix} b & 0 \\ 0 & 0 \end{bmatrix} T_T$. Since β is an equivalence relation and $X_T \beta Y_T$, $X_T \beta \begin{bmatrix} b & 0 \\ 0 & 0 \end{bmatrix} T_T$. Therefore, $\begin{bmatrix} c & 0 \\ 0 & 0 \end{bmatrix} T_T$ is G-extending.

Claim 2. $\begin{bmatrix} 0 & 0 \\ 0 & e \end{bmatrix} T$ is G-extending.

Proof of Claim 2. This is an immediate consequence of eS being G-extending by Corollary 5.91.

Claim 3. $\begin{bmatrix} c & 0 \\ 0 & 0 \end{bmatrix} T_T$ is $\begin{bmatrix} 0 & 0 \\ 0 & e \end{bmatrix} T_T$-ejective.

Proof of Claim 3. Let $X_T \leq \begin{bmatrix} 0 & 0 \\ 0 & e \end{bmatrix} T_T$ and $h : X_T \to \begin{bmatrix} c & 0 \\ 0 & 0 \end{bmatrix} T_T$ a homomorphism. Then $X = \begin{bmatrix} 0 & 0 \\ 0 & K \end{bmatrix}$ for some $K_S \leq eS_S$. Let $\begin{bmatrix} 0 & 0 \\ 0 & k \end{bmatrix} \in X$. Then $h\left(\begin{bmatrix} 0 & 0 \\ 0 & k \end{bmatrix}\right) = \begin{bmatrix} cr & ca \\ 0 & 0 \end{bmatrix}$, for some $r \in R$ and $a \in A$. So $\begin{bmatrix} cr & ca \\ 0 & 0 \end{bmatrix} = h\left(\begin{bmatrix} 0 & 0 \\ 0 & k \end{bmatrix}\right) = h\left(\begin{bmatrix} 0 & 0 \\ 0 & k \end{bmatrix}\begin{bmatrix} 0 & 0 \\ 0 & 1 \end{bmatrix}\right) = h\left(\begin{bmatrix} 0 & 0 \\ 0 & k \end{bmatrix}\right)\begin{bmatrix} 0 & 0 \\ 0 & 1 \end{bmatrix} = \begin{bmatrix} cr & ca \\ 0 & 0 \end{bmatrix}\begin{bmatrix} 0 & 0 \\ 0 & 1 \end{bmatrix} = \begin{bmatrix} 0 & ca \\ 0 & 0 \end{bmatrix}$. Hence $cr = 0$. Thus $h\left(\begin{bmatrix} 0 & 0 \\ 0 & K \end{bmatrix}\right) \subseteq \begin{bmatrix} 0 & cA \\ 0 & 0 \end{bmatrix} \subseteq \begin{bmatrix} 0 & A \\ 0 & 0 \end{bmatrix}$. Define $\varphi : K \to A$ by $\varphi(k) = ca$, where $h\left(\begin{bmatrix} 0 & 0 \\ 0 & k \end{bmatrix}\right) = \begin{bmatrix} 0 & ca \\ 0 & 0 \end{bmatrix}$. Then φ is an S-homomorphism. Observe that $\varphi(K) \subseteq cA$. Since A_S is S_S-ejective, there exist an $N_S \leq_{\rm e} K_S$ and an S-homomorphism $\overline{\varphi} : S_S \to A_S$ such that $\varphi(n) = \overline{\varphi}(n) = \varphi'(n)$ for all $n \in N$, where φ' is the restriction of $\overline{\varphi}$ to eS. Let $\pi : A \to cA$ be the projection defined by $\pi(a) = ca$, for all $a \in A$. Then define $\theta : eS_S \to cA_S$ by $\theta = \pi \circ \varphi'$. Hence $\varphi(n) = \theta(n)$ for all $n \in N$. Observe that $\begin{bmatrix} 0 & 0 \\ 0 & N \end{bmatrix}_T \leq_{\rm e} X_T$. Define $\overline{h} : \begin{bmatrix} 0 & 0 \\ 0 & e \end{bmatrix} T \to \begin{bmatrix} c & 0 \\ 0 & 0 \end{bmatrix} T$ by $\overline{h}\left(\begin{bmatrix} 0 & 0 \\ 0 & es \end{bmatrix}\right) = \begin{bmatrix} 0 & ca_1 \\ 0 & 0 \end{bmatrix}$, where $\theta(es) = ca_1$. Then $\overline{h}$ is a T-homomorphism with $h\left(\begin{bmatrix} 0 & 0 \\ 0 & n \end{bmatrix}\right) = \overline{h}\left(\begin{bmatrix} 0 & 0 \\ 0 & n \end{bmatrix}\right)$, for all $n \in N$. Therefore, $\begin{bmatrix} c & 0 \\ 0 & 0 \end{bmatrix} T_T$ is $\begin{bmatrix} 0 & 0 \\ 0 & e \end{bmatrix} T_T$-ejective.

Now by Claims 1–3 and Theorem 5.82, we deduce that $Z_2(T_T)_T = \begin{bmatrix} c & 0 \\ 0 & e \end{bmatrix} T_T$ is G-extending. We accomplish this through a series of cases depending on c and e.

Case 1. $1 - c = 0$ and $1 - e = 0$. Then $T' = 0$, hence T'_T is extending.

Case 2. $1 - c = 0$ and $1 - e \neq 0$. By Proposition 5.81 and Corollary 5.91, $(1 - e)S_S$ is extending. Since $T' = \begin{bmatrix} 0 & 0 \\ 0 & (1-e)S \end{bmatrix}$, T'_T is extending.

Case 3. $1 - c \neq 0$ and $1 - e = 0$. Hence $T' = \begin{bmatrix} 1-c & 0 \\ 0 & 0 \end{bmatrix} T = \begin{bmatrix} (1-c)R & (1-c)A \\ 0 & 0 \end{bmatrix}$. Let $X_T \leq T'_T$ and $Y = X \cap \begin{bmatrix} 0 & (1-c)A \\ 0 & 0 \end{bmatrix} = \begin{bmatrix} 0 & K \\ 0 & 0 \end{bmatrix}$. By Proposition 5.81, Corollary 5.91, and condition (i), there exists $f = f^2 \in R$ such that $K_S \beta f A_S$. So $(K \cap fA)_S \leq_{\rm e} K_S$ and $(K \cap fA)_S \leq_{\rm e} fA_S$. Since $1 - c \in \mathcal{S}_{\rm l}(R)$, then $(1 - c)f = (1 - c)f(1 - c) = [(1 - c)f]^2$. Hence $(K \cap fA)_S \leq_{\rm e} (1 - c)fA_S \subseteq (1 - c)A$. Since $(1 - c)A_S$ is nonsingular, each submodule has unique closure, and so $K_S \leq_{\rm e} (1 - c)fA_S$. Observe that $\begin{bmatrix} 0 & (1-c)A \\ 0 & 0 \end{bmatrix}_T \leq_{\rm e} T'_T$. Hence $\begin{bmatrix} 0 & K \\ 0 & 0 \end{bmatrix}_T \leq_{\rm e} X_T$.

Also, $\begin{bmatrix} 0 & K \\ 0 & 0 \end{bmatrix}_T \leq_{\rm e} \begin{bmatrix} (1-c)f & 0 \\ 0 & 0 \end{bmatrix} T_T$. Thus $X_T \beta Y_T$ and $Y_T \beta \begin{bmatrix} (1-c)f & 0 \\ 0 & 0 \end{bmatrix} T_T$. So $X_T \beta \begin{bmatrix} (1-c)f & 0 \\ 0 & 0 \end{bmatrix} T_T$. Since β is an equivalence relation, T'_T is G-extending. By Proposition 5.81, T'_T is extending.

Case 4. $1-c \neq 0$ and $1-e \neq 0$. By Proposition 5.144, T' is a ring with unity, and $(1-c)A$ is an (R', S')-bimodule, where $R' = (1-c)R(1-c)$ and $S' = (1-e)S(1-e)$. So $T' = \begin{bmatrix} R' & (1-c)A \\ 0 & S' \end{bmatrix}$. By Proposition 5.81, and Corollary 5.91, $(1-e)S_S$ is extending. Next, by Proposition 5.144, $S'_{S'}$ is nonsingular and extending. Since $A/Z_2(A_S) \cong (1-c)A_S$, $(1-c)A$ is a nonsingular injective S'-module thanks to Proposition 2.61.

Let $K_{S'}$ be a complement submodule of $(1-c)A_{S'}$. Then $K_{S'}$ is a direct summand of $(1-c)A_{S'}$. So K_S is a direct summand of $(1-c)A_S$. Hence there exists $f = f^2 \in R$ such that $K_S \beta f A_S$. As in Case 3, $K_S \leq_{\rm e} (1-c)fA_S$ where $(1-c)f = ((1-c)f)^2 \in R'$. Then $K_{S'} \leq_{\rm e} (1-c)fA_{S'}$. Since $K_{S'}$ is a complement, $K_{S'} = (1-c)fA_{S'}$. By Theorem 3.88, $T'_{T'}$ is nonsingular and extending. Proposition 5.144 yields that T'_T is extending.

Therefore in all cases, T'_T is extending. From Proposition 5.81 and Corollary 5.91, it follows that T_T is G-extending. □

Theorem 5.153. *T_T is G-extending if and only if all of the following conditions are satisfied.*

(i) *If D_S is a direct summand of A_S, then there is an $f = f^2 \in R$ such that $D_S \beta f A_S$.*

(ii) *A_S and S_S are G-extending.*

(iii) *A_S is S_S-ejective.*

(iv) *Let $N = \ell_R(A)$. There exists $x = x^2 \in R$ such that $xR = N$ and xR_R is G-extending.*

Proof. If T_T is G-extending, then conditions (i)–(iv) are satisfied by Lemma 5.151 and Proposition 5.145. Conversely, assume conditions (i)–(iv) are satisfied.

Claim 1. $\begin{bmatrix} N & 0 \\ 0 & 0 \end{bmatrix} = \begin{bmatrix} x & 0 \\ 0 & 0 \end{bmatrix} T_T$ is G-extending.

Proof of Claim 1. Clearly, $\begin{bmatrix} N & 0 \\ 0 & 0 \end{bmatrix} = \begin{bmatrix} x & 0 \\ 0 & 0 \end{bmatrix} T$. Let $X_T \leq \begin{bmatrix} xR & 0 \\ 0 & 0 \end{bmatrix}$. Then there exists $X'_R \leq xR$ such that $X = \begin{bmatrix} X' & 0 \\ 0 & 0 \end{bmatrix}$. By condition (iv), there is a $c = c^2 \in xR$ such that $X' \beta cR$. Hence $X_T \beta \begin{bmatrix} c & 0 \\ 0 & 0 \end{bmatrix} T_T$. Therefore, $\begin{bmatrix} x & 0 \\ 0 & 0 \end{bmatrix} T_T$ is G-extending.

Claim 2. $B = \begin{bmatrix} (1-x)R & A \\ 0 & S \end{bmatrix}$ is a ring with unity $\begin{bmatrix} (1-x) & 0 \\ 0 & 1 \end{bmatrix}$ and B_T is G-extending.

Proof of Claim 2. Since $\begin{bmatrix} N & 0 \\ 0 & 0 \end{bmatrix} \trianglelefteq T$, $\begin{bmatrix} x & 0 \\ 0 & 0 \end{bmatrix} \in S_l(T)$. By Proposition 5.144, B is a ring with unity $\begin{bmatrix} (1-x) & 0 \\ 0 & 1 \end{bmatrix}$. Let $R' = (1-x)R(1-x)$. Then ${}_{R'}A$ is faithful and A is an (R', S)-bimodule. By Lemma 5.149, B satisfies condition (i) of Lemma 5.152. Trivially, B satisfies conditions (ii) and (iii) of Lemma 5.152. Hence B_B is G-extending. By Proposition 5.144, B_T is G-extending. Now by Claims 1 and 2, Lemma 5.150, and Theorem 5.82, T_T is G-extending. □

Note that if R is a subring of $\text{End}(A_S)$ which contains all idempotents of $\text{End}(A_S)$ (e.g., $R = \text{End}(A_S)$), then conditions (i) and (iv) of Theorem 5.153 are satisfied. The following result generalizes one of the main results of Section 3.3 (see Theorem 3.88), by removing the condition that ${}_RA$ is faithful.

Corollary 5.154. *T_T is nonsingular and extending if and only if all the following conditions are satisfied.*

(i) *For every complement K in A_S, there exists $f = f^2 \in R$ such that $K = fA$.*
(ii) *S_S is nonsingular and extending.*
(iii) *A_S is nonsingular and injective.*
(iv) *Let $N = \ell_R(A)$. There exists $x = x^2 \in R$ such that $xR = N$ and xR_R is extending.*
(v) $\{c \in R : cJ = 0 \text{ for some } J_S \leq_{\text{e}} A_S\} \cap \{c \in R : cK = 0 \text{ for some } K_R \leq_{\text{e}} N_R\} = 0$.

Proof. The proof is a consequence of Proposition 5.142, Theorem 5.153, Proposition 5.81, [Go76, p. 113, Exc. 3], and the fact that each submodule of a nonsingular module M has a unique essential closure in M. □

By Theorem 3.88, if $T_n(S)$ is right extending, then S is right self-injective. If $Z(S_S) = 0$, the converse is true by Theorem 3.91. However, in general, the converse is false, as can be seen by considering $T_2(\mathbb{Z}_4)$ (see Example 3.93). Naturally, one can ask the following question: How "much" of the extending condition passes from S to $T_n(S)$ when S is right self-injective? Our next result provides an answer.

Corollary 5.155. *Let W be a ring with unity.*

(i) *$T_2(W)$ is right G-extending if and only if W is right G-extending and W_W is W_W-ejective.*
(ii) *For $n > 2$, $T_n(W)$ is right G-extending if and only if W is right G-extending and $cT_{n-1}(W)$ is $T_{n-1}(W)$-ejective, where $c \in T_{n-1}(W)$ with 1 in the $(1,1)$-position and zero elsewhere.*
(iii) *If W is right self-injective, then $T_n(W)$ is right G-extending for all $n > 0$.*

Proof. (i) is an immediate consequence of Theorem 5.153.

(ii) Let M be the 1-by-$(n-1)$ row matrix over W, $R = W$, and $S = T_{n-1}(W)$. Then $T_n(W) = T$. Also, $M_S \cong cS$, so $\text{End}(M_S) \cong cSc \cong W = R$. If $T_n(W)$ is right G-extending, then by Theorem 5.153 $cT_{n-1}(W)$ is $T_{n-1}(W)$-ejective. Note that $T_n(W)$ can be reblocked with M as an $(n-1)$-by-1 column matrix over W, $R = T_{n-1}(W)$, and $S = W$. Now, again by Theorem 5.153, W is right G-extending.

Conversely, assume W is right G-extending and $cT_{n-1}(W)$ is $T_{n-1}(W)$-ejective. Block $T_n(W)$ with M the 1-by-$(n-1)$ row matrix over W, $R = W$, and $S = T_{n-1}(W)$. Then conditions (i) and (iv) of Theorem 5.153 are satisfied since $\text{End}(M_S) \cong R = W$. By induction and part (i), S_S is G-extending. By Theorem 5.88 (ii), M_S is G-extending. So condition (ii) of Theorem 5.153 holds. Condition (iii) of Theorem 5.153 is satisfied by hypothesis. Therefore, $T_n(W)$ is right G-extending.

(iii) Clearly, W is right G-extending. Again block $T_n(W)$ as in the converse of part (ii). Observe that $M_{n-1}(W) = E(S_S)$. Since $M_S \cong cM_{n-1}(W)$ and $c = c^2$, M_S is injective. Hence $cT_{n-1}(W)$ is $T_{n-1}(W)$-ejective. Therefore, by part (ii), $T_n(W)$ is right G-extending. □

Example 5.156.

(i) Let S be any commutative subdirectly irreducible ring. Then S is extending and, by Corollary 2.57, S-ejective. Therefore, $T_2(S)$ is right G-extending by Corollary 5.155 (i). Since there exist commutative subdirectly irreducible rings that are not self-injective (Example 2.58 (ii)), the converse of Corollary 5.155 (iii) does not hold.

(ii) There exist self-injective rings S such that $T_2(S)$ is not right extending. Let $S = D/M^n$, where D is a Dedekind domain and M is a maximal ideal of D. By Corollary 5.155, $T_2(S)$ is right G-extending. For $S = \mathbb{Z}/\mathbb{Z}p^n$, $T_2(S)$ is not extending for $n > 1$. To see this, note that the right ideal generated by $\begin{bmatrix} 0 & \overline{q} \\ 0 & \overline{p} \end{bmatrix}$, where q is a prime such that $q \neq p$ and $\overline{p}, \overline{q} \in S$, is not essential in an idempotent generated right ideal of $T_2(S)$. Note that D/M^n is also a commutative subdirectly irreducible ring.

(iii) Let $R = N \oplus E$, where N and E are rings such that N_N is G-extending, E_E is injective, $A = E \trianglelefteq R$, and $S = E$ (the ring). Then by Theorem 5.153, T_T is G-extending and $\ell_R(A) = N$.

Example 5.157. Let $R = S = \mathbb{Z}_4$ and $A = 2\mathbb{Z}_4$. Then S_S is extending (in fact injective), $S/Z_2(S_S) = 0$, A_S is extending, and if D_S is a direct summand of A_S, then there is $f = f^2 \in R$ such that $D_S = fA_S$. Thus conditions (i)–(iii) of Theorem 5.153 are satisfied. But $T = \begin{bmatrix} \mathbb{Z}_4 & 2\mathbb{Z}_4 \\ 0 & \mathbb{Z}_4 \end{bmatrix}$ is not G-extending. Thus (iv) is not a superfluous condition in our results.

Corollary 5.158. *Assume that $Z(S_S) = 0$ and $Z_2(A_S) = A_S$. Then T_T is G-extending if and only if all of the following conditions are satisfied.*

(i) *If* D_S *is a direct summand of* A_S*, then there is an* $f = f^2 \in R$ *such that* $D_S \beta f A_S$.
(ii) A_S *and* S_S *are* G*-extending.*
(iii) *Let* $N = \ell_R(A)$*. There exists* $x = x^2 \in R$ *such that* $xR = N$ *and* xR_R *is* G*-extending.*

Proof. This result is a consequence of Theorem 5.153 and Proposition 2.61 (v). □

Our next result provides a wealth of examples of G-extending that are not extending.

Corollary 5.159. *Let* S *be a Dedekind domain that is not a field,* A_S *be a finitely generated torsion module, and* R *be a subring of* $\operatorname{End}(A_S)$ *such that for each direct summand* D_S *of* A_S*, there is an* $f = f^2 \in R$ *such that* $D_S \beta f A_S$ *(e.g.,* $R = \operatorname{End}(A_S)$*). Then* T_T *is* G*-extending, but* T_T *is not extending.*

Proof. By Theorem 5.147, A_S is a G-extending. Next, by Proposition 2.61 (v), A_S is S_S-ejective. Therefore, either Theorem 5.153 or Corollary 5.158 yield that T_T is G-extending. Since A_S is finitely generated, it is not injective. By Theorem 3.88, T_T is not extending. □

Recall from [Lam99, pp. 508–509] that if $\{X_i : i \in I\}$ is a set of representatives of the isomorphism classes of all simple S-modules and $C = \bigoplus_{i \in I} E(X_i)$, then C is called a canonical cogenerator of Mod-S.

Corollary 5.160. *Assume that* $\operatorname{soc}(S_S) = 0$ *and that* A_S *is semisimple module or a canonical cogenerator of Mod-*S*. Then* T_T *is* G*-extending if and only if* S_S *is* G*-extending and conditions (i) and (iv) of Theorem 5.153 are satisfied (e.g.,* $R = \operatorname{End}(A_S)$*).*

Proof. This result is a consequence of Theorem 5.153, Proposition 2.61 (vi), and Theorem 5.110 (ii). □

Our final theorem in this section complements Corollary 5.154 and Theorem 3.91 by providing a characterization of a large class of not nonsingular rings T for which T_T is extending.

Theorem 5.161. *Assume that* $Z(S_S) = 0$*, and either* $Z(A_S) = 0$ *or* $Z(A_S) \leq_{\mathrm{e}} A_S$*. If* T_T *is extending, then conditions* (i)–(iv) *are satisfied. Conversely, if conditions* (i)–(v) *are satisfied, then* T_T *is extending.*

(i) *If* D_S *is a direct summand of* A_S*, then there is an* $f = f^2 \in R$ *such that* $D_S = fA_S$.
(ii) S_S *is extending.*
(iii) A_S *is injective.*
(iv) *Let* $N = \ell_R(A)$*. There exists* $x = x^2 \in R$ *such that* $xR = N$ *and* xR_R *is extending.*
(v) $x \in \operatorname{Cen}(R)$.

Proof. Assume T_T is extending. By Theorem 3.88, conditions (i)–(iii) are satisfied and N_R is extending. By Lemma 5.151, $N = xR$, where $x = x^2 \in R$.

Conversely, assume conditions (i)–(v) hold. First, suppose that ${}_RA$ is faithful. By Proposition 5.148, $T = Z_2(T) \oplus T'$ (right ideal decomposition), where $\begin{bmatrix} 1 & 0 \\ 0 & 0 \end{bmatrix} T = \begin{bmatrix} R & A \\ 0 & 0 \end{bmatrix} = Z_2(T_T)$ and $T' = \begin{bmatrix} 0 & 0 \\ 0 & 1 \end{bmatrix} T = \begin{bmatrix} 0 & 0 \\ 0 & S \end{bmatrix} \cong T/Z_2(T_T)$. First, we show that $\begin{bmatrix} R & A \\ 0 & 0 \end{bmatrix}_T$ is extending. Let $X_T \le \begin{bmatrix} R & A \\ 0 & 0 \end{bmatrix}_T$. Clearly, $\begin{bmatrix} 0 & K \\ 0 & 0 \end{bmatrix} = X_T \cap \begin{bmatrix} 0 & A \\ 0 & 0 \end{bmatrix}_T \le_{\mathrm{e}} X_T$. Since A_S is injective, there exists $e = e^2 \in R$ such that $K_S \le_{\mathrm{e}} eA_S$. Let $0 \ne \begin{bmatrix} r & a \\ 0 & 0 \end{bmatrix} \in X$ and $x \in A$. Then $\begin{bmatrix} r & a \\ 0 & 0 \end{bmatrix} \begin{bmatrix} 0 & x \\ 0 & 0 \end{bmatrix} = \begin{bmatrix} 0 & rx \\ 0 & 0 \end{bmatrix}$, so $rx \in K$. Hence $erx = rx$, for all $x \in A$, whence $er = r$. Since $\begin{bmatrix} 0 & K \\ 0 & 0 \end{bmatrix} = X \begin{bmatrix} 0 & 0 \\ 0 & 1 \end{bmatrix}$, $X \subseteq \begin{bmatrix} e & 0 \\ 0 & 0 \end{bmatrix} T = \begin{bmatrix} eR & eA \\ 0 & 0 \end{bmatrix}$. Clearly, $\begin{bmatrix} 0 & eA \\ 0 & 0 \end{bmatrix} = \begin{bmatrix} e & 0 \\ 0 & 0 \end{bmatrix} \begin{bmatrix} 0 & A \\ 0 & 0 \end{bmatrix}_T \le_{\mathrm{e}} \begin{bmatrix} e & 0 \\ 0 & 0 \end{bmatrix} T_T$. Let $0 \ne \begin{bmatrix} 0 & ea \\ 0 & 0 \end{bmatrix} \in \begin{bmatrix} 0 & eA \\ 0 & 0 \end{bmatrix}$. Then there exists $s \in S$ such that $0 \ne \begin{bmatrix} 0 & ea \\ 0 & 0 \end{bmatrix} \begin{bmatrix} 0 & 0 \\ 0 & s \end{bmatrix} = \begin{bmatrix} 0 & eas \\ 0 & 0 \end{bmatrix} \in \begin{bmatrix} 0 & K \\ 0 & 0 \end{bmatrix}$. So $\begin{bmatrix} 0 & K \\ 0 & 0 \end{bmatrix}_T \le_e \begin{bmatrix} e & 0 \\ 0 & 0 \end{bmatrix} T_T$. Hence $X_T \le_{\mathrm{e}} \begin{bmatrix} e & 0 \\ 0 & 0 \end{bmatrix} T_T$. Therefore, $\begin{bmatrix} R & A \\ 0 & 0 \end{bmatrix}_T$ is extending. Since S_S is extending, $\begin{bmatrix} 0 & 0 \\ 0 & S \end{bmatrix}$ is extending. So we only need to show that $\begin{bmatrix} R & A \\ 0 & 0 \end{bmatrix}$ is $\begin{bmatrix} 0 & 0 \\ 0 & S \end{bmatrix}$-injective, by Exercise 3.12 (i). Let $\begin{bmatrix} 0 & 0 \\ 0 & V \end{bmatrix}_T \le \begin{bmatrix} 0 & 0 \\ 0 & S \end{bmatrix}_T$ and $g : \begin{bmatrix} 0 & 0 \\ 0 & V \end{bmatrix} \to \begin{bmatrix} R & A \\ 0 & 0 \end{bmatrix}$ be a T-homomorphism. Observe that if $g\left(\begin{bmatrix} 0 & 0 \\ 0 & v \end{bmatrix}\right) = \begin{bmatrix} r & a \\ 0 & 0 \end{bmatrix}$, then

$$\begin{bmatrix} r & a \\ 0 & 0 \end{bmatrix} = g\left(\begin{bmatrix} 0 & 0 \\ 0 & v \end{bmatrix} \begin{bmatrix} 0 & 0 \\ 0 & 1 \end{bmatrix}\right) = \begin{bmatrix} r & a \\ 0 & 0 \end{bmatrix} \begin{bmatrix} 0 & 0 \\ 0 & 1 \end{bmatrix} = \begin{bmatrix} 0 & a \\ 0 & 0 \end{bmatrix}.$$

Hence $g\left(\begin{bmatrix} 0 & 0 \\ 0 & V \end{bmatrix}\right) \subseteq \begin{bmatrix} 0 & A \\ 0 & 0 \end{bmatrix}$.

Define the S-homomorphism $h : V \to A$ by $h(v) = a$, where a is the $(1, 2)$-entry in $g\left(\begin{bmatrix} 0 & 0 \\ 0 & v \end{bmatrix}\right)$. Since A_S is injective, there is an S-homomorphism $\overline{h} : S \to A$ such that $\overline{h}|_V = h$. Now define the T-homomorphism $\overline{g} : \begin{bmatrix} 0 & 0 \\ 0 & S \end{bmatrix} \to \begin{bmatrix} R & A \\ 0 & 0 \end{bmatrix}$ by $\overline{g}\left(\begin{bmatrix} 0 & 0 \\ 0 & s \end{bmatrix}\right) = \begin{bmatrix} 0 & \overline{h}(s) \\ 0 & 0 \end{bmatrix}$. Then the restriction of $\overline{g}$ to $\begin{bmatrix} 0 & 0 \\ 0 & V \end{bmatrix}$ is g. So $\begin{bmatrix} R & A \\ 0 & 0 \end{bmatrix}$ is $\begin{bmatrix} 0 & 0 \\ 0 & S \end{bmatrix}$-injective. Therefore, T_T is extending when ${}_RA$ is faithful.

Now assume $N = \ell_R(A) \neq 0$. A proof similar to that used for the converse of Theorem 5.153 yields that $T = \begin{bmatrix} x & 0 \\ 0 & 0 \end{bmatrix} T \oplus B$, where $\begin{bmatrix} x & 0 \\ 0 & 0 \end{bmatrix} T_T$ and $B = \begin{bmatrix} (1-x)R & A \\ 0 & S \end{bmatrix}_T$ are extending and B_T is $\begin{bmatrix} x & 0 \\ 0 & 0 \end{bmatrix} T_T$-injective. By Lemma 5.150, $\begin{bmatrix} x & 0 \\ 0 & 0 \end{bmatrix} T_T$ is B_T-injective. By Exercise 3.12 (ii), T_T is extending. □

Note that if in Theorem 5.161 condition (iv) holds and R is semiprime, then condition (v) also holds.

Corollary 5.162. *Assume that $Z(S_S) = 0$, ${}_RA$ is faithful, and either $Z(A_S) = 0$ or $Z(A_S) \leq_e A_S$. Then T_T is extending if and only if conditions* (i)–(iii) *of Theorem 5.161 are satisfied.*

In our next result $\widehat{E}(X_S)$ denotes the quasicontinuous hull (also called π-injective hull) of X_S (see [JJ78]).

Corollary 5.163. *Let $Z(S_S) = 0$, X a nonsemisimple S-module, and $L_S \leq_e X_S$. Take $A_S = E(X/L)$ ($A_S = \widehat{E}(X/L)$) and R a subring of* End (A_S). *Then T_T is extending* (G-*extending*) *if and only if S_S is extending* (G-*extending*) *and condition* (i) *of Theorem* 5.161 (*Corollary* 5.158) *is satisfied.*

Proof. This result is a consequence of Corollaries 5.158 and 5.162. □

Example 5.164. Let S_S be nonsingular and extending and M a right S-module. Take $A_S = E(E(M_S)/M_S)$ and let R be a subring of End (A_S) such that if D_S is a direct summand of A_S, then there is an $f = f^2 \in R$ such that $D_S = fA_S$. Then T_T is extending by Corollary 5.162. If S is right hereditary, then $E(E(M_S)/M_S) = E(M_S)/M_S$. In particular, let D be a Dedekind domain with F its field of fractions. Then $T = \begin{bmatrix} \text{End}\,(F/D) & F/D \\ 0 & D \end{bmatrix}$ is such a ring.

Corollary 5.165. *Assume S_S is a nonsingular extending ring, $E(A_S)$ is indecomposable, and R is the subring of* End $(E(A_S)$ *generated by the identity endomorphism. Then*

$$H = \begin{bmatrix} R & E(A_S) \\ 0 & S \end{bmatrix}$$

is minimal among the right essential right extending overrings of T (i.e., H is a right extending right ring hull of T, see [BPR06]).

Proof. This result is an immediate consequence of Corollary 5.162. □

Example 5.166. For parts (i) and (ii), let R be as indicated in Corollary 5.165.

(i) Let D be a Dedekind domain. Take $T = \begin{bmatrix} R & D/P^n \\ 0 & D \end{bmatrix}$, where P is a maximal ideal. Then $H = \begin{bmatrix} R & E(D/P^n) \\ 0 & D \end{bmatrix}$. Note that T_T is G-extending by Corollary 5.159.

(ii) Let S be a right Ore domain and $Q(S)$ its classical right ring of quotients. Take $T = \begin{bmatrix} R & S \\ 0 & S \end{bmatrix}$. Then $H = \begin{bmatrix} R & Q(S) \\ 0 & S \end{bmatrix}$.

(iii) Let R, A, and S be as in Example 5.156 (iii) with N_N extending and either $Z(E_E) = 0$ or $Z(E_E) \leq_e E_E$. Then, by Theorem 5.161, T_T is extending and $\ell_R(A) = N$.

The next result and example make a connection to operator theory. Recall that an *AW^*-algebra* is a C^*-algebra that is a Baer ring. A commutative AW^*-algebra is mapped isometrically and $*$-isometrically onto $C(T)$, the $*$-algebra of continuous, complex-valued functions on a compact Hausdorff extremely disconnected space T vanishing at infinity (see [B72]). A ring S is *right cononsingular* if $\ell_S(X) = 0$ implies $X_S \leq_e S_S$, where X is a right ideal of S. Note that every commutative AW^*-algebra is right (and left) cononsingular. By [CK80, Thm. 2.1], a commutative C^*-algebra is extending if and only if it is an AW^*-algebra.

Corollary 5.167. *Let S be an AW^*-algebra that is right cononsingular, R a subring of* End(A_S)*, and either $Z(A_S) \leq_e A_S$, or $Z(A_S) = 0$. Then T_T is extending* (G-*extending*) *if and only if A_S is injective* (G-*extending*) *and condition* (i) *of Theorem* 5.161 (*Corollary* 5.158) *is satisfied.*

Proof. Since S is a Baer ring, $Z(S_S) = 0$. By [CK80, Thm. 2.1], S_S is extending. The result is now a consequence of Corollary 5.158 and Theorem 5.161. □

Example 5.168. Let S be a commutative AW^*-algebra, A_S a simple module, and R the field of complex numbers. Then T is a Banach algebra that is right G-extending. If S is (von Neumann) regular, then A_S is injective (since S is a V-ring); so T_T is extending. In fact, T_T is extending for all simple A in Mod-S if and only if S is regular.

Recall that a ring S is said to be a right SI-*ring* (SC-*ring*) if every singular right S-module is injective (continuous) (see [Goo72] and [RY90]).

Corollary 5.169. *Let S be a right* SI-*ring* (SC-*ring*), *$A = Z(A_S)$, and $_RA$ be faithful. Then T_T is extending* (G-*extending*) *if and only if S_S is extending* (G-*extending*) *and condition* (i) *of Theorem* 5.161 (*Corollary* 5.158) *is satisfied.*

Proof. Note that if S is a right SI-ring, then $Z(S_S) = 0$. Now the proof follows from Corollaries 5.158 and 5.162. □

We close this section with the following result which characterizes right extending right SI-rings.

Corollary 5.170. *S is a right extending right* SI*-ring if and only if T_T is extending for all singular modules A_S with $R = \text{End}(A_S)$.*

Proof. This is a consequence of Theorem 3.88 and Corollary 5.169. □

Exercises

5.18. Write out the proof of Proposition 5.145.

5.19. Prove that a right SI-ring is right nonsingular.

Remarks. C_{12}-modules were introduced and investigated by Smith and Tercan [ST93]. Theorem 5.15 appeared in [ST93]. Theorems 5.8 and 5.12 are due to Takıl and Tercan [TT09]. The gap in the proof of Theorem 5.8 was fixed and its weak C_{12} version was obtained by Kara and Tercan [KaT16]. Examples 5.38, 5.39 and Theorem 5.40 were included in Kara and Tercan [KaT16], which provide counterexamples to a question posed in [Er99].

Recall that P_1 condition is a proper generalization of extending property. Modules satisfying P_1 condition are also known as self-c-injective modules in the terminology of [SCS00]. To this end, Theorems 5.48, 5.51, and 5.55 were proved by Santa Clara and Smith [SCS00]. CLS-modules were introduced and investigated by Tercan [Ter95]. Amongst other results Theorems 5.63 and 5.64 are due to Tercan [Ter95].

E. Akalan, G.F. Birkenmeier, and A. Tercan introduced G-extending concept in [ABT09]. They considered the decomposition theory for G-extending modules and characterized Abelian groups, which are G-extending. In contrast to the characterization of extending Abelian groups, they obtained that all finitely generated Abelian groups are G-extending. They proved that a minimal cogenerator for R-Mod is G-extending, but not, in general, extending. It was also shown that if a module A is G-extending, then so is its rational hull. To this end, Theorems 5.82, 5.88, 5.96, 5.106, 5.110, 5.120 and Corollary 5.121 were obtained by Akalan, Birkenmeier, and Tercan [ABT09]. C_3 version of Proposition 5.100 is due to Wu and Wang [WuW12]. G-extending condition under various ring extensions was investigated in [ABT12]. Corollary 5.126 was sugguested by Jae K. Park and Syed T. Rizvi, thanks to personel communication. Theorems 5.122, 5.125, 5.128, and 5.130 were obtained by Akalan, Birkenmeier, and Tercan [ABT12]. Moreover, they characterized G-extending modules over a Dedekind domain and used the G-extending condition to characterize modules over a principal ideal domain, whose pure submodules are direct summand in [ABT11]. The latter result is presented as Theorem 5.137. Theorems 5.153, 5.161, and Corollary 5.154 are due to Akalan, Birkenmeier and Tercan [AkBT12], which improve and generalize the existing results on right extending generalized triangular matrix rings.

Chapter 6

Dual Goldie and EC-complement Versions of the Extending Property

It is well known that lifting modules are dual to extending modules, and there are plentiful of works on this subject. In this chapter, we consider modules with a property that is dual to Goldie extending property. Such modules are called Goldie*-supplemented modules. We investigate this new class of modules and the class of H-supplemented modules. Using another equivalence relation which is based on the radical of a module, we introduce the classes of Goldie-rad-supplemented and rad-H-supplemented modules. Finally, we discuss extending and C_{11} properties on ec-complement submodules and obtain fundamental properties of these classes of modules.

6.1 Goldie*-supplemented modules

In this section, we will study modules with property that is dual to Goldie extending property. Such modules will be called Goldie*-supplemented modules. This new class of modules and the class of H-supplemented modules are among various well-known classes of modules related to the class of lifting (dual to extending) modules.

Recall that a module A is H-*supplemented* if for every submodule N there is a direct summand D of A such that $N + X = A$ if and only if $D + X = A$ [MM90, p. 95]. In [MM90, p. 56, Def. 4.4], a module A is called *supplemented* if for any two submodules N and K with $N + K = A$, K contains a supplement of N. This definition of supplemented is equivalent to the more recent terminology of amply supplemented (see [CLVW06, p. 237] or [Wis91, p. 54]).

Recall from [CLVW06] that a submodule $N \leq A$ is called a *supplement* (*weak supplement*) of a submodule L of A if $N + L = A$ and $N \cap L \leq_{\mathrm{s}} N$ ($N \cap L \leq_{\mathrm{s}} A$). The module A is called (*weakly*) *supplemented* if every submodule N of A has a (weak) supplement. A is called *lifting* if every submodule N of A contains a direct summand D of A such that $N/D \leq_{\mathrm{s}} A/D$. A submodule N of A has *ample supplements* in A if for every $L \leq A$ with $A = N + L$, there is a supplement L' of N with $L' \leq L$. The module A is called *amply supplemented* if all submodules have ample supplements in A. A is called $\oplus$*-supplemented* if every submodule of A has a supplement that is a direct summand. A supplemented module A is called *strongly* $\oplus$*-supplemented* if every supplement submodule of A is a direct summand. Other terminology and notation can be found in [MM90, CLVW06, Wis91].

From the previous chapter, for submodules X, Y of A, $X\beta Y$ if $X \cap Y \leq_e X$ and $X \cap Y \leq_e Y$. Since this relation has proved fruitful in various applications such as characterizing when a pure subgroup of an Abelian group is a direct summand, it is natural to consider its dual relation.

Definition 6.1. We define the relation "β^*" on the set of submodules of A by $X\beta^*Y$ if $(X+Y)/X \leq_s A/X$ and $(X+Y)/Y \leq_s A/Y$.

Let us present the basic properties of the β^* relation on the set of submodules of A. These properties will be used in our ensuing results.

Lemma 6.2. *β^* is an equivalence relation.*

Proof. The reflexivity and symmetry properties are clear. For transitivity, assume $X\beta^*Y$ and $Y\beta^*Z$. So $(X+Y)/X \leq_s A/X$ and $(X+Y)/Y \leq_s A/Y$ and $(Y+Z)/Y \leq_s A/Y$ and $(Y+Z)/Z \leq_s A/Z$. Assume $B/X \leq A/X$ such that $(X+Z)/X + B/X = A/X$. Then $(X+Z+B)/X = (Z+B)/X = A/X$, so $Z+B = A$. Hence $A/Y = (Z+Y+B)/Y = (Z+Y)/Y + (Y+B)/Y$. Since $(Z+Y)/Y \leq_s A/Y$, $(Y+B)/Y = A/Y$. Hence $Y+B = A$. Then $A/X = (Y+B)/X = (X+Y)/X + B/X$. Since $(X+Y)/X \leq_s A/X$, $B/X = A/X$. Therefore $B = A$, so $(X+Z)/X \leq_s A/X$. Similarly, $(X+Z)/Z \leq_s A/Z$. □

Observe that the zero submodule is β^* equivalent to any small submodule. Also, note that two submodules may be isomorphic, but not β^* equivalent. For example, let F be a field and $R = \begin{bmatrix} F & F \\ 0 & F \end{bmatrix}$, $X = \begin{bmatrix} 0 & F \\ 0 & 0 \end{bmatrix}$ and $Y = \begin{bmatrix} 0 & 0 \\ 0 & F \end{bmatrix}$. Then X is R-isomorphic to Y, but X is not β^* equivalent to Y, as can be seen by taking $B = \begin{bmatrix} F & F \\ 0 & 0 \end{bmatrix}$ in item (iii) of the following result. Moreover, if $A = \mathbb{Z}_{\mathbb{Z}}$, then $m\mathbb{Z}\beta^*n\mathbb{Z}$ if and only if m and n are divisible by the same primes (Exercise 6.1).

Given submodules $K \leq L \leq A$, the inclusion $K \subseteq L$ is called *cosmall* in A if $L/K \leq_s A/K$, and then denoted is by $K \overset{cs}{\hookrightarrow} L$ (see [CLVW06, p. 20]). We have the following result.

Theorem 6.3. *Let $X, Y \leq A$. The following statements are equivalent.*

(i) $X\beta^*Y$.
(ii) $X \overset{cs}{\hookrightarrow} X+Y$ *and* $Y \overset{cs}{\hookrightarrow} X+Y$.
(iii) *For each $B \leq {}_RA$ such that $X+Y+B = A$ we have $X+B = A$ and $Y+B = A$.*

(iv) *If $K \leq A$ with $X+K=A$, then $Y+K=A$, and if $H \leq A$ with $Y+H=A$, then $X+H=A$.*

Proof. (i) $\Rightarrow$ (ii) Clear.

(ii) $\Rightarrow$ (iii) Let $B \leq A$ such that $X+Y+B=A$. Then $(X+Y)/Y+(Y+B)/Y = A/Y$ implies that $(Y+B)/Y = A/Y$. Hence $Y+B=A$. Similarly, $X+B=A$.

(iii) $\Leftrightarrow$ (iv) Let $K \leq A$ such that $X+K=A$. Then $X+K+Y=A$. By (iii), $Y+K=A$. Let $H \leq A$ such that $Y+H=A$. Then $X+Y+H=A$. By (iii), $X+H=A$. Conversely, assume $X+Y+B=A$. Then $X+(Y+B)=A$, so $Y+(Y+B)=A$. Hence $Y+B=A$. Similarly, $X+B=A$.

(iii) $\Rightarrow$ (i) Let $B/Y \leq A/Y$ such that $(X+Y)/Y+B/Y=A/Y$. Then $X+Y+B=A$. Hence $Y+B=B=A$ (since $Y \subseteq B$), so $(X+Y)/Y \leq_{\mathrm{s}} A/Y$. Similarly, $(X+Y/X) \leq_{\mathrm{s}} A/X$. □

Corollary 6.4. *Let $X, Y \leq A$ such that $X \subseteq Y+B$ and $Y \subseteq X+C$, where $B, C \leq_{\mathrm{s}} A$, then $X\beta^*Y$.*

Proof. Let $X+Y+K=A$ for some $K \leq A$. Then $(Y+B)+Y+K=A$. So $Y+B+K=A$, whence $Y+K=A$. Similarly, $X+K=A$. □

Note that there are modules A with $K, X, Y \leq A$ such that $A = X+K = Y+K$, but X is not β^* related to Y. For example, take $R=A=\mathbb{Z}$, $K=3\mathbb{Z}$, $X=2\mathbb{Z}$ and $Y=5\mathbb{Z}$.

Proposition 6.5. *If $X, Y, K \leq A$ are such that $A=X+K=Y+K$, $Y\cap K \subseteq X \cap K$ and $Y \overset{\mathrm{cs}}{\hookrightarrow} X+Y$ (i.e., $(X+Y)/Y \leq_{\mathrm{s}} A/Y$), then $X \overset{cs}{\hookrightarrow} X+Y$ (i.e., $(X+Y)/X \leq_{\mathrm{s}} A/X$), so $X\beta^*Y$.*

Proof. There exist canonical isomorphisms

$$K/(Y\cap K) \overset{\overset{\theta}{\leftarrow}}{\simeq} (Y+K)/Y = A/Y, \quad K/(X\cap K) \overset{\overset{\varphi}{\rightarrow}}{\simeq} (X+K)/X = A/X$$

and

$$f : K/(Y\cap K) \overset{\mathrm{epi}}{\to} K/(X\cap K),$$

defined by $f(k+Y\cap K) = k+X\cap K$. Define $h : A/Y \to A/X$ by $h = \varphi f \theta$. Let $a \in A$. Then $a = y+k$ for some $y \in Y$ and $k \in K$. Hence $\theta(a+Y) = \theta(y+k+Y) = k+Y\cap K$, $f(k+Y\cap K) = k+X\cap K$, and $\varphi(k+X\cap K) = k+X = k+x+X$ for any $x \in X$. Thus $h(a+Y) = k+x+X$.

We claim that $h((X+Y)/Y) = (X+Y)/X$. To see that $h((X+Y)/Y) \subseteq (X+Y)/X$, let $x+y+Y \in (X+Y)/Y$. Note that one can write $y_1+k = x+y$, where $y_1 \in Y$ and $k \in K$. Hence $h(x+y+Y) = x+y-y_1+X \in (X+Y)/X$. Now assume $x_2+y_2+X \in (X+Y)/X$. There exists $x_3 \in X$ and $k_3 \in K$ such that $y_2 = x_3+k_3$. Hence $x_2+y_2+X = k_3+X$. Now $k_3 = -x_3+y_2 \in X+Y$. So $h(k_3+Y) = \varphi f\theta(k_3+Y) = \varphi f(k_3+Y\cap K) = \varphi(k_3+X\cap K) = k_3+X = x_2+y_2+X$.

Thus $h((X+Y)/Y) = (X+Y)/X$. Since $(X+Y)/Y \leq_s A/Y$ we conclude that $(X+Y)/X \leq_s A/X$, by [CLVW06, p. 11, 2.2 (5)]. □

Theorem 6.6. *Let $X, Y \leq A$ such that $X\beta^*Y$. Then*

(i) *$X \leq_s A$ if and only if $Y \leq_s A$.*

(ii) *X has a (weak) supplement C in A if and only C is a (weak) supplement for Y.*

Proof. (i) ($\Rightarrow$) Assume $X \leq_s A$. Let $K \leq A$ such that $Y + K = A$. Then $X + Y + K = A$. By Theorem 6.3, $X + K = A$. Since $X \leq_s A$, $K = A$, whence $Y \leq_s A$.

($\Leftarrow$) The converse is true because β^* is symmetric (Lemma 6.2).

(ii) Assume C is a supplement for X. Then $A = X + C = X + Y + C$. By Theorem 6.3, $Y + C = A$. Assume $K \subseteq C$ and $Y + K = A$. Then $X + Y + K = A$. By Theorem 6.3, $X + K = A$. By the minimality of C, $K = C$. Thus C is a supplement for Y. The converse is true because β^* is symmetric (Lemma 6.2). Therefore, X has a supplement C if and only if C is a supplement for Y.

Now assume C is a weak supplement for X. Then $X + C = A$ and $X \cap C \leq_s A$. By Theorem 6.3, $Y + C = A$. We need to show that $Y \cap C \leq_s A$. Let $K \leq A$ such that $Y \cap C + K = A$. Since $Y \cap C \subseteq Y$, $Y + K = A$ and $C + K = A$. By Theorem 6.3, $X + K = A$. Since $Y \cap C \subseteq C$, the modular law yields that $C = C \cap A = (Y \cap C) + (C \cap K)$. Then

$$A = Y + C = Y + Y \cap C + C \cap K = Y + C \cap K$$

and hence $X + Y + C \cap K = A$. By Theorem 6.3, $X + C \cap K = A$, so

$$K = K \cap A = K \cap (C \cap K + X) = (C \cap K) + X \cap K,$$

again by the modular law. Now

$$A = C + K = (Y \cap C) + (C \cap K) + (X \cap K) \subseteq C + X \cap K \subseteq A.$$

Hence $A = C + (X \cap K)$. By the modular law,

$$X = X \cap A = X \cap ((X \cap K) + C) = X \cap K + X \cap C.$$

Thus $A = X + K = X \cap C + X \cap K + K = X \cap K + K = K$, since $X \cap C \leq_s A$. Therefore, $Y \cap C \leq_s A$. The converse holds by the symmetry of the β^* relation. □

Corollary 6.7. *Let $X, Y \leq A$ be such that $X \subseteq Y$ and X has a weak supplement C in A. Then $X\beta^*Y$ if and only if $Y \cap C \leq_s A$.*

Proof. Assume $X\beta^*Y$. By Theorem 6.6, C is a weak supplement of Y. Hence $Y \cap C \leq_s A$. Conversely, assume $Y \cap C \leq_s A$. Let $K \leq A$ such that $X + Y + K = A$. Since $X \subseteq Y$, $Y + K = A$. Since $X + C = A$ and $X \subseteq Y$, the modular law yields $Y = X + Y \cap C$. Then

$$A = Y + K = X + Y \cap C + K = X + K,$$

since $Y \cap C \leq_s A$. By Theorem 6.3, $X\beta^*Y$. Note that from the hypothesis and the modular law, $Y = X + Y \cap C$ and $X \cap C \leq_s A$. □

Observe that for any minimal right ideal X of R there is a direct summand D in R such that $X\beta^*D$. To see this, note that either $X = eR$ for some idempotent (in which case X is a direct summand), or $X^2 = 0$ (in which case $X\beta^*0$).

Corollary 6.8. *Let $A = C \oplus D$ and $L, S, X \leq A$, where S is a weak supplement of L.*

(i) *If $X\beta^*S$, then $X/(X \cap L) \cong S/(S \cap L)$. If S is a supplement, there is a small cover $f : S \to X/(X \cap L)$. If X is projective, then there is an epimorphism $h : X \to S$. Finally, if S is projective, then there is a projective cover $g : S \to X$.*
(ii) *If $X\beta^*D$, then $X/X \cap C \cong D$, $X \cap C \leq_s C$ and there is a small cover $h : A/D \to A/X$. If A is H-supplemented, then for each X there is a direct summand D such that $X\beta^*D$, an epimorphism $f : X \to D$, and a small cover $h : A/D \to A/X$.*
(iii) *Assume $X \subseteq D$. Then $X\beta^*D$ if and only if $X = D$. In particular, $X\beta^*A$ if and only if $X = A$.*
(iv) *Assume $D \subseteq X$. Then $X\beta^*D$ if and only if $X \cap C \leq_s A$.*

Proof. (i) Since $X\beta^*S$, Theorem 6.3 yields $X/(X \cap L) \cong (X + L)/L = A/L = (S + L)/L \cong S/(S \cap L)$. The remainder of the proof of this part follows from properties of projective modules.

(ii) is a consequence of (i) and Theorem 6.6 which yields that C is a supplement of X. Hence $X \cap C \leq_s C$. Then there is a small cover $k : C \to C/(X \cap C)$. Consequently, there is a small cover $h : A/D \to A/X$.

(iii) ($\Rightarrow$) Assume $X\beta^*D$. Then $X+D+C = A$. By Theorem 6.3, $X \oplus C = A$. Thus $X = D$.

($\Leftarrow$) Since β^* is reflexive (Lemma 6.2), $X = D$ implies that $X\beta^*D$.

(iv) is a consequence of Corollary 6.7. □

Note that in Corollary 6.8 (ii), if $\operatorname{rad} A = 0$, then $X \cap C = 0$, so $A = C \oplus X$.

Proposition 6.9. *Let $f : A \to N$ be an epimorphism. Then*

(i) *If $X, Y \leq A$ are such that $X\beta^*Y$, then $f(X)\beta^*f(Y)$.*
(ii) *If $X, Y \leq N$ are such that $X\beta^*Y$, then $f^{-1}(X)\beta^*f^{-1}(Y)$.*
(iii) *If f is a small cover, $X \leq A$ and $K \leq N$ are such that $f(X)\beta^*K$, then $X\beta^*f^{-1}(K)$.*

Proof. (i) Assume that $f(X) + f(Y) + K = N$. Then $X + Y + f^{-1}(K) = A$. To see this, let $a \in A$. There exists $x \in X$, $y \in Y$, and $k \in K$ such that $f(a) = f(x) + f(y) + k$. Hence $f(a - x - y) = k$, so $a - x - y \in f^{-1}(K)$. Thus $a \in X + Y + f^{-1}(K)$, so $X + Y + f^{-1}(K) = A$. Hence $A = X + f^{-1}(K) = Y + f^{-1}(K)$. Consequently, $N = f(X) + K = f(Y) + K$. Therefore, by Theorem 6.3, $f(X)\beta^*f(Y)$.

(ii) Let $f^{-1}(X)+f^{-1}(Y)+H=A$. Then $X+Y+f(H)=N$, so, by Theorem 6.3, $X+f(H)=N=Y+f(H)$. Let $a\in A$. Then $f(a)=x+f(h)$ for some $x\in X$ and $h\in H$. Hence $f(a-h)=x$, and $a-h\in f^{-1}(X)$. Thus $a\in f^{-1}(X)+H$. We see that $A=f^{-1}(X)+H$. Similarly, $A=f^{-1}(Y)+H$. Therefore, by Theorem 6.3, $f^{-1}(X)\beta^* f^{-1}(Y)$.

(iii) Assume that $A=X+f^{-1}(K)+B$. Then $N=f(X)+K+f(B)$. Since $f(X)\beta^*K$, Theorem 6.3 yields $N=f(B)+f(X)$ and $N=f(B)+K$. Hence $N=f(B+X)=f(A)$ and $N=f(A)=f(B)+K$. Let $a\in A$. Since $N=f(B)+f(X)$, there exist $b\in B$, $x\in X$ such that $f(a)=f(b)+f(x)$. Hence $f(a-b-x)=0$. Then $a-b-x\in\ker f$ and $a\in B+X+\ker f$. So $A=B+X+\ker f$. Since $\ker f\leq_s A$, $A=B+X$.

Since $N=f(B)+K$, there exist $b\in B$, $k\in K$ such that $f(a)=f(b)+k$. Since f is an epimorphism, there exists $y\in A$ such that $f(y)=k$. Then $f(a-b-y)=0$. Hence $a-b-y\in\ker f$, thus $a\in\ker f+B+f^{-1}(K)$. So $A=\ker f+B+f^{-1}(K)$, and since $\ker f\leq_s A$, $A=B+f^{-1}(K)$. □

Proposition 6.10. *Let $X\leq A$ and K a maximal submodule of A.*

(i) *Let $C_1,C_2\leq A$ such that $C_1+C_2=A$, $C_2\neq A$ and $X\beta^*C_1$. Then $X\nsubseteq C_2$.*
(ii) *If $X\beta^*Y$ and $X\subseteq K$, then $Y\subseteq K$.*
(iii) *If $X\beta^*K$, then $X\subseteq K$. Hence, if $X\beta^*Y$, then $X\subseteq \operatorname{rad}A$ if and only if $Y\subseteq\operatorname{rad}A$.*
(iv) *If $X\beta^*K$ and $X+W=A$ with $X\cap W\leq_s A$, then $K=X+(K\cap W)$ and $K\cap W\leq_s A$.*

Proof. (i) Assume $X\subseteq C_2$. Then $C_1+X+C_2=A$, hence $C_1+C_2=A$ and $X+C_2=A$. But $X\subseteq C_2$ implies that $C_2=A$, a contradiction.

(ii) Assume $Y\nsubseteq K$. Then $Y+K=A$ and $Y+K+X=A$. Hence $K+X=A$. But $X\subseteq K$ implies that $K=A$, a contradiction.

(iii) follows from (ii) using the symmetry of β^* with K and X replacing X and Y, respectively.

(iv) follows from (iii) and Theorem 6.6. □

Note that from Proposition 6.10 (i), if $X,Y\leq A$ such that $X\subsetneqq A$ and $X\beta^*Y$, then $X+Y\neq A$.

Proposition 6.11. *Let $X_1,X_2,Y_1,Y_2\leq A$ such that $X_1\beta^*Y_1$ and $X_2\beta^*Y_2$. Then $(X_1+X_2)\beta^*(Y_1+Y_2)$ and $(X_1+Y_2)\beta^*(Y_1+X_2)$. In particular, $X_1+X_2=A$ if and only if $Y_1+Y_2=A$, and $X_1+Y_2=A$ if and only if $Y_1+X_2=A$.*

Proof. Let $K\leq A$ such that $X_1+X_2+Y_1+Y_2+K=A$. Then $X_2+Y_1+Y_2+K=A$ and $X_1+X_2+Y_2+K=A$, because $X_1\beta^*Y_1$. Moreover, $Y_1+Y_2+K=A$ and $X_1+X_2+K=A$, because $X_2\beta^*Y_2$. By Theorem 6.3, $(X_1+X_2)\beta^*(Y_1+Y_2)$. Using Lemma 6.2, we obtain $(X_1+Y_2)\beta^*(Y_1+X_2)$. By Corollary 6.8 (iii), $X_1+X_2=A$ if and only if $Y_1+Y_2=A$, and $X_1+Y_2=A$ if and only if $Y_1+X_2=A$. □

Corollary 6.12. *Let $X, Y \leq A$ and $J \leq_s A$. Then $X\beta^* Y$ if and only if $X\beta^*(Y+J)$.*

Proof. ($\Rightarrow$) This implication follows from Proposition 6.11 and the fact that $0\beta^* J$.

($\Leftarrow$) As above, $Y\beta^*(Y+J)$. Now the implication follows from the transitivity of the β^* relation. □

Corollary 6.13. *Let $X, Y_1, Y_2, \dots, Y_n \leq A$. If $X\beta^* Y_i$ for each i, then $X\beta^* Y$, where $Y = \sum_{i=1}^{n} Y_i$.*

Definition 6.14. Let $X \leq A$. We denote $X_{\beta^*} = \sum\{N \leq A : N\beta^* X\}$.

Observe that $\operatorname{rad} A = 0_{\beta^*}$ and that if $X \leq K$ then $X_{\beta^*} \leq K$, where K is a maximal submodule of A by Proposition 6.10. Moreover, if $X\beta^* Y$, then $X_{\beta^*} = Y_{\beta^*}$. Finally, if A is Noetherian, Corollary 6.13 yields that $X\beta^* X_{\beta^*}$. However, this is not true in general, as the following example shows.

Example 6.15. Proposition 6.11 can be extended to finite, but not infinite sums. Let $R = \mathbb{Z}$ and $A = \mathbb{Q}$. Then $\mathbb{Q} = \operatorname{rad}\mathbb{Q} = \sum_{m\in\mathbb{Z}} \mathbb{Z}\frac{1}{m}$ but each $\mathbb{Z}\frac{1}{m} \leq_s \mathbb{Q}$. Hence $\mathbb{Z}\frac{1}{m}\beta^* 0$ for each m. If Proposition 6.11 was true for even countably infinite sums, then $\mathbb{Q}\beta^* 0$, a contradiction since $\mathbb{Q}$ is not small in $\mathbb{Q}$.

Proposition 6.16. *Let A be a Noetherian module which is weakly supplemented and $X \leq A$. Then $X_{\beta^*} = X + \operatorname{rad} A$.*

Proof. By Corollary 6.13, $X\beta^* X_{\beta^*}$. The result is now a consequence of Corollary 6.7 and Corollary 6.12. □

Observe that, by Proposition 6.9, if $f : A \to N$ is an epimorphism and $X \leq A$, then $f(X_{\beta^*}) \leq [f(X)]_{\beta^*}$. Moreover, if f is a small cover, then $f(X_{\beta^*}) = [f(X)]_{\beta^*}$.

Proposition 6.17. *Let $S \leq A$ and $I \leq R$. If $X \leq R$ is such that $I^n \leq X \leq I$ for some positive integer n, then $IS\beta^* XS$. In particular, $I\beta^* X$.*

Proof. Clearly, the statement is true for $n = 1$. Let $n > 1$. Let $B \leq A$ be such that $IS + B = A$. Then $I^2S + IB = IA, \dots, I^nS + I^{n-1}B = I^{n-1}A$. Hence $A = IS+B \leq IA+B = I^2S+IB+B = I^2S+B \leq I^2A+B \leq \cdots \leq I^nS+B \leq A$. Thus $I^nS + B = A$. So $XS + B = A$. By Theorem 6.3, $IS\beta^* XS$. □

Example 6.18. Let $A = R = \mathbb{Z}$ and $K = p\mathbb{Z}$, for some prime p. Take $X = p^2\mathbb{Z}$. Then $X_{\beta^*} = K \neq X = X + \operatorname{rad} A$. Therefore the condition that X has a weak supplement is not superfluous in Proposition 6.10 (iv) and Proposition 6.16.

Now we use the β^* equivalence relation to define the class of G*-*lifting* modules and the class of G*-*supplemented* modules. Some basic properties, including behaviour with respect to direct sums and direct summands are presented for these classes. We locate these classes of modules between the class of lifting (which is dual to extending) modules and the class of supplemented modules.

Definition 6.19.

(i) We say A is *Goldie*-lifting*, or *G*-lifting*, if for each $X \leq A$ there exists a direct summand D of A such that $X\beta^* D$.

(ii) We say A is *Goldie*-supplemented*, or G*-*supplemented*, if for each $X \leq A$ there exists a supplement submodule S of A such that $X\beta^* S$.

Theorem 6.20. *A is* G*-*supplemented* (G*-*lifting*) *if and only if for each $X \leq A$ there exists supplement S (direct summand D) and a small submodule H of A such that $X + H = S + H = X + S$ ($X + H = D + H = X + D$).*

Proof. Assume that A is G*-supplemented. There exists a supplement S such that $X\beta^* S$. Hence there exists $W \leq A$ such that $S + W = A$ and $S \cap W \leq_s S$. By Proposition 6.11, $X\beta^*(X + S)$ and $S\beta^*(X + S)$. By Theorem 6.6, W is a weak supplement for S, X, and $X + S$. By the modular law, $X + H = S + H = X + S$, where $H = (X + S) \cap W \leq_s A$. The converse follows from Corollary 6.4. The proof is similar for G*-lifting. □

Corollary 6.21.

(i) *If for each $X \leq A$ there exist a supplement S and $H \leq_s A$ such that $X = S + H$, then A is* G*-*supplemented. The converse holds if A is also distributive.*

(ii) *Let A be* G*-*supplemented and $X \leq A$ such that* $\operatorname{rad} A \leq X$. *Then $X = S + H$, where S is a supplement and $H \leq_s A$.*

Proof. (i) By Theorem 6.20, the hypothesis implies that A is G*-supplemented. Assume that A is G*-supplemented and distributive. Let $X \leq A$. Then there are $S, L \leq A$ such that $X\beta^* S$, $S + L = A$, and $S \cap L \leq_s S$. By Theorem 6.3, $X + L = A$. So $S = S \cap (X + L) = S \cap X + S \cap L = S \cap X$. Hence $S \leq X$. By Theorem 6.6, L is a weak supplement of X, so $X \cap L \leq_s A$. Thus $X = X \cap (S + L) = S + H$, where $H = X \cap L$.

(ii) follows from Theorem 6.20. □

Corollary 6.22. *Assume* $\operatorname{rad} A$ *is small in A (e.g., A is finitely generated). Then A is* G*-*supplemented if and only if for each $X \leq A$ there exists a supplement submodule S of A such that $S + \operatorname{rad} A = X + \operatorname{rad} A$.*

Proof. This result is a consequence of Theorem 6.20 and Corollary 6.4. □

The following theorem generalizes and extends Theorem 3.16 of [KK07]. To see this, observe that our result holds if every submodule of A is projection invariant.

Theorem 6.23. *Assume that $A = B \oplus C$, where $B = Ab$, $C = Ac$, $\{b, c\}$ is a set of orthogonal idempotents of* $\operatorname{End}({}_R A)$ *and $U = Ub + Uc$ for each $U \leq A$ (e.g., each U is fully invariant). Then A is* G*-*supplemented* (G*-*lifting*) *if and only if B and C are* G*-*supplemented* (G*-*lifting*).

Proof. (⇒) Let $X \leq B$. Then there exist $S, L \leq A$ such that $S+L = A$, $S \cap L \leq_s S$ and $X\beta^* S$. We claim that $X\beta^* Sb$ as submodules of B. To see this, suppose that $X + Sb + K = B$ for some $K \leq B$. Then $X + S + K + L = A$. By Theorem 6.3, $X + K + L = A$. Then $X + K + Lb = B$ and $Lc = C$. Thus $Sc \leq Lc$, so $Sb + L = A$. Since S is a supplement of L in A, $S = Sb$. Therefore $X\beta^* Sb$. Now $Sb + Lb = B$ and $Lb \cap Sb \leq L \cap S \leq_s S = Sb$. Hence Sb is a supplement in B. Therefore, B is G*-supplemented. Similarly, C is G*-supplemented. The proof for A being G*-lifting is similar.

(⇐) Let $U \leq A$, $U_1 = Ub$, and $U_2 = Uc$. There exist $L_1, S_1 \leq B$ such that $U_1\beta^* S_1$, $L_1 + S_1 = B$, and $L_1 \cap S_1 \leq_s S_1$. Likewise, there exist $L_2, S_2 \leq C$ such that $U_2\beta^* S_2$, $L_2 + S_2 = C$, and $L_2 \cap S_2 \leq_s S_2$. By Proposition 6.11, $U\beta^*(S_1 + S_2)$. Moreover, $S_1 + S_2 + L_1 + L_2 = A$ and $(S_1 + S_2) \cap (L_1 + L_2) = (S_1 \cap L_1) + (S_2 \cap L_2)$. Assume that $(S_1 + S_2) \cap (L_1 + L_2) + K = S_1 + S_2$ for some $K \leq A$. Then $(S_1 \cap L_1) + (S_2 \cap L_2) + Kb + Kc = S_1 + S_2$. Hence $(S_1 \cap L_1) + Kb = S_1$. But $S_1 \cap L_1 \leq_s S_1$, so $Kb = S_1$. Similarly, $Kc = S_2$. Hence $S_1 + S_2$ is a supplement in A. Therefore A is G*-supplemented. The proof for B and C being G*-lifting is similar. □

Theorem 6.24. *Let A be a module and consider the following conditions.*

(i) *A is lifting.*
(ii) *A is* G*-*lifting*
(iii) *A is* H-*supplemented.*
(iv) *A is* G*-*supplemented.*
(v) *A is supplemented.*

Then (i) ⇒ (ii) ⇔ (iii) ⇒ (iv) ⇒ (v).

Proof. (i) ⇒ (ii) This implication follows from [CLVW06, p. 266, 22.3 (a) ⇔ (b)] and Corollary 6.8 (iv).

(ii) ⇔ (iii) This equivalence follows from Theorem 6.3.

(ii) ⇒ (iv) This implication follows from the fact that every direct summand is a supplement.

(iv) ⇒ (v) Let X be a submodule of A. Then $X\beta^* S$, where S is a supplement. So there exists $W \leq A$ such that S is a supplement of W. There exists a supplement T such that $W\beta^* T$. Hence S is a supplement of T, by Theorem 6.6. By [CLVW06, p. 234, 20.4 (9)], T is a supplement of S. By Theorem 6.6, T is a supplement of X. Therefore A is supplemented. □

Proposition 6.25.

(i) *Let A be* G*-*supplemented and K a maximal submodule of A. Then, $K = S + (K \cap T)$, where S is a supplement of T, T is a supplement of K, T is a local module and $K\beta^* S$.*
(ii) *Let A be* G*-*lifting and K a maximal submodule of A. There exist $C, D \leq A$ such that $A = C \oplus D$, $K\beta^* D$, $K = D \oplus (K \cap C)$, and C is a local module.*

Proof. (i) Since A is G*-supplemented, there exists a supplement S such that $K\beta^* S$. By Theorem 6.24, K has a supplement T. By Theorem 6.6, T is a supplement of S. By [CLVW06, p. 234, 20.4 (9)], S is a supplement of T. Proposition

6.10 yields that $K = S + (K \cap T)$. Then, $K \cap T$ is a maximal submodule of T. Let $t \in T$ such that $t \notin K$. Then, $K + Rt = A$. Since T is a supplement of K, $T = Rt$. Therefore, T is a local module.

(ii) Since A is G*-lifting there exist $C, D \leq A$ such that $A = C \oplus D$ and $K\beta^*D$. By Theorem 6.6, C is a supplement of K. By Proposition 6.10, $K = D \oplus (K \cap C)$. As in part (i), C is a local module. □

Using the previous result, we obtain a structure theorem for G*-lifting Noetherian modules.

Theorem 6.26. *Let A be a Noetherian module such that each submodule is projection invariant. If A is* G*-*lifting, then A is a finite direct sum of local modules.*

Proof. Since A is Noetherian, it is a finite direct sum of indecomposable Noetherian modules. By [KK07, Cor. 2.4], each indecomposable direct summand of A is G*-lifting. By Proposition 6.25 (ii), each indecomposable direct summand of A is local. □

Corollary 6.27.

(i) *A is a lifting module if and only if A is amply supplemented and strongly $\oplus$-supplemented.*

(ii) *If A is* G*-*lifting, then A is* G*-*supplemented and $\oplus$-supplemented.*

(iii) *If A is* G*-*supplemented and strongly $\oplus$-supplemented, then A is* G*-*lifting.*

Proof. (i) This part is in [CLVW06, p. 266].

(ii) follows from Theorem 6.6 and [MM90, pp. 95–96].

(iii) Let $X \leq A$. Then there exists a supplement submodule S of A such that $X\beta^*S$. Since A is strongly $\oplus$-supplemented, S is a direct summand of A. Hence A is G*-lifting. □

Proposition 6.28. *Let A be a module. Consider the following conditions.*

(i) *A is amply supplemented.*

(ii) *For each $X \leq A$ there exist a supplement S and $L \leq A$ such that $A = S + L = X + L$, $S \cap L \subseteq X \cap L$, and $S \overset{\text{cs}}{\hookrightarrow} X + S$.*

(iii) *A is* G*-*supplemented.*

Then (i) ⇒ (ii) ⇒ (iii).

Proof. (i) ⇒ (ii) If X is small, then take $S = 0$ and $L = A$. So now assume X is not small. Since A is weakly supplemented, there exists L such that $X + L = A$ and $X \cap L \leq_s A$. Also, there exists a supplement S of L such that $S \subseteq X$. Hence $S + L = A$, $S \cap L \subseteq X \cap L$ and $S \cap L \leq_s S$. Thus L is a weak supplement for both S and X. By Corollary 6.7, $X\beta^*S$. By Theorem 6.3, $S \overset{\text{cs}}{\hookrightarrow} X + S$.

(ii) ⇒ (iii) By Proposition 6.5, $X\beta^*S$. Thus A is G*-supplemented. □

Note that if A is amply supplemented, then its submodules that have the same supplements in A are β^* equivalent, by [CLVW06, p. 254, Exc. 8].

Proposition 6.29. *If A is a projective module, then the following conditions are equivalent.*

(i) *A is supplemented.*

(ii) *A is* G*-*supplemented.*

(iii) *A is amply supplemented.*

(iv) *A is lifting.*

(v) *A is* G*-*lifting.*

(vi) *A is semiperfect.*

Proof. This result follows from Theorem 6.24, Proposition 6.28, and [CLVW06, 27.21]. □

The next result shows that in general a weakly supplemented module need not be G*-supplemented.

Proposition 6.30. *Let R be a commutative domain with only two maximal ideals. Then R is a weakly supplemented module which is not* G*-*supplemented.*

Proof. By [CLVW06, p. 210, 17.9], R_R is weakly supplemented. Let M_1 and M_2 be maximal ideals of R. Note that R_R is not supplemented by [CLVW06, 27.21]. Also, observe that if $Y \leq R_R$, then either $Y \leq M_1$ or $Y \leq M_2$ and that $\operatorname{rad} R = M_1 \cap M_2 \leq_s R_R$. Now

Claim 1. Let $X \leq R_R$ such that X_R is not small in R_R. Then $X \leq M_i$ if and only if $X\beta^* M_i$ where $i \in \{1, 2\}$.

Proof of Claim 1. Assume that $i = 1$. Since R_R is weakly supplemented, there exists $W \leq R_R$ such that $X + W = R$ and $X \cap W \leq_s R_R$. First assume $X \leq M_1$. Then $W \leq M_2$. By the modular law, $M_1 = X + (M_1 \cap W)$ and $M_1 \cap W \leq \operatorname{rad} R \leq_s R$. Let $K \leq R_R$ such that $X + M_1 + K = R_R$. Since $X \leq M_1$, $M_1 + K = R_R$. So $R_R = X + (M_1 \cap W) + K = X + K$. By Theorem 6.3, $X\beta^* M_1$. Conversely, assume $X\beta^* M_1$. Suppose to the contrary that X is not a submodule of M_1. Then $X\beta^* M_2$. It follows that $M_1\beta^* M_2$. Then $R_R = M_1 + M_2 + M_1$. By Lemma 6.2, $M_1 + M_1 = M_1 = R_R$, a contradiction. Thus $X \leq M_1$.

Claim 2. There exists no supplement $S \leq R_R$ such that $M_2\beta^* S$.

Proof of Claim 2. Assume, to the contrary, that $M_2\beta^* S$ for some supplement $S \leq R_R$. By Claim 1, $S \leq M_2$. Hence there exists $V \leq R_R$ such that $V + S = R_R$ and $V \cap S \leq_s S$. Then $V \leq M_1$. By Claim 1, $V\beta^* M_1$. Since $X \leq M_1$, $X\beta^* M_1$, by Claim 1. By Lemma 6.2, $X\beta^* V$, a contradiction. Thus Claim 2 is proved. It follows that R_R is not G*-supplemented. □

For a concrete example to Proposition 6.30, let $R = \{\frac{m}{n} \in \mathbb{Q} : p \nmid n,\, q \nmid n\}$ (see [S90, p. 60, Exc. 3.67]) where p and q are distinct primes. Then R is a commutative domain with exactly two maximal ideals.

We close this section by two examples. In the first, we show that the one-way implications of Theorem 6.24 cannot be reversed. On the other hand, the second example shows that there is a H-supplemented (and hence G*-supplemented) module which is not amply supplemented. Thus, in Proposition 6.28, (iii) ⇒ (ii) does not hold and [MM90, Open prob. 18, p. 106] fails.

Example 6.31.

(i) Let $R = \mathbb{Z}_8$ and $A = \mathbb{Z}_2 \oplus \mathbb{Z}_4 \oplus \mathbb{Z}_8$. By [MM90, p. 97], A is G*-lifting (i.e., H-supplemented), but not lifting.

(ii) Let R be a commutative local ring which has two incomparable ideals I and J. Let $A = (R/I) \oplus (R/J)$. By [MM90, p. 97, Lem. A.4 (1)], A is amply supplemented, hence G*-supplemented by Proposition 6.28, but A is not G*-lifting by [MM90, p. 97, Lem. A.4 (3)] and Theorem 6.24. For a concrete example, let F be a field and $T = F[x]/\langle x^4\rangle = \{a\overline{1} + b\overline{x} + c\overline{x}^2 + d\overline{x}^3 : a, b, c, d \in F \text{ and } \overline{x} = x + \langle x^4\rangle\}$. Let $R = \{a\overline{1} + c\overline{x}^2 + d\overline{x}^3 \in T\}$. Thus R is a subring of T. Moreover, R is a commutative local Kash ring (where a Kash ring has the property that every maximal ideal is an annihilator). The ideals of R are: $0, R, F\overline{x}^2, F\overline{x}^3, F\overline{x}^2 + F\overline{x}^3$. Note that $F\overline{x}^2 + F\overline{x}^3$ is maximal and $F\overline{x}^2 \cap F\overline{x}^3 = 0$. Then $A = R/F\overline{x}^2 \oplus R/F\overline{x}^3$ is $\oplus$-supplemented and amply supplemented, but not G*-lifting [MM90, p. 97].

(iii) (See [CLVW06, p. 279, Exa. 23.7]). Let K be the quotient field of a discrete valuation domain R which is not complete. Let $A = K \oplus K$. Then A is supplemented, but not G*-supplemented. To see this, assume that A is G*-supplemented. Let $X \leq A$. Then $X\beta^*S$, where S is a supplement in A. By [CLVW06, p. 233, 20.2], S is coclosed in A. From [CLVW06, p. 279, Exa. 23.7], S is a direct summand of A. Thus A is G*-lifting. By Theorem 6.24 and [MM90, pp. 97–98], A is not G*-lifting, a contradiction.

Example 6.32 (See [WW12, Exa. 3.11]). Let R be a discrete valuation ring with field of fractions K. Assume ${}_RK$ is not quasi-projective. Let $A = K \oplus K$. By [WW12, Exa. 3.11], A is H-supplemented (and hence G*-supplemented) but not amply supplemented.

Exercises

6.1. Let ${}_{\mathbb{Z}}A = {}_{\mathbb{Z}}\mathbb{Z}$, $m, n \in \mathbb{Z}$. Prove that $\mathbb{Z}m\beta^*\mathbb{Z}n$ if and only if m and n are divisible by the same primes.

6.2. Write out in detail the proof of Corollary 6.8 (i).

6.3. Prove Theorems 6.20, 6.23 for G*-lifting cases.

6.4. Let R be a discrete valuation ring with field of fractions K. Let P be the unique maximal ideal of R. Let ${}_RA = (K/R) \oplus (R/P)$. Show that A is G*-lifting, but not lifting.

6.5. Let A be the $\mathbb{Z}$-module $(\mathbb{Z}/\mathbb{Z}2) \oplus (\mathbb{Z}/\mathbb{Z}8)$. Prove that every direct summand of ${}_{\mathbb{Z}}A$ is G*-lifting, but ${}_{\mathbb{Z}}A$ is not lifting.

6.6. Let ${}_RM$ be a uniserial module with unique composition series $0 \neq V \subset U \subset M$. Let ${}_RA = M \oplus (U/V)$. Prove that A is G*-lifting, but not lifting.

6.2 Goldie-rad-supplemented modules

We introduce the β^{**} relation on the lattice of submodules of a module A. We show that β^{**} is an equivalence relation. Then we investigate some general properties of this relation. Moreover, β^{**} is used to define and study classes of Goldie-rad-supplemented and rad-H-supplemented modules.

Recall from [AS91] that a module A is said to have *P^*-property*, or to be a *P^*-module*, if for any submodule N of A there exists a direct summand D of A such that $D \subseteq N$ and $N/D \in \operatorname{rad}(A/D)$; equivalently, for every submodule N of A there exists a decomposition $A = K \oplus K'$ such that $K \subseteq N$ and $(N \cap K') \subseteq \operatorname{rad}(K')$. Let $K, L \leq A$. K is called a (*weak*) rad-*supplement* of L in A if whenever $A = K + L$, we have $(K \cap L \subseteq \operatorname{rad} A)$ $K \cap L \subseteq \operatorname{rad} K$. Furthermore, a module A is called (*weakly*) rad-*supplemented* if every submodule of A has a (weak) rad-supplement (see [THT14]). Recall that for submodules X, Y of A, $X\beta^*Y$ if and only if $(X+Y)/X \leq_s A/X$ and $(X+Y)/Y \leq_s A/Y$. In the same manner, we have the following.

Definition 6.33. Let A be a module and $X, Y \leq A$. We say X and Y are β^{**} equivalent, and write $X\beta^{**}Y$, if $(X+Y)/X \subseteq (\operatorname{rad} A + X)/X$ and $(X+Y)/Y \subseteq (\operatorname{rad} A + Y)/Y$.

Let us present some basic properties of β^{**} relation on the set of submodules of A.

Lemma 6.34. *β^{**} is an equivalence relation.*

Proof. The reflexivity and symmetry properties are clear. For transitivity, assume $X\beta^{**}Y$ and $Y\beta^{**}Z$. So $(X+Y)/X \subseteq (\operatorname{rad} A + X)/X$ and $(X+Y)/Y \subseteq (\operatorname{rad} A + Y)/Y$, $(Y+Z)/Y \subseteq (\operatorname{rad} A + Y)/Y$ and $(Y+Z)/Z \subseteq (\operatorname{rad} A + Z)/Z$. Then we have $X + Y \subseteq \operatorname{rad} A + X$ and $X + Y \subseteq \operatorname{rad} A + Y$; $Y + Z \subseteq \operatorname{rad} A + Y$ and $Y + Z \subseteq \operatorname{rad} A + Z$. It is easy to see that $X + Z \subseteq \operatorname{rad} A + X$ and $X + Z \subseteq \operatorname{rad} A + Z$. Thus $X\beta^{**}Z$. □

It is clear that any submodule contained in $\operatorname{rad} A$ is β^{**} equivalent to the zero submodule. Also, note that two submodules may be isomorphic, but not β^{**} equivalent. For example, let F be a field and $R = \begin{bmatrix} F & F \\ 0 & F \end{bmatrix}$, $X = \begin{bmatrix} 0 & F \\ 0 & 0 \end{bmatrix}$, and $Y = \begin{bmatrix} 0 & 0 \\ 0 & F \end{bmatrix}$. Since $\operatorname{rad}(R_R) = X$, X and Y are not β^{**} equivalent, but they are R-isomorphic. Also, in $A = {}_{\mathbb{Z}}\mathbb{Z}$, $\mathbb{Z}m\beta^{**}\mathbb{Z}n$ if and only if $m = n$ (see Exercise 6.1).

Proposition 6.35. *Let $f : A \to N$ be an epimorphism. The following statements hold.*

(i) *If $X, Y \leq A$ are such that $X\beta^{**}Y$, then $f(X)\beta^{**}f(Y)$*
(ii) *If $X, Y \leq N$ are such that $X\beta^{**}Y$, then $f^{-1}(X)\beta^{**}f^{-1}(Y)$*
(iii) *If $X \leq A$ is such that $X \subseteq \operatorname{rad} A$, $K \leq N$, and $f(X)\beta^{**}K$, then $X\beta^{**}f^{-1}(K)$.*

Proof. (i) Suppose that $X\beta^{**}Y$ for submodules X, Y of A. Then $X+Y \subseteq \operatorname{rad} A+X$ and $X+Y \subseteq \operatorname{rad} A+Y$. Thus we have $f(X)+f(Y) \subseteq \operatorname{rad} N+f(X)$ and $f(X)+f(Y) \subseteq \operatorname{rad} N+f(Y)$. This implies that $f(X)\beta^{**}f(Y)$.

(ii) Let $X\beta^{**}Y$ for submodules X, Y of N. Then $X+Y \subseteq \operatorname{rad} N+X$ and $X+Y \subseteq \operatorname{rad} N+Y$. Since f is an epimorphism, $f^{-1}(X)+f^{-1}(Y) \subseteq \operatorname{rad} A+X$ and $f^{-1}(X)+f^{-1}(Y) \subseteq \operatorname{rad} A+Y$. It follows that $f^{-1}(X)\beta^{**}f^{-1}(Y)$.

(iii) Assume that $f(X)\beta^{**}K$, $X \subseteq \operatorname{rad} A$, and $K \leq N$. Then $f(X)+K \subseteq \operatorname{rad} N+f(X)$ and $f(X)+K \subseteq \operatorname{rad} N+K$. Since f is an epimorphism and $X \subseteq \operatorname{rad} A$, we get $f^{-1}(K)+X \subseteq \operatorname{rad} A+f^{-1}(K)$ and $f^{-1}(K)+X \subseteq \operatorname{rad} A+X$. Therefore, $X\beta^{**}f^{-1}(K)$. □

Proposition 6.36. *Let $X \leq A$ and K a maximal submodule of A.*

(i) *Suppose $C_1, C_2 \leq A$, $\operatorname{rad} A \subseteq C_2$ are such that $C_1+C_2=A$, $C_2 \neq A$ and $X\beta^{**}C_1$. Then $X \nsubseteq C_2$.*

(ii) *If $X\beta^{**}Y$ and $X \subseteq K$, then $Y \subseteq K$.*

Proof. (i) Assume that $X \subseteq C_2$. Since $\operatorname{rad} A \subseteq C_2$, we have $X+C_2=A$. By assumption, $C_2=A$, a contradiction.

(ii) Assume that $Y \nsubseteq K$. Then $Y+K=A$. Since $X\beta^{**}Y$ and $\operatorname{rad} A \subseteq K$, we obtain $K+X=A$. But $X \subseteq K$ implies that $K=A$, a contradiction. □

Proposition 6.37. *Let $X_1, X_2, Y_1, Y_2 \leq A$ such that $X_1\beta^{**}Y_1$ and $X_2\beta^{**}Y_2$. Then $(X_1+X_2)\beta^{**}(Y_1+Y_2)$ and $(X_1+Y_2)\beta^{**}(Y_1+X_2)$.*

Proof. Suppose that $X_1\beta^{**}Y_1$ and $X_2\beta^{**}Y_2$. Then $X_1+Y_1 \subseteq \operatorname{rad} A+X_1$ and $X_1+Y_1 \subseteq \operatorname{rad} A+Y_1$, $X_2+Y_2 \subseteq \operatorname{rad} A+X_2$ and $X_2+Y_2 \subseteq \operatorname{rad} A+Y_2$. Hence by using the above inequalities, we can easily see that $(X_1+X_2)\beta^{**}(Y_1+Y_2)$ and $(X_1+Y_2)\beta^{**}(Y_1+X_2)$. □

Corollary 6.38. *Let $X, Y \leq A$ and $K \subseteq \operatorname{rad} A$. Then $X\beta^{**}Y$ if and only if $X\beta^{**}(Y+K)$.*

Proof. ($\Rightarrow$) This implication follows from Proposition 6.37 and the fact that $0\beta^{**}K$.

($\Leftarrow$) Since $K \subseteq \operatorname{rad} A$, we have $Y\beta^{**}(Y+K)$. Now the implication follows from the transitivity of the β^{**} relation. □

Corollary 6.39. *Let $X, Y_1, Y_2, \ldots, Y_n \leq A$. If $X\beta^{**}Y_i$, for $i=1,2,\ldots,n$, then $X\beta^{**}\sum_{i=1}^{n} Y_i$.*

In the preceding section, we defined and studied the β^* relation. Using β^*, we introduced two new classes of modules, namely G*-lifting and G*-supplemented. Now we introduce two new classes of modules which are based on the β^{**} relation.

Definition 6.40. Let A be a module.

(i) We say A is *Goldie*-rad-*supplemented* (or G-rad-*supplemented*) if for every submodule N of A, there exists a rad-supplement submodule S in A such that $N\beta^{**}S$.

(ii) We say A is rad-H-*supplemented* if for every submodule N of A, there exists a direct summand D of A such that $N\beta^{**}D$.

Observe that if A is a module with the property that every rad-supplement submodule is a direct summand, then for A being G-rad-supplemented is equivalent to being rad-H-supplemented. In fact, for a module A, we have the following implications:

$$P^* \Rightarrow \text{rad-H-supplemented} \Rightarrow \text{G-rad-supplemented}.$$

Next we give a general example of modules which are rad-H-supplemented (Goldie-rad-supplemented), but not G*-supplemented. This example also shows that rad-H-supplemented modules (Goldie-rad-supplemented modules) are a proper generalization of H-supplemented modules (G*-supplemented modules).

Example 6.41.

(i) A radical module A (i.e., $\operatorname{rad} A = A$) is rad-H-supplemented and hence G-rad-supplemented. This yields that any non-supplemented module A with $\operatorname{rad} A = A$ is rad-H-supplemented, but not H-supplemented (i.e., G*-lifting). Hence, all injective non-supplemented modules over a Dedekind domain (e.g., the quotient field of a non-local Dedekind domain) (see [MM90, Prop. A.8]) are rad-H-supplemented (hence G-rad-supplemented) but not G*-supplemented (H-supplemented) by Theorem 6.24. In particular, ${}_{\mathbb{Z}}\mathbb{Q}$ is G-rad-supplemented, but not G*-supplemented.

(ii) The $\mathbb{Z}$-module $\mathbb{Z}$ is neither rad-H-supplemented, nor G-rad-supplemented. In fact, an (indecomposable) rad-H-supplemented module with zero radical is (local) semisimple.

Proposition 6.42. *Let A be a* H-*supplemented module. Then A is* rad-H-*supplemented. If* rad $\leq_s A$*, then the converse holds.*

Proof. Let $N \leq A$. By assumption, A has a decomposition $A = D \oplus D'$ such that $(N + D)/N \leq_s A/N$ and $(N + D)/D \leq_s A/D$. Then $A = D + D' = N + D'$ and $(N + D)/D \subseteq (\operatorname{rad} A + D)/D$. Let $\theta : (D + D')/D \to D'$, $\varphi : D'/N \cap D' \to (N + D')/N$ be the natural isomorphisms and $f : D' \to D'/(N \cap D')$ be the natural epimorphism. Set $h = \varphi f \theta$. By a similar argument to Proposition 6.5, $(N + D)/N = h((N + D)/D)$. Since $(N + D)/D \subseteq (\operatorname{rad} A + D)/D$, we have $(N + D)/N \subseteq (\operatorname{rad} A + N)/N$. Hence A is rad-H-supplemented. For the converse, when $\operatorname{rad} A \leq_s A$, it is easy to check that A is H-supplemented. □

Proposition 6.43. *Let A be a module. Then A is* G-rad-*supplemented if and only if for every $X \leq A$ there exists a* rad-*supplement submodule S of A such that $S + \operatorname{rad} A = X + \operatorname{rad} A$.*

Proof. Let A be G-rad-supplemented and $X \leq A$. Then there is a rad-supplement submodule S of A such that $X + S \subseteq \operatorname{rad} A + X$ and $X + S \subseteq \operatorname{rad} A + S$. Then

$S + \operatorname{rad} A \subseteq X + \operatorname{rad} A$ and $X + \operatorname{rad} A \subseteq S + \operatorname{rad} A$. It follows that $S + \operatorname{rad} A = X + \operatorname{rad} A$. The converse is easy. □

Proposition 6.44. *Let A be a module. If for every $X \leq A$, there is a* rad*-supplement submodule S of A and a submodule $H \subseteq \operatorname{rad} A$ such that $X = S + H$, then A is* G-rad*-supplemented.*

Proof. We prove that $X\beta^{**}S$. Since $X + S = S + H \subseteq \operatorname{rad} A + S + H = \operatorname{rad} A + X$ and $X + S = S + H + S \subseteq \operatorname{rad} A + S$, then $(X + S)/X \subseteq (\operatorname{rad} A + X)/X$ and $(X + S)/S \subseteq (\operatorname{rad} A + S)/S$, as required. □

Proposition 6.45. *Let A be a* G-rad*-supplemented module. Then for each $X \leq A$ with $\operatorname{rad} A \subseteq X$, we have $X = S + H$, where S is a* rad*-supplement in A and $H \subseteq \operatorname{rad} A$.*

Proof. Let $X \leq A$ such that $\operatorname{rad} A \subseteq X$. By assumption, there exists a rad-supplement submodule S of A such that $X\beta^{**}S$. Then $S \subseteq X$ and $X = \operatorname{rad} A + (S \cap X) = \operatorname{rad} A + S$. This completes the proof. □

The following result is analogous to Theorem 6.23, but it asserts distributivity for the lattice of submodules. Recall that A is distributive if, for submodules K, L, N of A, $N + (K \cap L) = (N + K) \cap (N + L)$ or $N \cap (K + L) = (N \cap K) + (N \cap L)$.

Theorem 6.46. *Let $A = B \oplus C$ be a distributive module. Then A is* G-rad-*supplemented* (rad-H-*supplemented*) *if and only if B and C are* G-rad-*supplemented* (rad-H-*supplemented*).

Proof. ($\Rightarrow$) Let $X \leq B$. Then there exist submodules S, L of A such that $S + L = A$ and $S \cap L \subseteq \operatorname{rad} S$ and $X\beta^{**}S$. We prove that $X\beta^{**}(B \cap S)$. Since $X\beta^{**}S$, we have $X + S \subseteq \operatorname{rad} A + X$ and $X + S \subseteq \operatorname{rad} A + S$. Since $X \subseteq B$, we get $X + (B \cap S) \subseteq \operatorname{rad} B + X$ and $X + (B \cap S) \subseteq \operatorname{rad} B + B \cap S + C \cap S + \operatorname{rad} C \cap B$. By the modular law, $X + (B \cap S) \subseteq \operatorname{rad} B + X$ and $X + (B \cap S) \subseteq \operatorname{rad} B + (B \cap S)$. Thus $X\beta^{**}(B \cap S)$. By assumption, $(B \cap S) + (B \cap L) = B$ and $(B \cap S) \cap (B \cap L) = B \cap S \cap L \subseteq \operatorname{rad}(B \cap S) \oplus \operatorname{rad}(C \cap S)$. It follows that $B \cap S \cap L \subseteq \operatorname{rad}(B \cap S)$. So $(B \cap S)$ is a rad-supplement of $(B \cap L)$ in B. Therefore, B is G-rad-supplemented. Similarly, C is G-rad-supplemented.

($\Leftarrow$) Let $U \leq A$, $U_1 = B \cap U$ and $U_2 = C \cap U$. There exist $L_1, S_1 \leq B$ such that $U_1\beta^{**}S_2$, $L_1 + S_1 = B$, and $L_1 \cap S_1 \subseteq \operatorname{rad} S_1$. There also exist $L_2, S_2 \leq C$ such that $U_2\beta^{**}S_2$, $L_2 + S_2 = C$, and $L_2 \cap S_2 \subseteq \operatorname{rad} S_2$. By Proposition 6.37, $U\beta^{**}(S_1 + S_2)$. Moreover, $S_1 + S_2 + L_1 + L_2 = A$ and $(S_1 + S_2) \cap (L_1 + L_2) = (S_1 \cap L_1) + (S_2 \cap L_2) \subseteq \operatorname{rad} S_1 \oplus \operatorname{rad} S_2 \subseteq \operatorname{rad}(S_1 + S_2)$. This means that $S_1 + S_2$ is a rad-supplement submodule of A. Hence A is G-rad-supplemented. The proof that B and C are rad-H-supplemented is similar. □

Recall that a module A is called *semilocal* if $A/\operatorname{rad} A$ is semisimple (see [CLVW06]). It is clear that rad-supplemented modules are semilocal. Now we have the following fact.

Lemma 6.47. *Let A be a* rad-H-*supplemented module. Then $A/\operatorname{rad} A$ is semisimple.*

Proof. Let $N/\operatorname{rad} A \leq A/\operatorname{rad} A$. Since A is rad-H-supplemented, there exists a direct summand D of A such that $N\beta^{**}D$. So $(N+D)/N \subseteq (\operatorname{rad} A + N)/N$ and $(N+D)/D \subseteq (\operatorname{rad} A+D)/D$. Since $D \leq_{\mathrm{d}} A$, $A = D \oplus D'$ for some submodule D' of A. Then $A = D' + N$. It follows that $A/\operatorname{rad} A = N/\operatorname{rad} A + (D' + \operatorname{rad} A)/\operatorname{rad} A$. Since $N \cap D' \subseteq \operatorname{rad} D'$, $A/\operatorname{rad} A = (N/\operatorname{rad} A) \oplus ((D' + \operatorname{rad} A)/\operatorname{rad} A)$. Hence $A/\operatorname{rad} A$ is semisimple. □

Proposition 6.48. *Let A be a module. Then the following statements are equivalent.*

(i) *A is* rad-H-*supplemented.*

(ii) *A is semilocal and each direct summand of $A/\operatorname{rad} A$ lifts to a direct summand of A.*

Proof. (i) ⇒ (ii) By Lemma 6.47, it is enough to prove the last statement. Let $N/\operatorname{rad} A \leq A/\operatorname{rad} A$. Since A is rad-H-supplemented, there exists $D \leq_{\mathrm{d}} A$ such that $N\beta^{**}D$, i.e., $(N+D)/N \subseteq (\operatorname{rad} A+N)/N$ and $(N+D)/D \subseteq (\operatorname{rad} A+D)/D$. Thus $D \subseteq N$. Hence $N/\operatorname{rad} A = (D + \operatorname{rad} A)/\operatorname{rad} A$. This means that $N/\operatorname{rad} A$ lifts to D.

(ii) ⇒ (i) Let $N \leq A$. By hypothesis, $(N + \operatorname{rad} A)/\operatorname{rad} A = \overline{N}$ is a direct summand of $A/\operatorname{rad} A = \overline{A}$. Hence, by (ii), $\overline{N} = \overline{L}$ such that $L \leq_{\mathrm{d}} A$. Now it is straightforward to check that $N\beta^{**}L$. □

We note that a finite direct sum of H-supplemented modules need not be H-supplemented (see [TNT11]). The following example shows that a (finite) direct sum of rad-H-supplemented modules need not be rad-H-supplemented either.

Example 6.49. Let R be a commutative local ring and A a finitely generated R-module. Assume $A \cong \bigoplus_{i=1}^{n} R/I_i$ for ideals I_i $(1 \leq i \leq n)$. Since every I_i is fully invariant in R, each R/I_i is H-supplemented by [KK07, Thm. 2.3] and hence rad-H-supplemented by Proposition 6.42. By [MM90, Lem. A.4], A is rad-H-supplemented if $I_1 \leq I_2 \leq \cdots \leq I_n$. Otherwise, A is not rad-H-supplemented, in view of Proposition 6.42.

In contrast, we have the following result.

Theorem 6.50. *Let $A = \bigoplus_{i\in I} H_i$ be a direct sum of* rad-H-*supplemented modules H_i $(i \in I)$. Assume that each direct summand of $A/\operatorname{rad} A$ lifts to a direct summand of A. Then A is* rad-H-*supplemented.*

Proof. By Lemma 6.47, $A/\operatorname{rad} A$ is semisimple. Now Proposition 6.48 yields that A is rad-H-supplemented. □

A module A is called rad-⊕-*supplemented* if for every $B \leq A$, there exists a $C \leq_{\mathrm{d}} A$ such that $B + C = A$ and $B \cap C \subseteq \operatorname{rad} C$. Clearly, every P^*-module is rad-⊕-supplemented and every rad-⊕-supplemented module is rad-supplemented.

A module A is called *amply* rad-*supplemented* if for any submodules K and V of A such that $A = K + V$, there is a submodule U of V such that $K + U = A$ and $K \cap U \subseteq \operatorname{rad} U$. It is easy to show that every amply rad-supplemented module is weakly rad-supplemented (Exercise 6.7).

Proposition 6.51. *Every amply* rad-*supplemented module is* G-rad-*supplemented.*

Proof. Let A be amply rad-supplemented and $X \leq A$. Let $X \subseteq \operatorname{rad} A$. Then clearly $X\beta^{**}0$. So assume that $X \not\subseteq \operatorname{rad} A$. Since A is weakly rad-supplemented, there exists a submodule L of A such that $X + L = A$ and $X \cap L \subseteq \operatorname{rad} A$. By hypothesis, there is a rad-supplement S of L in X. So $A = S+L$ and $S \cap L \subseteq \operatorname{rad} S$. Since $S \subseteq X$, we have $X = S + (L \cap X) \subseteq \operatorname{rad} A + S$. It follows that $X\beta^{**}S$. Therefore, A is G-rad-supplemented. □

Example 6.52.

(i) Let $R = \mathbb{Z}_8$ and $A = \mathbb{Z}_2 \oplus \mathbb{Z}_4 \oplus \mathbb{Z}_8$. By Example 6.31 (i), A is an H-supplemented R-module and hence a rad-H-supplemented R-module by Proposition 6.42. A is not lifting and since it is finitely generated, A is not P^*.

(ii) Let R and A be as in Example 6.31 (ii). Then A is amply supplemented and hence amply rad-supplemented. By Proposition 6.51, A is G-rad-supplemented, but A is not H-supplemented. Now, by Proposition 6.42, A is not rad-H-supplemented.

Let A be any module. A submodule U of A is called *quasi strongly lifting* (QSL) in A if whenever $(B + U)/U$ is a direct summand of A/U, there exists a direct summand P of A such that $P \leq B$ and $P + U = B + U$ (see [A09]).

Lemma 6.53. *Let A be a module. Then the following statements are equivalent.*

(i) *A is a P^*-module.*

(ii) *A is* rad-H-*supplemented and* $\operatorname{rad} A$ *is* QSL *in A.*

Proof. By Lemma 6.47 and [A09, Lem. 3.5 and Prop. 3.6]. □

Lemma 6.54. *Let A be a projective module such that every* rad-*supplement submodule of A is a direct summand of A. Then the following statements are equivalent.*

(i) *A is* rad-*supplemented.*

(ii) *A is amply* rad-*supplemented.*

(iii) *A is P^*.*

(iv) *A is* rad-$\oplus$-*supplemented.*

Proof. (i) $\Leftrightarrow$ (ii) By [WD06, Thm. 2.15].

(i) $\Rightarrow$ (iii) Take $\tau = \operatorname{rad}$ in [A09, Lem. 3.2].

(iii) $\Rightarrow$ (i) and (i) $\Leftrightarrow$ (iv) are clear by the definitions and the assumption that every rad-supplement submodule of A is a direct summand of A. □

A module A is called *strongly* rad-$\oplus$-*supplemented* if A is rad-$\oplus$-supplemented and every rad-supplement submodule of A is a direct summand [THT14].

Proposition 6.55. *If A is* G-rad-*supplemented and strongly* rad-$\oplus$-*supplemented, then A is* rad-H-*supplemented.*

Proof. Let $N \leq A$. Then there exists a rad-supplement submodule S of A such that $N\beta^{**}S$. By hypothesis, S is a direct summand of A. Hence, A is rad-H-supplemented. □

The following result holds.

Theorem 6.56. *Let A be a projective module such that every* rad-*supplement submodule of A is a direct summand. Then the following are equivalent.*

(i) *A is* rad-*supplemented.*
(ii) *A is P^*.*
(iii) *A is amply* rad-*supplemented.*
(iv) *A is* rad-H-*supplemented and* $\operatorname{rad} A$ *is* QSL *in A.*
(v) *A is* rad-$\oplus$-*supplemented.*
(vi) *A is* G-rad-*supplemented and* $\operatorname{rad} A$ *is* QSL *in A.*

Proof. (i) $\Leftrightarrow$ (ii) $\Leftrightarrow$ (iii) $\Leftrightarrow$ (v) follows from Lemma 6.54.
(ii) $\Leftrightarrow$ (iv) by Lemma 6.53.
(iv) $\Leftrightarrow$ (vi) follows from Proposition 6.55. □

Recall that a module A is called *refinable* if whenever $A = B + C$ for submodules B, C, then there exists a direct summand D of A such that $D \subseteq B$ and $A = D+C$ (see [Wis91]). From [A09, Thm. 3.7], if A is refinable, then $\operatorname{rad} A$ is QSL in A. Also, by [A09, Cor. 3.21], if R_R is lifting, then for every finitely generated projective R-module A, $\operatorname{rad} A$ is QSL in A. Thus, we have the following corollary without any further proof.

Corollary 6.57. *Let A be a projective module such that every* rad-*supplement submodule is a direct summand. Then the following statements are equivalent in case A is refinable, or ${}_RR$ is lifting and A is finitely generated.*

(i) *A is* rad-*supplemented.*
(ii) *A is P^*.*
(iii) *A is amply* rad-*supplemented.*
(iv) *A is* rad-H-*supplemented.*
(v) *A is* rad-$\oplus$-*supplemented.*
(vi) *A is* G-rad-*supplemented.*

Over a left perfect ring every left R-module is G-rad-supplemented. If ${}_RR$ is rad-H-supplemented, then R is a semiperfect ring. So if every module over a ring R is rad-H-supplemented, then R is semiperfect. But there exists a semiperfect ring which has a module that is not rad-H-supplemented.

Example 6.58. Let $R = F[[x,y]]$ be the ring of formal power series over a field F in the indeterminates x and y. Then R is a commutative Noetherian local domain with maximal ideal $J = Rx + Ry$. Therefore the ring R is semiperfect. Since R is a domain, ${}_RJ$ is a uniform R-module. It followws that ${}_RJ$ is indecomposable. Now, assume that ${}_RJ$ is rad-H-supplemented and $N \subsetneqq J$ such that $N \not\subseteq \mathrm{rad}({}_RJ)$. Then $N\beta^{**}0$ or $N\beta^{**}J$. Thus $N \subseteq \mathrm{rad}({}_RJ)$ or $N = {}_RJ$. It follows that ${}_RJ$ is not rad-H-supplemented.

Exercises

6.7. Show that an amply rad-supplemented module is weakly rad-supplemented.

6.8. Find semilocal modules which are not semisimple.

6.3 ECS and EC_{11}-modules

In this section, we introduce ECS and EC_{11}-modules as weaker forms of CS and C_{11}-modules, respectively. A module A is called an ECS-*module* if every ec-*complement* submodule of A is a direct summand. A module A is said to satisfy condition EC_{11} (or EC_{11}-*module*) if every ec-submodule of A has a complement which is direct summand (see [CYT09, CYT10]).

Following [KE05], a (complement) submodule is called ec-(*complement*) *submodule* if it contains essentially a cyclic submodule. Recall that a module A is said to be *principally extending* (for short P-*extending*) if every cyclic submodule of A is essential in a direct summand. Observe that clearly CS-modules and (von Neumann) regular rings are ECS-modules. For instance, we have the following fact.

Proposition 6.59. *Let A be a module. Consider the following statements.*

(i) *A is* CS. (ii) *A is* ECS. (iii) *A is P-extending.*

Then (i) $\Rightarrow$ (ii) $\Rightarrow$ (iii). *In general, the converses to these implications do not hold.*

Proof. (i) $\Rightarrow$ (ii) This implication is clear.

(ii) $\Rightarrow$ (iii) Let Ra be any cyclic submodule of A. Then the closure of Ra in A, L say, is an ec-complement. By hypothesis, L is a direct summand of A. Thus A is P-extending.

Let $M_2(R)$ be the ring as in Example 4.77. Then $M_2(R)$ is a von Neumann regular ring which is neither left, nor right CS. Thus (ii)$\not\Rightarrow$(i). Finally, let G be the ring as in Exercise 3.34., i.e., $G = \begin{bmatrix} \mathbb{Z}_2 & \mathbb{Z}_2 \\ 0 & \mathbb{Z} \end{bmatrix}$. Then G is not right CS-ring, but it is P-extending. Since G_G has finite uniform dimension, it has a maximal uniform (and hence an ec-complement) submodule which is not a direct summand of G_G. So G is not a right ECS-module. □

Proposition 6.59 shows that the classes of modules with the CS, ECS and P-extending properties are different from each other. In [KE05], authors assumed that ECS and P-extending properties are equivalent and proved several results for P-extending modules. However, the proof of Proposition 6.59 provides a counterexample to these assertions by exhibiting a P-extending module which does not satisfy the ECS condition. Actually, most of the results in [KE05] which are stated for P-extending modules remain true only for ECS-modules.

Since the ECS property lies strictly between the CS and P-extending properties, it is natural to seek conditions which ensure that a P-extending module is ECS or that a ECS-module is CS. Such conditions are illustrated in our next result.

Proposition 6.60.

(i) *Let* ${}_RA$ *be a nonsingular module. Then* A *is* P*-extending if and only if* A *is* ECS.

(ii) *Let* A *be a left* R*-module with the property that a direct sum of any ec-complement with a direct summand of* A *is a complement in* A*. Then* A *is* P*-extending if and only if* A *is* ECS.

(iii) *Let* A *be a module with finite uniform dimension. Then* A *is* CS *if and only if* A *is* ECS.

Proof. (i) Assume A is a left P-extending R-module. Let X be any ec-complement submodule of A. Then Rx is essential in X for some $x \in X$. By hypothesis, there exists a direct summand L of A which contains Rx as an essential submodule. Since ${}_RA$ is nonsingular, $X = L$. Thus A is ECS. The converse follows from Proposition 6.59.

(ii) Assume A is P-extending. Let C be an ec-complement submodule of A such that Rc is essential in C. By hypothesis, there are submodules D, D' of A such that Rc is essential in D and $A = D \oplus D'$. It follows that $C \oplus D'$ is essential in A. By hypothesis, $A = C \oplus D'$. Hence A is ECS. The converse follows from Proposition 6.59.

(iii) Assume A is an ECS-module. Let N be any maximal uniform submodule of A. Clearly, N is an ec-complement in A. By hypothesis, N is a direct summand of A. Hence A is CS. The converse is clear by Proposition 6.59. □

Observe that the extending version of the following result is Corollary 3.20.

Theorem 6.61. *Let* A *be a* P*-extending module.*

(i) *If* A *is distributive, then every submodule of* A *is* P*-extending.*

(ii) *If* X *is a projection invariant submodule of* A*, then* X *is* P*-extending.*

(iii) *If* X *is a submodule of* A *such that the intersection of* X *with any direct summand of* A *is a direct summand of* X*, then* X *is* P*-extending.*

(iv) *If* A *has* SIP*, then every direct summand of* A *is* P*-extending.*

Proof. (i) Let X be any submodule of A and Rx be a cyclic submodule of X. Then there exists a direct summand D of A such that Rx is essential in D. Hence Rx is essential in $D \cap X$. Since $X = X \cap (D \oplus D') = (X \cap D) \oplus (X \cap D')$, where D' is a submodule of A, then $X \cap D$ is a direct summand of X. So, X is P-extending.

(ii) Let X be a projection invariant submodule of A. Let D be any direct summand of A and $\pi : A \to D$ be the projection. Then $\pi(X) \subseteq X$ and $\pi(X) = D \cap X$, which is a direct summand of X. By (iii), X is P-extending.

(iii) Let B be any cyclic submodule of X. Then $B = Rx$ for some $x \in X$. Then there exists a direct summand D of A such that B is essential in D. So B is essential in $D \cap X$ and $D \cap X$ is a direct summand of X. Thus X is a P-extending module.

(iv) is a direct consequence of (iii). □

By adapting the proof of [DS92, Lem. 1], we have the following corollary.

Corollary 6.62. *Let R be any ring and A a projective P-extending module which has* SIP. *Then there exists an index set I such that A is a direct sum $\bigoplus_{i \in I} A_i$ of submodules A_i ($i \in I$) of A such that each submodule A_i is an* ec-*complement in A.*

Proof. By Kaplansky's Theorem (see [Fai76, p. 120]), the module A is a direct sum of countably generated submodules. By Theorem 6.61 (iv), we may suppose that A is countably generated. There exists a countable set of elements $a_1, a_2, \dots$ in A such that $A = \sum_i Ra_i$. By hypothesis, there exist submodules A_1, N_1 of A such that $A = A_1 \oplus N_1$ and Ra_1 is essential in A_1. Let π_i be the projection of A_i in N_1 for all $i \geq 2$. By Theorem 6.61 (iv) again, there exists a direct summand A_2 of N_1 which contains Ra_2 as an essential submodule. Continuing in this manner we obtain a direct sum $A_1 \oplus A_2 \oplus \cdots$ of submodules in the module A such that $Ra_1 + Ra_2 + \cdots + Ra_k \subseteq A_1 \oplus A_2 \oplus \cdots \oplus A_k$, for all positive integers k. It follows that $A = \bigoplus_i A_i$. Moreover, by the construction, each submodule A_i is an ec-complement in A. □

Recall that CS-modules can be characterized by the lifting of homomorphisms from certain submodules to the module itself (see Corollary 3.98). This yields the following characterization for ECS-modules.

Corollary 6.63. *Let A be a module. Then A is* ECS *if and only if for each* ec-*complement submodule K of A there exists a complement L of K in A such that $K \oplus L \in \mathrm{Lift}_A(A)$.*

Proof. This is a direct consequence of Lemma 3.97. □

Recall that we defined the EC_{11} property as each ec-submodule having a complement which is a direct summand. For instance, the following holds.

Proposition 6.64. *Let A be a module. Then the following statements are equivalent.*

(i) *A has EC_{11}.*
(ii) *For any* ec-*complement submodule L in A, there exists a direct summand K of A such that K is a complement of L in A.*

(iii) *For any* ec-*submodule* N *in* A, *there exists a direct summand* K *of* A *such that* $K \cap N = 0$ *and* $K \oplus N$ *is an essential submodule of* A.

(iv) *For any* ec-*complement submodule* L *in* A, *there exists a direct summand* K *of* A *such that* $K \cap L = 0$ *and* $K \oplus L$ *is an essential submodule of* A.

Proof. The equivalences (i) $\Leftrightarrow$ (ii) and (iii) $\Leftrightarrow$ (iv) are obvious.

The equivalence (i) $\Leftrightarrow$ (iii) follows from Lemma 4.24. □

The next lemma and corollary provide relationships between the EC_{11} condition and various generalizations of the extending property.

Lemma 6.65. *Let* ${}_RA$ *be a module. Consider the following statements.*

(i) ${}_RA$ *is* ECS.

(ii) ${}_RA$ *is* P*-extending.*

(iii) ${}_RA$ *is an* EC_{11}*-module.*

(iv) ${}_RA$ *is a* C_{11}*-module.*

Then (i) $\Rightarrow$ (ii) $\Rightarrow$ (iii) *and* (iv) $\Rightarrow$ (iii). *In general, the converses to these implications do not hold.*

Proof. (i) $\Rightarrow$ (ii) By Proposition 6.59.

(ii) $\Rightarrow$ (iii) Let K be an ec-complement submodule of A. Then there exists $x \in K$ such that Rx is essential in K. Since A is P-extending, there exists a direct summand D of A such that Rx is essential in D. Now $A = D \oplus D'$ for some submodule D' of A. Then $K \cap D' = 0$ and $K \oplus D'$ is essential in A. By Lemma 4.24, A is an EC_{11}-module.

(iv) $\Rightarrow$ (iii) Clear.

Let $G = \begin{bmatrix} \mathbb{Z}_2 & \mathbb{Z}_2 \\ 0 & \mathbb{Z} \end{bmatrix}$. By the proof of Proposition 6.59, G_G is P-extending. However, G_G is not ECS. Thus (ii) $\not\Rightarrow$ (i). Now, let A be the $\mathbb{Z}[x]$-module $\mathbb{Z}[x] \oplus \mathbb{Z}[x]$. So A is an EC_{11}-module. But A is not P-extending, by Proposition 6.60. Thus (iii) $\not\Rightarrow$ (ii). Finally, let R be the ring as in [Lam99, Exa. 7.54]. Then R is a commutative, regular ring which is not Baer. Now by [BMR02, Thm. 4.7 (iii)], R_R is not FI-extending. Hence, Proposition 4.67 yields that R_R is not C_{11}-module. Thus, (iii) $\not\Rightarrow$ (iv). □

Corollary 6.66. *Let* ${}_RA$ *be an indecomposable module. Then the following statements are equivalent.*

(i) ${}_RA$ *is* ECS.

(ii) ${}_RA$ *is* P*-extending.*

(iii) ${}_RA$ *is an* EC_{11}*-module.*

(iv) ${}_RA$ *is uniform.*

Proof. (i) $\Rightarrow$ (ii) and (ii) $\Rightarrow$ (iii) follow from Lemma 6.65.

(iii) $\Rightarrow$ (iv) Let $0 \neq X \leq A$. Then there exists $0 \neq x \in X$. Let L be any closure of Rx in A. Then L is an ec-complement submodule of A. By hypothesis, there exists a direct summand D of A such that $L \cap D = 0$ and $L \oplus D$ is essential in A. It follows that L is essential in A. Since L is a complement of A, then $L = A$. Hence X is essential in A, and consequently $_RA$ is uniform.

(iv) $\Rightarrow$ (i) is obvious. □

Our next objective is to determine when the P-extending and EC_{11} conditions are equivalent. We have the following theorem.

Theorem 6.67. *Let A be an R-module such that* $\mathrm{End}\,({}_RA)$ *is Abelian and $X \leq A$ implies $X = \sum_{i\in I} h_i(A)$, where $h_i \in$* $\mathrm{End}\,({}_RA)$. *Then A is EC_{11}-module if and only if A is P-extending.*

Proof. Assume A is EC_{11}-module and X is a cyclic submodule of A. Let Y be a closure of X in A. Then X is essential in Y. So Y is an ec-complement submodule of A. Now $Y = \sum_{i\in I} h_i(A)$, where each $h_i \in \mathrm{End}\,({}_RA)$. By hypothesis, Ae is a complement of Y, where $e^2 = e \in \mathrm{End}\,({}_RA)$. Let $0 \neq y \in Y$. Then $y = ye + y(1-e)$. But $y = \sum_{i\in I} h_i(a_i)$, where $a_i \in A$. Thus $ye = (\sum_{i\in I} h_i(a_i))e = \sum_{i\in I} h_i(a_i e) \in Y \cap Ae = 0$, i.e., $y = y(1-e)$. Hence Y is essential in $A(1-e)$. Then $Y = A(1-e)$ is a direct summand of A. Hence $_RA$ is P-extending. The converse follows from Lemma 6.65. □

Corollary 6.68. *If A is an R-module satisfying any of the following conditions, then A is EC_{11}-module if and only if A is P-extending.*

(i) *$_RA = {}_RR$ and R is Abelian.*

(ii) *A is cyclic and R is commutative.*

(iii) *A is a multiplication module and R is commutative.*

Proof. By Theorem 6.67, the result is true for condition (i). Now assume that A is cyclic and R is commutative. There exists $_RB \leq {}_RR$ such that $_RA$ is isomorphic to R/B. Let Y/B be an R-submodule of R/B. So $Y/B = (\sum_{i\in I} Ry_i) + B = R(\sum_{i\in I} y_i + B)$, where each $y_i \in Y$. Define $h_i : R/B \to R/B$ by $h_i(r+B) = y_i + B$. Then $h_i \in \mathrm{End}\,({}_R(R/B))$. Hence $Y/B = \sum_{i\in I} h_i(R/B)$. Since R is commutative, $\mathrm{End}\,({}_R(R/B))$ is commutative. Thus Theorem 6.67 yields the result for condition (ii).

Finally, assume that A is a multiplication module and R is commutative. Let $X = IA$, where $_RI \leq {}_RR$. For each $s \in I$ define $h_s : A \to A$ by $h_s(a) = sa$ for $a \in A$. Then $X = IA = \sum_{s\in I} h_s(A)$. Observe that every submodule of a multiplication module is fully invariant. By [BMR02, Lem. 1.9], if $e^2 = e \in \mathrm{End}\,({}_RA)$, then e and $1-e \in S_{\mathrm{r}}(\mathrm{End}\,({}_RA))$ where $S_{\mathrm{r}}(\mathrm{End}\,({}_RA))$ is the set of all right semicentral idempotent elements of $\mathrm{End}\,({}_RA)$. Hence e is central. So $\mathrm{End}\,({}_RA)$ is Abelian. Again, Theorem 6.67 yields the result. □

Note that Example 4.33, or more generally Example 4.132, show that direct summands of an EC_{11}-module need not satisfy the EC_{11} property. Observe that the tangent bundles in the proofs of the aforementioned examples (for instance, the submodules ${}_RK$ and ${}_RD$, respectively) are complements which are not ec-complements (Exercise 6.9). In the rest of this section, we focus on direct summands of EC_{11}-modules.

Lemma 6.69. *Let A be an EC_{11}-module and X a submodule. If the intersection of X with any direct summand of A is a direct summand of X, then X is an EC_{11}-module.*

Proof. Clear. □

Corollary 6.70. *Let A be an EC_{11}-module.*

(i) *If X is a projection invariant submodule of A, then X is an EC_{11}-module. In particular, every fully invariant submodule of A is an EC_{11}-module.*
(ii) *If A has* SIP, *then every direct summand of A has EC_{11}.*

Proof. (i) Let D be a direct summand of A and $\pi : A \to D$ be the canonical projection. By Lemma 6.69, X is an EC_{11}-module.

(ii) is an immediate consequence of Lemma 6.69. □

Lemma 6.71. *Let $A = A_1 \oplus A_2$. Then A_1 satisfies EC_{11} if and only if for every* ec-*submodule N of A_1, there exists a direct summand K of A such that $A_2 \subseteq K$, $K \cap N = 0$, and $K \oplus N$ is an essential submodule of A.*

Proof. Modifying the proof of Lemma 4.40 and Proposition 6.64 yields the result. □

Theorem 6.72. *Let $A = A_1 \oplus A_2$ be an EC_{11}-module such that for every ec-submodule K of A with $K \cap A_2 = 0$, $K \oplus A_2$ is a direct summand of A. Then A_1 is an EC_{11}-module. In this case, A_1 is a P-extending module.*

Proof. By Lemma 6.71, A_1 is an EC_{11}-module. For the second part, let K be an ec-submodule of A_1. Hence K is an ec-submodule of A with $K \cap A_2 = 0$. By hypothesis, $K \oplus A_2$ is a direct summand of A. Therefore, K is a direct summand of A and hence also of A_1. It follows that A_1 is a P-extending module. □

Proposition 6.73. *Let A be an EC_{11}-module. If $\operatorname{soc} A$ is cyclic, then $A = A_1 \oplus A_2$, where A_1 is a submodule of A with essential socle and A_2 a submodule of A with zero socle.*

Proof. Let S denote the socle of A. By hypothesis, there exist submodules A_1 and A_2 of A such that $A = A_1 \oplus A_2$, $S \cap A_2 = 0$, and $S \oplus A_2$ is an essential submodule of A. So $S = \operatorname{soc} A = \operatorname{soc} A_1 \oplus \operatorname{soc} A_2$. Clearly $\operatorname{soc} A_2 = 0$, so that $S \leq A_1$. Now $S \oplus A_2$ essential in A implies S essential in A_1. Thus we have the required decomposition. □

The converse of Proposition 6.73 in not true, in general, as the next counterexample shows.

Example 6.74. With the notation of the construction carried out after Proposition 5.28, let $A_1 = R$, $A_2 = R/I$ and $A = A_1 \oplus A_2$. Note that $\operatorname{soc} A = I \oplus 0$, and $\operatorname{soc} A_2 = 0$. Now, let N be any simple submodule of A. It is clear that N is an ec-submodule of A. By the proof of Corollary 5.31, there is no direct summand L of A such that $L \cap N = 0$ and $L \oplus N$ is essential in A. Since this would imply that $L \oplus N$ contains $\operatorname{soc} A$, Lemma 5.30, combined with the fact that $\operatorname{soc} A$ is not simple shows that this is impossible. It follows that A is not an EC_{11}-module.

Exercises

6.9. Prove that the tangent bundle $_RD$ in Example 4.132 is not ec-complement.

6.10. Show that every von Neumann regular ring is an EC_{11}-module.

Remarks. It is well known that the existence of complements in a module is due to upper continuity of the lattice of submodules of a module. Since, in general, the lattice of submodules of a module is not lower continuous, the existence of supplements is not guaranteed. Recall that a module A is called lifting if every submodule N of A contains a direct summand X of A such that $N/X \leq_s A/X$. Lifting modules and related concepts were well-presented in [CLVW06] and [MM90]. It is known that, extending modules (or CS modules) are modules, whose complement submodules are direct summands (equivalently; every submodule is essential in direct summand), while lifting modules provide a dualization of extending modules.

G*-lifting modules and G*-supplemented modules were introduced and investigated by Birkenmeir, Mutlu, Nebiyev, Sökmez, and Tercan [BMNST10]. Theorems 6.20, 6.23, 6.24, and 6.26 were obtained in [BMNST10]. Example 6.32 is due to Wang and Wu [WW12], which shows that G*-lifting (or H-supplemented) property does not imply amply supplemented condition. On using β^{**} equivalence relation, Goldie-rad-supplemented and Goldie-H-supplemented modules were defined and investigated by Talebi, Hamzekolaee, and Tercan [THT14]. Theorems 6.46, 6.50, and 6.56 were proved in [THT14]. Celep Yücel and Tercan defined and investigated ECS and EC_{11}-modules as weaker types of CS and C_{11}-modules in [CYT09] and [CYT10], respectively. Amongst other results, Theorems 6.67 and 6.72 appeared in [CYT10].

Chapter 7

Open Problems and Questions

This chapter formulates a number of open problems and questions. We think of our problems are legitimate and lead further researches which greatly broadened the scope of the theory. To this end, we should mention that the latticial counterparts of conditions C_i $(i = 1, 2, 3)$ and C_{1j} $(j = 1, 2)$, for modules and applications to Grothendieck categories, module categories equipped with a torsion theory have been investigated in [AIT16].

1. Let A be a weak CS-module and $N \leq_{\text{d}} A$. Is N weak CS?
2. Find necessary and sufficient conditions to characterize when direct summands of G-extending, C_{11}, $\underline{P}\underline{I}$-extending, or FI-extending modules are G-extending, C_{11}, $\underline{P}\underline{I}$-extending, or FI-extending, respectively.
3. Is being C_{11} a Morita invariant property?
4. Characterize Abelian C_{11}-groups.
5. Investigate modules A in which $\langle \underline{H} \rangle$ is a clean subring (i.e., every element of $\langle \underline{H} \rangle$ is the sum of an idempotent and a unit) of $\underline{E}_A$ (see Corollary 4.120 (ii)).
6. Determine modules A such that $\underline{E}_A$ is a clean ring (see Corollary 4.140 (viii)).
7. Find conditions which ensure that any direct sum of CLS-modules is also a CLS-module.
8. Clarify whether (iii) $\Rightarrow$ (ii) holds in Proposition 4.108. If not, find conditions which make the former implication true.
9. Let A be a weak C_{11}-module and $N \leq_{\text{d}} A$. Is N weak C_{11}?
10. Find necessary and sufficient conditions to characterize (finite) direct sums of G-extending modules.
11. Characterize when a generalized extending module (i.e., an inner or outer generalization of extending modules) is a (finite) direct sum of uniform modules.
12. When does a module have a generalized extending hull?

13. How do generalized extending modules behave with respect to the tensor product or Hom functors?
14. Investigate how generalized extending conditions interact with torsion theory.
15. Characterize the rings such that every (cyclic, finitely generated, singular etc.) module is a generalized extending module for some fixed generalized extending property.
16. Characterize when $R[x]$ is a generalized extending ring.
17. Determine necessary and/or sufficient conditions when the homomorphic image of a module is generalized extending.
18. Determine necessary and/or sufficient conditions for a submodule to be generalized extending.
19. Must a finitely generated module over a Prüfer domain be G-extending?
20. If R is a Dedekind domain and A is a G-extending R-module, must every pure submodule of A be a direct summand?
21. If A is finitely generated and each maximal submodule is β^* equivalent to a supplement submodule, is A G*-supplemented?
22. Characterize those rings over which every (cyclic, finitely generated) module is G*-supplemented.
23. Characterize those rings over which every (cyclic, finitely generated) module is G*-lifting.
24. Determine when a full matrix ring and/or a generalized upper triangular matrix ring is G*-lifting (G*-supplemented).
25. Find conditions which ensure that a G-rad-supplemented module is a direct sum of hollow modules.
26. Determine if either A_1 or A_2 satisfies C_{11} property without additional conditions in Lemma 4.36.
27. Investigate the dual classes of $\mathrm{Lift}_X(A)$ and $\mathrm{Elift}_X(A)$ for modules X and A over a ring R?

Appendix

Construction of Rings of Quotients via Gabriel Topologies

Let R be a ring with identity 1. For any right ideal E of R and $a \in R$, define

$$a^{-1}E = \{r \in R : ar \in E\}.$$

Note that $a^{-1}E$ *is a right ideal* of R for each right ideal E and element a.

Let $\underline{F}$ be a collection of right ideals of R with the following properties:

(1) $R \in \underline{F}$,

(2) $E \in \underline{F}$, $E \subseteq F$, F a right ideal of $R \Rightarrow F \in \underline{F}$,

(3) $E, F \in \underline{F} \Rightarrow E \cap F \in \underline{F}$,

(4) $E \in \underline{F}$, $a \in R \Rightarrow a^{-1}E \in \underline{F}$, and

(5) E a right ideal of R, $F \in \underline{F}$, $a^{-1}E \in \underline{F}$ $(\forall a \in F) \Rightarrow E \in \underline{F}$.

(Such a collection $\underline{F}$ is called *a right Gabriel topology on* R.)

For any right ideal E of R, let $\mathrm{Hom}(E, R)$ denote the collection of R-homomorphisms $f : E \to R$. Let $S = \bigcup_{E \in \underline{F}} \mathrm{Hom}(E, R)$, the collection of all R-homomorphisms $f : E \to R$ with $E \in \underline{F}$.

Define a relation $\sim$ on S by

$$(f : E \to R) \sim (g : F \to R) \Leftrightarrow \exists G \subseteq E \cap F, G \in \underline{F} \text{ and } f|_G = g|_G.$$

Then $\sim$ *is an equivalence relation* on S; this is clear by property (3) above. For any $(f : E \to R)$ in S let $[f : E \to R]$ denote its equivalence class.

Fact: $E, F \in \underline{F}$, $f \in \mathrm{Hom}\,(E, R) \Rightarrow f^{-1}(F) \in \underline{F}$.

Proof. $f^{-1}(F) = \{r \in R : f(r) \in F\}$ is a right ideal of R. Let $a \in E$. Then $a^{-1}(f^{-1}(F)) = \{r \in R : ar \in f^{-1}(F)\} = \{r \in R : f(ar) \in F\} = \{r \in R : f(a)r \in F\}$, i.e.,

$$a^{-1}(f^{-1}(F)) = f(a)^{-1}F.$$

Thus $a^{-1}(f^{-1}(F)) \in \underline{F}$ ($\forall a \in E$), by property (4), and hence $f^{-1}(F) \in \underline{F}$ by property (5). Now define, for $E, F \in \underline{F}$ and $f \in \mathrm{Hom}(E, R)$, $g \in \mathrm{Hom}(F, R)$,

$$[f : E \to R] + [g : F \to R] = [f + g : E \cap F \to R]$$
$$[f : E \to R][g : F \to R] = [fg : g^{-1}(E) \to R].$$

Here $(f+g)(x) = f(x)+g(x)$ ($x \in E \cap F$) and $(fg)(x) = f(g(x))$ ($x \in g^{-1}(E)$). □

Fact: Addition and multiplication are well defined.

Proof. Suppose $[f : E \to R] = [f' : E' \to R]$ and $[g : F \to R] = [g' : F' \to R]$ for some $E, E', F, F' \in \underline{F}$. Then there exist $G \subseteq E \cap E'$, $H \subseteq F \cap F'$ with $G, H \in \underline{F}$ and $f|_G = f'|_G$, $g|_H = g'|_H$. Let $K = G \cap H$. Then $K \in \underline{F}$ by property (3) and

$$(f + g)(x) = f(x) + g(x) = f'(x) + g'(x) = (f' + g')(x) \ (\forall x \in K).$$

Thus $[f + g : E \cap F \to R] = [f' + g' : E' \cap F' \to R]$, and so

$$[f : E \to R] + [g : F \to R] = [f' : E' \to R] + [g' : F' \to R].$$

On the other hand, with the same notation, let $L = H \cap g^{-1}(G)$. Then $L \in \underline{F}$ (see above). Moreover, for any $x \in L$,

$$g(x) = g'(x) \in G \subseteq E \cap E',$$

so that $x \in g^{-1}(E) \cap g'^{-1}(E')$. Thus

$$L \subseteq g^{-1}(E) \cap g'^{-1}(E').$$

Also, if $x \in L$, then

$$fg(x) = f(g(x)) = f(g'(x)) = f'(g'(x)) = f'g'(x).$$

If follows that $[fg : g^{-1}(E) \to R] = [f'g' : g'^{-1}(E') \to R]$ and hence

$$[f : E \to R][g : F \to R] = [f' : E' \to R][g' : F' \to R].$$

Thus addition and multiplication are well defined. □

Now we define $R_{\underline{F}}$ to be the collection of all equivalence classes $[f : E \to R]$, where $E \in \underline{F}$, $f \in \mathrm{Hom}(E, R)$.

Fact: $R_{\underline{F}}$ is a ring with respect to the addition and multiplication defined above.

Proof. The proof is long and tedious. Define $z : R \to R$ by $z(r) = 0$ ($r \in R$). Then $[z : R \to R]$ is the zero of $R_{\underline{F}}$. For any $[f : E \to R]$, the negative is $[-f : E \to R]$, where $(-f)(x) = -f(x)$ ($x \in E$). Define $i : R \to R$ by $i(r) = r$ ($r \in R$). Then

$[i : R \to R]$ is the identity of $R_{\underline{F}}$. Let $[f : E \to R]$, $[g : F \to R]$, $[h : G \to R] \in R_{\underline{F}}$. Then

$$\begin{aligned}\{[f : E \to R] + [g : F \to R]\}[h : G \to R] &= [f + g : E \cap F \to R][h : G \to R] \\ &= [(f + g)h : h^{-1}(E \cap F) \to R]\end{aligned}$$

whereas

$$\begin{aligned}&[f : E \to R][h : G \to R] + [g : F \to R][h : G \to R] \\ &\quad= [fh : h^{-1}(E) \to R] + [gh : h^{-1}(F) \to R] = [fh + gh : h^{-1}(E) \cap h^{-1}(F) \to R] \\ &\quad= [(f + g)h : h^{-1}(E \cap F) \to R].\end{aligned}$$

Therefore,

$$\begin{aligned}&\{[f : E \to R] + [g : F \to R]\}[h : G \to R] \\ &\quad= [f : E \to R][h : G \to R] + [g : F \to R][h : G \to R].\end{aligned}$$

In this way it can be shown that $R_{\underline{F}}$ is a ring.

Let $a \in R$. Define $a_l : R \to R$ by $a_l(r) = ar$ $(r \in R)$. Then $a_l \in \text{Hom}(R, R)$. Define $\varphi : R \to R_{\underline{F}}$ by $\varphi(a) = [a_l : R \to R]$ $(a \in R)$. Let $a, b \in R$. Then

$$\begin{aligned}\varphi(a + b) &= [(a + b)_l : R \to R] = [a_l : R \to R] + [b_l : R \to R] = \varphi(a) + \varphi(b), \\ \varphi(ab) &= [(ab)_l : R \to R] = [a_l : R \to R][b_l : R \to R] = \varphi(a)\varphi(b).\end{aligned}$$

Thus φ is a ring homomorphism. □

Fact: $\ker \varphi = \{r \in R : rE = 0 \text{ for some } E \in \underline{F}\}$.

Proof. Suppose $a \in R$ and $aF = 0$ for some $F \in \underline{F}$. Then $a_l|_F = z|_F$, so that $[a_l : R \to R] = [z : R \to R]$. Thus $\varphi(a) = 0$. Conversely, suppose $\varphi(b) = 0$, where $b \in R$. Then $[b_l : R \to R] = [z : R \to R]$, so that $bG = 0$ for some $G \in \underline{F}$.

Therefore, there exists a ring monomorphism from $R/\ker\varphi$ to $R_{\underline{F}}$. □

Special cases

(i) $R_{\underline{T}}$ *where* $\underline{T} = \{R\}$. Note that $\underline{T} = \{R\}$ is a right Gabriel topology on R (this is an easy check). Now

$$(f : R \to R) \sim (g : R \to R) \Leftrightarrow f|_R = g|_R \Leftrightarrow f = g.$$

Thus every equivalence class contains a single element. Given $f \in \text{Hom}(R, R)$,

$$f(r) = f(1r) = f(1)r, \ (r \in R)$$

so that $f = f(1)_l$. Thus the above mapping φ is onto and $\ker\varphi = 0$. Thus $R_{\underline{T}} \cong R$.

(ii) $R_{\underline{U}}$, *where* $\underline{U}$ *is the collection of all right ideals of* R. Clearly $\underline{U}$ is a right Gabriel topology on R. Given any right ideals E, F of R and $f \in \mathrm{Hom}(E, R)$, $g \in \mathrm{Hom}(F, R)$, note that $0 \in \underline{U}$ and $f|_0 = g|_0$ implies

$$(f : E \to R) \sim (g : F \to R).$$

Thus $R_{\underline{U}}$ has a unique element, namely $[z : R \to R]$, i.e., $R_{\underline{U}} = 0$. Also $\ker \varphi = R$ (clearly).

Note: Since $R/\ker \varphi$ embeds in $R_{\underline{F}}$, we naturally want $\ker \varphi = 0$, in order not to lose information about R. This is not always possible of course. To have $\ker \varphi = 0$ we require $r \in R$, $E \in \underline{F}$, $rE = 0 \Rightarrow r = 0$. Given any right ideal I of R, the *left annihilator* $\ell(I)$ of I is defined by $\ell(I) = \{r \in R : rI = 0\} = \{r \in R : ra = 0\, (\forall a \in I)\}$. Recall that $\ell(I)$ is a left ideal of R. We might consider $\underline{L} = \{I : I \text{ is a right ideal of } R \text{ and } \ell(I) = 0\}$, but then we have the problem:

$$I \in \underline{L},\ r \in R \Rightarrow r^{-1}I \in \underline{L}?$$

A right ideal D of R is called *dense* if

$$\ell(r^{-1}D) = 0\ (\forall r \in R).$$

(iii) $R_{\underline{D}}$, *where* $\underline{D}$ *is the collection of all dense right ideals of* R.

We need to prove:

Proposition A.1. $\underline{D}$ *is a right Gabriel topology on* R.

Proof. For any r in R, $r^{-1}R = R$, and so $\ell(r^{-1}R) = 0$, i.e., $R \in \underline{D}$. Let $E \subseteq F$ be right ideals with $E \in \underline{D}$ and let $r \in R$. Then $r^{-1}E \subseteq r^{-1}F$ and hence $\ell(r^{-1}F) \subseteq \ell(r^{-1}E) = 0$. Thus $\ell(r^{-1}F) = 0$ $(\forall r \in R)$, i.e., $F \in \underline{D}$. Let $E, F \in \underline{D}$ and $r \in R$. Then $r^{-1}(E \cap F) = r^{-1}E \cap r^{-1}F$. Let $a \in \ell(r^{-1}E \cap r^{-1}F)$. For any $x \in r^{-1}E$, $xR \cap r^{-1}F \subseteq r^{-1}E \cap r^{-1}F$, so that $a(xR \cap r^{-1}F) = 0$. But $xR \cap r^{-1}F = x\{x^{-1}(r^{-1}F)\} = x\{(rx)^{-1}F\}$ and hence $ax\{(rx)^{-1}F\} = 0$, so that $ax = 0$ because $F \in \underline{D}$. Thus $a(r^{-1}E) = 0$ and hence $a = 0$, because $E \in \underline{D}$. It follows that $\ell(r^{-1}E \cap F)) = 0$ for all $r \in R$, and so $E \cap F \in \underline{D}$. Let $D \in \underline{D}$ and $a \in R$. Then $\ell(r^{-1}(a^{-1}D)) = \ell(ar)^{-1}D) = 0$ for all $r \in R$. Thus $a^{-1}D \in \underline{D}$. Finally, let E be a right ideal of R and $D \in \underline{D}$ such that $a^{-1}E \in \underline{D}$ $(\forall a \in D)$. Let $r \in R$ and $x \in \ell(r^{-1}E)$. Let $y \in r^{-1}D$. Then $ry \in D$ and hence $(ry)^{-1}E \in \underline{D}$. But $(ry)^{-1}E = y^{-1}(r^{-1}E)$. Let $s \in (ry)^{-1}E$. Then $rys \in E$, so $ys \in r^{-1}E$, and then $xys = 0$. Thus $xy\{(ry)^{-1}E\} = 0$ and $xy = 0$. We see that $x(r^{-1}D) = 0$ and hence $x = 0$. Thus $\ell(r^{-1}E) = 0$ $(\forall r \in R)$, and then $E \in \underline{D}$. This completes the proof of Proposition A.1. □

The ring $R_{\underline{D}}$ is called the *maximal* (or *complete*) *right ring of quotients,* or the *maximal right quotient ring, of* R. Note that in this case

$$\ker \varphi = \{r \in R : rE = 0 \text{ for some } E \in \underline{D}\} = 0,$$

and hence R is isomorphic to a subring of $R_{\underline{D}}$. Further, note that if I is a two-sided ideal of R, then $I \subseteq r^{-1}I$ for all $r \in R$, and hence I is a dense right ideal if and only if $\ell(I) = 0$. In checking whether a given collection $\underline{F}$ of right ideals is a right Gabriel topology conditions (2) and (3) are redundant, as the following result shows.

Proposition A.2. *A collection $\underline{F}$ of right ideals of a ring R is a right Gabriel topology on R if and only if*

(1) $R \in \underline{F}$,

(4) $E \in \underline{F}, a \in R \Rightarrow a^{-1}E \in \underline{F}$, *and*

(5) E *a right ideal of* R, $F \in \underline{F}$, $a^{-1}E \in \underline{F}$ $(\forall a \in F) \Rightarrow E \in \underline{F}$.

Proof. Suppose $\underline{F}$ satisfies (1), (4) and (5). We show that $\underline{F}$ satisfies (2) and (3).

(2) Suppose $E \in \underline{F}$, $E \subseteq F$, F a right ideal of R. Let $a \in E$. Then $aR \subseteq E \subseteq F$, so that $a^{-1}F = R \in \underline{F}$, by (1). Thus $a^{-1}F \in \underline{F}$ $(\forall a \in E)$. By (5), $F \in \underline{F}$.

(3) Let $E, F \in \underline{F}$. Let $a \in F$. Then $a^{-1}(E \cap F) = a^{-1}E \cap a^{-1}F = a^{-1}E \cap R = a^{-1}E \in \underline{F}$, by (4). Thus $a^{-1}(E \cap F) \in \underline{F}$ $(\forall a \in F)$. By (5), $E \cap F \in \underline{F}$. The result follows. □

We now turn to another example. Recall that a right ideal E of R is called *essential* if $E \cap I \neq 0$ for every non-zero right ideal I of R. Note the following basic facts:

(i) R is an essential right ideal of R.

(ii) E, F are essential right ideals $\Rightarrow E \cap F$ is an essential right ideal.

(iii) E is an essential right ideal, $E \subseteq F$, F is a right ideal $\Rightarrow F$ is essential.

(iv) E is an essential right ideal, $r \in R \Rightarrow r^{-1}E$ is an essential right ideal of R.

Proof. Assertions (i)–(iii) are obvious. (iv) Let $0 \neq I$ be a right ideal of R. If $rI = 0$, then $0 \neq I \subseteq I \cap r^{-1}E$, and hence $I \cap r^{-1}E \neq 0$. Suppose $rI \neq 0$. Since rI is a non-zero right ideal of R, it follows that there exists $a \in I$ with $0 \neq ra \in E$. Then $0 \neq a \in I \cap r^{-1}E$. Thus, in any case, $I \cap r^{-1}E \neq 0$. Consequently $r^{-1}E$ is an essential right ideal of R. □

Let $\underline{E}$ denote the collection of essential right ideals of R. Is $\underline{E}$ a right Gabriel topology on R? Certainly $\underline{E}$ satisfies (1), (2), (3) and (4). But $\underline{E}$ need not satisfy (5).

Example A.3. Let R be the trivial extension of $\mathbb{Z}$ with the $\mathbb{Z}$-module $\mathbb{Z} \oplus \mathbb{Z}$, i.e., $R = \begin{bmatrix} \mathbb{Z} & \mathbb{Z}\oplus\mathbb{Z} \\ 0 & \mathbb{Z} \end{bmatrix} = \left\{\begin{bmatrix} a & (b,c) \\ 0 & a \end{bmatrix} : a, b, c \in \mathbb{Z}\right\}$. Then R is a commutative ring. Let $B = \left\{\begin{bmatrix} 0 & (b,0) \\ 0 & 0 \end{bmatrix} : b \in \mathbb{Z}\right\}$, $C = \left\{\begin{bmatrix} 0 & (0,c) \\ 0 & 0 \end{bmatrix} : c \in \mathbb{Z}\right\}$, $N = \left\{\begin{bmatrix} 0 & (b,c) \\ 0 & 0 \end{bmatrix} : b, c \in \mathbb{Z}\right\}$. Note that B, C, N are all ideals of R, $N = B + C$, and $N^2 = 0$. Moreover, $N \in \underline{E}$. Also, $N \subseteq r^{-1}B$ $(\forall r \in N)$, so that $r^{-1}B \in \underline{E}$ $(\forall r \in N)$. But $B \notin \underline{E}$, since $B \cap C = 0$. Thus $\underline{E}$ does not satisfy property (5). However, $\underline{E}$ and $\underline{D}$ are related.

Proposition A.4. *For any ring* R, $\underline{D} \subseteq \underline{E}$. *Moreover* $\underline{D} = \underline{E}$ *if and only if* $\{r \in R : rE = 0$ *for some* $E \in \underline{E}\} = 0$.

Proof. Let $D \in \underline{D}$. Let I be a non-zero right ideal of R. Let $0 \neq a \in I$. Then $aR \cap D = a(a^{-1}D) \Rightarrow aR \cap D \neq 0$, because $\ell(a^{-1}D) = 0$. Thus $I \cap D \neq 0$. It follows that $D \in \underline{E}$.
Suppose $\underline{D} = \underline{E}$. Suppose $r \in R$ and $rE = 0$ for some $E \in \underline{E}$. Then $E \in \underline{D}$, $\ell(E) = 0$, and hence $r = 0$. Conversely, suppose $\{r \in R : rE = 0$ for some $E \in \underline{E}\} = 0$. Let $F \in \underline{E}$, and $a \in R$. Then $a^{-1}F \in \underline{E}$ (see (iv) above), hence $\ell(a^{-1}F) = 0$, by hypothesis. Thus $F \in \underline{D}$. □

Recall that for any ring R, $Z(R) = \{r \in R : rE = 0$ for some $E \in \underline{E}\}$ is called the *right singular ideal* of R, and R is called *right nonsingular* if $Z(R) = 0$. Note that there exists an example of a commutative ring R with $Z(R) \neq 0$, but for which $\underline{E}$ is a Gabriel topology on R. For example, let $R = \mathbb{Z}/4\mathbb{Z}$. Then $Z(R) \neq 0$, but clearly $\underline{E}$ is the set of all ideals of R, which is a Gabriel topology on R.

(iv) $R_{\underline{C}}$, where $\underline{C} = \{E : E$ is a right ideal of R containing a regular element$\}$.

Recall that an element $c \in R$ is *regular* if $cr \neq 0$ and $rc \neq 0$ $(\forall 0 \neq r \in R)$. The ring R satisfies the *right Ore condition* if $\forall r \in R$, $c \in R$, c regular $\exists r_1, c_1 \in R$ with c_1 regular and $rc_1 = cr_1$.

Proposition A.5. $\underline{C}$ *is a right Gabriel topology on* R *if and only if* R *satisfies the right Ore condition. In this case,* $\underline{C} \subseteq \underline{D}$.

Proof. Suppose $\underline{C}$ is a right Gabriel topology on R. Let $r, c \in R$ with c regular. Then $cR \in \underline{C}$ and hence $r^{-1}(cR) \in \underline{C}$. There exists $c_1 \in R$, c_1 regular, such that $rc_1 \in cR$. Thus R satisfies the right Ore condition.

Conversely, suppose R satisfies the right Ore condition. Clearly, $R \in \underline{C}$. Let $E \in \underline{C}$, $a \in R$. There exists $d \in E$, d regular. By hypothesis, there exist $a_1, d_1 \in R$ with d_1 regular such that $ad_1 = da_1$. Then $d_1 \in a^{-1}(dR) \subseteq a^{-1}E$ and hence $a^{-1}E \in \underline{C}$. Finally, let E be a right ideal of R, $F \in \underline{C}$ and $a^{-1}E \in \underline{C}$ $(\forall a \in F)$. Then F contains a regular element f. Since $f^{-1}E \in \underline{C}$, it follows that $f^{-1}E$ contains a regular element g. Then the regular element $fg \in E$ and hence $E \in \underline{C}$. We see that $\underline{C}$ is a right Gabriel topology on R.

Now suppose $\underline{C}$ is a right Gabriel topology on R. Let $E \in \underline{C}$, $r \in R$. Then $r^{-1}E \in \underline{C}$ implies that $r^{-1}E$ contains a regular element c, and hence

$$\ell(r^{-1}E) \subseteq \ell(c) = 0.$$

Thus E is a dense right ideal. □

Suppose R satisfies the right Ore condition. Then $R_{\underline{C}}$ is a subring of $R_{\underline{D}}$ by Proposition A.5. Let c be a regular element of R. Then $[c_l : R \to R]$ is a unit in $R_{\underline{D}}$ with inverse $[f_c : cR \to R]$ defined by $f_c(cr) = r$ $(\forall r \in R)$. Let $g : E \to R$, where $E \in \underline{C}$. Suppose $d \in E$, d regular, and $g(d) = a$. Then it can be checked that $[g : E \to R] = [a_l : R \to R][f_d : dR \to R]$.

Thus $R_{\underline{C}}$ has the following properties:

(i) $\varphi(R)$ is a subring of $R_{\underline{C}}$ and $\varphi(R) \cong R$,
(ii) every regular element of $\varphi(R)$ is a unit in $R_{\underline{C}}$, and
(iii) $\forall q \in R_{\underline{C}}\ \exists a, d \in \varphi(R)$, d regular, such that $q = ad^{-1}$.

The ring $R_{\underline{C}}$ is called the *classical right quotient ring of* R. As we have seen above, $R_{\underline{C}}$ is a subring of $R_{\underline{D}}$, the maximal right quotient ring of R.

Example A.6. Let K be a field and let R denote the free algebra $K\langle X, Y\rangle$ in the indeterminates X, Y. (Recall that $K\langle X, Y\rangle$ is like the polynomial ring $K[X, Y]$ except $XY \neq YX$.) Then R is a domain, i.e., every non-zero element is regular, but $XR \cap YR = 0$. Thus R does not satisfy the right Ore condition and $\underline{C}$ is not a right Gabriel topology on R.

We list a number of properties of $R_{\underline{F}}$:

Let $\underline{F}$ be a right Gabriel topology on a ring R, and let $\varphi : R \to R_{\underline{F}}$ be the canonical ring homomorphism. Then $\varphi(R)$ is a subring of $R_{\underline{F}}$.

Property 1. $\forall q \in R_{\underline{F}}\ \exists F \in \underline{F}$ such that $q\varphi(F) \subseteq \varphi(R)$.

Proof. There exist $F \in \underline{F}$ and $f \in \mathrm{Hom}(F, R)$ such that $q = [f : F \to R]$. Let $a \in F$. Then $q(\varphi(a)) = [f : F \to R][a_l : R \to R] = [g : a^{-1}F \to R]$ where, $\forall x \in a^{-1}F$, $g(x) = f(ax) = f(a)x = f(a)_l(x)$. Thus $q(\varphi(a)) = \varphi(f(a))$. It follows that $q\varphi(F) \subseteq \varphi(R)$. □

Property 2. $\forall F \in \underline{F}$ and $f \in \mathrm{Hom}(F, R)\ \exists q \in R_{\underline{F}}$ such that $q\varphi(a) = \varphi(f(a))$ $(\forall a \in F)$.

Proof. Let $q = [f : F \to R]$. By the above proof, $q\varphi(a) = \varphi(f(a))$ $(\forall a \in F)$. □

Property 3. $q \in R_{\underline{F}}$, $q\varphi(E) = 0$ for some $E \in \underline{F} \Rightarrow q = 0$.

Proof. Suppose $q = [f : F \to R]$, where $F \in \underline{F}$, $f \in \mathrm{Hom}(F, R)$. Let $K = \ker f$. Let $e \in E$. Then $q\varphi(e) = [f : F \to R][e_l : R \to R] = [g : e^{-1}F \to R]$, where $g(x) = f(ex)\ \forall x \in e^{-1}F$. There exists $G \in \underline{F}$ such that $g(y) = 0\ \forall y \in G$. Thus $f(ey) = 0$ for all $y \in G$ and hence $G \subseteq e^{-1}K$. It follows that $e^{-1}K \in \underline{F}$. Hence $e^{-1}K \in \underline{F}$ $(\forall e \in E)$ and thus $K \in \underline{F}$. Therefore, $q = 0$. □

Property 4. A commutative ring R is semiprime if and only if $R_{\underline{D}}$ is von Neumann regular. For R commutative, $\underline{D} = \{I : I \text{ is an ideal of } R \text{ and } \ell(I) = 0\}$. Note that if R is a semiprime ring, then

(i) $B \trianglelefteq R \Rightarrow \ell(B) \trianglelefteq R$ and $B \cap \ell(B) = 0$, and
(ii) $B \trianglelefteq R \Rightarrow \ell(B \oplus \ell(B)) = 0 \Rightarrow B \oplus \ell(B) \in \underline{D}$.

Note further that for any commutative ring R, $R_{\underline{D}}$ is also commutative. For, let $q_1, q_2 \in R_{\underline{D}}$. There exist $A_1, A_2 \in \underline{D}$, $f_i \in \mathrm{Hom}(A_i, R)$ $(i = 1, 2)$ such that $q_i = [f_i : A_i \to R]$ $(i = 1, 2)$. Now $\forall a_i \in A_i$ $(i = 1, 2)$, $f_1 f_2(a_1 a_2) = f_1(f_2(a_2)a_1) =$

$f_1(a_1)f_2(a_2) = f_2f_1(a_1a_2)$. But $A_1A_2 \in \underline{D}$ and $A_1A_2 \subseteq f_2^{-1}(A_1) \cap f_1^{-1}(A_2)$. Thus $q_1q_2 = q_2q_1$, and so $R_{\underline{D}}$ is commutative.

Proof of Property 4. Suppose $R_{\underline{D}}$ is regular. Then $R_{\underline{D}}$ is semiprime, and hence so is R. Conversely, suppose R is semiprime. Let $q \in R_{\underline{D}}$. Then $q = [f : I \to R]$ for some $I \in \underline{D}$ and $f \in \mathrm{Hom}(I, R)$. Let $K = \ker f$, $B = f(I \cap \ell(K))$, and $C = B \oplus \ell(B) \in \underline{D}$. Define $g : C \to R$ as follows: $g\{f(x) + y\} = x$ $(\forall x \in I \cap \ell(K), y \in \ell(B))$. Suppose $f(x) + y = f(x') + y'$ for some $x, x' \in I \cap \ell(K)$, $y, y' \in \ell(B)$. Then $f(x) = f(x')$ and hence $f(x-x') = 0$, i.e., $x-x' \in K \cap \ell(K) = 0$. Thus $x = x'$ and g is well defined. Put $D = I \cap (K \oplus \ell(K))$. Then $D \in \underline{D}$ and $D = K \oplus I \cap \ell(K)$. Let $d \in D$. Then $d = x + z$ for some $x \in I \cap \ell(K)$, $z \in K$. Then $[fgf - f](d) = fgf(x+z) - f(x+z) = fgf(x) - f(x) = f(x) - f(x) = 0$. Thus $fgf(d) = f(d)$ $(\forall d \in D)$. Let $q' = [g : C \to R] \in R_{\underline{D}}$. Then $qq'q = q$. It follows that $R_{\underline{D}}$ is von Neumann regular. □

Bibliography

[AB96] Abulkheir, H.E., Birkenmeier, G.F.: Prime ideals in right complement bounded rings. Kyungpook Math. J., **36**, 129–141 (1996)

[ABT09] Akalan, E., Birkenmeier, G.F., Tercan, A.: Goldie extending modules. Comm. Algebra, **37**, 663–683 (2009). Corrigendum **38**, 4747–4748 (2010). Corrigendum **41**, (2013)

[ABT11] Akalan, E., Birkenmeier, G.F., Tercan, A.: A characterization of Goldie extending modules over Dedekind domains. J. Algebra Appl., **6**, 1291–1299 (2011)

[ABT12] Akalan, E., Birkenmeier, G.F., Tercan, A.: Goldie Extending rings. Comm. Algebra, **40**, 423–428 (2012) Corrigendum **43**, No:3, 1366–1368 (2015)

[AkBT12] Akalan, E., Birkenmeier, G.F., Tercan, A.: Characterizations of extending and G-extending generalized triangular matrix rings. Comm. Algebra, **40**, 1069–1085 (2012)

[AIT16] Albu, T., Iosif, M., Tercan, A.: The conditions (C_i) in modular lattices, and applications. J. Algebra Appl. **40** (2016)

[A09] Alkan, M.: On τ-lifting and τ-semiperfect modules. Turkish J. Math., **33**, 117–130 (2009)

[AS91] Al-Khazzi, I., Smith, P.F.: Modules with chain conditions on superfluous Submodules. Comm. Algebra, **19**, 2331–2351 (1991)

[AF74] Anderson, F.W., Fuller, K.R.: Rings and Categories of Modules. New York, Springer-Verlag (1974)

[AH90] Arnold, M., Hausen, J.: A characterization of modules with the summand Intersection property. Comm. Algebra, **18**, 519–528 (1990)

[AMV75] Azumaya, G., Mbuntum, F., Varadarajan, K.: On M-projective and M-injective modules. Pacific J. Math., **59**, 9–16 (1975)

[Bas63] Bass, H.: Big projective modules are free. Illinois J. Math., **7**, 24–31 (1963)

[Bas68] Bass, H.: Algebraic K-Theory. New York, Benjamin (1968)

[BKN82] Bican, L., Kepka, T., Nemec, P.: Rings, Modules and Preradicals. New York, Marcel Dekker (1982)

[B72] Berberian, S.K.: Baer*- Rings. Berlin, Springer-Verlag (1972)

[Birk76] Birkenmeier, G.F.: Self-injective rings and the minimal summand containing the nilpotents. Comm. Algebra, **4**, 705–721 (1976)

[Birk81] Birkenmeier, G.F.: Baer rings and quasi-continuous rings have a MDSN. Pacific J. Math., **97**, 283–292 (1981)

[Birk83] Birkenmeier, G.F.: Idempotents and completely semiprime ideals. Comm. Algebra, **11**, 567–580 (1983)

[Birk90] Birkenmeier, G.F.: Split-null extensions of strongly right bounded rings. Publ. Mat., **33**, 37–44 (1989)

[BCFG01] Birkenmeier, G.F., Calugareanu, G., Fuchs, L., Goeters, H.P.: The fully invariant extension property for Abelian groups. Comm. Algebra, **29**, 673–685 (2001)

[BKT06] Birkenmeier, G.F., Karabacak, F., Tercan, A.: When is the SIP(SSP) property inherited by free modules. Acta Math. Hungar., **112**, 103–106 (2006)

[BKP01] Birkenmeier, G.F., Kim, J.Y., Park, J.K.: Semicentral reduced algebras. In: Birkenmeier, G.F., Park, J.K., Park, Y.S., eds., The International Symposium on Ring Theory, Trends in Math., Boston, Birkhäuser, 67–84, (2001)

[BL14] Birkenmeier, G.F., Lennon, M.J.: Dense intrinsic extensions. Houston J. Math., **40**, 21–42 (2014)

[BiL14] Birkenmeier, G.F., Lennon, M.J.: Extending sets of idempotents to ring extensions. Comm. Algebra, **42**, 5134–5151 (2014)

[BMR02] Birkenmeier, G.F., Müller, B.J., Rizvi, S.T.: Modules in which every fully invariant submodule is essential in a direct summand. Comm. Algebra, **30(3)**, 1395–1415 (2002)

[BKP99] Birkenmeier, G.F., Park, J.Y., Park, J.K.: When is the CS condition hereditary? Comm. Algebra, **27**, 3875–3885 (1999)

[BPR02] Birkenmeier, G.F., Park, J.K., Rizvi, S.T.: Modules with fully invariant sumodules essential in fully invariant sumodules. Comm. Algebra, **30**, 1833–1852 (2002)

[BPR06] Birkenmeier, G.F., Park, J.K., Rizvi, S.T.: Ring hulls and applications. J. Algebra, **304**, 633–665 (2006)

[BPR13] Birkenmeier, G.F., Park, J.K., Rizvi, S.T.: Extensions of Rings and Modules. Springer, New York (2013)

[BT07] Birkenmeier, G.F., Tercan, A.: When some complement of a submodule is a summand. Comm. Algebra, **35(2)**, 597–611 (2007). Addendum and Corrigendum **43**, 1364–1365, (2015)

[BMNST10] Birkenmeier, G.F., Mutlu, F.T., Nebiyev, C., Sökmez, Tercan, A.: Goldie*-supplemented modules. Glasgow Math. J., **52**, 41–52 (2010). Corrigendum **54**, 479–480 (2012)

[BiTY14] Birkenmeier, G.F., Tercan, A., Yaşar, R.: A survey of some generalized extending conditions. Palestine J. Math., **3**, 518–530 (2014)

[BT77] Birkenmeier, G.F., Tucci, R.P.: Does every right ideal of a matrix ring contain a nilpotent element? Amer. Math. Monthly, **84**, 631–633 (1977)

[BTY14] Birkenmeier, G.F., Tercan, A., Yücel, C.C.: The extending condition relative to sets of submodules. Comm. Algebra, **42**, 764–778 (2014)

[BTY16] Birkenmeier, G.F., Tercan, A., Yücel, C.C.: Essential Extensions with the C-extending property. Quaestiones Math., In press.

[BiTY16] Birkenmeier, G.F., Tercan, A., Yücel, C.C.: Projection invariant Extending Rings. J. Algebra Appl., **15(6)** 1650121 (2016)

[BL04] Blecher, D.P., Le Merdy, C.: Operator Algebras and their Modules. Oxford University Press (2004)

[Bre93] Bredon, G.E.: Topology and Geometry. Springer-Verlag, New York (1993)

[BR93] Burgess, W.D, Raphael, R.: On modules with the absolute direct summand property. In: Jain, S.K., Rizvi, S.T., eds., Proceding of the Biennial Ohio-State-Denison Conference 1992, in Ring Theory, Singapore: World Scientific, 137–148 (1993)

[CY91] Camillo, V., Yousif, M.F.: CS-modules with acc or dcc on essential submodules. Comm. Algebra, **19**, 655–662, (1991)

[CHN07] Camillo, V., Herzog, J., Nielsen, P., Pace: Non-self-injective hulls with compatible multiplication. J. Algebra, **314**, 471–478 (2007)

[CYT09] Celep Yücel, C., Tercan, A.: Modules whose ec-closed submodules are direct summand. Taiwanese J. Math., **13**, 1247–1256 (2009)

[CYT10] Celep Yücel, C., Tercan, A.: When some complement of an ec-submodule is a direct summand. Kyungpook Math. J., **50**, 101–107 (2010)

[CC07] Charalambides, S., Clark, J.: CS modules relative to a torsion theory. Mediterr. J. Math., **4**, 291–308 (2007)

[CH77] Chatters, A.W., Hajarnavis, C.R.: Rings in which every complement right ideal is a direct-summand. Quart. J. Math. Oxford, **28(2)**, 61–80 (1977)

[CH80] Chatters, A.W., Hajarnavis, C.R.: Rings with Chain Conditions. Pitman, Boston (1980)

[CK80] Chatters, A.W., Khuri, S.M.: Endomorphism rings of modules over non-singular CS-rings. J. London Math., Soc., **21**, 434–444 (1980)

[CLVW06] Clark, J., Lomp, C., Vanaja, N., Wisbauer, R.: Lifting Modules: Supplements and Projectivity in Module Theory. Birkhäuser Verlag, Basel (2006)

[Co59] Cohn, P.M.: On the free product of associative rings. Math. Z., **71**, 380–398 (1959)

[CS14] Crivei, S., Sahinkaya, S.: Modules whose closed submodules with essential socle are direct summands. Taiwanese J. Math., **18**, 989–1002 (2014)

[DT10] Dancheng, L., Tongsuo, W.: Topological characterizations of the extending property of rings. Acta Math., Scientia, 1086–1092 (2010)

[DY12] Dejun, W., Yongduo, W.: Two open questions on Goldie extending modules. Comm. Algebra, **40**, 2685–2692 (2012)

[DS98] Dogruoz, S., Smith, P.F.: Modules which are extending relative to module classes. Comm. Algebra, **26**, 1699–1721 (1998)

[DHSW94] Dung, N.V., Huynh, D.V, Smith, P.F., Wisbauer, R.: Extending Modules. Pitman Research Notes in Mathematics Series, Longman Scientific and Technical, Harlow, Essex, England, **313**, (1994)

[DS92] Dung, N.V., Smith, P.F.: Hereditary CS-modules. Math. Scand., **71**, 173–180 (1992)

[Er99] Er, N.: Direct sums and summands of weak CS-modules and continuous modules. Rocky Mount. J. Math., **29**, 491–503 (1999)

[Fai76] Faith, C.: Algebra II; Ring Theory. Springer-Verlag, Berlin (1976)

[Fai95] Faith, C.: The minimal regular ideal of self-injective and continuous rings splits off. Arch. Math., **44**, 511–521 (1995)

[FU64] Faith, C., Utumi, Y.: Intrinsic extensions of rings. Pacific J. Math. **14**, 505–512 (1964)

[Fu70] Fuchs, L.: Infinite Abelian Groups. Vol. I, Academic Press (1970)

[Fu73] Fuchs, L.: Infinite Abelian Groups. Vol. II, Academic Press (1973)

[Gar89] Garcia, J.L.: Properties of direct summands of modules. Comm. Algebra, **17**, 73–92 (1989)

[Gil72] Gilmer, R.: Multiplicative Ideal Theory. Marcel Dekker, New York (1972)

[GW89] Goodearl, K.R., Warfield, R.B.: An Introduction to Noncommutative Noetherian Rings. Lecture Note Series. London Math. Soc., Cambridge University Press (1989)

[Goo72] Goodearl, K.R.: Singular torsion and the splitting properties. Mem. Amer. Math. Soc., **124**, (1972)

[Go76] Goodearl, K.R.: Ring Theory: Nonsingular Rings and Modules. Marcel, Dekker, New York (1976)

[Goo79] Goodearl, K.R.: Von Neumann Regular Rings. Pitman, London (1979)

[Gol60] Goldie, A.W.: Semi-prime rings with maximum condition. Proc. London Math. Soc., **10**, 201–220 (1960)

[Hau89] Hausen, J.: Modules with the summand intersection property. Comm. Algebra, **17**, 135–148 (1989)

[Har82] Harada, M.: On modules with lifting properties. Osaka J. Math., **19**, 189–201 (1982)

[Ha82] Harada, M.: On modules with extending properties. Osaka J. Math., **19**, 203–215 (1982)

[HS93] Harmancı, A., Smith, P.F.: Finite direct sums of CS-modules. Houston J. Math., **19**, 523–532 (1993)

[HSTT96] Harmancı, A., Smith, P.F., Tercan, A., Tıraş, Y.: Direct sums of CS-modules. Houston J. Math., **22**, 61–71, (1996)

[HT00] Heatherly, H.E., Tucci, R.P.: Central and semicentral idempotents. Kyungpook Math. J., **40**, 255–258 (2000)

[HO87] Hodges, T.J., Osterburg, J.: A rank two indecomposable projective module over a Notherian domain of Krull dimension one. Bull. London Math. Soc., **19**, 139–144 (1987)

[Hod89] Hodges, T.J., Stafford, J.T.: Noetherian rings with big indecomposable projective modules. Bull., London Math., Soc., **21**, 249–254 (1987)

[Jat98] Jategaonkar, A.V.: Localization in Noetherian Rings. Lecture Note Series. London Math. Soc., Cambridge University Press (1998)

[JJ78] Joel, V.K., Jain, S.K.: π-injective modules and rings whose cyclics are π-injective. Comm. Algebra, **6**, 59–73 (1978)

[KE05] Kamal, M.A., Elmnophy, O.A.: On P-extending modules. Acta Math. Univ. Comenianae, **74**, 279–286 (2005)

[KM88] Kamal, M.A., Müller, B.J.: The structure of extending modules over Noetherian rings. Osaka J. Math., **25**, 539–551 (1988)

[KaM88] Kamal, M.A., Müller, B.J.: Extending modules over commutative domains. Osaka J. Math., **25**, 531–538 (1988)

[KaT16] Kara, Y., Tercan, A.: Modules whose certain submodules are essentially embedded in direct summands. Rocky Mountain J. of Math., In press.

[KT03] Karabacak, F., Tercan, A.: Matrix rings with summand intersection property. Czech. Math. Journal, **53** (128), 621–626 (2003)

[Kap68] Kaplansky, I.: Rings of Operators. Benjamin, New York, (1968)

[Kap69] Kaplansky, I.: Infinite Abelian Groups. University of Michigan Press (1969)

[Kap52] Kaplansky, I.: Modules over Dedekind rings and valuation rings. Trans. Amer. Math. Soc. **72**, 327–340 (1952)

[KK07] Koşan, T., Keskin, D.: H-supplemented duo modules. J. Algebra Appl., **6**, 965–971 (2007)

[Lam78] Lam, T.Y.: Serre's Conjecture. Lecture Notes in Math., Springer-Verlag, Berlin, **635** (1978)

[Lam99] Lam, T.Y.: Lectures on Modules and Rings. Springer-Verlag, New York (1999)

[Lamb86] Lambek, J.: Lectures on Rings and Modules. 3rd ed., Chelsea, New York (1986)

[Lev66] Levy, L.S.: Commutative rings whose homomorphic images are self-injective. Pacific J. Math., **18**, 149–153 (1966)

[Lev89] Levy, L.S.: Projectives of large uniform-rank in Krull dimension 1. Bull., London Math., Soc., **21**, 57–64 (1987)

[LZ88] Li, M.S., Zelmanowitz, J.M.: On the generalizations of injectivity. Comm. Algebra, **16**, 483–491 (1988)

[LOR98] Lopez-Permouth, S.R., Oshiro, K., Rizvi, S.T.: On the relative (quasi-) continuity of modules. Comm. Algebra, **26**, 3497–3510 (1998)

[Ma74] Marubayashi, H.: Modules over Dedekind prime rings II. Osaka J. Math., **11**, 517–545 (1974)

[Me77] Megibben, C.: Projection-invariant subgroups of abelian groups. Tamkang J. Math. **8**, 177–182 (1977)

[MM90] Mohamed, H.S., Müller, B.J.: Continuous and Discrete Modules. London Math. Soc. Lecture Note Series, Cambridge Univ. Press., **147** (1990)

[MR83] Müller, B.J., Rizvi, S.T.: On injective and quasi-continuous modules. J. Pure Appl. Algebra, **28**, 197–212 (1983)

[NY03] Nicholson, W.K., Yousif, M.F.: Quasi Frobenius Rings. Cambridge Univ. Press (2003)

[Nu62] Nunke, R.J.: On direct products of infinite cyclic groups. Proc. Amer. Math. Soc., **13**, 66–71 (1962)

[Oshi91] Oshiro, K.: Minimal cogenerators need not be unique. Comm. Algebra, **19**, 2071–2080 (1991)

[Oshi83] Oshiro, K.: Continuous modules and quasi-continuous modules. Osaka J. Math., **20**, 681–694 (1983)

[Oshi84] Oshiro, K.: Lifting modules, extending modules and their applications to generalized uniserial rings. Hokkaido Math. J., **13**, 339–346 (1984)

[Osof64] Osofsky, B.L.: Homological properties of rings and modules. Ph.D thesis, Rutgers University, New Jersey (1964)

[Osof67] Osofsky, B.L.: On ring properties of injective hulls. Canad. Math. Bull., **10**, 275–282 (1967)

[Osof91] Osofsky, B.L.: Minimal cogenerators need not be unique. Comm. Algebra, **19**, 2071–2080 (1991)

[PS88] Parmenter, M.M., Stewart, P.N.: Excellent extensions. Comm. Algebra, **16** (4), 703–713 (1988)

[PZ03] Parmenter, M.M., Zhou, Y.: Fintely $\sum$-CS property of excellent extensions of rings. Algebra Colloq., **10**, 17–21 (2003)

[Pas77] Passman, D.S.: The Algebraic Structure of Group Rings. Wiley Interscience, New York (1977)

[Ri72] Ribenboim, P.: Algebric Numbers. Wiley (1972)

[RR04] Rizvi, S.T., Roman, C.S.: Baer and Quasi-Baer modules. Comm. Algebra, **32**, 103–123 (2004)

[RR07] Rizvi, S.T., Roman, C.S.: On $\mathcal{K}$-nonsingular modules and applications. Comm. Algebra, **35**, 2969–2982 (2007)

[RY90] Rizvi, S.T., Yousif, M.F.: On continuous and singular modules. In: Noncommutative Ring Theory, Springer Lecture Notes in Math. **1448**, 116–124 (1990) lean

[Sand68] Sandomierski, F.L.: Nonsingular rings. Proc. Amer. Math., Soc., **19**, 225–230 (1968)

[SCS98] Santa-Clara, C., Smith, P.F.: Extending modules with injective or semisimple summands. J. Pure. Appl. Algebra, **127**, 193–203 (1998)

[SCS00] Santa-Clara, C., Smith, P.F.: Modules which are self-injective relative to closed submodules. Contemp. Math., **259**, 487–499 (2000)

[S90] Sharp, R.Y.: Steps in Commutative Algebra. London Math. Soc., **19**, (1990)

[SV72] Sharpe, D.W., Vamos, P.: Injective Modules. Cambridge University Press (1972)

[Smi79] Smith, P.F.: On the structure of certain pp-rings. Math. Z., **166**, 147–157 (1979)

[Smi90] Smith, P.F.: CS-modules and weak CS-modules, In: Non-commutative ring theory. Springer Lecture Notes in Mathematics, Springer-Verlag, **1448**, 99–115 (1990)

[Smi93] Smith, P.F.: Modules for which every submodule has a unique closure in ring theory. (Proc. Briennial Ohio State-Denison Conf., May. 1992), eds., Jain, S.K., Rizvi, S.T., World Sciebntific, Singapore, 302–313 (1993)

[Smi94] Smith, P.F.: Lectures on CS-modules. unpublished (1994)

[Smi97] Smith, P.F.:Nonsingular extending modules. Advances in Ring Theory (Granville, OH, 1996), Trends Math., Birkhäuser Boston, Boston, MA, 305–311 (1997)

[SHD90] Smith, P.F., Huynh, N.V., Dung, N.V.: A characterization of Noetherian modules. Quart. J. Math. Oxford **41**, 225–235 (1990)

[ST92] Smith, P.F., Tercan, A.: Continuous and quasi-continuous modules. Houston J. Math., **18**, 339–348 (1992)

[ST93] Smith, P.F., Tercan, A.: Generalizations of CS-modules. Comm. Algebra, **21**, 1809–1847 (1993)

[ST04] Smith, P.F., Tercan, A.: Direct summands of modules which satisfy C_{11}. Algebra Colloq., **11**, 231–237 (1993)

[Ste75] Stenström, B.: Rings of Quotients. Springer-Verlag, New York (1975)

[Swan62] Swan, R.G.: Vector bundles and projective modules. Trans. Amer. Math. Soc., **105**, 264–277 (1962)

[THT14] Talebi, Y., Hamzekolace, A.R.M., Tercan, A.: Goldie-Rad-Supplemented Modules. An. Şt. Univ. Ovid. Const., **22**, 205–218 (2014)

[TT09] Takıl, F., Tercan, A.: Modules whose submodules are essentially embedded in direct summands. Comm. Algebra, **37**, 460–469 (2009)

[Ter92] Tercan, A.: Generalizations of CS-modules. Ph. D. thesis., University of Glasgow (1992)

[Ter94] Tercan, A.: On the endomorphism rings of modules with C_{11} and C_2, Hacettepe Bulletin of Math. Sci. and Engineering, **21**, 3887–3901 (1994)

[Ter95] Tercan, A.: On CLS-modules. Rocky Mount. J. Math., **25**, 1557–1564 (1995)

[Te95] Tercan, A.: On certain CS-rings. Comm. Algebra, **23**, 405–419 (1995)

[Ter01] Tercan, A.: Weak C_{11}^{+}-modules with ACC and DCC on essential submodules. Taiwanese J. Math., **5**, 731–738 (2001) Addendum and Corrigendum **8**, 557–558 (2004)

[Ter03] Tercan, A.: Eventually weak C_{11}-modules and matrix C_{11}-rings. Southeast Asian Bull. Math., **27**, 729–737 (2003)

[Ter04] Tercan, A.: Weak C_{11}-modules and algebraic topology type examples. Rocky Mount. J. Math., **34**, 783–792 (2004)

[TNT11] Tütüncü, D.K., Nematollahi, M.J., Talebi, Y.: On H-supplemented modules. Algebra Colloq., **18**, 915–924 (2011)

[WD06] Wang, Y., Ding, N.: Generalized supplemented Modules. Taiwanese J. Math., **10**, 1589–1601 (2006)

[WW12] Wang, Y., Wu, D.: On H-supplemented Modules. Comm. Algebra, **40**, 3679–3689 (2012)

[Wil86] Wilson, G.V.: Modules with the summand intersection property. Comm. Algebra, **14**, 21–38 (1986)

[Wis88] Wisbauer, R.: Grundlagen der Modul- und Ringtheorie. Verrlag Reinhard Fischer (1988)

[Wis91] Wisbauer, R.: Foundations of Module and Ring Theory. Gordon and Breach., Philadelphia (1991)

[Wis96] Wisbauer, R.: Modules and Algebras. Bimodule Sructure and Group Actions on Algebras: Addison-Wesley Longman, Essex (1996)

[WuW12] Wu, D., Wang, Y.: Two open questions on Goldie extending modules, Comm. Algebra, **40**, 2685–2692 (2012)

[Yu94] Yu, H.Y.: On modules for which the finite exchage property implies the countable exchange property. Comm. Algebra, **22**, 3887–3901 (1994)

[ZS58] Zariski, O., Samuel, P.: Commutative Algebra. Vol. I, Van Nostrand, Princeton (1958)

[Zh99] Zhou, Y.: Examples of rings and modules as trivial extensions. Comm. Algebra, **27**, 1997–2001 (1999)

[ZH84] Zimmermann, W., Husgen, Z.B.: Classes of modules with exchange property. Journal of Algebra, **88**, 416–434 (1984)

Index

MIX
Papier aus verantwortungsvollen Quellen
Paper from responsible sources
FSC® C105338
FSC
www.fsc.org

If you have any concerns about our products,
you can contact us on
ProductSafety@springernature.com

In case Publisher is established outside the EU,
the EU authorized representative is:
Springer Nature Customer Service Center GmbH
Europaplatz 3, 69115 Heidelberg, Germany

Printed by Libri Plureos GmbH
in Hamburg, Germany